DATE DUE FOR RETURN 2011
2013

HYDROLOGY: WATER QUANTITY AND QUALITY CONTROL

HYDROLOGY:
WATER QUANTITY
AND QUALITY CONTROL

SECOND EDITION

Martin Wanielista
University of Central Florida

Robert Kersten
University of Central Florida

Ron Eaglin
University of Central Florida

John Wiley & Sons, Inc.
New York · Chichester · Brisbane · Toronto · Singapore · Weinheim

To Betty, Sue, and Linda

Acquisitions Editor	Cliff Robichaud
Marketing Manager	Debra Reigert
Senior Production Editor	Cathy Ronda
Designer	Ann Marie Renzi
Manufacturing Manager	Dorothy Sinclair
Illustration	Eugene Aiello
Cover Photo	J. D. Marston
Cover Design	Nancy Field

This book was set in Times Ten by Bi-Comp, and printed and bound by Hamilton. The cover was printed by Lehigh Press.

Recognizing the importance of preserving what has been written, it is a policy of John Wiley & Sons, Inc. to have books of enduring value published in the United States printed on acid-free paper, and we exert our best efforts to that end.

100 1717603

The paper in this book was manufactured by a mill whose forest management programs include sustained yield harvesting of its timberlands. Sustained yield harvesting principles ensure that the number of trees cut each year does not exceed the amount of new growth.

Library of Congress Cataloging in Publication Data:
Wanielista, Martin P.
   Hydrology : water quantity and quality control / Martin
Wanielista, Robert Kersten, Ron Ealgin. — 2nd ed.
      p.   cm.
   Includes bibliographical references.
   ISBN 0-471-07259-1 (alk. paper)
      1. Water-supply—Management.   2. Water quality management.
   3. Hydrology.   I. Kersten, Robert.   II. Ealgin, Ron.   III. Title.
   TD353.W337   1997
   551.48—dc20   96-22390
                                                              CIP

Printed in the United States of America

10 9 8 7 6 5 4 3

Overview

Why are hydrology and water controls important? Hydrology provides a basic understanding of the occurrences and distribution of waters above, on, and below the earth. As development occurs, the quantity and quality of water may not be sufficient to provide for all intended economical uses. Thus, we must be able to control waters within their intended risk and for beneficial uses. Major disasters are caused by below normal or excessive rainfall and snowfall. Droughts cause populations to be displaced and human suffering. Or at the other extreme, streams and rivers overflow into populated areas. Wildlife is displaced or destroyed. Damage to transportation systems occurs. And, unfortunately there is a long list of human sufferings resulting in excessive government expenditures when we do not understand hydrology and the management of waters. We should be able to design structures to control water related events at a risk that is acceptable to the peoples of an area and within budget expenditures.

Hydrology is the major discipline used to understand and design water management systems. Principles and concepts related to basic hydrologic processes and their use in analysis and design form a major part of this text. The hydrologic systems that we seek to identify are processes that change over time. Some can be identified by deterministic relationships, but others must be described using the laws and concepts of probability and statistics. We strive to put together a blend of both theory and practical applications of empirical equations. There are many solved problems in each chapter that are useful in expanding the calculation methods.

Hydrology: Water Quantity and Quality Control was developed in response to the educational needs of scientists, engineers, planners, and environmentalists to aid in the understanding and solving of water control problems. As the demand for quantitative solutions increase, we find that both the scientific and lay communities can benefit from the many example problems of quantitative nature in the text. We believe the text can also be used to aid in the design of systems using hydrologic data. The text has been used in engineering curriculums as a total design content course. We incorporate hydrologic concepts, ideas, and models into comprehensive water control studies. Comprehensive water quality and quantity studies are necessary when water supplies are developed or land use changes are contemplated. The emphasis of the text materials are placed on (1) measurement and interpretation of hydrologic cycle data, such as precipitation, evaporation, infiltration, and runoff, and (2) control of runoff water quality and quantity.

This new edition of **Hydrology: Water Quantity and Quality Control** is a major revision of our previously successful text. The first edition was published in 1990, and comments from those using the text in an academic environment provided the interest for rewriting. This new version of the text includes expanded treatment of water quality and control, thus the term *water quality* has been included in the title. Additional problems have been added to the end of each chapter. Case studies relating to design problems have been incorporated to better explain the assumptions and choice of values for the parameters of models. An additional appendix was added to suggest classroom or laboratory problems that support and help strengthen lecture and reading materials. Also, new computer programs to demonstrate solutions to complex problems otherwise not solved by hand calculations were added. The programs, unlike those of the first edition, can be copied for individual use as many times as needed. The programs run and are resident in either **Windows 3.1 or Windows 95 or Windows NT** environments. Additional text materials were added throughout the book to expand topic areas and introduce new materials. We hope the users of the last edition will be pleased to find additional solved problems in the text and as before, a solution manual is available to those adopting the text.

Organization

This book is primarily used by advanced level undergraduate and graduate students. It is used in its total content in a three semester hour course that typically meets for 30 hours of lecture, 30 hours of laboratory with assisted problem solving, and about 6 hours for evaluation or testing. Professionals also use the text because of the need to understand the materials and the value of the computer programs. Chapter 1 is introductory material, and Chapter 4, Evapotranspiration has relatively less material, thus, proportionally less time is spent on them. In general, equal time is spent on the remaining chapters. Because of the greater volume of material and computational level, Chapter 6, *Hydrographs,* may be given more time relative to the others.

We have rearranged materials in some chapters and rearranged some chapters to increase flexibility in the presentation and to maintain greater continuity of materials. Spreadsheet calculations and computer programs are introduced early in Chapter 1, followed in Chapter 2 by calculations based on methods from probability and statistics. Computational methods were introduced earlier in this edition because of their frequent use in the remaining chapters. For the less mathematical treatment of the subject, Chapter 2 can be left to the end of the course or lightly treated in the order as presented.

Features

This revised edition of **Hydrology: Water Quantity and Quality Control** contains some unique features. The features may be beneficial when considering its use:

- Practical and realistic case study (design) related problems.
- Presentation of basic principles and use in analysis and design.
- End of chapter problems with a solution manual.
- Example problems in each chapter to reinforce computational skills.
- Computer programs useful to explain complex concepts and solve large data based problems. The computer programs were written to solve actual problems.
- Water quality control methods and solved problems.
- Coverage of hydrograph topics, including synthetic and convolution methods.
- Empirical frequency and theoretical probability distribution methods with computer programs to ease the calculation burden.

- Integration of economic, hydraulic, and risk concepts into design.
- Example problems from all parts of the world.

Many of the problems used as case studies and examples come from the consulting experiences of the authors. Thus, the data are real and the solution procedures are complex and in-depth, but by the nature of learning, concise solutions and number "rounding" are used to reduce the volume of work. Many of the problems are solved more accurately and faster with the aid of the computer programs. The solutions using the computer programs can frequently lend a better understanding of the nature of input model changes, and thus the computer-assisted problems are also valuable in demonstrating ideas and concepts as well as problem solving. The computer programs are included on a diskette at the end of the book. The computer programs are very valuable to the professional doing larger problems. In fact, the programs were developed to solve real "honest-to-goodness" problems. However, the materials in the book can be learned without using the computer programs.

We have tried to write the text in an easy to read style with sequential materials at a level of understanding for most college students. In the past, we have encouraged students and faculty using this book to return suggestions on materials and use of the computer programs; we will continue that policy for this edition as well. An INTERNET address, namely *HTTP://BADER.ENGR.UCF.EDU/HYDRO.HTM*, with a course syllabus, laboratory exercises, updated materials, and a web users group is available. Also, as we develop new and improved computer programs, they can be downloaded from our University network. These activities are also helpful to exchange views and improve learning capabilities on the part of the users. Contact Dr. Ron Eaglin, *EAGLIN@CEE.ENGR.UCF.EDU*, for additional on-line help and information.

Marty Wanielista
Robert Kersten
Ron Eaglin

ACKNOWLEDGMENTS

We thank all the individuals who provided written materials in the form of consulting reports, example learning ideas, and reviews. Professor Anderson, Memphis State University; Professor Appleman; CPT Ed Gully, U.S. Military Academy; Professor William Hughes, University of Colorado, Denver; Phil King, New Mexico State University; Professor Mersky, Widener University; Dr. Miller, Brigham Young University; Dr. Robert Pitt, University of Alabama; Professor Ernest Pogge, University of Kansas; Dr. Paul Ruff, Arizona State University; and Dr. Tyagi, Oklahoma State University. Drs. J. Paul Hartman, Yousef A. Yousef, and Ola Nnadi at the University of Central Florida also provided valuable reviews in the early stages of the text development. Without the input of all reviewers, the practical application of the theory would not have been possible and the style of writing would not have addressed the intended audiences.

A special thanks to the students of Hydrology classes that reviewed and endured the early writings for the text. Also, we recognize those that helped with the typing, early manuscript review, and drafting, namely Saro Mayo, Betty Wanielista, Linda Eaglin, Dawn Fetter, Cheryl Brooks, Jackie Sullivan, John Florio, Marlo Wanielista, Denver Stutler, and many others. This was a "team" effort.

Marty Wanielista
Robert Kersten
Ron Eaglin

CONTENTS

CHAPTER 10   STORMWATER MANAGEMENT   365

INTRODUCTION

The study and practice of hydrology aids in explaining and quantifying the occurrence of water on, under, and over the earth's surface. Hydrology is both a scientific and engineering field of study. The subject area is derived from many basic sciences such as mathematics, physics, meteorology, and geology. In this text, methods for measurement and methodologies for the description and prediction of hydrologic processes are presented. Hydrologic processes vary with time and also change with geographic location. The quantities (volume and rate) of each process and the quality associated with each quantity measure resulting from modifications to other hydrologic processes and land use conditions are the fundamental considerations in this text. Both theoretical and experimental descriptions are presented.

1.1
ORGANIZATION AND CONTENT OF THE TEXT

This chapter presents information concerning the content and organization of the text. Next, the social importance of hydrology, the value of accounting for the location of water (water budget), units of measurement, and computation aids are briefly introduced. To enforce the concepts and ideas of each chapter, some example problems are solved and then additional problems are found at the end of each chapter. The end-of-chapter problems recognize the complexity of hydrologic systems and provide repetition to learn problem-solving procedures. The computer programs on the diskette included with this book serve as teaching aids in addition to being computation aids.

Summary statements are also used to reinforce the material of each chapter. The summaries are listed before the problem sections of each chapter to highlight important ideas, issues, and concepts.

The text material is primarily used for the study of water quantity and quality for basic hydrologic processes: precipitation, streamflow, runoff, evaporation, transpiration, infiltration, and storage. Both volumes and rates of flow are included for water quantity studies. Measurement techniques for each process are developed. Information to relate one process to other common processes of the hydrologic cycle are presented. Watershed and meteorological factors that affect these processes are developed. Quantification of the hydrologic processes is done by relating each process to easily available watershed and meteorological data. Concepts of probability and statistics are introduced as they relate to hydrology and water quantity management. The example problems and end-of-chapter problems are developed to aid in understanding hydrologic processes and the relationships among watershed, meteorological, and hydrologic processes.

Water quality is a relatively new consideration for the management of surface and ground waters. Water quality issues are included as some of the basic concepts of this text. Hydrologic and quality considerations are applied to stormwater management. Other texts are available (Novotny and Chesters, 1981; Wanielista et al., 1984; Wanielista and Yousef, 1993).

1.2
SOCIAL IMPORTANCE AND ENGINEERING RELEVANCE

Many aspects of social life are dependent on the economic availability and acceptable quality of water. The availability of water determines the basic existence of society. Thus, people educated and trained in a variety of water-related jobs are necessary. The role of water in sustaining life on the planet is vital.

The use of the concepts, ideas, and methods of hydrology are found in both the public sector of governments and the private sector of an economy. Individuals with hydrologic backgrounds can perform design calculations, collect hydrologic data, inspect construction, and conduct operations and maintenance activities. There are many types of hydrologic-related work including flood mitigation, roadway drainage, irrigation systems, navigation, water supply, pollution control, hydropower development, and ecological protection. Certainly, there is a tendency toward specialization, because details in design and operation are necessary. But also, there are needs for the individual with general knowledge about the many interactions of water projects. The material in this text introduces broad concepts, interactions, and design specifics related to the quantity and quality of water.

1.2.1 Water Quantity

Modern society appears to be very dependent on water projects. Early societies used water experts to help plan, design, and build canals for irrigation and transportation. Modern society has shifted the primary use of canals from transporting people and goods to advanced irrigation schemes, drainage, potable water supply and a host of other water-based activities. With this shift comes greater interest in water quantity management. For example, along our highways there exists the need to manage waters to prevent flooding of adjacent lands. Also, as population increases, the required volume of potable water increases. Thus, emphasis is on the volume of water and the rate of flow.

Planning is an important part of any society. Furthermore, all planning must conform to reasonable economic forecasts and have a solid technological base. Rate of flow, volume projections, and management methods must be planned. Plans must be completed for future water requirements, but the plans must conform to some economic and technical model requiring different levels of benefits to be estimated.

These benefits usually can be expressed in both quantitative and qualitative ways. To warrant investment, costs are generally required to be less than benefits expected. As a simple example, peak streamflow varies from year to year. Is it reasonable to provide protection against the very rare streamflow? The cost of this protection may be much greater than the losses from a flood. Thus, individuals in the planning area must be aware of the risk due to chance events associated with probable outcomes and incorporate economic considerations into the analysis.

1.2.2 Water Quality

The quality of water has become an issue of increasing importance over the years. Information in this text is primarily related to water pollution resulting from stormwater runoff and the prevention of pollution found in stormwaters. Many water quality issues are difficult to define in economic terms, such as the degradation of the recreational use of water and the loss of wildlife habitat due to pollution. Other water quality issues have a direct economic impact such as the quality of source waters that must be treated before being used.

Stormwater runoff has recently been identified as a significant source of pollutants in the natural environment (Yousef et al., 1985; Betson, 1978; Wanielista and Yousef, 1993). In many cases the water quality control systems can be easily adapted to include the prevention of pollutants in stormwater runoff. In some cases more costly measures need to be taken in the prevention of pollution.

1.3

WATER BUDGET AND MASS BALANCE

A water budget is an accounting of the volume of flow rate of water in all possible locations. Since density is constant it may be interpreted as a mass balance. One has to focus interest on a region and determine how the quantity of water in the region can be changed. The regional boundaries have to be determined across which water may move or be confined. Also, a time period must be specified.

A simple example of a budget is water from a parking lot. First, one must determine the surface boundaries of the parking lot that can contribute water to a collection point. The boundary may be defined on the surface as an imaginary line bordering the surface area from which precipitation can be accumulated and routed to some control point. At the control point, a decision may be made on the volume and rate of discharge. The control point can be an inlet grate at the lowest elevation of the parking lot. If the parking lot is constructed with curbs to contain the water on site, the boundary is easily determined. A parking lot is shown schematically in Figure 1.1. A water budget for that parking lot is helpful to present the concepts of a mass balance. Assume that all precipitation remains on the surface of the parking lot and is routed to a control point. This water volume is given the term rainfall excess. The rate at which rainfall excess appears over time at a discharge (control) point is called *runoff*. A water budget can be written in volume terms (mass balance):

$$\text{Inputs} - \text{outputs} \pm \text{accumulation} = 0$$

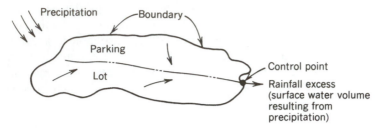

FIGURE 1.1 Parking lot water budget.

If water is not stored on the parking surface, the accumulation term is zero with input equal to precipitation and output equal to rainfall excess, or:

$$\text{Rainfall excess} = \text{volume of precipitation} \qquad (1.1)$$

If precipitation is abstracted by depression storage in the parking lot, then the water budget in volume terms is altered as shown in Equation 1.2.

$$\text{Rainfall excess} = \text{precipitation} - \text{depression storage} \qquad (1.2)$$

The complexity of a budget depends on the physical system and the ultimate use of the budget. Water budgets for large areas are complex with many parameters.

1.3.1   Global Water Budget

Available freshwater on earth comprises only about 3% of the total supply (Miller, 1986). More than 97% of the water on earth contains dissolved solids, which result in a salty taste. Direct human use and most industrial uses of salt water are possible only after some or most of the solids are removed. Other waters become unfit for municipal and industrial use when unwanted chemicals or rubbish are disposed of in them. Also biological changes can take place within the water resulting in a need for treatment before use.

Much of the freshwater is not readily available for use due to its remote locations. A majority of the freshwater (90%) is found in glaciers and polar ice caps. The small fraction of the total supply available for human consumption is estimated at about 4 billion billion liters. This is 400 million liters per person at a population of 10 billion people. It would appear then that water supply would not be a problem if remote sources could be made economically available and pollution could be minimized. Making water economically available is the major task.

On a global scale, the quantity of water is finite and its usable fraction can be altered (Wanielista et al., 1984). With a global water budget, one can establish water quality and quantity levels. Thus, the availability of water can be a matter of record, which provides responsible people with the database for decision making. This decision making can affect both water and land uses in a large region or a small council district. Useful storage must be monitored to ensure that there are no significant losses over time. When one of the inputs or outputs to the usable storage is significantly altered, changes will take place. These changes may be costly and life-styles may have to be altered. Life-style changes are already evident in places where usable water is in short supply. People must limit their consumption, which changes life-style.

As will be demonstrated in later chapters, the nature of the hydrologic problem becomes one of thoughtful consideration of the variability of water supply, i.e., distribution in time and space. Alterations of this naturally occurring scheme typically require engineering works and should be undertaken with extreme care. Complete hydrologic investigations will involve aspects of many other supporting sciences.

To illustrate variability of a hydrologic process, consider a flowing stream, creek, or river. Rarely, if ever, over a one-year period of observation can the observer record a constant flow rate or depth of flow. A listing of these flow rates over time or a graphical display (chart a figure) will more vividly illustrate the variability. These flow rates over time are called hydrographs. The hydrograph obtained from empirical observations or generated using computer programs is one of the more frequently used hydrologic descriptions. A plot of a hydrograph is shown as Figure 6.1.

Pollutants entering the water course directly affect water quality. These pollutants may be natural or man-induced. Quality also influences availability of supply depending on necessity. That is, it may be better to have poor or marginal quality water than none at all. Quality standards are important, even necessary, but not always sufficient to ensure adequate supplies.

1.3.2 Geographic Area Water Budgets

On a smaller scale, examples exist that indicate the value of a water budget. One such example is Amboseli National Park in southern Kenya. The park is located just north of Mount Kilimanjaro. Highly saline groundwater percolates from Kilimanjaro to Amboseli. The climate is arid, with on the average only 400 mm (16 in.) of rain per year.

After the early 1960s, the vegetation in the park changed along with the animal life that fed on the vegetation. What caused the changes? After an examination of a water budget for the area, it was found that precipitation on Kilimanjaro had increased dramatically. This caused groundwater levels to increase, which raised the groundwater levels beneath the lake beds in the National Park and caused a shift in the surface vegetation.

There are also examples of precipitation greatly exceeding watershed storage. In many cases, floods and subsequent damages can be reduced simply by storing water for release during dry periods. These examples show that simple water budgets are valuable and can provide operating guidelines for reservoir sizing.

Excessive precipitation is not the only problem. There are examples of too little precipitation causing drought conditions and rapid depletion of groundwater. This in turn causes an increased cost of water treatment. Subsidence, or sinkhole activity, has also been related to groundwater withdrawal. Possibly one of the most famous subsidence studies was conducted near Venice (Gambolati et al., 1974). Again, a water budget was used. The budget indicated a reduction in subsurface water occurred from pumping activities and could have caused the subsidence. Sinkhole activity in Florida and Texas has been related to groundwater withdrawal. A relatively complete history of sinkhole activities and related groundwater levels have been reported for the central Florida area. In the Houston–Galveston area, up to 3 m (9.8 ft) of subsidence has been noted and related to groundwater withdrawals. As pumps in wells remove great quantities of water from one of the United States most plentiful aquifers, about 1700 mi^2 of land has subsided. To prevent this subsidence, the source of water supply is being changed from groundwater to surface water and to practice more conservative use of water (*U.S. Water News*, 1985).

In Pennsylvania and other areas, increases in stormwater discharges to groundwaters were related to sinkhole activity. Possibly the increased volume and flow rates
of stormwater being discharged to the aquifer caused dissolving of the limestone rock
structure. Again, a water budget was used to estimate groundwater volume increases.

1.4

HYDROLOGIC CYCLE

The hydrologic cycle is a simplified accounting of the complex interactions of meteorological, biological, chemical, and geological phenomena. It is the movement of water
from surface water, groundwater, and vegetation to the atmosphere and back to the
earth in the form of precipitation. The transfer of water from plant tissues to the
atmosphere is called transpiration. Plants absorb water from the soil through the root
system. Rainfall can be abstracted onto vegetation, intentionally stored in ponds, be
abstracted by depression storage (unintentional small volumes), infiltrated into the
soil, or be available for discharge (rainfall excess). Rainfall that infiltrates into the
soil, moves or percolates to the water table. Some of this groundwater may help
recharge the aquifers. Some infiltrated waters may evaporate or flow in the direction
of surface waters. This is a simplified accounting of a very dynamic process, shown in
Figure 1.2.

Some surface water will remain after infiltration, abstraction by vegetation, and
depression storage has been filled. An example of depression storage is a hole in
pavement that stores water. This depression storage is different from the intentional
design of surface storage ponds to reduce rainfall excess and runoff.

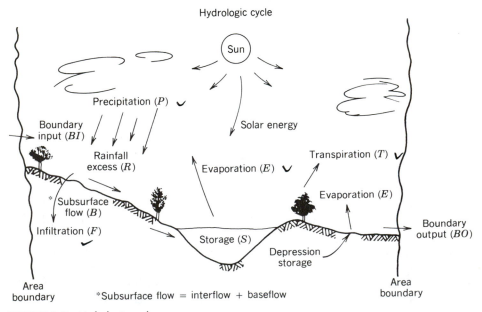

FIGURE 1.2 Hydrologic cycle.

1.4.1 The Hydrologic Cycle as a Mass Balance

Using a water budget to represent the hydrologic cycle, an equation to estimate surface water volumes can be developed. This equation is elementary but useful for surface water accounting where inputs and outputs vary with time. Volume units are used and time is fixed for most elementary applications of the water budget. All elements of the hydrologic cycle are dependent in some way on each other. A schematic representation of a pond is helpful to develop a mass balance equation, shown in Figure 1.3. All variables that cross the boundary must be accounted for in equation form. Rainfall excess is usually separated from surface storage to denote its significance in stormwater management studies. A mass balance equation for Figure 1.3 is

$$\text{Inputs} - \text{outputs} = \text{change in storage}$$

$$P + R + B - F - E - T = \Delta S \tag{1.3}$$

where ΔS = change in storage volume. A boundary transfer can be net human consumptive uses from surface volumes and/or groundwater volumes. In the water budget, volumes are measured in units of cubic meters, liters, acre-feet, cubic feet, gallons, or inches and centimeters over the watershed area. Also, a common way to express quantities of surface waters is in discharge units (volume per unit time). In the United States, data on discharge are available from the U.S. Geological Survey (USGS, all years) and other state and federal environmental departments. Other countries publish similar data.

It is often desired to perform a water budget for a specific time and on a specific watershed. The watershed of concern must be delineated to determine significant inputs over the time period. The time periods for analysis are chosen to be consistent with desired accuracy or storm event. Surface storage will change with time and is dependent on meteorological, geological, topographical, and human consumption factors.

1.4.2 Surface Water Supplies

One of the uses of the hydrologic cycle and a mass balance equation is in the estimation of surface storage. As an example, an area has an insufficient supply of water. Net input boundary exchanges may be increased by constructing reservoirs (water impoundments), restricting consumptives uses and importing water. Weather modifications are possible (however, not extensively used) for either increasing or decreasing precipitation. Geological surveys may indicate ways of directing subsurface runoff so

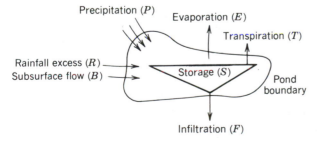

FIGURE 1.3 Schematic of hydrologic cycle.

that groundwater supplies are supplemented or transferred to surface waters. Thus, surface water inventories can be affected by various engineering projects. In addition to volume of inventory analysis of the hydrologic cycle, flow rates are important for establishing transport systems and quality degradation.

Storing and transferring a sufficient quantity of water from one location to another has been one of the major problems of society. Some questions related to surface storage and the hydrologic cycle are: What volume of water is stored in a surface reservoir and how does the volume change over time? What causes the water supply to be depleted or increased? How are the storage and releases managed?

Surface flows into a reservoir are necessary to maintain the beneficial uses of the reservoir. Adequate descriptions of these flow rates would be helpful to define flood levels, wildlife management, irrigation volumes, water-based recreation, and other social needs. Water quantity would be less of a problem if the source of usable, unpolluted water were located close to the users. Water is not always economically available at a particular location because rainfall quantities vary from one area to another. In addition, for one specific area, the stochastic (time-varying) aspects of rainfall must be considered for more reliable predictions of storage levels. Precipitation, evaporation, infiltration, and streamflows will be the material for latter chapters. The details for measurement and interpretation are necessary to use the basic mass balance of the hydrologic cycle as developed within this chapter.

1.4.3 Rainfall Excess

During a precipitation event, a mass balance of the total volume of rainfall onto and flow from an area is helpful to understand rainfall excess. Consider as variables the volume of precipitation P, rainfall excess R, infiltration F, evaporation E, transpiration T, and initial abstraction I_A. Initial abstraction is water intercepted by vegetation and stored in surface depressions. A mass balance of a simplified water budget for a fixed time period, considering negligible boundary transfers and no storage change, is written as

$$\text{Rainfall excess} = \text{precipitation} - \text{outputs}$$

$$R = P - E - T - F - I_A \qquad (1.4)$$

Equation 1.4 is illustrated by the schematic in Figure 1.4.

In many locations it is difficult to separate evaporation and transpiration. Thus, the variables are considered together and most likely can be estimated as one value, identified as evapotranspiration (ET). For a short time period (hours-day) each of

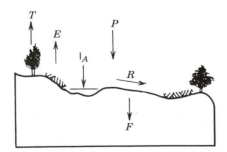

FIGURE 1.4 Water budget schematic.

the above variables can be considered constant, and since evapotranspiration (ET) may be considered negligible during a precipitation event, Equation 1.4 is rewritten as

$$R = P - F - I_A \tag{1.5}$$

If the volume of infiltration and initial abstraction is proportional to the precipitation volume, the quantity of rainfall excess can be expressed as a fraction of precipitation:

$$R = CP \tag{1.6}$$

where C = runoff coefficient (dimensionless) such that $0 \leq C \leq 1$

The runoff coefficient, as defined here, is the ratio of rainfall excess to precipitation. The runoff coefficient can be determined from extensive rainfall and runoff studies, or from published values. In later chapters, the relationship of runoff (flow rate) to precipitation intensity and the runoff coefficient is developed.

❏ EXAMPLE PROBLEM 1.1

Over a two-month period of time, a catchment in the Pontypridd area of Wales is expected to receive 254 mm of rain with an expected evapotranspiration estimated at 85 mm and that lost to groundwater storage of 20 mm. There is no other significant storage in the watershed. What is the expected rainfall excess to a reservoir storage area if the catchment area is 65 km²? Express your answer in cubic meters and liters. Also, determine how many people can be serviced by this water if the per person per day water use rate is 160 liters.

Solution

Solving for rainfall excess using a mass balance:

$$R = P - ET - F$$

$$R = 254 - 85 - 20 = 149 \text{ mm (over the catchment)}$$

or in m³

$$R = \frac{149 \text{ mm}}{1000 \text{ mm/m}} \times 65 \text{ km}^2 \times 10^6 \text{m}^2/\text{km}^2 = 9.685 \, (10^6) \text{ m}^3$$

or in liters

$$R = 9.685(10^6)\text{m}^3 \times 10^3 \text{L/m}^3 = 9.685(10^9) \text{ liters}$$

Solving for the per person use rate:

$$\text{Usage/2 mo} = 160 \text{ liters/person-day} \times 30 \text{ d/mo} \times 2 \text{ mo}$$

$$= 9600 \text{ liters/person}$$

$$\text{and people served} = \frac{9.685(10^9) \text{ liters}}{9.6(10^3) \text{ liters/person}} = 10^6 \text{ people} \quad ❏$$

1.5

UNITS OF MEASUREMENT

Consistency among individuals in the reporting of data would increase the rate of knowledge transfer. The transfer of data is enhanced by modern definitions of primary and secondary units of measurement. The primary quantities are defined as mass, length, time, and temperature. A secondary quantity is one defined in terms of primary quantities. Examples of secondary quantities are volume in cubic meters (m^3), density in kilograms per cubic meter (kg/m^3), and rainfall intensity (mm/s). The U.S., or English customary System, and the International System of Units (SI) are both used in this text. It is important for hydrologists, planners, and engineers to understand and calculate with a knowledge of both systems of units. Thus, a mixture of SI and customary units is found in the text with conversion factors included. Some of the more common notation and dimensions for basic units of hydrology are shown in Table 1.1. A more complete notation list is shown in Appendix A.

It should be noted that not all equations used in hydrologic analyses are dimensionally homogeneous (e.g., Manning's equation), and therefore numerical coefficients in the equations vary with the set of units used. Great care should be exercised in all calculations to make sure the dimensions and units are consistent.

A large part of hydrology involves measurements of various parameters. Since measured data are inherently not exact, it is essential that methods of manipulating the data be carefully examined so that information derived therefrom can be evaluated properly (Beakley, 1967).

A *significant figure* in a number is defined as a figure that may be considered reliable as a result of measurement or computation. For example, any instrument can be assumed to be accurate only to one-half of the smallest scale division that has been marked on the instrument by the manufacturer. Consider a thermometer so marked

TABLE 1.1 Some Common Hydrologic Symbols and Dimensions

PROPERTY	SYMBOL	U.S. CUSTOMARY	SI UNITS
Length	L	ft	m
Mass	M	slug	kg
Time	t	s	s
Temperature	T	°F or °R	°C or K
Precipitation volume	P	in.	mm
Precipitation rate	i	in./hr	mm/hr
Rainfall excess	R	in. (ft^3)[a]	mm (m^3)[a]
Infiltration rate	f	in./hr	cm/hr
Velocity	v	ft/s	m/s
Area	A	ft^2	m^2
Volume	V	ft^3	m^3
Flow rate	Q	ft^3/s (CFS)	m^3/s
Concentration	C	lb/gal[b]	mg/L

[a]Volume (depth over an area).
[b]lb used as a mass unit.

that the smallest graduation is 1°. If the liquid column rises to a point between 70° and 71°, we would normally include an estimated 0.5 in our reading. Therefore, the reading would be recorded as 70.5 and would contain three significant figures. In some cases, we could estimate by eye beyond the accuracy obtained from the graduations, say 0.6° or 0.7°. It should be recognized that we still have three significant figures and the last is an estimated figure.

For engineering calculations, we should guard against introducing a false accuracy by including more digits in the answer than a *significant figure* analysis would warrant. In adding or subtracting numbers, the sum should not be written to more digits than the number which has the least number of significant figures. In multiplication and division, the result should only be carried to the same number of significant figures that there are in the quantity entering into the calculation with the least number of significant figures. Last, since we often have results including a number multiplied by a power of ten (i.e., scientific notation), it should be kept in mind that this does not change the number of significant figures but only moves the decimal point.

❑   **EXAMPLE PROBLEM 1.2**

From basic units, develop conversion factors for volumetric flow rate and area loadings.

Solution

For flow rate:

$$Q = \text{flow rate} = L^3/t \qquad (\text{from ft}^3/\text{s to m}^3/\text{s})$$

$$= 1 \text{ ft}^3/\text{s} \times (0.3049 \text{ m/ft})^3 = 0.0283 \text{ m}^3/\text{s}$$

and area loadings:

$$\text{loadings} = M/L^2 \qquad (\text{from lb/ac to kg/ha})$$

$$1 \text{ lb/ac} \times 0.454 \text{ kg/lb} \times 2.471 \text{ ac/ha} = 1.121 \text{ kg/ha}$$

A listing of conversion factors is presented in Appendix B, and an abbreviated list is found within the front and back covers.   ❑

❑   **EXAMPLE PROBLEM 1.3**

Using the conversion factors of Appendix B, convert 1 in. of rainfall over a 1 mi^2 watershed to rainfall excess expressed as acre-feet, cubic feet, gallons, and cubic meters. Assume that all the rainfall is available as rainfall excess.

Solution

From a mass balance, rainfall excess is equal to rainfall, thus using the conversion tables of Appendix B:

a.   Conversion to acre-feet:

$$1 \text{ in.-mi}^2 \times 1 \text{ ft/12 in.} \times 640 \text{ acres/mi}^2$$
$$= 53.33 \text{ acre-feet/in.-mi}^2$$

which agrees with the conversion factor under Miscellaneous in Appendix B.

b.   Conversion to cubic feet:
There are 43,560 ft^2/acre, thus $(53.33 \times 43{,}560)/1$ in.-mi^2 = 2.323 (10^6) ft^3

c.   Conversion to gallons:
There are 7.48 gal/ft^3, thus $(7.48 \times 2.323 \ (10^6))/1$ in.-mi^2 = 17.38 (10^6) gal

d.   Conversion to cubic meters:
There are 35.314 ft^3/m^3, thus $(2.323 \ (10^6)/35.314)/1$ in.-mi^2 = 65,781 m^3, which
agrees with the conversion factor under Miscellaneous in Appendix B.   ❑

1.6
COMPUTATION AIDS

Problems in hydrology frequently require repetitive and precise calculations. Some
are solved by trial-and-error methods. Others require repeating the basic solution
algorithm many times. Many aids for solving these problems have been developed
over the years. Nomographs and coaxial graphs were popular about 35 years ago
(1950–1965) and are still used today. These graphs make the computation procedure
less complex and save time. Proper interpretation of the graphs is necessary and can
be learned by those educated and trained in the subject matter. Starting in the 1960s,
calculators of various types further streamlined the repetitive and routine calculations.
Also, around the same time, computers were gaining popularity. In the early 1980s,
the personal computer (PC) gained greater recognition as an aid to solving complex
repetitive problems in hydrology. As more software developed, the availability of PCs
increased and cost decreased. In the late 1980s, many researchers, planners, engineers,
and hydrologists increased their dependency on computers and other electronic devices
for solving equations.

1.7
COMPUTER PROGRAMS

Mathematical equations used in Hydrology have been adapted to software used on
computers extremely well. As computers have become faster and more powerful,
hydrologists have utilized the computer's potential to solve more complex problems.
Today, hundreds of computer programs and computer tools exist which are useful for
solving hydrology and related problems. The back of this book contains a selection
of programs written to accompany the text and assist the student in solving hydrology
problems. Appendix I contains a discussion of each of the included programs.

1.7.1   Spreadsheets

One of the most useful tools for the solution of hydrologic equations is the spreadsheet.
In its most simplistic form the spreadsheet is a computer program which stores, displays,
and performs calculations with rows and columns of numbers. Also, most spreadsheets
contain programming or macro languages, 'what if' scenario ability, table look-up,
trial and error solutions, and array manipulation; all of which are useful in hydrology.

Almost all spreadsheets today are capable of generating a variety of graphs for displaying data. Spreadsheets are used in example problems throughout the text and the computer problems section of most chapters contain problems which require the use of a spreadsheet.

❏ **EXAMPLE PROBLEM 1.4**

Using an incremental precipitation of 0.5, 1.0, and 0.2 in. given at 1-h intervals calculate cumulative rainfall excess using the equations

$$R = \frac{(P - 0.8)^2}{P + 3.2} \qquad P > 0.8$$

$$R = 0 \qquad\qquad P \leq 0.8$$

where

R = cumulative rainfall excess (inches)
P = cumulative precipitation (inches)

A spreadsheet should be used to perform the calculations.

Solution

Set up a spreadsheet as follows:

	A	B	C	D
1	Time	Incremental	Cumulative	Rainfall
2		Precipitation	Precipitation	Excess
3	(hours)	(in)	(in)	(in)
4	1	0.50	0.50	0.000
5	2	1.00	1.50	0.104
6	3	0.20	1.70	0.165

Cell C5   +C4 + B5
Cell D4   IF(C4 > 0.8, (C4 - 0.8)^2/(C4 + 3.2), 0)

Cell C5 equation should be copied to C6, and cell D4 equation should be copied to D5 and D6. Since the equations use *relative addresses* the equation will automatically adjust to use the numbers from the correct cells in the calculations.   ❏

1.7.2 Computer Drafting

Drafting and engineering drawing used to be performed primarily by hand. Today, nearly all drafting is done on the computer, primarily because drafting software has become more diversified and less expensive. A popular computer aided design (CAD) program is AutoCAD® from AutoDesk, Inc. For most hydrology projects, however, less expensive CAD programs will suffice.

CAD offers the advantage of being able to produce high quality engineering drawings very quickly and easily with changes to the drawing being fairly routine. New CAD users may find the ruler and compass at first quicker, but with a little experience they will find the computer much faster. CAD also offers the advantage of saving the drawing electronically and permits modifying the drawing easily. Throughout the text, guidelines will be offered on how to perform different tasks using CAD. The text will use AutoCAD® commands; however, the user should realize that other CAD programs can typically perform the tasks described.

1.7.3   Hydrology Computer Programs

A number of programs that have been written to assist in the solution of hydrology problems are described in this section. Some of these programs are included on the disk that accompanies this book. Programs included with this text are SMADA, DISTRIB, REGRESS, OPSEW, EZMAT, TCCALC and PLOAD. Further documentation for these programs is included in Appendix I.

1.7.3A   HEC-1
HEC-1 is a program written by the United States Army Corps of Engineers to perform hydrologic routing (U.S. Army Corps of Engineers, 1990). HEC-1 generates hydrographs from both rainfall and snowmelt and adds, diverts, or routes them to stream reaches, reservoirs, or ponds. The program was originally written in FORTRAN. A number of preprocessors (Haestad Methods, 1992) and help manuals (U.S. Army Corps of Engineers, 1991) exist to simplify this input.

1.7.3B   SMADA
Stormwater Management and Design Aid (SMADA) is a computer program for the Microsoft Windows™ operating system. The program allows the user to generate hydrographs using rainfall and watershed characteristics. Users can also enter or calculate stage-storage-discharge relationships for ponds and route hydrographs to these ponds. SMADA was written by Marty Wanielista and Ron Eaglin at the University of Central Florida and is available with this text.

1.7.3C   SWMM
Stormwater Management Model (SWMM) is a program originally produced by United States Environmental Protection Agency. The program can generate hydrographs and estimate pollutant runoff, sediment loading, and sewage loads. SWMM is designed with an open architecture allowing for the addition of more modules. SWMM is primarily maintained by Wayne Huber (Huber et al., 1988) at Oregon State University.

1.7.3D   DISTRIB
DISTRIB is a Microsoft Windows™ program which fits univariate data to a probability distribution and allows predictions based on these data. It performs method of moments analysis using normal, log normal, Pearson, Log Pearson, and Gumbel type distributions. DISTRIB was written by Ron Eaglin and Marty Wanielista at the University of Central Florida and is available with this text.

1.7.3E   REGRESS
REGRESS is a Microsoft Windows™ program which performs linear least squares regression. It uses linearization techniques to fit data to a number of forms. XY data pairs are entered in a spreadsheet, and a curve form is chosen from a number of choices. REGRESS also includes lookup tables for t statistics and F statistics. REGRESS was written by Marty Wanielista and Ron Eaglin at the Univeristy of Central Florida and is available with this text.

1.7.3F PLOAD

PLOAD is a Microsoft Windows™ program which estimates pollutant loading on a time basis using typical watershed land uses and total rainfall. The program allows for estimation of loading rate and modification of loading rate parameters. Land use information can also be entered along with pollutants. PLOAD was written by Ron Eaglin and Marty Wanielista at the University of Central Florida and is available with this text.

1.7.3G EZMAT

EZMAT is a matrix calculator which is useful in performing matrix multiplication, addition, subtraction, inversion, and other matrix operations. EZMAT was written by Ron Eaglin at the University of Central Florida and is available with this text.

1.7.3H OPSEW

OPSEW is a Microsoft Windows™ program which assists in the design of storm sewer systems. It was written by Ron Eaglin at the University of Central Florida and is available with this text.

1.7.3I TCCALC

TCCALC is a Microsoft Windows™ program which calculates the time of concentration using a number of different methods given watershed parameters. It was written by Ron Eaglin at the University of Central Florida and is available with this text.

1.7.4 Programming Languages

At one time hydrologists and engineers relied heavily on programming to solve routing and other hydrologic problems. Today hundreds of different programs exist to perform nearly any task a hydrologist might require. Use of programming in solving hydrology problems has become less common as these software tools have become widely available. Occasionally, however, the hydrologist might have a need to solve a problem which requires the use of a program. A number of languages are typically available to the hydrologist to perform this task.

1.7.4A BASIC

Versions of BASIC are standard with nearly all versions of DOS. BASIC is primarily a DOS-based language and is not typically available in other operating systems. A major exception to this is Visual BASIC, which is available for the Microsoft Windows™ operating system. A detailed discussion of the BASIC language is beyond the scope of this text. Hundreds of texts address programming in BASIC. BASIC is an interpreted language, meaning that the program can be run without compiling it to an executable program. BASIC programs can also be compiled.

1.7.4B FORTRAN

FORTRAN is a procedural language that is very similar to BASIC. FORTRAN is a portable language that typically adheres to standards allowing programs written in FORTRAN to be compiled for many operating systems, including UNIX and DOS. FORTRAN is a compiled language, meaning the programmer must compile the program to an executable file before running and a FORTRAN compiler is required to do this. Many hydrology and hydraulic programs written in FORTRAN are available in compiled and uncompiled source code format. HEC-1 is an example of one such program.

1.7.4C   C and C++
The C language and its object-oriented counterpart C++ are two extremely powerful languages. These languages allow for both high- and low-level coding, pointers, and data structures. For professional programmers, C and C++ are often the language of choice. It would not be time effective, however, for us to learn the C language to solve a simple problem. C and C++ both offer the advantage of portability. They are both compiled languages.

1.7.4D   Others
Other languages exist, including but not limited to ADA, COBOL, Pascal, Clipper, SmallTalk, Prolog, etc. When undertaking a large project which requires programming it would be advisable to investigate the advantages and disadvantages of each language.

1.7.5   Other Programs

There are many programs on the market that perform most of the calculations done in Hydrology. As these programs become more sophisticated, the users must understand their capabilities and limitations. Some programs require other programs to run, such as AutoCAD Development System (ADS) programs which run within and require AutoCAD®. Many programs also have specific hardware requirements, and it is advisable to be sure that the hardware is in place before purchasing these programs. ICPR® is a commonly used program to assist in hydraulic and hydrologic analysis. Another package is Pond Pack® available from Haestad Methods.

1.7.5A   ICPR®
The Interconnected Channel and Pond Routing Model is a comprehensive one-dimensional unsteady hydrodynamic model used to route storm hydrographs through complex systems of ponds, nonprismatic channels, storm sewers, culverts, horizontal and vertical weirs, orifices, gates, pump stations, drop structures, operable structures, bridges, and dam breaches. Hydrograph methods include the SCS unit hydrograph, the Santa Barbara urban hydrograph, and a finite element overland flow procedure. The program has been approved by the Federal Emergency Management Agency for use on flood insurance studies. Extensive user interfaces, including a full AutoCAD® graphical interface, are available. This program was written by Peter J. Singhofen, P. E., Peter J. Singhofen, Jr., and Gary R. Griffin. ICPR® is available from Streamline Technologies, Inc.

1.7.5B   TR-55
TR-55 stands for Technical Release 55 available from the Natural Resources Conservation Service. The actual title of the publication is "Urban Hydrology for Small Watersheds." The document and accompanying computer program contain methodologies (worksheets) for the calculation of time of concentration, curve number, watershed peak flow, and watershed hydrographs for small urban watersheds.

1.7.5C   Watershed Modelling System (WMS)
WMS is a computer program which performs all of the basic function of HEC-1 providing for comprehensive hydrologic analysis of watersheds. The program was

TABLE 1.2 Selected Government Worldwide Web Addresses

AGENCY	WWW Address
United States Geological Survey (USGS)	http://www.usgs.gov/
Environmental Protection Agency (EPA)	http://www.epa.gov/
National Technical Information Service (NTIS)	http://www.fedworld.gov/ntis/ntishome.html
Natural Resource Conservation Service (NRCS)	http://www.ncg.nrcs.usda.gov/
Federal Emergency Management Agency (FEMA)	http://www.fema.gov/

produced by the Engineering Computer Graphics Laboratory at Brigham Young University. Information about the program is available via the worldwide web at address http://www.et.byu.edu/~geos/software/wms/wmsinfo.html.

Readers who have a computer with a modem or have other access to the internet can view and retrieve a tremendous amount of hydrologic data and programs. Using any browser for the worldwide web (WWW) the user may wish to browse the sites listed in Table 1.2. This list is not to be considered a complete list of sites, but a starting point for the reader to access information. Updates of the computer programs found with this text will also be accessible via the University of Central Florida home page (http://www.ucf.edu/) link to the Civil and Environmental Engineering Department page.

1.8

SUMMARY

Brief summary statements to highlight important concepts, ideas, problems, and formulas are provided at the end of each chapter to reinforce material. Details related to the material must be obtained by reading the complete text and solving both the example problems in the chapter and problems at the end of the text. The depth of understanding can be increased by reading some of the references at the end of each chapter.

- Hydrology is the study of waters on, under, and over the earth's surface.
- Water quantity control can be understood knowing hydrologic principles. Both peak rates and volume control are studied in this text.
- The basic concept of a water budget is necessary to understand and interpret hydrologic data. A volume/mass balance of rainfall and rainfall excess for a parking lot helps to illustrate the concept.
- Rainfall excess is the volume of water from rainfall that remains on the surface and is available for runoff, while runoff is the rate at which water is discharged from a watershed.
- Rainfall excess can be assumed to be directly related to precipitation, resulting in a coefficient relating both. This coefficient is called the runoff coefficient and

can be defined as the ratio of rainfall excess to precipitation. In later chapters, the runoff coefficient will be used in another formula for estimating runoff.

- A mass balance of the hydrologic cycle results in useful equations for estimating one of the parameters of the hydrologic cycle.
- A notation list is found in Appendix A.
- Conversion tables are found in Appendix B and on the inside of the front and back cover of the book.
- To expand an understanding of hydrologic processes, computer programs are helpful and are provided with this text.

1.9

PROBLEMS

1.9.1   Hand Problems

1. What is the hydrologic cycle? Explain in your own words how it is related to the study of meteorology.

2. List at least five hydrologic processes.

3. Explain in your own words and by means of an example how you can use hydrology and economics to aid in solving a problem related to a hydrologic process.

4. Convert and show all basic units for:
   a. Pounds/acre-day to kilograms/hectare-year
   b. Acre-feet to million gallons
   c. Cubic feet/second to cubic meters/second
   d. Inches/hour to millimeters/hour
   e. Liters/second to gallons/minute

5. What is a water budget? Explain at least three applications of a water budget to hydrologic systems.

6. Write an equation of a water budget for a parking lot that has on-site storage. Using common notation for precipitation volume, rainfall excess, and storage volume, show an equation to solve for rainfall excess. Using both the SI and U.S. customary set of units for the variables of your equation, define at least four consistent sets of units.

7. If the rainfall on a watershed is 3 in., and the rainfall excess is 21,780 ft³, what is an estimate of the runoff coefficient if the watershed area is 4 acres?

8. For the hydrologic cycle of Figure 1.2, write the equation for estimating the ending storage after 12 one-month periods considering monthly hydrologic data that are available for evaporation, transpiration, infiltration, subsurface runoff, boundary output, precipitation, and initial storage.

9. Estimate the amount of depression storage (inches and cubic feet) in a 6-acre parking lot using the following rainfall and discharge data. The discharge data were measured at the only outlet in the parking lot. Rainfall = 0.88 in. Rainfall excess = 5 cfs average for one hour, the total duration of runoff. Also list assumptions for the other variables of the hydrologic cycle.

10. It rains 3.0 in. on a 20.0-acre watershed. What is the volume of rainfall excess if 30% of the watershed area is a lake and infiltration on the soil is estimated at 1.0 in.? What is the total volume of runoff? Express your answer in cubic feet. Assume that the lake level is low and does not contribute to rainfall excess. Hint: Lakes are treated as initial abstraction.

11. Estimate the depth of depression storage (units of inches and cubic feet) for a 12-acre parking lot given site specific measures for discharge and rainfall. The rainfall volume was 0.90 in. and the average runoff rate was 10 cfs with a duration of one hour.

12. Average runoff from a 6.7-acre parking lot in Chicago, Illinois, was measured as 20 cfs for a 2-hr period of time. If the depression storage were 0.08 inches, what is the volume of rainfall (express your answer in inches of water)?

13. You must calculate the depth of water in a stormwater pond resulting from a two-hour rainfall event. The rainfall occurs over a 24-acre watershed at an average rate of 1 inch/hour. The runoff coefficient is 0.5. The average area of the pond is 2 acres and the depth at the beginning of rainfall is 5 ft. The pond water exfiltrates at a rate of 4 in./hr through the pond side walls (area $= 2100$ ft^2) when the water level is above 5 ft. What is the pond depth above the 5-ft level 8.2 hr after the rainfall ends?

1.9.2 Computer-Assisted Problems

1. Initiate the computer program diskette provided with this book. Obtain the menus of programs and locate at least one chapter of this book where the programs can be used. This exercise will start to familiarize you with the computer programs and the subject material.

2. Using your own disk, develop a computer program to convert any inputed data from pounds/acre-day to kilograms/hectare-year, from acre-feet to million gallons, from cubic feet/second to cubic meters/second, from inches/hour to millimeters/hour, and from liter/second to gallons/minute. Be sure to document the program well and make it user friendly. Print out the inputed data as well as the converted data along with the appropriate units.

1.9.3 Case Studies

1. A given storage reservoir has a normal pool area of 4000 acres and serves a drainage area of 200 mi^2. During an average year rainfall is 37 in., rainfall excess is 10 in., and reservoir evaporation is 30 in. Determine the net change in storage during the year if a local municipality is allowed to take 100 mgd from the reservoir. You may neglect infiltration and seepage from the reservoir.

2. Little Joe Creek, a small tributary of the Arkansas River near Tulsa, Oklahoma has a 640-acre surface reservoir planned for construciton. The following data are known for the worst (lowest) 12-month period of record: precipitation is 20 in., lake evaporation is 58 in., and controlled outflow to satisfy water rights downstream is 5.0 cfs.

   The stream gaging record for the year indicates an average discharge into the reservoir of 3.0 cfs, i.e., 0.5 cfs/mi^2 of drainage area. Several farmers will be allowed to irrigate 700 acres (typical consumptive use will be 3.5 ft/yr). Adjacent to and feeding from the reservoir water are 100 acres of wetlands, which are estimated to use 9.6 ft per year.

There are small ungaged areas (totaling 2000 acres and including a small spring) draining into the reservoir. Determine the annual volume of ungaged flows (i.e., surface flow and flow from the spring) if the reservoir elevation is unchanged during the year.

1.10

REFERENCES

Betson, Roger. 1978. "Bulk Precipitation and Streamflow Quality Relationships in Urban Areas," *Water Resources Research,* **14** (6), pp. 1165–1169.

Engineering Computer Graphics Laboratory, 1996. *Watershed Modeling System,* Brigham Young University, Utah, USA.

Gambolati, G., Gatto, P., and Freeze, R. A. 1974. "Predictive Simulation of the Subsidence of Venice," *Science* **183,** pp. 849–851.

Haestad Methods, 1992. *HEC PACK,* 37 Brookside Rd., Waterbury, CT (phone 1-800-727-6555).

Haestad Methods, 1993. *POND PACK,* 37 Brookside Rd., Waterbury, CT (phone 1-800-727-6555).

Huber, W. C. and Dickinson, R. E., 1988. *Storm Water Management Model User's Manual,* Version 4 with addendums, EPA/600/3-88/001, U.S. Environmental Protection Agency, Athens, GA.

Miller, G. T., Jr. 1986. *Living in the Environment—Concepts, Problems and Alternatives.* Wadsworth Publishing Co., Inc., Belmont, CA.

Natural Resources Conservation Service (formerly Soil Conservation Service), 1986. "Urban Hydrology for Small Watersheds," Technical Release 55, Washington, D.C.

Novotny, V., and Chesters, G. 1981. *Handbook of Nonpoint Pollution: Sources and Management.* Van Nostrand-Reinhold, New York.

Streamline Technologies Inc., 7125 University Blvd., Winter Park, FL, 32792 (phone 1-407-679-1696).

U.S. Army Corps of Engineers, 1990. *HEC-1 Flood Hydrograph Package,* Hydrologic Engineering Center, Davis, CA.

U.S. Army Corps of Engineers, 1991. *Using HEC-1 on a Personal Computer,* Training Document No. 32, Hydrologic Engineering Center, Davis, CA.

U.S. Water News, May 1985, "Houston's Subsidence Problem," **1** (11), pp. 1 and 17.

Wanielista, M. P., and Yousef, Y. A. 1993. *Stormwater Management.* John Wiley & Sons, New York.

Wanielista, M. P., Yousef, Y. A., Taylor, J. S., and Cooper, C. D. 1984. *Engineering and The Environment.* Brooks/Cole Engineering Division, Wadsworth, Belmont, CA.

Yousef, Y. A. et al. 1985. *Consequential Species of Heavy Metals,* BMR-85-286 (FL-ER-29-85), Florida Department of Transportation, Tallahassee, FL.

CHAPTER 2

COMPUTATIONAL METHODS

In this chapter computational methods frequently used to solve hydrology problems are described. Probability and statistics are mathematical tools which are often used to describe and predict the variable nature of hydrologic events. Since hydrology often deals with large arrays of numbers, matrix mathematics are also often used. These are described in this chapter.

2.1

PROBABILITY CONCEPTS

When one accurately predicts future water flow rates and volumes with knowledge of the uncertainties of meteorologic, hydrologic, and watershed changes, design and operation of projects cost less and are more effective. However, a designer or planner is not certain of the future rate and volume that will affect a project. To predict these flow rates and volumes of hydrological processes, probabilistic and statistical models can be designed and used. This section presents probability and statistics concepts for rainfall and runoff events, especially the extreme events of droughts and floods. The concepts of this section are used in many water resources related areas.

Hydrologic and meteorologic information is sometimes available at a location where a change is anticipated. This availability of data is very helpful and statistical descriptions from the data can be developed. However, data information at a location is frequently incomplete or available for some other location. Statistics still can be used to correlate one location to another in hopes of determining "best" predictive and descriptive mathematical models.

2.1.1 Terminology

Some hydrology and stormwater management terms that use concepts from probability and statistics may be misleading or not completely understood. In many cases, data are collected over a short period of time in hopes of estimating the results of processes or driving forces (hydrologic, hydraulic, biological, etc.) that are not well defined. Thus, samples are taken and inferences about the longer time period for driving forces are made. There exists terminology that relates to these estimates, some of the terminology is shown in Table 2.1.

TABLE 2.1   Terminology Related to Statistical Observations

TERM	DESCRIPTION
Accuracy	The closeness of a measurement to the true value of the quantity being measured or to an accepted reference value.
Bias	A systematic variation or lack of randomness in a set of observations that results from a systematic error in data collection or analysis.
Confidence level	A quantitative expression of the reliability of an estimated value. The expression is usually stated in probability terms.
Consistency	A property of numbers that is related to sample size. As sample size increases, a consistent sample does not deviate from the population mean more or less than a fixed amount.
Efficiency	A measure of the quality of an estimator or set of observations. Efficiency is inversely proportional to the variance. A sample is more efficient if it has a smaller variance than another sample.
Independent	One event (i.e., rainfall) is independent of another (pipe roughness) when the occurrence of the second event has no bearing whatsoever on the occurrence or nonoccurrence of the first event.
Mutually exclusive	The occurrence of one event (i.e., rainfall) precludes the occurrence of another (i.e., evaporation).
Precision	The variation in an observation or set of observations due to random error. It is the measure of the repeatability of a series of observations or measurements.
Random error	Chance fluctuations in the value of a variable that occur when a series of measurements are taken under the exact same conditions.
Reliability	The expression of how well a model or other predictor technique measures what it is supposed to measure, including both accuracy and precision properties. Refers also to the consistency and precision of an instrument or measurement technique.
Repeatability	The precision associated with an individual observer taking several measurements of the same variable under the same conditions but at different times using the same equipment or measurement techniques.
Sample	A small number of observations from a larger number or potential observations. A random sample is one of which each observation has the same probability of being chosen.
Sensitivity	The ability of an instrument (or instrument plus observer) to measure changes in a variable or the ability of a model or other mathematical predictor to produce realistic responses in output variables when levels of input variables are changed.
Significance	In data analysis, significance refers to the relationship of a number of some reference value. In statistical analyses the term should be qualified with a probability statement.
Sufficiency	The degree to which a parameter derived from a given sample represents or extracts information from the corresponding population.
Tolerance	The allowable deviation from a numerical standard or the range of variation permitted a given number.
Uncertainty	The variation of a variable or the tendency of outcomes to vary when repeated measurements are made under identical conditions. Uncertainty consists of both random and systematic error.

2.1.2 Definition of Probability

Probability is empirically defined as the number of times a specific event occurs from the total number of events measured, or

$$Pr(X) = x/n; \qquad \lim n \to \infty \tag{2.1}$$

where

$Pr(X)$ = probability of event X
    x = number of occurrences of X event
    n = total number of recorded outcomes

2.1.3 Probability and Statistics Concepts

Because of the extreme variability and lack of deterministic relationships for rainfall, runoff, and other hydrologic processes, it is frequently necessary to use probability and statistics concepts to aid in defining and predicting these events. Some of the basic ideas and concepts of probability and statistics are defined in this section.

2.1.3A Independence

When the occurrence of one event does not affect the occurrence of another, the two events are independent. As an example, the percent impervious area of a watershed is independent of volume of rainfall per year. Rainfall is independent of watershed characteristics. However, runoff is dependent on watershed characteristics. The probability of the occurrence of two independent events is the product of both probabilities or their intersection (symbolized by ∩).

$$Pr\{A \cap B\} = P\{A\} \cdot P\{B\} \tag{2.2}$$

Another probability statement is one for which either one event or both can occur. It is called a union (symbolized by ∪). An example would be the probability of either runoff or rainfall or both occurring.

2.1.3B Conditional Probability

A conditional probability is one which depends on the occurrence of some other event. Almost all probabilities are conditional in a space or time relationship. Example statements are (1) the probability of a 5-in. rainfall event given data for the midwestern United States, (2) the probability of at least 1-in. of runoff given a 10-in. snowfall in Alaska during December, or (3) the probability of a 3-in. rainfall given a dry period of 72 hr following a storm event. In these three statements, specific quantities of the hydrologic variable are specified and after the word "given" is the conditional statement.

2.1.3C Empirical Probabilities

It is most likely that exact probabilities will never be obtained because the sample size hardly ever approaches the total number of outcomes (trials) for hydrologic processes. However, samples of the recorded outcomes can be tabulated with associated empirical probabilities.

Most hydrologic processes occur in patterns; and the engineer, hydrologist, and planner must determine and use this pattern. One way of describing a pattern is to develop a frequency distribution for the values of the hydrologic process. A frequency distribution is a plot of the values of a hydrologic process as the abscissa and the frequency of occurrence as the ordinate. If the frequency is divided by the total number of samples, the ordinate is expressed as empirical probability. To use a probability or frequency distribution effectively, one must understand fundamental probability properties and the development of a distribution.

□ **EXAMPLE PROBLEM 2.1**

Consider calculation of a probability for average yearly stream flow greater than 40 m^3/sec, given stream flow records for 50 years. As an additional qualification, the users of this probability wish to know if this flow rate is exceeded in a year and how many times per year. On examining the stream flow data, it was determined that 40 m^3/sec was exceeded in 10 yearly recordings. Therefore, the empirical probability based on a sample size of 50 is:

Solution

$$\Pr(X > 40 \text{ m}^3/\text{sec}) = 10/50 = .20$$

This estimate for Example Problem 2.1 was made from a number of possible outcomes and is a sample of that population. As more data are collected, the empirical probability will become more accurate.   □

2.1.3D Probability Properties

Hydrologic events can be assumed to be discrete but often are continuous variables. Regardless of the assumption, the properties of probability distributions are analogous for discrete, $\Pr(x)$, and continuous, $f(x)$, cases:

$$\Pr(x) \quad \text{or} \quad f(x) \geq 0$$

and

$$\sum_{(all\,x)} \Pr(x) = 1 \quad \text{or} \quad \int_{-\infty}^{+\infty} f(x)\,dx = 1 \tag{2.3}$$

The sum of probabilities is referred to as the cumulative distribution function and is written as:

1. Less than type:

$$F(x) = \Pr(X \leq x) = \sum_{z \leq x} \Pr(z) \tag{2.4}$$

$$F(x) = \Pr(X \leq x) = \int_{-\infty}^{x} f(z)\,dz \tag{2.5}$$

2. Greater than type (exceedence):

$$G(x) = \Pr(X \geq x) = 1 - \sum \Pr(z) = 1 - F(x) \tag{2.6}$$

$$G(x) = \Pr(X \geq x) = \int_{x}^{\infty} f(z)\,dz \tag{2.7}$$

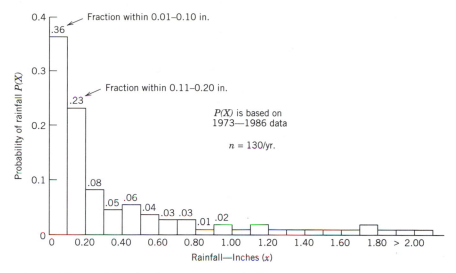

FIGURE 2.1 Rainfall probability histogram—Orlando Jetport. *Source:* U.S. Department of Commerce, 1973–1986, interevent dry period of 4 hr.

2.1.3E Graphical Presentations of Probability

Since there exists a need to determine the shape (or type) of distributions, there is a need for graphic presentations of hydrological data. One such graphic is the histogram—a plot of empirical probabilities corresponding to process values. An example is shown here to illustrate its use.

A histogram of rainfall data using 14 yr of rainfall event data is shown in Figure 2.1. The rainfall quantity is the total rain per storm event, where a storm event is defined by any uninterrupted rainfall preceded by 4 hr or more of no rainfall. Cumulative frequency distributions of rainfall are shown for the less than type for Austin, Texas; and Baltimore, Maryland in Figure 2.2. Note that the histograms have a similar shape. These figures illustrate the estimated probability of rainfall per storm event being less than or equal to a stated volume. For example, the probability of rainfall being less than or equal to 0.50 in. is approximately 80%. Local climatological data can be used to construct rainfall, temperature, wind speed, or other types of histograms. Once the histogram has been determined, a theoretical probability distribution can be assigned to the random events.

2.1.3F Central Tendency and Distribution Moments

A distribution may be described by a central value (mean), variability, and skewness.

The most common measure of central tendency is the mean of the empirical (observed) data or first moment about the origin, and is defined by

$$\bar{x} = \frac{1}{n} \sum_{i=1}^{n} x_i \tag{2.8}$$

where

$\bar{x}$ = mean value sample estimate
x_i = data points for all measurements i
n = total number of data points

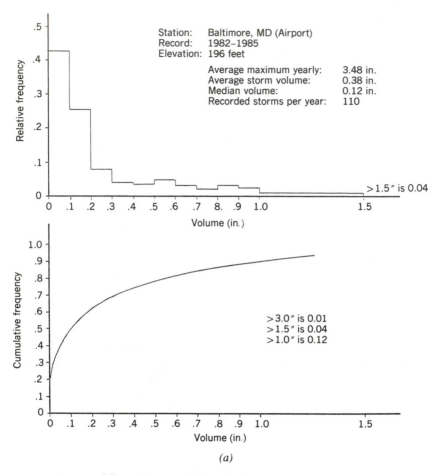

FIGURE 2.2   Rainfall cumulative probability distributions.

2.1.3G   Variability

Other measures of central tendency are the median or middle value of the ordered empirical data and the mode or value that appears most frequently (peak of a distribution).

The most common measure of variability of a sample is the standard deviation (s) and the square of the standard deviation is known as the variance (s^2). The standard deviation has the units of the variable itself, while the variance is the square of the values and possibly more difficult to interpret. The variance of the sample is calculated from:

$$s^2 = \left[\frac{1}{(n-1)} \right] \sum_{i=1}^{n} (x_i - \bar{x})^2 \tag{2.9}$$

The estimate for the variance of the sample is based on ($n - 1$) data points, reportedly because $\bar{x}$ is used in the estimate and replaces one of the data points, x_i. This is called a reduction in the degree of freedom or an unbiased estimate. The variance also is known as the second moment of the distribution. Relative variability among sample data sets can be compared using the coefficient of variability (C_v), as defined by

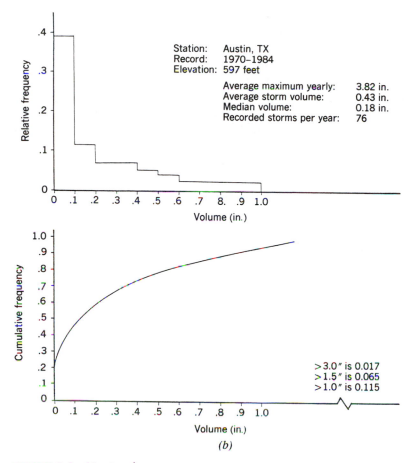

FIGURE 2.2 (Continued)

$$C_v = \frac{s}{\bar{x}} \tag{2.10}$$

A measure of statistical reliability is the standard error of estimate, or just standard error, expressed by

$$s_e = \delta \frac{s}{\sqrt{n}} \tag{2.11}$$

where
s = standard deviation
n = sample size
δ = standard error parameter dependent on type of distribution

The proportion of the total variation that can be explained by the relationship between the sample parameters is often referred to as the correlation coefficient (see Equation 2.39). That which is due to unexplained factors is related to the standard error of estimate (see Equation 2.41). It should be noted that the standard error parameter depends on the type of distribution:

normal δ is a function of return period, T_r
log normal δ is a function of T_r and C_v (coefficient of variability)
Gumbel δ is a function of T_r and n
log Pearson δ is a function of T_r and G (skewness).

Skewness is a measure for the degree of asymmetry. Right skewness indicates a distribution with greater variability to the right (typically, the distribution is "stretched" to the right). Similarly, there is a left skewness and a perfectly symmetrical distribution about the central value. The untransformed data of Figure 2.1 indicate a right skewed distribution.

The third moment of a distribution would estimate the skewness or weights on either side of the mean. An estimator for the true population skewness (G) using the sample observation is:

$$G = \frac{n \sum_{i=1}^{n} (x_i - \bar{x})^3}{(n-1)(n-2)s^3} \tag{2.12}$$

❑ **EXAMPLE PROBLEM 2.2**

For the following data recorded in mm-annual rainfall, calculate the mean, standard deviation, coefficient of variability, and skewness.

YEAR	SCRANTON, PA	ORLANDO, FL	YEAR	SCRANTON, PA	ORLANDO, FL
	LOCATION			LOCATION	
1950	602	1150	1968	680	1252
1951	541	1035	1969	480	892
1952	732	1341	1970	922	1494
1953	446	860	1971	710	1200
1954	584	1196	1972	454	1026
1955	690	1244	1973	635	1162
1956	724	1299	1974	699	1227
1957	832	1363	1975	703	1150
1958	717	1248	1976	814	1143
1959	659	1190	1977	710	1190
1960	571	1040	1978	784	1340
1961	692	1274	1979	507	994
1962	840	1401	1980	512	1043
1963	730	1296	1981	852	1350
1964	630	1162	1982	671	1109
1965	618	1100	1983	784	1206
1966	687	1291	1984	842	1304
1967	777	1340	1985	649	1156

TABLE 2.2 Results of Example Problem 2.2

	A	B	C	D	E	F
1	Year	Data		$(x - \bar{x})^2$		$(x - \bar{x})^3$
2	1950	602		6084		−474552
3	1951	541		19321		−2685619
4	1952	732		2704		140608
5	1953	446		54756		−12812904
6	1954	584		9216		−884736
7	1955	690		100		1000
8	1956	724		1936		85184
9	1957	832		23104		3511808
10	1958	717		1369		50653
11	1959	659		441		−9261
12	1960	571		11881		−1295029
13	1961	692		144		1728
14	1962	840		25600		4096000
15	1963	730		2500		125000
16	1964	630		2500		−125000
17	1965	618		3844		−238328
18	1966	687		49		343
19	1967	777		9409		912673
20	1968	680		0		0
21	1969	480		40000		−8000000
22	1970	922		58564		14172488
23	1971	710		900		27000
24	1972	454		51076		−11543176
25	1973	635		2025		−91125
26	1974	699		361		6859
27	1975	703		529		12167
28	1976	814		17956		2406104
29	1977	710		900		27000
30	1978	784		10816		1124864
31	1979	507		29929		−5177717
32	1980	512		28224		−4741632
33	1981	852		29584		5088448
34	1982	671		81		−729
35	1983	784		10816		1124864
36	1984	842		26244		4251528
37	1985	649		961		−29791
38	Total	24480	Total	483924	Total	−10943280
39	Number Pts	36	s^2	13826.4	G	−0.2036
40	Average	680	s	117.59		
41			C_v	0.173		

Solution

STATISTICAL MEASURE	SCRANTON		ORLANDO	
	mm	in.	mm	in.
Mean	680	26.77	1196	47.07
Standard deviation	118	4.64	140	5.50
Skewness	−0.19	−0.19	−0.37	−0.37
Coefficient of variability	0.17	0.17	0.12	0.12

This problem can be solved using a programmable calculator or a spreadsheet. A sample spreadsheet showing the calculations involved is given. A computer program could also be written to solve for these statistics.

```
Cell B38    @SUM(B2..B37)               Cell D40    +D39^.5
Cell B39    @COUNT(B2..B37)             Cell D41    +D40/B40
Cell B40    +B38/B39                    Cell F38    @SUM(F2..F37)
Cell D2     +(B2-$B$40)^2               Cell F2     +(B2-$B$40)^3
Cell D39    +D38/(B39-1)                Cell D38    @SUM(D2..D37)
Cell F39    +(B39*F38)/((B39-1)*(B39-2)*D40^3)
```

The results indicate a higher average in Orlando than in Scranton, 1196 mm and 680 mm, respectively. The variability in the yearly averages is relatively lower in Orlando as measured by the coefficient of variability although the standard deviation is higher in Orlando. The distributions both have a slight negative (left) skewness.   ❑

2.2

PROBABILITY DISTRIBUTIONS

Seven probability distributions are discussed. The first two are discrete. The remaining ones are continuous. All are useful to describe hydrologic events. The last three are needed most frequently for flood flow events.

2.2.1 Binomial Distribution

A random variable (X), that can assume two values, where the probability of one value, p, has a binomial distribution if its probability distribution function can be written as

$$\Pr(x) = \binom{n}{x} p^x (1 - p)^{n-x} \qquad \text{for } x = 0, 1, \ldots, n \qquad (2.13)$$

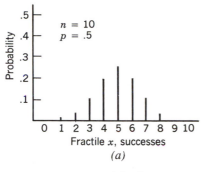
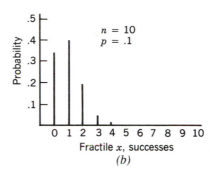

FIGURE 2.3 Binomial distribution.

where

$$\binom{n}{x} = \frac{n!}{x!(n-x)!}$$

p = probability of one value, usually defined as a success, $0 \le p \le 1$
n = number of trials
x = number of successes

A graph of the binomial distribution with fixed n and p would be similar to that shown in Figure 2.3.

❑ **EXAMPLE PROBLEM 2.3**

You are in charge of a field crew measuring stage downstream from a regulated reservoir. Every time the stage is greater than 5 m the regulatory agency has successfully determined a violation of state and federal standards. If the probability of violation over a long period of time is 0.20, what are some conclusions when four out of six stage measures are greater than 5 m during runoff conditions?

Solution

The probability of four out of six violations (successes) is very rare or $\Pr(x = 4) = \binom{6}{4} (.2)^4 (.8)^2 = .015$. Thus, the violations have increased or the sample is characteristic of direct runoff rather than general stream flow data. Possible mixed populations are involved, that is, runoff (nonpoint), reservoir releases (point), and groundwater conditions. The random or nonexplained nature of hydrologic events can be analyzed using the binomial distribution. ❑

2.2.2 Poisson Distribution

Stream flow can be characterized by a Poisson distribution when it is input to a reservoir. Over a fixed period of time, stream flow is considered an "arrival" at the contemplated reservoir site. These arrivals characteristically follow a Poisson distribution. The random variable (X) in these situations has a Poisson distribution if its probability density function can be written as

$$\Pr(x) = \frac{\lambda^x e^{-\lambda}}{x!} \qquad \text{for } x = 0, 1, 2, \ldots \tag{2.14}$$

where
λ = positive constant (mean value of arrivals)
x = any nonnegative integer
e = a constant = 2.7182 . . .

A graph of the Poisson distribution is typically right skewed as in Figure 2.4.

The Poisson distribution is useful for design of reservoirs when surface water inputs of seepage are not described by a deterministic (nonrandom) relationship. It is possible that the random portion of inputs or outputs can be described by the Poisson distribution.

❏ **EXAMPLE PROBLEM 2.4**
_____

Assume that the water level change in a reservoir per month not explained by deterministic equations will average 1 cm and from a graphical plot the changes can be described by a Poisson distribution. Unit changes in depth are the closest measurements possible and are considered discrete. What is the probability of a 2-cm change in elevation during one month?

Solution

$$\Pr(x = 2) = \frac{1^2 e^{-1}}{2!} = .184$$

What is the probability of _at most a_ 2-cm change? This is the cumulative probability or:

$$\Pr(x = 0) + \Pr(x = 1) + \Pr(x = 2)$$

or

$$F(x \le 2) = .368 + .368 + .184 = .92 \quad ❏$$

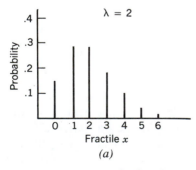

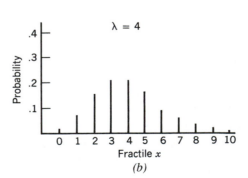

FIGURE 2.4   Poisson distribution.

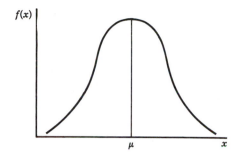

FIGURE 2.5 Normal distribution.

2.2.3 Normal Distribution

One of the most important continuous probability distributions is the normal. Its distribution function is

$$f(x) = \left(\frac{1}{\sigma\sqrt{2\pi}}\right) \exp\left[-\frac{1}{2}\left(\frac{x-\mu}{\sigma}\right)^2\right] \quad \text{for } -\infty < x < \infty \tag{2.15}$$

where

σ = standard deviation or s, the sample standard deviation, which is an estimate of σ.

x = any value

μ = population mean or $\bar{x}$, the sample mean, which is an estimate of μ.

A graph of the typical normal density function is shown in Figure 2.5.

☐ **EXAMPLE PROBLEM 2.6**

Suppose that the average yearly concentration of total phosphorus in a stream is 1.65 mg/L with a variance of 0.64 mg²/L². These averages are based on weekly samples averaged for one year. Thus, the sample size for the average yearly values is 52. Twelve years of data are available. During a rainy year (25% more than average), the yearly average concentration for that year (1 of 12 years) was 2.20 mg/L. What is the probability that yearly average concentrations will be equal to or greater than 2.20 mg/L?

Solution

The response to this question relies on the assumption of the application of the central limit theorem, which states that the distribution of mean values is normal. Solving Equation 2.15 is usually done by tables or the use of standard deviates (number of standard deviations on either side of the mean value). Statistical tables are presented in Appendix D, and are used with the standard normal deviate, calculated as: $z = (x - \bar{x})/s$. For this problem:

$$z = \frac{(2.20 - 1.65)}{.8} = .6875$$

Using the Table D.1, Appendix D, the probability of being less than 2.20 mg/L is .7541, or 75% of the time the average yearly concentrations will be less than

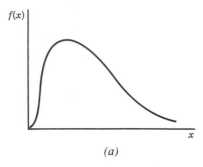 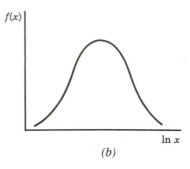

(a)   (b)

FIGURE 2.6 Log-normal distribution.

2.20 mg/L and 25% of the time the average yearly concentrations will be greater than 2.20 mg/L. ❑

2.2.4 Log-Normal Distribution

A plot of average weekly flow rates is generally right skewed and does not appear to be a normal distribution. However, if the log of weekly values is plotted as a histogram, the resulting distribution may be normal. The characteristic shapes are shown in Figure 2.6. The distribution function is

$$f(x) = \frac{1}{\sigma x \sqrt{2\pi}} \exp\left[-\frac{1}{2}\left(\frac{\ln x - \mu}{\sigma}\right)^2\right] \qquad \text{for } x > 0 \qquad (2.16)$$

Note that μ and σ are the mean and standard deviation of $\ln(x)$. If the probability of being between two values is required, it can be calculated using:

$$\Pr(a \leq X \leq b) = \int_a^b \frac{1}{\sigma \sqrt{2\pi}\, x} \exp\left[-\frac{1}{2}\left(\frac{\ln X - \mu}{\sigma}\right)^2\right] dx \qquad (2.17)$$

or with the use of normal tables:

$$\Pr(a \leq X \leq b) = F\left(\frac{\ln b - \mu}{\sigma}\right) - F\left(\frac{\ln a - \mu}{\sigma}\right) \qquad (2.18)$$

2.2.5 Exponential Distribution

The exponential distribution is used in the study of reservoir holding volumes and rainfall intensities. It is a continuous random variable having the following form:

$$f(x; \lambda) = \lambda e^{-\lambda x} \qquad \text{for all } x \geq 0$$

and

$$F(x; \lambda) = 1 - e^{-\lambda x} \qquad (2.19)$$

where λ = constant > 0 and the inverse of the average.

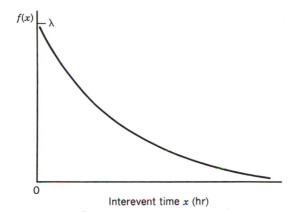

$f(x)$

λ

0

Interevent time x (hr)

FIGURE 2.7 Exponential distribution.

The interarrival time distribution for two successive arrivals of flows to a detention pond will characteristically follow an exponential distribution as shown in Figure 2.7. If the average interevent time is 84 hr, $\lambda = \frac{1}{84}$ and the probability of an interevent time less than 5 days (120 hr) is .76.

2.2.6 Gumbel Distribution

Gumbel or Type I distributions (Gumbel, 1945) have been used to describe annual flow events. A generalized flood flow diagram is shown as Figure 2.8. Most flood flow probability distributions have similar shapes. Flood flows can be approximated by a Gumbel distribution to estimate flood probabilities. The Gumbel distribution has its skewness fixed at $+1.14$. The largest flood flow/year over 365 daily values is picked for each year. The distribution of yearly flood flows is then plotted and an exceedence probability calculated from

$$1 - F(x) = 1 - e^{-e^{-b}} \tag{2.20}$$

where
$b = (1/0.7797s)(x - \bar{x} + 0.45s)$
x = magnitude of the flood (m³/sec)
$\bar{x}$ = average flood magnitude (m³/sec)
s = standard deviation of flood magnitudes (m³/sec)

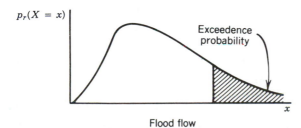

$p_r(X = x)$

Exceedence
probability

x

Flood flow

FIGURE 2.8 Flood peak distributions.

TABLE 2.3 Gumbel Probability and Return Period

REDUCED VARIATE b	PROBABILITY OF EXCEEDENCE	RETURN PERIOD (RECURRENCE INTERVAL)
0.367	.50	2.0
1.367	.20	5.0
2.250	.10	10.0
2.970	.05	20.0
3.199	.04	25.0
3.902	.02	50.0
4.601	.01	100.0
5.296	.005	200.0

☐ **EXAMPLE PROBLEM 2.6**

If flood flows on a large watershed in Pennsylvania has an average value of 1200 m^3/sec with a variance of 62,500 $(m^3/sec)^2$, what is the probability that a flood will be equal to or exceed 2000 m^3/sec using the Gumbel distribution?

Solution

Calculating the reduced variate, b, then the probability one obtains

$$b = \frac{1}{0.7797(250)} [(2,000 - 1,200 + .45(250)]$$
$$= 4.68$$

and

$$1 - F(x) = 1 - e^{-e^{-4.68}} = .0092$$

which means that this flood has a chance of occurring approximately one time each 100 years.   ☐

Since the exceedence probability, and hence the return period, is a direct function of the value of b, Table 2.3 may be of some help in determining the probabilities.

2.2.7 Log-Pearson Type III

The log Pearson Type III distribution is used as the acceptable procedure for estimating flood flows (Work Group on Flood Flow Frequency, 1977). If a series of floods for each year are available, these are converted to logarithm values from which the skewness, mean and standard deviation are estimated. Exceedence probabilities are estimated using a normal distribution with adjustments for the skewness of the flood flows (Beard, 1962). Therefore, the skewness of the distribution is estimated from the transformed logarithm values using Equation 2.12.

From the Water Resources Council Hydrology Committee (Work Group on Flood Flow Frequency, 1977), alternate computing formulas for skewness and standard deviation are

$$G = \frac{N^2(\Sigma X^3) - 3N(\Sigma X)(\Sigma X^2) + 2(\Sigma X)^3}{N(N-1)(N-2)s^3} \qquad (2.21)$$

where

$$s = \left[\frac{(\Sigma X^2) - (\Sigma X)^2/N}{N-1} \right]^{.5} \qquad (2.22)$$

Knowing the skewness coefficient and the exceedence probability, it is possible to calculate the value of the corresponding flood using

$$X = \overline{X} + Ks \qquad (2.23)$$

where

$$K = \frac{2}{G} \left[\left[\left(K_n - \frac{G}{6} \right) \left(\frac{G}{6} \right) + 1 \right]^3 - 1 \right] \qquad \text{for } -1.0 < G < 1.0 \qquad (2.24)$$

and

$$K = \text{Type III reduced deviate}$$

$$K_n = \text{standard normal deviate for that probability}$$

Table D.3, Appendix D, can be used for estimating K.

It should be noted that Equation 2.23 is generally applicable to any distribution. The reduced variate, K, sometimes called the frequency factor, may be computed with the aid of appropriate tables (e.g., D.1, D.3, or D.9) of Appendix D.

The skew coefficient (G) is sensitive to extreme events and thus difficult to estimate using small samples (less than 100 yr). Adjustments for the skew coefficient are suggested as follows:

RECORD YEARS, N	SKEW VALUE	
≥ 100	G	(2.25)
$26 - 100$	$\left(\dfrac{N-25}{75} \right) G + \left(1 - \dfrac{N-25}{75} \right) G_g$	(2.26)
≤ 25	G_g	(2.27)

where
G_g = generalized skew coefficient for an area (Hardison, 1974)

2.2.8 Recurrence Interval and Risk

The recurrence interval (T_r) is the average interval of time within which an event will be equaled or exceeded or $(1/T_r)$ the probability of occurrence in any one year. The recurrence interval should be based on a long-time history of the event. It is one of the more significant statistics and is easy to understand by a nontechnical person. Of course it must be emphasized that a long period of record is necessary because a rainfall that occurs once per 10-yr interval (10-yr recurrence interval) may occur with low probability 2 years in a row.

If an event is designated by $(X \geq x)$ and its probability of occurrences by $\Pr(X \geq x)$, the recurrence interval is the inverse of the probability, or

$$T_r = \frac{1}{\Pr(X \geq x)} = \frac{1}{G(x)}$$

$$T_r = \frac{1}{1 - \Pr(X \leq x)} = \frac{1}{1 - (1 - G(x))} \tag{2.28}$$

The probability that this event will not occur is:

$$1 - \Pr(X \geq x) = 1 - G(x) \tag{2.29}$$

The probability that this event will not occur in n time periods is calculated from the binomial distribution (see Section 2.2.1) as

$$\Pr(x = 0: n, p) = \binom{n}{0}(1 - p)^n(p)^0$$

or

$$\Pr(0, n, p) = (1 - p)^n \tag{2.30}$$

The probability (R) that this event, x, will occur at least once (one or more times) in n time periods is expressed by a term called risk, or

$$R = \sum_{z=1}^{n} \Pr(z; n, p) = 1 - \Pr(0; n, p) = 1 - (1 - p)^n \tag{2.31}$$

If the time period is equal to one year, the recurring interval has been defined as the return period. This is the same definition of return period as used in the intensity–duration–frequency curves for rainfall.

❑   **EXAMPLE PROBLEM 2.7**

What is the probability of no rainfall volume during the next 3 yr greater than or equal to the one in 20-yr rainfall volume?

Solution

The probability of a 1 in 20-yr rainfall is the inverse of the return period or 1/20. Using Equation 2.30:

$$\Pr(0; 3, .05) = (1 - .05)^3 = .86 \quad ❑$$

□   ***EXAMPLE PROBLEM 2.8***

Given the annual peak gage heights of a canal, calculate the empirical probability and the return period of an annual peak greater than or equal to 17 m. The gage heights are

YEAR	ANNUAL HEIGHT (m)
1968	17
1969	21
1970	12
1971	10
1972	19
1973	14
1974	16
1975	12
1976	15

Solution

$$\Pr(X \geq 17) = \frac{x}{n} = \frac{3}{9} = 0.33 = G(x)$$

and

$$T_r = \frac{1}{G(x)} = 3 \text{ yr} \quad \square$$

2.2.9   Evaluation of Risk

Rewriting Equation 2.31 in terms of the return period (recurrence interval), the probability of a stated event (design discharge, rainfall) being exceeded at least once in a design life is a measure of the risk one takes. If n is the design life in years, and T_r is the recurrence interval, a tabulation of risk using Equation 2.31 is developed and shown in Table 2.4.

□   ***EXAMPLE PROBLEM 2.9***

For a structure on a highway, it was decided a design life of 50 yr should be used. If the structure transports runoff, what is the risk of at least one exceedence if the runoff quantity is associated with the 100-yr discharge?

Solution

From Table 2.4, there is a 39% chance that the 100-yr discharge will occur during the 50-yr life of the project.   □

□   ***EXAMPLE PROBLEM 2.10***

If the designer decides that a 10% risk of failure of the transport structure is acceptable, what is the recurrence interval for a design life of 50 yr?

TABLE 2.4 Approximate Risk of At Least One Exceedence During a Design Life for a Recurrence Interval or Exceedence Probability

EXCEEDENCE PROBABILITY (G)	RECURRENCE INTERVAL (T_r)	DESIGN LIFE—YEARS (n)					
		2	5	10	25	50	100
.5	2	.75	.97	≅1.00	≅1.00	≅1.00	≅1.00
.2	5	.36	.67	.89	≅1.00	≅1.00	≅1.00
.1	10	.19	.41	.65	.93	.99	≅1.00
.04	25	.08	.18	.34	.64	.87	.98
.02	50	.04	.10	.18	.40	.64	.87[a]
.01	100	.02	.05	.10	.22	.39	.63
.002	500	.004	.01	.02	.05	.10	.18
.001	1000	.002	.005	.01	.03	.05	.10

[a] Example calculation:
$R = 1 - (1 - 1/T_r)^n$
$R = 1 - (1 - 1/50)^{100}$
$R = 1 - (.98)^{100} = 1 - .13 = .87$

Solution

Rearranging Equation 2.31 to solve for T_r.

$$T_r = \frac{1}{1 - (1 - R)^{1/n}}$$

$$= \frac{1}{1 - (1 - .10)^{1/50}} = 475 \text{ yr}$$

From Table 2.4, the approximate value is 500 yr. To reduce the risk using the data of Example Problem 2.9 from 39% to 10%, the transport structure must be designed with a peak flow associated with the 1/500-yr storm event. Thus, for very low levels of risk, accurate measures must be used to predict the values associated with the high-return period storm. ❑

2.2.10 Limited Samples and Uncertainty

When repeated measurements are made of the same hydrologic process there is variability in the estimate of the hydrologic process. Given a return period for rainfall, estimates for the volume of rainfall can and do vary for the same location. Uncertainty in the estimates can be reduced by eliminating sources of data collection and analyses errors. How accurate are these estimates? How accurate do they have to be for a given application? All these questions have to be answered with a limited database.

The accuracy of the data depends on the length of record available and on the assumed probability relationships (frequency distribution). The reliability of an estimate includes both the accuracy and the precision at which the event is measured. The reliability of the estimate can be measured by the confidence limit for that frequency. As the length of record increases, the reliability increases. Approximate values of reliability (percent chance) can be calculated for different exceedence probabilities (return periods). Approximate values are shown in Table 2.5. The results of being 10%, 25%, or 50% greater than or less than an estimate should be evaluated. In fact, there is approximately a 100% chance that the range of values of the estimate

TABLE 2.5 Approximate Reliabilities (% chance) as a Function of confidence Limit, Return Period, and Record Length

RETURN PERIOD (yr)	RECORD LENGTH (yr)	CONFIDENCE LIMIT (% ERROR)		
		±10%	±25%	±50%
2	10	47	88	99
	25	68	99	100
	100	96	100	100
10	10	46	77	97
	25	50	93	99
	100	85	100	100
50	10	37	70	91
	25	46	91	97
	100	73	99	100
100	10	35	66	90
	25	45	89	98
	100	64	99	100

for the 2-yr return period with 25 yr of data will be within 50% of the estimated values. There is a 68% chance that the range of values will be within 10% of the estimate.

□ **EXAMPLE PROBLEM 2.11**

If the frequency distribution estimate of the 100-yr storm event is 28 cm (11.0 in.) using 25 yr of data, and we wish to be 98% reliable in the estimate, what is the range of storm event values to satisfy this criteria?

Solution

From Table 2.5, there is a 98% chance that the estimated values will have a confidence interval of 50%, or the storm event can range from 14 cm (5.5 in.) to 42 cm (16.5 in.).   □

2.3

EMPIRICAL FREQUENCY DISTRIBUTION ANALYSIS

When hydrologic data are available for a location, then a statistical analysis of the data using frequency distributions is possible. Theoretical frequency distributions were defined in Section 2.2. More detail on the procedures for developing frequency distributions and for estimating the parameters of the distributions are presented in this section.

Before an analysis can be performed, a specific frequency distribution must be chosen with reasons for selection. Past experiences with rainfall and flood data generally indicate that these data will follow extreme event distributions, such as two parameter log-normal, log Pearson Type III, Weibull, Gumbel, or the three parameter log-normal. A calculation of the skewness parameter (third moment) helps to identify the shape as an extreme event (positive skewness).

TABLE 2.6 Maximum Annual Flow Rate Histogram Using Class Intervals

(1) FLOW RATE INTERVAL	(2) TALLY OF FREQUENCIES	(3) NUMBER	(4) RELATIVE FREQUENCY (3) ÷ 29	(5) CUMULATIVE FREQUENCY Σ(4)
41–60	//	2	.069	.069
61–80	///	3	.103	.172
81–100	7₩ //	7	.242	.414
101–120	7₩	5	.172	.586
121–140	7₩ /	6	.207	.793
141–160	///	3	.103	.896
161–180	//	2	.069	.965
181–200	/	1	.035	1.000
	Totals	29	1.000	

2.3.1 Histogram Development

A common way of arranging flow data is by order of magnitude (largest to smallest or smallest to largest). The arrangement is identified as a histogram or an empirical frequency distribution. One presentation is by classes or categories with associated frequencies for each class. If a large number of data points are available that permits grouping the data into 7 to 15 intervals (classes), then it is possible to represent the shape of a frequency distribution as a histogram. The histogram can be developed by noting the range of magnitudes and dividing the range into class intervals. There are no absolute rules concerning histogram construction. However, the following guidelines are helpful for tabulation (Table 2.6).

1.   Class intervals should not overlap (i.e., 1–25 mm, 26–50 mm, 51–75 mm, of rainfall, etc.).
2.   The intervals should include all the data points.
3.   The intervals should be of uniform size except the beginning or ending ones that may be open (i.e., greater than or less than a given value).
4.   The number of intervals should be changed and another graphic done for comparison to the previous one.

Using these simple guidelines, the flow rates of Table 2.7 can be used to develop a simple histogram. The data of Table 2.7 include the day of the year on which the maximum annual flows occurred. A histogram for the runoff data of Table 2.7 by class interval is shown in Table 2.6. From the histogram of Table 2.6, the shape of the distribution is noted as somewhat symmetrical but it is difficult to determine an exact theoretical distribution to fit the data. Also, there may be a skewness to the data.

2.3.2 Plotting Position

Another way to develop an empirical distribution is to use a method called "plotting position." One can arrange the data by order of magnitude (sort the data) and calculate

TABLE 2.7 A Listing by Month, Day, and Year for the Banana River Watershed

YEAR	MONTH–DAY	RUNOFF (m³/s)
1956	04–24	0.141E + 03
1957	05–16	0.900E + 02
1958	11–08	0.500E + 02
1959	03–25	0.109E + 03
1960	04–21	0.103E + 03
1961	12–25	0.142E + 03
1962	11–30	0.111E + 03
1963	06–01	0.750E + 02
1964	05–12	0.850E + 02
1965	03–16	0.630E + 02
1966	04–26	0.144E + 03
1967	05–11	0.135E + 03
1968	04–30	0.410E + 02
1969	07–24	0.123E + 03
1970	09–20	0.106E + 03
1971	06–08	0.121E + 03
1972	11–25	0.189E + 03
1973	07–22	0.640E + 02
1974	05–30	0.128E + 03
1975	04–04	0.176E + 03
1976	05–29	0.131E + 03
1977	09–11	0.164E + 03
1978	04–01	0.860E + 02
1979	05–09	0.137E + 03
1980	09–04	0.910E + 02
1981	08–23	0.990E + 02
1982	04–07	0.940E + 02
1983	05–28	0.920E + 02
1984	07–12	0.118E + 03

Average is .111E + 03 m³/s.
Standard deviation is 0.357E + 02 m³/s.

an empirical cumulative probability of either a "less than" a certain value (smallest to largest) or "greater than" a certain value type (largest to smallest). The probability of a value being greater than is called the exceedence probability. There are various ways of determining the empirical distribution by plotting position formulas. Once all the data equal to a number n are ordered, an order number (identified as m) is assigned to the position value. There are at least four plotting formulae in general use (Table 2.8). All have special advantages, however, the one used most frequently in stormwater management is Weibull (Benson, 1962), probably because it does not tend to over or underestimate the true probability at extreme points. Usually, greater than 25 data

TABLE 2.8   Less Than or Equal to Probabilities Using Plotting Formulas

SORTED FLOW RATES (cfs)	m PLOTTING POSITION	PLOTTING POSITION FORMULA PROBABILITIES			
		WEIBULL $m/(n+1)$	CALIFORNIA m/n	FOSTER $(2m-1)/2n$	EXCEEDENCE $(m-1)/n$
6	1	.125	.143	.071	0
14	2	.250	.286	.214	.143
36	3	.375	.429	.357	.286
42	4	.500	.571	.500	.429
58	5	.625	.714	.643	.571
90	6	.750	.857	.786	.714
99	7	.875	1.000	.929	.857

points are used. For a small number of data points (< 25) and for high-return periods, the Foster formula gives a larger value than the Weibull formula and smaller than the California formula. The California and exceedence formulas generally can be used with a large number of data points but rarely with a small number of points. The exceedence formula is best for estimating the largest values (upper limit) while the California formula is best for estimating the lowest values (lower limit). The spread of the estimates among the four procedures is very small for shorter return periods but can be large toward the upper end (longer return periods). As the sample size increases, the probability estimates by the four plotting formulae tend to converge.

Using the Weibull formula, the data of Table 2.7 are ordered and shown in Table 2.9. The plot position is converted to a less than or equal to probability in column 3. The exceedence probability is one minus the less than probability and the return period is the inverse of the exceedence probability. Table 2.9 was obtained using a computer program; however, the calculations are relatively simple.

2.3.3   Annual and Partial Duration Series

The data of Table 2.9 were arranged as the maximum per time period (annual peak discharge values), thus it can be called an annual series. Another ordering is by size regardless of the time in which they occur. This is called a partial duration series or basic stage ordering. This ordering uses all data above a base flow. Over a 20-yr period, one may then elect to have 30 or more data points. This method may eliminate peak values for a given time because of their low values. Differences between the partial and annual series probability distributions tend to be greater at lower values which correspond to lower return periods. Beard (1962) indicated that the partial duration series is most useful for determining flood flows for those lower return periods (less than 10 yr). When using partial series of flood flows, one must be careful to insure independence. One peak flow may be influenced by another in a previous time period. One must also carefully choose a lower limit (cut off) because it affects the parameters of the resulting distribution. In many cases, it is best to use an annual series to compute a probability distribution and then convert the return periods to partial series return periods using Equation 2.32 (U.S. Department of Transportation, 1984).

$$T_p = 1/[\ln(T_A) - \ln(T_A - 1)] \qquad T_A > 1 \qquad (2.32)$$

TABLE 2.9 Weibull Order for Empirical Distribution Function

EVENTS: X(l) FLOW RATE (m^3/s)	m PLOT POSITION	$Pr = \dfrac{m}{(n+1)}$ PROBABILITY	1 − Pr EXCEEDENCE PROBABILITY	T_r (yrs) RETURN PERIOD
41.00	1	.0333	.967	1.034
50.00	2	.0667	.933	1.071
63.00	3	.1000	.900	1.111
64.00	4	.1333	.867	1.154
75.00	5	.1667	.833	1.200
85.00	6	.2000	.800	1.250
86.00	7	.2333	.767	1.304
90.00	8	.2667	.733	1.364
91.00	9	.3000	.700	1.429
92.00	10	.3333	.667	1.500
94.00	11	.3667	.633	1.579
99.00	12	.4000	.600	1.667
103.00	13	.4333	.567	1.765
106.00	14	.4667	.533	1.875
109.00	15	.5000	.500	2.000
111.00	16	.5333	.467	2.143
118.00	17	.5667	.433	2.308
121.00	18	.6000	.400	2.500
123.00	19	.6333	.367	2.727
128.00	20	.6667	.333	3.000
131.00	21	.7000	.300	3.333
135.00	22	.7333	.267	3.750
137.00	23	.7667	.233	4.286
141.00	24	.8000	.200	5.000
142.00	25	.8333	.167	6.000
144.00	26	.8667	.133	7.500
164.00	27	.9000	.100	10.000
176.00	28	.9333	.067	15.000
189.00	29	.9667	.033	30.000

where

T_P = return period partial series (years)
T_A = return period annual series (year)

2.3.4 Daily and True Interval Comparisons

Rainfall volumes are commonly reported on a daily basis (midnight to midnight). However, the maximum 24-hr rainfall volume may occur over another 24-hr period. It can be shown that for longer return periods (> 10 yr) the daily data can be used to estimate 24-hr rainfall events. A plot showing the ratio of 24 hr rainfall volumes (true interval) to maximum daily rainfall volume is shown in Figure 2.9 for Brooksville,

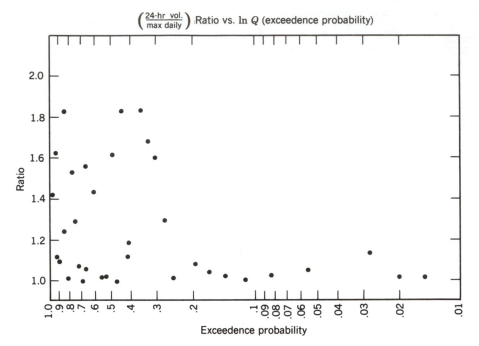

FIGURE 2.9 Ratio of partial to daily rainfalls as a function of exceedence probability.

Florida. The x axis is the exceedence probability and the return period is the inverse of the exceedence probability. Note the scatter of the data at low-return periods (high exceedence) and the convergence of the data to a 1.0 ratio at higher-return periods (low exceedence). Similar results are available for other locations.

□ **EXAMPLE PROBLEM 2.12**

Using the data of Table 2.10 for Bushnell, Florida, order the data from smallest to largest and estimate the empirical frequency distribution using the Weibull plotting formula.

Solution

Using computer programs or simple calculations, Table 2.11 results. The reader should verify some of the calculations.   □

2.4
REGRESSION AND CORRELATION ANALYSIS

Cause and effects relationships among hydrologic (discharge, volume runoff, mass of pollutants) and meterological parameters are frequently needed for engineering and planning investigations. Usually, a quantitative relationship within the investigative limits is desired. A variable for prediction depends on other independent variables. Many times it is assumed that the independent variables are measured with little or

TABLE 2.10 A Listing by Month, Day, and Year for Maximum Daily Storms for Each Year at Bushnell

YEAR	MONTH–DAY	RAINFALL (in.)
1918	09–27	0.200E + 01
1937	07–30	0.300E + 01
1938	06–22	0.380E + 01
1939	07–7	0.281E + 01
1940	07–4	0.290E + 01
1941	04–3	0.357E + 01
1942	02–24	0.155E + 01
1943	04–9	0.300E + 01
1944	10–19	0.760E + 01
1945	06–24	0.832E + 01
1946	06–27	0.253E + 01
1947	10–24	0.365E + 01
1948	07–2	0.350E + 01
1949	04–5	0.390E + 01
1950	09–6	0.908E + 01
1951	11–16	0.304E + 01
1952	05–20	0.309E + 01
1953	12–23	0.269E + 01
1954	07–26	0.394E + 01
1955	11–10	0.309E + 01
1956	10–16	0.413E + 01
1957	12–26	0.470E + 01
1958	03–2	0.308E + 01
1959	07–17	0.270E + 01
1960	07–29	0.527E + 01
1961	02–7	0.173E + 01
1962	05–28	0.280E + 01
1963	11–10	0.310E + 01
1964	09–11	0.382E + 01
1965	08–5	0.311E + 01
1966	05–8	0.302E + 01
1967	06–11	0.329E + 01
1968	10–19	0.379E + 01
1969	03–17	0.317E + 01
1970	02–3	0.397E + 01
1971	02–8	0.400E + 01
1972	03–31	0.665E + 01
1973	02–15	0.241E + 01
1974	06–25	0.690E + 01
1975	10–29	0.298E + 01
1976	6–5	0.298E + 01
1984	04–4	0.258E + 01
1985	8–18	0.515E + 01

Average is 0.378E + 01 in.
Standard deviation is 0.166E + 01 in.

TABLE 2.11 Empirical Cumulative Distribution Function

DATA TITLE NUMBER OF DATA PTS		:MAXIMUM DAILY STORMS FOR EACH YEAR AT BUSHNELL :43		
EVENTS: X(l) (in.)	PLOT POSITION	$Pr = \dfrac{m}{(n+1)}$ PROBABILITY	$1 - Pr$ EXCEEDENCE PROBABILITY	T_r (yr) RETURN PERIOD
1.55	1	.0227	.977	1.023
1.73	2	.0455	.955	1.048
2.00	3	.0682	.932	1.073
2.41	4	.0909	.909	1.100
2.53	5	.1136	.886	1.128
2.58	6	.1364	.864	1.158
2.69	7	.1591	.841	1.189
2.70	8	.1818	.818	1.222
2.80	9	.2045	.795	1.257
2.81	10	.2273	.773	1.294
2.90	11	.2500	.720	1.333
2.98	12	.2727	.727	1.375
2.98	12	.2727	.727	1.375
3.00	14	.3182	.682	1.467
3.00	14	.3182	.682	1.467
3.02	16	.3636	.636	1.571
3.04	17	.3864	.614	1.630
3.08	18	.4091	.591	1.692
3.09	19	.4318	.568	1.760
3.09	19	.4318	.568	1.760
3.10	21	.4773	.523	1.913
3.11	22	.5000	.500	2.000
3.17	23	.5227	.477	2.095
3.29	24	.5455	.455	2.200
3.50	25	.5682	.432	2.316
3.57	26	.5909	.409	2.444
3.65	27	.6136	.386	2.588
3.79	28	.6364	.364	2.750
3.80	29	.6591	.341	2.933
3.82	30	.6818	.318	3.143
3.90	31	.7045	.295	3.385
3.94	32	.7273	.273	3.667
3.97	33	.7500	.250	4.000
4.00	34	.7727	.227	4.400
4.13	35	.7955	.205	4.889
4.70	36	.8182	.182	5.500
5.15	37	.8409	.159	6.286
5.27	38	.8636	.136	7.333
6.65	39	.8864	.114	8.800
6.90	40	.9091	.091	11.000
7.60	41	.9318	.068	14.667
8.32	42	.9545	.045	22.000
9.08	43	.9773	.023	44.000

no error. The variability in the estimates for the independent variables is minimal. For example, the measurement of precipitation on a small watershed (< 5 ha) is generally representative of the watershed. Others, such as temperature, humidity and solar radiation, can be measured with a small error. However, the runoff from the watershed depends on the rainfall, soil moisture conditions and chance variation. Thus, the independent variable, x, may be fixed and with repeated values produce different values of the dependent variable, y. Some of the variability in the y variable may be explained by another independent variable, x_2, and the remaining variability is by chance. Thus, if this chance variability can be minimized, the estimate of the dependent variable can be improved.

2.4.1 Bivariate Case

For two variable situations, the x and y values are measured and a relationship between these two is determined. This relationship can be linear or nonlinear. Assume a linear relationship does exist and is given by:

$$y' = a + bx + \varepsilon$$

where

y' = predicted dependent variable
x = independent variable
a, b = constants
ε = error with mean zero

Consider a situation in which runoff is estimated from rainfall. Data on runoff and rainfall are shown plotted in Figure 2.10. Runoff is dependent on rainfall.

We wish to minimize the variability in the estimate of runoff; therefore, a criterion for minimizing the variance of runoff (y variable) appears reasonable.

$$\text{Minimize} \sum_i (y_i - y')^2$$

and

$$\text{Minimize} \sum_i (y_i - a - bx_i)^2 \tag{2.33}$$

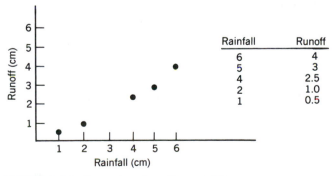

Rainfall	Runoff
6	4
5	3
4	2.5
2	1.0
1	0.5

FIGURE 2.10   Regression of runoff on rainfall.

Taking the partial derivatives of this equation with respect to the parameters a and b and setting these equal to zero, one obtains the "least square" estimators of a and b or those that minimize the variability in the estimate of the y variable. The partial derivatives result in normal equations, which are

$$\sum y_i = an + b \sum x_i$$
$$\sum x_i y_i = a \sum x_i + b \sum x_i^2 \tag{2.34}$$

These normal equations are linear with two unknowns, a and b, and reduce to

$$a = \frac{(\sum x^2)(\sum y) - (\sum x)(\sum xy)}{n(\sum x^2) - (\sum x)^2}$$
$$b = \frac{n(\sum xy) - (\sum x)(\sum y)}{n(\sum x^2) - (\sum x)^2} \tag{2.35}$$

□ **EXAMPLE PROBLEM 2.13**

For the data of Figure 2.10, what is the best estimate of the linear regression equation for runoff by measuring rainfall?

Solution

Regression Coefficients Tabulation

x	x^2	y	y^2	$x \cdot y$
6	36	4	16	24
5	25	3	9	15
4	16	2.5	6.25	10
2	4	1	1	2
1	1	.5	.25	.5
Σ 18	82	11	32.5	51.5

$$a = \frac{(82)(11) - (18)(51.5)}{(5)(82) - 18^2}$$

$$a = -0.29$$

$$b = \frac{(5)(51.5) - (18)(11)}{(5)(82) - 18^2}$$

$$b = .69$$

The linear equation of best fit is

$$\text{Runoff} = -.29 + .69 \text{ (rainfall)} \qquad (1.0 < \text{rainfall} < 6.0 \text{ cm})$$

It is noted in the above equation that the limits of fit are rainfall quantities between 1 and 6 cm. These are the limits of the raw data and should be presented with an equation. □

A linear equation of best fit has been assumed in the above example; however, if a curvilinear equation is assumed, the least squares estimators above can be used if the curvilinear equation can be transformed into a linear equation. Some example transformations are

$$y = \alpha \cdot \beta^x \tag{2.36}$$

and

$$\log y = \log \alpha + x \log \beta$$

also:

$$y = ae^{-bx} \qquad (\text{note } \alpha = a, \quad \beta = b) \tag{2.37}$$

and

$$\ln y = \ln a - bx$$

Multiple (more than two variables) regression is derived by a similar procedure as the bivariate case. For the three-variable case, the equation for minimum variance is

$$\sum (Y_i - a - bX_{1i} - bX_{2i})^2 \tag{2.38}$$

and the resulting equations are

$$\sum y = an + b_1 \sum x_1 + b_2 \sum x_2$$

$$\sum x_1 y = a \sum x_1 + b_1 \sum x_1^2 + b_2 \sum x_1 x_2$$

$$\sum x_2 y = a \sum x_2 + b_1 \sum x_1 x_2 + b_2 \sum x_2^2$$

As before, there are three equations and three unknowns, which are easily solved using the calculators and computers of today.

❑ **EXAMPLE PROBLEM 2.14**

Duration (min)	Intensity (in/hr)
10	4.0
15	3.2
20	2.7
30	1.9
60	1.2
120	0.6

Fit the following data to the curve form $i = \dfrac{a}{b + D}$

where

i = intensity (in/hr)
D = Duration (min)

Solution

To use least squares linear regression we must put the equation into the form:

$$Y = m'X + b'$$

this can be done by inverting both sides of the equation

$$\frac{1}{i} = \frac{b}{a} + \frac{D}{a}$$

and setting

$$Y = 1/i, \qquad b' = b/a, \qquad m' = 1/a, \qquad \text{and } X = D.$$

Using a spreadsheet you can set the columns so that the data appears as shown.

	A	B	C	D	E	F	G	H
1	Duration	Intensity	1/Intensity	X	Y	X²	Y²	XY
2	10	4	0.250	10	0.250	100	0.063	2.500
3	15	3.2	0.313	15	0.313	225	0.098	4.688
4	20	2.7	0.370	20	0.370	400	0.137	7.407
5	30	1.9	0.526	30	0.526	900	0.277	15.789
6	60	1.2	0.833	60	0.833	3600	0.694	50.000
7	120	0.6	1.667	120	1.667	14400	2.778	200.000
8			Sum	255	3.959	19625	4.047	280.384

Linear least squares regression of x (independent) and y (dependent) with the form $Y = m'X + b'$ yields the results $b' = b/a = 0.11761$ and $m' = 1/a = 0.012759$. Solving for b and a gives: $a = 78.38$ and $b = 9.22$. The solution is then (with i in inches per hour and D in minutes):

$$i = \frac{78.38}{9.22 + D}$$

The results of this equation should be plotted against the original data as shown in Figure 2.11. The solution of the problem using the REGRESS computer program is also shown in Figure 2.12.

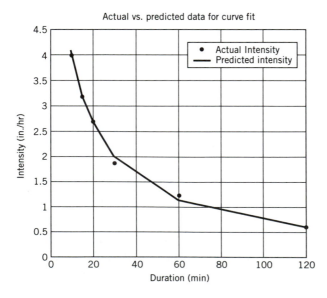

FIGURE 2.11   Actual versus curve fit data for Example Problem 2.14.

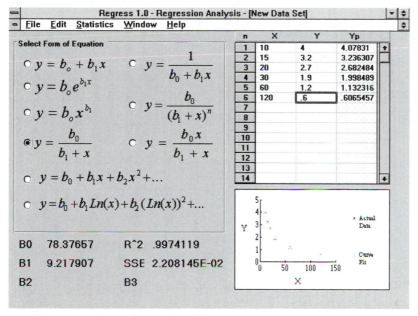

FIGURE 2.12 Solution of Example Problem 2.14 using REGRESS.

2.4.2 Correlation

Up to this point, it has been assumed that the regression equations were "best fit" estimators of the dependent variable. This best fit has to be quantified in terms of the degree of variability explained in the dependent variable by measuring the independent variability and using the "least squares" estimator. If the total variance in y is expressed by s^2 and the unexplained variance is given by s_u^2, the fraction of variance explained is

$$r^2 = 1 - \frac{s_u^2}{s^2} \tag{2.39}$$

and the sample correlation coefficient or measure of goodness of fit is given by

$$r = \pm \sqrt{1 - \frac{s_u^2}{s^2}} \tag{2.40}$$

When the explained variance approaches the total variance, the correlation coefficient approaches either $+1$ or -1. Perfect correlation exists when $r = +1$ or -1. Inverse correlation is a negative value of r and results when the variables are inversely related to one another. The computation formula for r is derived as follows:

$$r^2 = \text{explained fraction} = 1 - \text{unexplained fraction}$$

$$r = \pm \sqrt{1 - \frac{\sum (y - y')^2}{\sum (y - \bar{y})^2}} \tag{2.41}$$

which reduces to

$$r = \frac{n(\sum xy) - (\sum x)(\sum y)}{\sqrt{n(\sum x^2) - (\sum x)^2}\sqrt{n(\sum y^2) - (\sum y)^2}}$$
(2.42)

❑ *EXAMPLE PROBLEM 2.15*
_____

Calculate the correlation coefficient for the data of Example Problem 2.13. What percentage variation is explained?

Solution

$$r = \frac{5(51.5) - (18)(11)}{\sqrt{5(82) - 18^2}\sqrt{5(32.5) - 11^2}}$$

$$= .996$$

$$r^2 = .992 \quad \text{or} \quad 99.2\% \text{ variation explained} \quad ❑$$

The significance of this value is whatever one's intuition tells us about the standard deviation as a measure of dispersion or repeatability.

2.5

MATRIX METHODS

From our previous discussion of spreadsheets, hydrology problems can be solved using calculations of rows and columns of numbers. Some hydrology problems require a special form of column and row calculations that follows specific rules and is classified into mathematical techniques called matrix methods.

2.5.1 Matrix Multiplication

A *matrix* is a rectangular array of numbers arranged in rows and columns. Matrices can be expressed as a single variable. The use of matrices and matrix notation can be used to simplify calculations that involve many numbers. There are certain matrix multiplication terms with which the reader should be familiar. These terms are defined in this section.

Inverse   If matrix **A** is a rectangular matrix with m rows and n columns, then the *inverse matrix* of **A** is denoted as $\mathbf{A}^{-1}$ and is the matrix such that

$$\mathbf{A} \times \mathbf{A}^{-1} = \mathbf{I}$$

where: $\mathbf{I} =$ The unity (or identity) matrix which has 1 along the diagonal and 0 elsewhere.

Methods used to determine the invert of a matrix are beyond the scope of this text; however, a program (EZMAT.EXE Matrix Calculator) included in the back of this book can be used to find the invert of any matrix.

☐ **EXAMPLE PROBLEM 2.16**

Find the inverse of the following matrix **A**.

$$
\begin{array}{ccc}
11.0 & 3.00 & 1.00 \\
2.00 & 2.00 & 14.0 \\
19.0 & 1.00 & 3.00
\end{array}
$$

Solution

Using the matrix calculator the inverse $\mathbf{A}^{-1}$ can be found as

$$
\begin{array}{ccc}
-0.0122 & -0.0122 & 0.0610 \\
0.396 & 0.0213 & -0.232 \\
-0.0549 & 0.0701 & 0.0244
\end{array}
$$

Transpose   The *transpose* of matrix **A** is denoted as $\mathbf{A}^{\mathbf{T}}$ and is found by interchanging the rows and columns of the original matrix.   ☐

☐ **EXAMPLE PROBLEM 2.17**

Find the transpose of the matrix **A** from Example Problem 2.16.

Solution

The transpose can be found as

$$
\begin{array}{ccc}
11.0 & 2.00 & 19.0 \\
3.00 & 2.00 & 1.00 \\
1.00 & 14.0 & 3.00
\end{array}
$$ ☐

Multiplication of Two Matrices

Matrix **A** can be multiplied by matrix **B** with the product denoted as $\mathbf{C} = \mathbf{A} \times \mathbf{B}$ if the number of rows in matrix **B** is equal to the number of columns in matrix **A**. The product matrix will have a number of rows equal to the number of rows in matrix **A** and a number of columns equal to the number of columns in matrix **B**.

2.5.2 Solutions to Simultaneous Equations

Matrices can be used to simplify the solutions of many types of equations. A common use of matrix notation is the solution of linear simultaneous equations with n variables and n unknowns. These equations are typically expressed as

$$
\begin{aligned}
a_{11}X_1 + a_{12}X_2 + a_{13}X_3 + \cdots + a_{1n}X_n &= b_1 \\
a_{21}X_1 + a_{22}X_2 + a_{23}X_3 + \cdots + a_{2n}X_n &= b_2 \\
a_{31}X_1 + a_{32}X_2 + a_{33}X_3 + \cdots + a_{3n}X_n &= b_3 \\
\cdots & \\
a_{m1}X_1 + a_{m2}X_2 + a_{m3}X_3 + \cdots + a_{mn}X_n &= b_m
\end{aligned}
$$

by defining matrix **A, X,** and **B** as

$$
\mathbf{A} = \begin{matrix} a_{11} & a_{12} & \cdots & a_{1n} \\ a_{21} & a_{22} & \cdots & a_{2n} \\ & & \cdots & \\ a_{m1} & a_{m2} & \cdots & a_{mn} \end{matrix} \qquad \mathbf{X} = \begin{matrix} X_1 \\ X_2 \\ \cdots \\ X_n \end{matrix} \qquad \mathbf{B} = \begin{matrix} b_1 \\ b_2 \\ \cdots \\ b_m \end{matrix}
$$

the series of linear equations can be expressed as

$$\mathbf{A} \times \mathbf{X} = \mathbf{B}$$

Solving for the vector matrix X gives

$$\mathbf{A}^{-1} \times \mathbf{A} \times \mathbf{X} = \mathbf{A}^{-1} \times \mathbf{B}$$
$$\mathbf{I} \times \mathbf{X} = \mathbf{A}^{-1} \times \mathbf{B}$$
$$\mathbf{X} = \mathbf{A}^{-1} \times \mathbf{B}$$

❑ **EXAMPLE PROBLEM 2.18**
_____

Using Matrix methods solve the following set of equations:

$$11x_1 + 3x_2 + 1x_3 = 20$$

$$2x_1 + 2x_2 + 14x_3 = 48$$

$$19x_1 + 1x_2 + 3x_3 = 30$$

Solution

Using the Matrix Calculator, enter the following two matrices:

$$
\mathbf{A} = \begin{matrix} 11 & 3 & 1 \\ 2 & 2 & 14 \\ 19 & 1 & 3 \end{matrix} \qquad \mathbf{B} = \begin{matrix} 20 \\ 48 \\ 30 \end{matrix}
$$

the inverse of matrix **A** is

$$
\mathbf{A}^{-1} = \begin{matrix} -0.0122 & -0.0122 & 0.0610 \\ 0.396 & 0.0213 & -0.232 \\ -0.0549 & 0.0701 & 0.0244 \end{matrix}
$$

which when multiplied by matrix **B** gives the matrix $\mathbf{X}(\mathbf{X} = \mathbf{A}^{-1} \times \mathbf{B})$:

$$
\mathbf{X} = \begin{matrix} -0.0122(20) + -0.0122(48) + 0.0610(30) = 1.0 \\ 0.396(20) + 0.0213(48) + -0.232(30) = 2.0 \\ -0.0549(20) + 0.0701(48) + 0.0244(30) = 3.0 \end{matrix}
$$

thus $X_1 = 1, X_2 = 2, X_3 = 3$ ❑

❑ **EXAMPLE PROBLEM 2.19**
_____

Given the rainfall excess volumes of Example Problem 1.4 and the resulting flow rates from 1 in. of rainfall excess, what are the cumulative flow rates resulting from all the rainfall excess ($\mathbf{R}$). The flow rates resulting from 1 in. of rainfall excess ($\mathbf{U}$) are; @ 1 hr U_1 = 6 cfs, @ 2 hr U_2 = 18 cfs, @ 3 hr U_3 = 10 cfs, @ 4 hr U_4 = 4 cfs, @ 5 hr U_5 = 0 cfs. Solve for the resultant flow rates by setting up a matrix formulation of rainfall excess ($\mathbf{R}$) and solving for $\mathbf{Q} = \mathbf{R} \times \mathbf{U}$.

Solution

To develop the matrix $\mathbf{R}$, the number of rows in matrix $\mathbf{R}$ must be equal to the sum of the number of steps in the rainfall excess plus the number of steps in the unit hydrograph. The number of columns in the matrix $\mathbf{R}$ are equal to the number of steps in the unit hydrograph. The time steps for the rainfall excess and the unit hydrograph must be equal.

$$\mathbf{R} = \begin{matrix}
R_1 & 0 & 0 & 0 & 0 & 0 & 0 \\
R_2 & R_1 & 0 & 0 & .104 & 0 & 0 & 0 \\
R_3 & R_2 & R_1 & 0 & .165 & .104 & 0 & 0 \\
R_4 & R_3 & R_2 & R_1 = 0 & & .165 & .104 & 0 \\
R_5 & R_4 & R_3 & R_2 & 0 & 0 & .165 & .104 \\
R_6 & R_5 & R_4 & R_3 & 0 & 0 & 0 & .165 \\
R_7 & R_6 & R_5 & R_4 & 0 & 0 & 0 & 0
\end{matrix}$$

The matrix $\mathbf{U}$ is expressed as

$$= \begin{matrix} U_1 \\ U_2 \\ U_3 \\ U_4 \end{matrix} = \begin{matrix} 6 \\ 18 \\ 10 \\ 4 \end{matrix}$$

therefore the matrix $\mathbf{Q} = \mathbf{R} \times \mathbf{U}$ can be calculated as

$$\mathbf{Q} = \begin{matrix}
R_1 U_1 & 0 \\
R_2 U_1 + R_1 U_2 & .624 \\
R_3 U_1 + R_2 U_2 + R_1 U_3 & 2.862 \\
R_4 U_1 + R_3 U_2 + R_2 U_3 + R_1 U_4 & = 4.01 \\
R_5 U_1 + R_4 U_2 + R_3 U_3 + R_2 U_4 & 2.066 \\
R_6 U_1 + R_5 U_2 + R_4 U_3 + R_3 U_4 & .99 \\
R_7 U_1 + R_6 U_2 + R_5 U_3 + R_4 U_4 & 0
\end{matrix} \quad \square$$

2.6
SUMMARY

- Many new tools are available to engineers and hydrologists in the form of computer software. Spreadsheets are useful for dealing with large amounts of numbers and calculations. Computer-aided design programs simplify watershed analysis. Many programs exist to solve specific hydrology problems.
- Many meteorologic and hydrologic processes can be defined using probability and statistics concepts and formulas. Empirical data form the basis for determining the type of theoretical distribution that "best" fits the situation. In hydrologic

studies empirical data form the basis for determining frequency distributions. Computer programs can aid in this determination.

- All distributions are defined by measures of (1) central tendency, (2) variability, and (3) skewness. These are also referenced as the first, second, and third moment of a distribution.

- Return period or recurrence interval is the inverse of exceedence probability. It is based on the assumption of a long period of time over which changes will not affect the average time interval.

- A histogram can describe graphically the shape of a frequency distribution. There are useful guidelines for developing histograms in the text.

- There are plotting position formulas for empirical data. The Weibull formula is widely used.

- The ordering or ranking of data by magnitude is called a partial series. If ordered by time, the series is called annual.

- The exceedence probability can be developed from a cumulative distribution function.

- The risk involved with a hydrologic event is defined as the probability of at least one exceedence during a design life. It can be quantified using Equation 2.31.

- The Gumbel distribution and log-Pearson Type III are extreme event distributions used extensively for the description and analyses of hydrologic events.

- The exponential distribution is being used to aid in the description of rainfall and runoff events.

- Bivariate regression analysis can be used to determine the "best" fit line between two variables. A correlation coefficient is used to measure the "goodness of fit" of the dependent variable explained using the "best" equation.

- Matrix methods are useful in performing calculations which involve large amounts of data which can be arranged in columns and rows. Matrices can be used to solve series of linear equations.

2.7
PROBLEMS

2.7.1 Hand Calculations

1. a. What is the approximate reliability at which you can estimate a hydrologic event with a return of 10 yr if you are using 25 yr of data and you wish to be within $\pm 25\%$ of the true estimate?
   b. What is your record length if you wish to be 100% reliable?

2. a. For the data of Example Problem 2.8, calculate a measure of central tendency and a measure of variability.
   b. If the years 1977 through 1980 were added as 16, 14, 22, and 18 m, respectively, how does skewness, mean, and the coefficient of variability change?

3. Explain what is meant by skewness and give an example and show a frequency distribution for a right-skewed distribution.

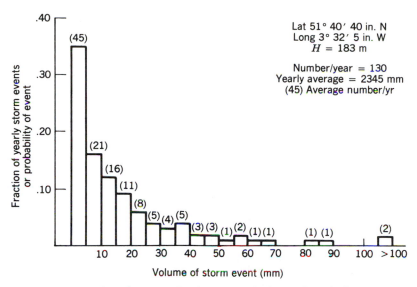

FIGURE 2.13 Volume frequency distributions—Treherbert Park, Mid-Glamorgan, U.K.

4. A flood flow analysis is completed on a tributary to the Delaware River. A Gumbel-type distribution is believed to be representative of annual flood peaks. If the average flood peak is 4000 m³/sec and the standard deviation is 600 m³/sec, what is the magnitude of a flood with a recurrence interval of 20 yr?

5. What is the probability that a drought that is more severe than a drought of a return period of once in 50 yr will occur only once in 100 yr?

6. For the histograms shown in Figures 2.13 and 2.14, develop a cumulative frequency distribution on the fraction of storms with volume less than or equal to a stated

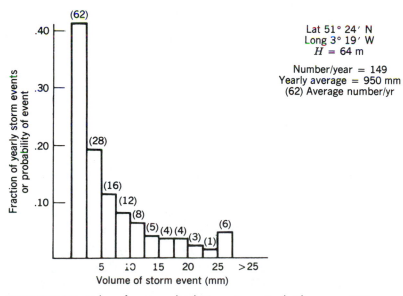

FIGURE 2.14 Volume frequency distribution—Barry, South Glamorgan, U.K.

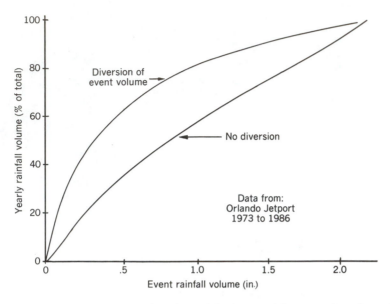

FIGURE 2.15   Percentage of yearly rainfall volume and diversion volume for each and every storm.

volume. Compare these results for rainfall in Wales to that of the Orlando Jetport in Figure 2.15. Comment on all statistical measures.

7. Using the cumulative distribution function of Figure 2.1, a cumulative rainfall volume distribution function was developed as presented in Figure 2.15. The lower curve, labeled "no diversion," was developed assuming that the rainfall volume was accumulated up to and including the stated event rainfall volume. The upper curve assumes that the stated rainfall event volume will be diverted for treatment and then the remaining amount is discharged. If 1 in. of every rainfall is diverted for treatment, 80% of the yearly rainfall volume will be treated. Using data for your area, develop a similar curve. (*Hints:* Volume of rainfall per year is the product of the frequency of that event interval times the average rainfall for that interval and the diversion volume is the sum of rainfall up to the diversion volume plus the sum of diversion volume and the frequency of exceedence.)

$$F(\text{Vol}|\text{Diversion Vol.}) = \sum_{i=0}^{\text{Diversion Vol.}} P(i)_i \bar{x}_i n$$

$$+ \sum_{i=\text{Diversion Vol.}}^{\infty} P(i)_i \, \text{Diversion Vol.} n$$

8. For the data for Figure 2.10, determine an equation of the form runoff = K precipitation, or determine K. This assumes that the intercept is equal to zero.

9. Using the 1950–1985 data of Example Problem 2.2 for one of the locations, develop a histogram and comment on the shape of the distribution (use six class intervals). Also, calculate the average, standard deviation, and skewness if the first 5 yr of

data were missing. Comment on the results as compared to the calculations using all the data.

10. In the analyses of annual flood peaks on a stream with 60 yr of records, it was determined that the annual flood peaks follow a Gumbel-type distribution with an annual mean peak of 83,333 cfs.
    a. What is the probability that an annual flood magnitude greater than 110,000 cfs will not occur in 25 yr? The sample standard deviation is 10,000.
    b. What is the probability that this 110,000 cfs flood flow will occur at least once in 25 yr?
    c. What is the probability that exactly two floods equal to or greater than 110,000 cfs will occur in the next 25 yr?

11. For the following-flow rate data, use the four plotting position formulae of Table 2.8 and plot the annual series empirical data on probability paper. Comment on the results.

YEAR	PEAK FLOW (m³/s)	YEAR	PEAK FLOW (m³/sec)
1970	602	1979	178
1971	214	1980	249
1972	106	1981	365
1973	312	1982	250
1974	280	1983	912
1975	143	1984	404
1976	190	1985	136
1977	236	1986	101
1978	737		

12. Using the maximum yearly rainfall data of Table 2.9 construct a histogram of rainfall. Comment on the shape of the histogram.

13. Cumulative probability distributions for rainfall intensity (i), duration (D), and interevent time (Δ) are used for the following estimates.

$$Pr(i \leq 0.40 \text{ in./hr}) = 0.98$$

$$Pr(D \geq 6 \text{ hr}) = 0.05$$

$$Pr(\Delta \geq 92 \text{ hr}) = 0.10$$

    a. If there were 100 events/year, calculate the return period and the number of events per year for each rainfall description (i, D, Δ).
    b. Also, what is the return period and number of events per year for a storm of intensity greater than 0.40 in./hr and a duration greater than 6 hr?

14. From the following independent flow rate data, develop an empirical probability distribution using annual series, then a distribution using partial series data for flows over 200 m³/s. What is the flow rate estimate for the 1 in 10-yr storm using

both the annual and partial series data? Use the Weibull method for the empirical distributions. The highest two peak flows per year are as follows:

YEAR	PEAK FLOW (m³/s)		YEAR	PEAK FLOW (m³/s)	
1970	602	390	1979	178	120
1971	214	174	1980	249	145
1972	106	88	1981	365	204
1973	312	140	1982	250	170
1974	280	210	1983	912	350
1975	143	138	1984	404	100
1976	190	108	1985	136	135
1977	236	190	1986	101	99
1978	737	304			

15. A water resources project has a design life of 50 yr. The hydrologic variable of interest follows a log normal distribution with a mean equal to 10 in. and a standard deviation of 2.5 in. If hydrologic design criteria calls for 12.5-in value, what is exceedence probability and the risk involved over a 4-yr period with this design criteria? Does the risk change if the distribution is a Gumbel type?

16. A 1 in 100-yr rainfall volume for a 24-hr time period is 10 in.
    a. What is the probability that this rainfall volume will not occur in the next 5 yr?
    b. What is the risk you assume if there is at least one exceedence in the design life of 5 yr?
    c. Assuming that the distribution can be approximated by a Gumbel one, what is the average if the standard deviation is 1.5 in.?

17. Transpose the following matrix:

$$\begin{matrix} 12 & 15 & 21 \\ 9 & 14 & 45 \end{matrix}$$

18. Use matrix methods to solve the linear set of equations:

$$4x_1 - 3x_2 = 5$$
$$2x_1 + 2x_2 = 6$$

19. Given an annual series of flood peak flows for the Suwannee River, determine the mean and standard deviation for the data set. What is your estimate for the once in 50 years flood using the Gumbel distribution? Note that Equation 2.20 may be written as $b = -\ln \left[\ln \dfrac{T_r}{T_r - 1} \right]$, and hence $K = \dfrac{b - b_n}{s_n}$ (see Table D.9).

```
Peak Flows
Suwannee River
at Branford, Fla.
station number 2320500
```

Month	Day	Year	Flow (cfs)
DEC	31	1972	4280
APR	18	1973	54700
SEP	21–22	1974	12500
APR	28	1975	24900
DEC	24–26	1976	20000
JAN	19	1977	21800
MAR	24–26	1978	18600
MAR	8–9	1979	15500
APR	17–18	1980	23300
APR	9	1981	8170
FEB	26	1982	10200
APR	24–26	1983	26900
APR	13–14	1984	41400
DEC	25–26	1985	11300
FEB	23	1986	38500
MAR	13–15	1987	27100
MAR	15–16	1988	20400
JUL	31	1989	6830
MAR	4–5	1990	12900
MAR	19	1991	37700
JAN	31	1992	6410

20. How would you improve the reliability of your estimate in the previous problem? Hint: Compute the skewness and find the frequency factor from the appropriate Log Pearson distribution (see Table D.3).

21. Utilize the information of Example Problem 2.2 and (a) compare the standard error parameters for normal, log normal, and log Pearson distributions, assuming a return period of 25 years. (b) Which of these gives the best statistical estimate for the expected precipitation?

22. Determine the standard error parameters for the data in problems 2.18 and 2.19 for Gumbel and log Pearson type distributions. Based on these factors and the estimates made for the 50-yr flood, estimate the relative reliability of the two methods and indicate the range of the expected values.

2.7.2 Computer-Assisted Problems

1. For the following rainfall intensity and time data, estimate a relationship between intensity and time to minimize the estimate of intensity. Use least squares linear regression for the equation $i = b/(a + t)$.

t-TIME (min)	*i*-INTENSITY (in./hr)
5	6.72
10	5.16
15	4.26
30	3.02
60	1.95

2. Using the data of Problem 1, what is your estimate using two other equation forms? Comment on the results.

3. Write a simple computer program to calculate the mean, standard deviation, coefficient of variability, and skew of a series of numbers. The program should enter the data in a numerical array and allow at least 30 numbers to be input from the keyboard. Use the computer program with the data from Problem 11 (Section 2.7.1) with your program.

4. Use a spreadsheet to calculate the mean, standard deviation, coefficient of variability, and skew of the data from Problem 11 (Section 2.7.1). Use the spreadsheet to sort the data and assign a Weibull probability to each ordered data. Plot the data on an *XY* plot versus Weibull probability. Use the plot to estimate the 50-yr return period value.

5. Use REGRESS to determine a "best fit" mathematical relationship between the rainfall data of Example Problem 2.2. Select one location as the dependent variable and the other location as the independent variable. Would you expect to find a relation between the two sites?

6. Using the Intensity-Duration-Frequency curve for Orange County, Florida (Appendix C); fit the 10-yr return period IDF. The curve fit should be done with intensity as a function of duration. The REGRESS program can be used to try different curve fit forms. The resulting curve should be plotted against the original curve on a log–log plot using a spreadsheet or plotting program.

7. Use DISTRIB to determine which distribution best fits the data of Problem 11 (Section 2.7.1). Using the best distribution, estimate the flow values for the 2, 3, 5, 10, 25, 50, and 100-yr return periods.

8. Use the EZMAT program to solve the following set of linear equations:

$$12x_1 + 5x_2 - 3x_3 = 51$$
$$15x_1 - 4x_2 + 7x_3 = 46$$
$$7x_1 + 19x_2 - 8x_3 = 39$$

2.7.3   Case Studies—Comparison of Distribution Fits for Rainfall Data

The following Rainfall data are for a NOAA site in Apalachicola, Florida. The data were collected in hourly increments. By adding successive time increments and comparison it is possible to determine the maximum volume of rainfall for given durations for each year of record. These records are referred to as *annual series maximum volumes*. These records are shown in the Table 2.12.

TABLE 2.12 Volumes (in inches) for Given Storm Durations for Rainfall Station in Apalachicola, Florida

YEAR	1	2	4	6	10	12	24	YEAR	1	2	4	6	10	12	24
	DURATION (hr)								DURATION (hr)						
1942	2.2	2.4	3.1	3.3	3.8	3.8	5	1963	2.4	2.7	4.2	4.5	4.8	4.9	5.5
1943	2	2	3	4.1	5.8	5.9	5.9	1964	2.5	3.3	4.2	5.3	5.4	5.5	7.9
1944	2.1	3.1	4.2	4.7	4.9	4.9	5.1	1965	5	5	5.1	5.5	5.7	5.8	6.3
1945	1.4	1.6	1.6	1.7	1.8	2.1	2.3	1966	3.4	3.8	4.4	4.6	6.6	7	7.5
1946	4.9	6.9	8.9	9.1	9.4	9.6	10.1	1967	2.9	2.9	3.1	3.1	3.2	3.3	4.1
1947	2.2	3	3.2	3.7	4	4	4.7	1968	2.8	2.8	3.8	4.2	4.4	4.4	4.5
1948	2.1	2.7	3.9	5.6	7.1	7.7	8.2	1969	2.3	2.6	2.9	3.3	4.4	5.1	7.1
1949	3.5	4.2	5.3	5.3	5.3	5.3	5.3	1970	5	5.5	5.8	6.1	6.1	6.1	6.1
1950	4	4.1	4.4	4.6	4.7	5.1	6	1971	1.4	1.7	2	2.1	2.7	3.1	3.2
1951	4.8	6.1	8.5	8.7	8.8	8.8	8.8	1972	1.5	1.6	1.9	2.3	2.8	2.8	3.1
1952	1.9	2.1	2.2	2.6	2.8	2.8	3.7	1973	2.8	3.3	3.4	3.4	3.5	3.5	3.5
1953	2.6	2.6	3.4	4.4	6.8	7.7	9	1974	3.1	4.6	5.4	5.9	6.9	7	8.2
1954	4.3	4.3	4.3	4.4	4.4	4.4	5	1975	1.9	2.8	3.8	4.7	6.1	6.2	6.8
1955	2.6	2.6	2.6	2.6	2.6	2.6	4.3	1976	1.4	1.8	2	2.1	2.5	2.6	3.7
1956	3.3	3.5	3.8	4	4.8	4.9	5.3	1977	1.8	1.9	2.8	3.1	3.2	3.5	3.7
1957	2.9	3.5	3.5	3.6	3.6	3.6	3.7	1978	1.9	2.5	2.6	2.7	2.7	2.7	3
1958	2.9	3.5	3.7	3.9	5.7	6	7.6	1979	2.7	2.9	3.7	3.9	4.7	4.8	6.5
1959	3.1	4.1	5.7	6.2	6.8	7	7.3	1980	1.6	2.1	2.5	3	3.9	4	4.6
1960	1.7	1.9	2.5	3.3	4.2	4.6	6.4	1981	3.8	4.3	4.8	4.8	4.8	4.8	5.2
1961	2.2	2.7	2.9	3.1	3.3	3.3	3.8	1982	4.5	6.2	7.2	7.3	7.6	7.6	8
1962	2.1	2.1	2.2	2.2	2.2	2.5	3.4	1983	4.3	4.8	4.8	5	5.1	5.1	5.1

The DISTRIB program is used to fit each of these series to a probability distribution. The probability distributions which will be considered are the 2 Parameter Log Normal, 3 Parameter Log Normal, Pearson Type III, Log Pearson Type III, and the Gumbel (Extremal Type I) Distribution. The program will be used to determine the probability distribution that best fits each set of data (for each duration). Use of the DISTRIB program is outlined in Appendix I. A plot of each distribution is shown in Figure 2.16 for the 1-hr duration along with the predictions for 2-, 3-, 5-, 10-, 25-, 50-, 100-, and 200-yr return periods for each distribution (see Table 2.13). From a visual observation of the graphical comparison, the log Pearson Type III and the log normal may be the better fits.

TABLE 2.13 Predictions (in inches) for each Distribution Type for Selected Return Periods (in years) using 1-hr Duration Data

RETURN PERIOD	NORMAL	2 PARAMETER LOG NORMAL	3 PARAMETER LOG NORMAL	PEARSON	LOG PEARSON	GUMBEL
200	5.48	6.67	6.24	6.43	7.39	6.59
100	5.22	6.09	5.79	5.93	6.62	6.03
50	4.93	5.51	5.32	5.43	5.87	5.47
25	4.61	4.93	4.84	4.91	5.16	4.91
10	4.12	4.14	4.16	4.17	4.24	4.14
5	3.65	3.53	3.59	3.57	3.54	3.54
3	3.22	3.03	3.10	3.07	3.01	3.06
2	2.77	2.59	2.65	2.60	2.55	2.63

Statistics, such as the chi-squared (χ^2) statistic, the sum-squared error, and the Kolmogorov-Smirnov number exist to allow a quantitative comparison between distributions. The best method for comparison of statistics is still visual. When choosing a "best fit" distribution you should be careful to consider the range in which the distribution is going to be used for predictions. If you are predicting high return period values, the distribution should fit well in the higher probability range.

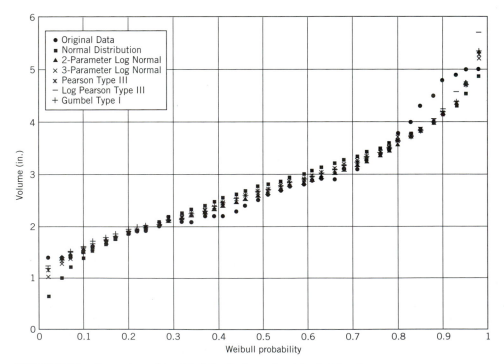

FIGURE 2.16 Comparison of empirical distributions using 1-hr duration data.

2.8
REFERENCES

Beard, L. R. 1962. "Statistical Methods in Hydrology," U.S. Army Corps of Engineers, Civil Works Project CW-151.

Benson, M. A. 1962. "Plotting Positions and Economics of Engineering Planning," *ASCE, Journal of the Hydraulics Division,* November, No. 88, pp. 57–71.

Gumbel, E. J. 1945. "Floods Estimated by the Probability Method," *Engineering News Record,* **134,** pp. 833–837.

Hardison, C. H. 1974. "Generalized Skew Coefficient of Annual Floods in the United States and Their Application," *Water Resources Research,* **10**(5), pp. 745–752.

U.S. Department of Commerce, 1973–1986. "Local Climatological Data," National Oceanic and Atmospheric Administration, Asheville, NC.

U.S. Department of Transportation. 1984. *Hydrology.* FHWA Report No. 1P-84-15, Federal Highway Administration, McLean, VA, pp. 45–52.

Wanielista, M. P. and Yousef, Y. A. 1993. Stormwater Management, John Wiley & Sons, New York.

Work Group on Flood Flow Frequency—Hydrology Committee. 1977. "A Uniform Technique for Determining Flood Flow Frequency," Bulletin No. 15, Thomas S. Kleppe, Chairman, Water Resources Council.

CHAPTER 3

PRECIPITATION

Some moisture is always present in the atmosphere. It is stored in the atmosphere awaiting forces that can cause precipitation. As moisture content increases, the chances of precipitation increase. This chapter discusses the formation, classification, and measurement of precipitation. Once the data on precipitation depth and rate (intensity) are available, interpretation of data and various methods for presentation are developed. These presentations form some of the input data required in latter chapters to estimate and predict precipitation and runoff quantities.

3.1

METEOROLOGY

Meteorology is the study of the atmosphere with special interest in weather and climate conditions. Weather conditions are those existing in a watershed or area at a specific time. Climate is the average of prevailing conditions over a period of years and is defined by measures of central tendency (average, median, etc.) or variability (standard deviation, range, etc.).

Weather conditions with ground cover, topography, and geology of an area determine surface water storage volumes and runoff flow rates to a great extent. Description or prediction of storage and flow rates must begin with, or at least be related to, meteorological and geological conditions.

3.1.1 The Atmosphere

The atmosphere is the gaseous envelope surrounding the earth. It sustains life by cycling chemicals and water. For the study of hydrology, the complex atmosphere can be divided into three parts: dry air, water vapor, and impurities related to the hydrologic cycling of water.

The dry portion of air is a mixture of gases. If the gases were separated and brought to the same temperature and pressure, the relative volumes that the four principal gases would occupy are shown in Table 3.1. These four gases make up all but 0.003% of the atmosphere 15 miles (25 km) above the earth. Some of the other trace gases are helium, ozone, hydrogen, radon, neon, and krypton. Ozone is an

68

TABLE 3.1 Volume of Four Gases in the Atmosphere

GAS	PERCENTAGE
Nitrogen (N_2)	78.09
Oxygen (O_2)	20.95
Argon (Ar)	0.93
Carbon dioxide (CO_2)	0.03

important trace gas because it prevents ultraviolet radiation from reaching the earth's surface. The carbon dioxide and water vapor in the air cause radiation from the earth to be absorbed. This absorption causes temperature increases. The quantity of carbon dioxide is changing because of its consumption by vegetation, absorption by oceans, and production by animals, volcanoes, and the burning of fossil fuels. Changes in atmospheric gases, especially ozone and carbon dioxide will continue to be examined to determine effects on the amount and distribution of water on the earth. Some investigators (Hansen, et al., 1981) predict dramatic effects on the earth's climate and weather which could change precipitation and evaporation rates.

3.1.2 Water Vapor

In the atmosphere of the earth, water vapor by weight is about 1.5×10^{13} metric tons. The atmosphere weight is 5.6×10^{15} metric tons. Water vapor must be present with temperature differences and other impurities for clouds and precipitation to form. Generally, water vapor occurs in the atmosphere about 18,000 ft above the earth (Petterssen, 1964). Within the earth's atmosphere, as distance from the earth increases, temperature in general will decrease. As temperature decreases, water vapor content will decrease. The amount of water vapor can be expressed as the pressure that vapor would exert in the absence of other gases, and is known as vapor pressure. The vapor pressure of water vapor saturating the air at 86°F (30°C) is about 1.18 in. of mercury (40 millibars [mbar]) while at 32°F (0°C), the vapor pressure is about 0.20 in. of mercury (7 mbar).

The atmosphere is infrequently saturated. The degree of saturation is expressed as the ratio of actual vapor pressure to that at saturation for a given temperature. This ratio is expressed as a percentage and is called relative humidity, or

$$f = 100 \frac{e}{e_s} \tag{3.1}$$

where

f = relative humidity (%)
e = actual vapor pressure (mbar)
e_s = saturated vapor pressure (mbar)

The variation of relative humidity with temperature and wet-bulb depression is shown in table form in Appendix B. As the relative humidity increases and approaches 100%, the chances for precipitation increases.

3.1.3 Solar Energy

Solar energy initiates the hydrologic cycle and influences climate. On the average, solar energy reaching the earth is approximately 0.5 Ly/min* with the planet absorbing about 0.22 Ly/min. There are wide geographic differences in the net rate at which solar energy is received at the earth's surface. Snow, for example, can reflect about 80% while dark soil can absorb about 90%. These differences aid in air movements and serve to redistribute energy. The energy produces mass movements in the atmosphere and oceans and is the energy source for evaporation and transpiration. Evaporation occurs from water surfaces while transpiration is the loss of water from plant life. Of special interest for the understanding of the hydrologic cycle are the fundamental processes of conduction, convection, and radiation. Solar energy enters the hydrologic cycle by the radiation process. The redistribution of energy is done by conduction from the land and convection in the water bodies and atmosphere.

Conduction is the transport of air between adjacent layers if the layers are at different temperatures. A net transport of heat will result. Also, the amount of heat transfer is dependent on velocity and concentration. The degree of proportionality can be determined by laboratory measures and is expressed by the following.

For thermal conductivity:

$$q_x = K_T[(\Delta T)/x] \tag{3.2}$$

where

q_x = rate of transport of heat per unit area in the x direction (cal/cm²-sec)
K_T = thermal conductivity constant at steady state (cal/deg-cm-sec)
ΔT = temperature differential (degrees centigrade)
x = distance (cm)

For concentration, Fick's law of diffusion can be used, and for velocity, Newton's law of viscosity can be used.

Convection is the movement of heat by the mass movement of air or water. It is one of the results of instability of the air or water masses, frequently caused by the potential energy in latent heat being converted into air currents. The rate at which the energy is released determines the meteorological conditions of rainfall and wind. Usually, stronger updrafts of air produce shorter rainfall durations. Another convective process—advection—is the transfer of heat by horizontal movement. The large-scale advective circulations of the atmosphere and the oceans are under the control of solar activity and the rotation of the earth.

3.1.4 Wind

Wind is the movement of air. The measures of wind are speed and direction. Wind speed is important because it can be related to water losses and precipitation events. In order for precipitation to occur, a sustained inflow of moist air is required. Winds provide the forces to sustain the moist air flow. Instruments for measuring wind speed are called anemometers. Common measures are kilometers per hour, miles per hour, meters per second, or knots.

If the world did not rotate, wind patterns would be established solely by thermal circulation. Winds would move toward the equator as warmer and lighter air would

---

*One Langley (Ly) equals one calorie per square cm.

rise and be replaced by cooler dense air. Because of the earth's rotation, air mass (frontal) movements are from west to east. Solar energy and the earth's rotation are superimposed on thermal circulation. When air masses with different temperatures meet, precipitation can occur at the boundary (front).

During a day, wind speed and direction may change. These are called diurnal changes and are significant only near the ground. Examples are changes that result from temperature contrasts between land and water. Winds tend to blow from cooler bodies toward warmer ones. In the summer afternoons, winds tend to blow toward the warmer land from the cooler water bodies. Mountains also influence wind currents because of diurnal heating of the sides of mountains, which can cause unequal heating of air masses, causing wind currents. Heating loss from other heat sources (sand, pavement, buildings) also can set up wind movement.

Over any one of the major continents, there will exist prevailing winds, either called trade or westerly winds. Between the trade and westerlies are calm or lightly variable winds. Nevertheless, pressure systems make the direction and speed of winds generally variable with time. In the northern hemisphere, these pressure systems move across a continent spinning counterclockwise if a low pressure system and clockwise if a high pressure system.

3.1.5 Temperature

Temperature influences the form of precipitation and the rates of evaporation, transpiration, and snowmelt. It is one measure that has been related to the prediction and explanation of the occurrence and distribution of waters on the earth. The measurement of temperature and factors influencing the measurement will be discussed first.

The measurement of temperature requires consideration of air circulation and surfaces in the immediate vicinity of the measuring device. There are over 8500 stations in the United States for which records are compiled by a government agency. The recording stations are constructed of louvered, wooden shelters positioned about 1.5 m (4–5 ft) above the ground. The shelters are protected from direct sunlight, wind, and precipitation. The recording station will list daily maximum and minimum temperature values. Temperatures are expressed using the Celsius (°C), Fahrenheit (°F), or absolute scales. Mean daily temperatures are expressed in terms of the interval of data collection. There are a few hundred hourly recording stations in the United States. For hourly temperature data:

$$T_{avg} = \sum_{i=1}^{24} T_i / 24 \tag{3.3}$$

where
T_{avg} = average daily temperature, °F or °C
T_i = hourly temperature, °F or °C

If maximum and minimum data are collected, then the average temperature is defined as

$$T_{avg} = (T_{max} + T_{min})/2 \tag{3.4}$$

where

T_{max} = maximum daily temperature, °C
T_{min} = minimum daily temperature, °C

The person using temperature data should know how averages were calculated because the average will vary with the method used.

Temperatures exhibit diurnal, seasonal, geographical, and elevation variations. Consequently, average conditions must be defined by time and location. An example of monthly variability with yearly averages is shown in Table 3.2 for a semitropical area in the northern hemisphere.

Under normal atmospheric conditions, temperatures will decrease with elevation. The mean decrease with increasing elevation is 0.7°C/100 m (3.8°F/1,000 ft). Thus, applying temperature data from one location (elevation) to a second location (elevation) may not be accurate. Tests of comparability would be advisable to establish deviations and levels of acceptable accuracy.

The temperature associated with saturated air is referred to as dewpoint temperature (T_D). Dewpoint temperature is easily obtained in the field or available from Weather Bureau records. Knowing the air and dewpoint temperature, an estimate of relative humidity can be obtained using

$$f = (112 - 0.1T + T_D)/(112 + 0.9T) \tag{3.5}$$

where

f = relative humidity,
T = temperature, °C
T_D = dewpoint temperature, °C

The diurnal variation of atmospheric moisture is normally small with relative humidity being a maximum early in the morning.

3.1.6 Variability of Meteorological Data

Meteorological data change frequently, certainly daily and seasonally, but sometimes instantaneously. As an example of change, consider the local climatological data in Table 3.3. Note the statistical data, such as averages and departures from normal, which indicate variability in the data. For some studies, site-specific meteorological

TABLE 3.2 Yearly and Monthly Temperature Averages

	AVERAGE TEMPERATURE °F												
YEAR	JAN	FEB	MAR	APR	MAY	JUNE	JULY	AUG	SEPT	OCT	NOV	DEC	ANNUAL
1934	62.2	59.8	65.4	70.8	76.4	60.0	81.8	82.0	79.6	75.0	67.2	60.4	71.7
1935	61.2	60.0	70.7	72.7	78.0	81.2	80.6	82.3	78.6	73.8	64.8	51.8	71.3
1936	58.6	56.9	63.9	69.8	74.5	77.8	81.6	81.0	79.4	76.8	64.4	62.8	70.6
1937	69.4	62.2	64.6	69.7	75.6	80.5	81.2	81.4	78.8	72.0	63.8	58.4	71.5
1938	59.3	65.2	70.2	70.4	77.0	77.8	79.6	81.2	78.3	70.2	66.6	57.8	71.1
1939	60.3	67.0	69.0	72.2	76.2	81.0	82.0	80.4	81.2	75.4	63.6	59.0	72.3

TABLE 3.2 (*Continued*)

						AVERAGE TEMPERATURE °F						
YEAR	JAN	FEB	MAR	APR	MAY	JUNE	JULY	AUG	SEPT	OCT	NOV	DEC	ANNUAL
1940	50.4	57.5	64.8	68.4	73.8	80.0	82.0	82.3	78.2	70.9	64.6	63.8	69.8
1941	57.0	55.8	60.2	70.0	72.9	81.4	80.6	82.8	79.2	76.2	65.4	62.3	70.3
1942	56.6	57.0	65.1	70.0	76.6	80.8	84.5	82.4	81.3	74.1	68.2	63.2	71.7
1943	62.8	59.4	66.4	70.6	78.2	82.8	82.8	88.0	80.4	70.6	65.2	62.0	71.9
1944	58.8	68.2	69.6	72.9	75.2	82.8	82.1	83.0	82.5	72.6	65.4	58.3	72.6
1945	59.8	66.2	72.8	75.7	76.8	82.0	81.8	82.1	81.0	75.0	65.4	59.4	73.2
1946	62.3	63.3	68.8	72.5	78.4	79.8	82.6	82.8	80.9	76.0	73.9	66.8	74.0
1947	69.0	54.2	62.2	76.4	77.8	80.4	80.0	82.0	80.2	76.0	68.8	64.0	72.6
1948	58.0	66.0	71.5	74.1	79.0	82.6	82.0	82.0	80.6	73.2	73.7	67.4	74.2
1949	64.4	69.6	67.4	72.5	77.8	80.6	82.6	81.8	81.1	77.8	62.7	65.2	73.8
1950	68.2	65.1	66.5	67.1	78.0	83.0	81.5	82.3	80.2	77.0	63.5	57.7	72.5
1951	60.0	59.8	66.1	68.5	75.7	80.5	81.8	83.8	81.4	76.1	63.7	65.1	71.9
1952	62.5	61.1	67.3	67.8	77.4	83.0	82.5	82.5	80.5	73.3	65.6	58.0	71.8
1953	60.2	63.5	70.0	71.0	79.7	81.1	82.8	81.5	80.0	71.7	66.1	52.7	72.5
1954	61.9	62.2	64.1	74.1	75.3	81.1	81.4	83.5	80.9	72.1	63.2	57.3	71.4
1955	58.4	62.2	67.4	72.3	77.8	79.3	81.2	82.1	81.4	72.4	65.6	60.8	71.7
1956	53.1	65.5	65.7	70.2	77.8	79.9	82.1	82.5	78.3	73.4	63.8	64.5	71.6
1957	64.7	67.5	65.8	72.3	76.9	80.7	82.6	81.3	81.0	72.1	69.4	59.1	72.8
1958	52.9	52.4	63.7	71.4	75.6	82.0	82.6	82.7	81.9	72.7	71.3	60.6	70.8
1959	58.3	67.6	63.7	70.9	78.0	81.0	81.3	81.7	79.7	78.5	66.9	60.7	72.3
1960	60.5	59.7	60.9	72.0	76.0	79.5	83.1	83.3	80.3	77.2	69.9	57.6	71.1
1961	56.9	64.7	70.6	69.4	77.0	80.6	83.3	82.9	81.1	73.6	69.4	63.9	72.8
1962	60.9	68.4	63.7	70.3	79.8	81.6	83.9	82.8	80.6	75.0	63.0	57.7	72.3
1963	59.5	57.2	69.5	73.6	77.9	82.4	82.9	83.9	80.5	73.8	65.4	56.5	71.9
1964	58.5	58.3	68.1	74.1	77.1	82.4	81.6	82.8	79.8	72.5	70.5	64.4	72.5
1965	60.0	64.1	67.0	74.8	77.1	79.0	80.5	82.2	80.8	74.2	69.5	62.6	72.7
1966	58.7	62.3	64.4	70.5	77.4	78.0	82.3	82.3	80.1	75.8	65.8	60.1	71.5
1967	63.2	60.0	68.3	74.3	78.3	80.2	82.4	82.0	79.7	74.0	67.5	65.9	73.0
1968	59.6	54.8	61.4	73.5	76.6	78.8	81.3	82.3	80.3	74.3	63.4	58.7	70.4
1969	59.8	57.8	60.4	72.5	76.9	82.9	84.4	82.2	81.2	77.9	64.0	58.7	71.6
1970	55.1	58.7	67.0	75.8	77.7	81.8	83.8	82.3	83.6	77.0	63.4	64.6	72.6
1971	62.0	64.1	64.8	72.1	78.2	81.7	83.1	83.3	81.8	79.0	69.5	71.4	74.2
1972	68.9	62.0	68.7	72.7	77.4	82.2	83.2	82.8	81.8	76.8	68.9	66.1	74.3
1973	62.4	59.7	71.1	71.1	78.3	83.1	84.2	81.8	81.4	75.6	70.9	60.4	73.3
						RECORD							
Mean	60.8	62.1	66.6	72.1	77.5	81.2	82.4	82.5	80.8	74.8	66.9	61.9	72.5
Max	71.7	73.3	77.8	83.3	88.4	91.0	91.8	91.6	89.3	83.7	77.3	72.7	82.7
Min	49.9	50.9	55.4	60.9	66.5	71.3	72.9	73.4	72.2	65.8	56.5	51.0	62.2

Latitude: 29 degrees 26 minutes north.
Longitude: 81 degrees 19 minutes west.
Source: National Oceanic and Atmospheric Administration. 1975. Annual Report, National Climatic Center, Asheville, NC.

TABLE 3.3 Example Climatological Data

LOCAL CLIMATOLOGICAL DATA

STATION: MCO
MONTH: MAR
YEAR: 1986

LATITUDE 28 DEGREES 26 MINUTES NORTH LONGITUDE 81 DEGREES 19 MINUTES WEST ELEVATION 96 FEET TIME ZONE: EASTERN

| | TEMPERATURE DEGREES F | | | | DEGREE DAYS (BASE 65) | | PRECIPITATION | | | WIND | | | SUNSHINE | | SKY COVER | WEATHER | PEAK |
| | | | | | | | | | | | FASTEST MILE | | | | | | | |
DAY	MAXI-MUM	MINI-MUM	AVER-AGE	DEPARTURE FROM NORMAL	HEATING	COOLING	TOTAL WATER EQUIV	SNOW-FALL, ICE PELLETS	SNOW, ICE PELLETS OR ICE ON GROUND AT	AVG SPEED (MPH)	SPEED (MPH)	DIREC-TION	TOTAL (MIN)	% POS-SIBLE	SUNRISE -SUNSET	OCCUR-RENCES	WIND
1	53	42	48	−16	17	0	0.03	0.0	0	11.2	18	26	0	0	5	1, 8	W 21
2	60	35	49	−16	17	0	0.00	0.0	0	9.1	14	29	0	0	0		NW 13
3	72	38	55	−9	10	0	0.00	0.0	0	8.5	14	25	0	0	5		W 22
4	73	52	63	−1	2	0	0.12	0.0	0	10.1	20	24	0	0	9	1	SW 25
5	66	48	57	−8	8	0	0.00	0.0	0	6.3	13	28	0	0	1		W 17
6	71	40	56	−9	9	0	0.00	0.0	0	9.9	18	28	0	0	5	1, 8	SW 25
7	72	50	61	−4	4	0	0.00	0.0	0	7.6	10	29	0	0	0	1	NW 20
8	76	48	62	−3	3	0	0.00	0.0	0	7.6	14	08	0	0	9		E 13
9	79	59	69	3	0	4	0.00	0.0	0	10.1	15	12	0	0	10		SE 17
10	77	64	71	5	0	6	0.00	0.0	0	9.9	15	12	0	0	9		E 19
11	85	64	75	9	0	10	0.00	0.0	0	7.2	12	24	0	0	7	8	W 14
12	88	64	76	10	0	11	0.00	0.0	0	8.7	14	11	0	0	4	1, 8	S 14
13	88	69	79	13	0	14	T	0.0	0	13.3	21	18	0	0	6		S 19
14	79	60	70	3	0	5	0.37	0.0	0	11.6	17	21	0	0	8	1, 3	NW 24
15	77	61	69	2	0	4	0.90	0.0	0	8.2	20	31	0	0	10	1, 3	NW 24
16	82	62	72	5	0	7	0.66	0.0	0	8.1	14	18	0	0	7	3	SW 16
17	82	63	73	6	0	8	0.00	0.0	0	5.9	12	07	0	0	5	1	NE 11
18	82	61	72	5	0	7	0.00	0.0	0	9.4	15	13	0	0	5	1	SE 15
19	87	68	78	11	0	13	0.15	0.0	0	10.9	18	18	0	0	8	1	S 18
20	88	60	74	6	0	9	0.00	0.0	0	11.9	17	20	0	0	7	1	SW 21
21	66	49	58	−10	7	0	0.15	0.0	0	9.6	14	33	0	0	7		N 16

Day															Weather		Weather	Wind	
22	59	44	52	-16	13	0	0.01	0.0	0	0	12.1	16	36	0	0		1		N	20
23	67	42	55	-13	10	0	0.00	0.0	0	0	9.5	13	02	0	0		0		N	15
24	73	47	60	-8	5	0	0.00	0.0	0	0	8.2	14	06	0	0		0		E	13
25	77	54	66	-3	0	1	0.00	0.0	0	0	9.1	16	06	0	0		6		NE	20
26	79	61	70	1	0	5	T	0.0	0	0	8.5	14	07	0	0		7		E	14
27	78	60	69	0	0	4	0.02	0.0	0	0	7.1	12	06	0	0		9	1	NE	12
28	81	57	69	0	0	4	0.00	0.0	0	0	7.7	13	06	0	0		3	1, 8	NE	15
29	79	59	69	0	0	4	0.14	0.0	0	0	7.8	13	07	0	0		6	1, 3, 8	NE	13
30	78	59	69	0	0	4	0.00	0.0	0	0	7.8	13	05	0	0		4	1	NE	14
31	79	59	69	0	0	4	0.00	0.0	0	0	7.2	13	07	0	0		4	1	NE	12
Sum	2353	1699			105	124	2.63	0.0			280.1						167			
Avg	75.9	54.8									9.0						5.4			

Misc. 21 18 31 0

TEMPERATURE DATA

Average monthly	65.4
Departure from normal	-1.4
Highest 88 on 12, 13, 20	
Lowest 35 on 2	
Number of days with	
Max 32 or below	0
Max 90 or above	0
Min 32 or below	0
Min 0 or below	0
Heating degree days (base 65)	
Total this month	105
Departure from normal	37
Seasonal total	609
Departure from normal	-47
Cooling degree days (base 65)	
Total this month	124
Departure from normal	0
Seasonal total	218
Departure from normal	-53

PRECIPITATION DATA (INCHES)

Total for the month	2.63
Departure from normal	-0.57
Snowfall, ice pellets	none
Total for the month	0.0 in.
T-Trace	< 0.01 in.

WEATHER

Number of days-

Clear (scale 0–3)	7
Partly cloudy (scale 4–7)	16
Cloudy (scale 8–10)	8
With 0.01 inch or more precip	10
With 0.10 inch or more precip	7
With 0.50 inch or more precip	2
With 1.00 inch or more precip	0

WEATHER SYMBOLS

1 = Fog
2 = fog w/visibility 1/4 mile or less
3 = Thunder
4 = Ice pellets
5 = Hail
6 = Glaze or rime
7 = Duststorm or sandstorm
8 = Smoke or haze
9 = Blowing snow
X = Tornado

Source: National Oceanic and Atmospheric Administration. 1986. Local Climatic and Data, March. National Climatic Center, Asheville, NC.

data are not available. Thus, data from the Weather Bureau or similar organizations located some distance from the study must be used and significant errors may be introduced. Equations used for description or prediction of hydrologic events could be tested for sensitivity of meteorological inputs. If necessary, field meteorologic data collected at a study site may be used, or at least statistically compared to longer records at permanent sites.

3.2

WEATHER SYSTEMS

The state of the atmosphere as measured by its temperature, water vapor, wind, and pressure determine the weather of a region. An understanding of the weather in a region will help explain some of the reasons for the existence of the hydrologic cycle and the measurement of the parameters of the cycle. The weather systems are related to types of precipitation events and air masses.

An air mass is a large air body whose physical properties (temperature, vapor, wind, pressure) are approximately constant in a horizontal plane. But violent changes occur on the border of the mass. These air masses can be found in the Arctic and Antarctic regions, subtropical ocean areas, and arid subtropical lands. The air masses will move from these regions. If the air mass is colder than the surface while in motion, precipitation will usually occur.

The border between air masses is called a front. Fronts are classified by the displacement of air. If cold air replaces warm air, a cold front results. A warm front results when warmer air moves into an area. When warm air masses rise, cooling takes place and precipitation can result. Storm systems can be classified by the factors responsible for the lifting of air masses (wind or solar heating). There are four major types of storms:

1. Convective storms
2. Orographic storms
3. Cyclonic storms
4. Tropical cyclones

3.2.1 Convective Storms

In the atmosphere, convection enables winds to be maintained by an upward and downward transfer of air masses of different temperatures. Convective storms result as warm, humid air rises into cooler overlying air (Figure 3.1). A common form of convective precipitation is the summer thunderstorm. The earth's surface is warmed by mid to late afternoon on a hot summer day. The surface imparts heat to the air mass directly above. The warmed air rises through the overlying air, and if the air mass has a moisture content equal to the condensation level, moisture will be condensed from the rising, rapidly cooling air. This may often result in a large volume of rain from a single thunderstorm.

Unequal cooling of air masses associated with large water bodies or "heat-islands"

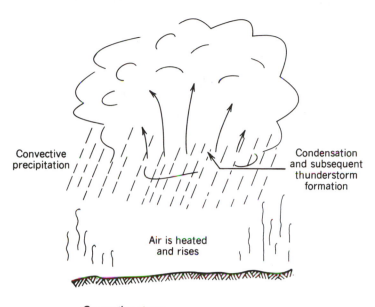

Convective storm.

FIGURE 3.1 Convective storm.

around metropolitan areas also can cause warm air to rise. Typically, these storms are very intense, are of short duration, and have wide area distribution.

3.2.2 Orographic Storms

Orography is the study of elevation relief between highlands (mountains) to other land and water features. Orographic precipitation results as warmer air rises over a high geographic feature such as a range of mountains and meets cooler air, as shown in Figure 3.2. Precipitation results if the rising air mass has a condensation level of moisture. Consequently, mountain slopes facing prevailing winds get more precipita-

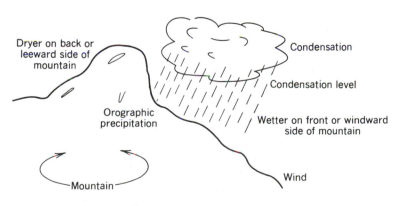

Orographic storm.

FIGURE 3.2 Orographic storm.

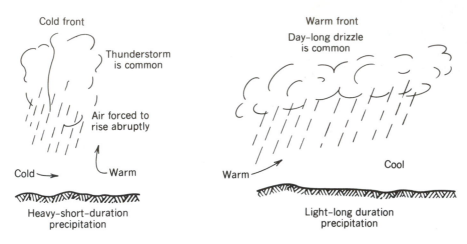

Cyclonic storms in mid-latitude.

FIGURE 3.3   Cyclonic storms in mid-latitude.

tion than the back, or leeward, slopes. The precipitation patterns of the Pacific coastal areas of North America are the result of significant orographic influences.

3.2.3   Cyclonic Storms

Cyclonic storms are caused by the rising or lifting of air as it converges on an area of low pressure. The movement of air is from high to low pressure areas and the boundary between air masses of different pressure is called a front. Frontal precipitation is formed from the lifting of warm air over cold air. Cold fronts are formed by cold air advancing under warmer air; a warm front is formed by warm air advancing over colder air (Figure 3.3). The intensity of precipitation associated with a cold front is usually heavy and covers a relatively small area, whereas less intense precipitation is associated with a warm front, but it covers a much larger area. Tornadoes and other violent weather phenomena are associated with cold fronts.

Cold fronts move faster than warm fronts, and, thus, warm air is lifted at a faster rate causing a higher intensity of precipitation. Cyclonic precipitation can also be associated with low pressure areas in the absence of frontal movements.

3.2.4   Tropical Cyclones

A tropical cyclone is an intense cyclone with its source in the tropic regions where the surface water temperature is generally greater than 85°F (29°C). The wind speed is 75 mph and above. An example of a tropical cyclone in North America is shown in Figure 3.4. Note the center, or eye, and the counterclockwise wind movement. It is called a hurricane, but in the Far East it is termed a typhoon, and in the Indian Ocean it is called a cyclone. The waters must remain warm to sustain the winds and rain. A tropical cyclone must remain in warm waters to sustain the high winds and rain. Tropical cyclones have been known to cause more than 25 cm (10 in.) of rain on an area in a relatively short time period (12–24 hr). In most regions, tropical cyclones cause the greatest volume of rainfall per storm event.

FIGURE 3.4 Tropical cyclone in North America, commonly called a hurricane (from U.S. National Environmental Satellite Service, NOAA).

3.3
PRECIPITATION CONCEPTS

3.3.1 The Formation of Precipitation

In general, there are four conditions that must be present for the production of precipitation: (1) condensation onto nuclei, (2) cooling of the atmosphere, (3) growth of water droplets, and (4) mechanisms to cause a sufficient density of the droplets. These conditions can occur in a relatively short time period and may be observed simultaneously.

Vapors are present in the atmosphere and can change to a liquid. The process by which vapor changes to a liquid or solid form is called condensation. In the atmosphere, cloud droplets form on condensation nuclei. These nuclei are usually sea salts and combustion by-products. The size of the nuclei are generally less than 1 micron in diameter. Prior to precipitation, most water droplets and ice crystals in clouds are less than 10 microns. During the condensation process, the water droplets and ice crystals will tend to enlarge because of vapor pressure differences. However, without any other factors present, it takes about one or two days for the particles of water and ice to reach the size of a small raindrop, which is about 3000 microns (3 mm). Thus, other factors are more important if precipitation is to occur.

One of these factors is the collision of particles. Collisions occur because of differences in rising and falling velocities. Particles that collide usually coalesce to form larger particles. Gravity acts on the particle to increase its falling velocity, but the friction drag causes a terminal velocity that depends on temperature, pressure, and size of the raindrop. At around 7 mm in diameter, the raindrop travels at about 10 m/sec (30 ft/sec) and usually breaks up into smaller drops.

Another factor is the ice crystal growth process. When ice elements form a vapor pressure imbalance with the water, larger water drops are created. The equilibrium vapor pressure over the water droplets is higher than over the ice elements. This causes the water vapor to evaporate and condense on the ice. Larger particles are formed and precipitation may result.

Precipitation formation can be modified using cloud seeding. Dry ice and silver iodide deposited in clouds can increase the possibility of rainfall from that cloud formation. The effectiveness depends on many factors, but the idea is to increase the number of condensation nuclei and promote ice crystal growth.

3.3.2 Classification of Precipitation

A meteor is a small particle of matter in the atmosphere. Any formation that results from the condensation process is called a hydrometeor. Fog, haze, frost, and blowing snow are hydrometeors of some concern in the science of hydrology. Related to the hydrologic cycle are hydrometeors that fall to the earth, which are important for the study and application of hydrology. The general classes of precipitation are as follows:

- Snow is complex ice crystals. A snowflake consists of agglomerated ice crystals. The average water content of snow is assumed to be about 10% of an equal volume of water.
- Hailstones are balls of ice that are about 5 to over 125 mm in diameter. Their specific gravity is about 0.7 to 0.9. Thus, hailstones have the potential for agricultural and other property damage.
- Sleet results from the freezing of raindrops and is usually a combination of snow and rain.
- Rain consists of liquid water drops of a size 0.5 mm to about 7 mm in diameter. Drizzle refers to small water drops less than 0.5 mm in diameter. The settling velocity is slow, with the intensity rarely exceeding 1 mm/hr (0.04 in./hr).

3.3.3 Water Quality of Precipitation

The aqueous environment is one of the ultimate repositories for trace amounts of all the synthetic chemicals made by man (Metcalf, 1977). Pollutants which are transported in stormwater runoff are typically considered the major source of pollution in our natural waters; however, precipitation itself can also be a major source of pollution. *Dustfall,* or dry precipitation, is a continuous process of atmospheric deposition of pollutants. These pollutants are then later picked up by stormwater runoff and contribute to the total pollutant concentration within stormwater. *Rainfall* or wet precipitation is not totally pure. Many pollutants have been reported in rainfall, including nutrient forms (nitrogen and phosphorus), metals, organics, and suspended solids. A listing is shown in Table 3.4.

TABLE 3.4 Rainfall Constituents (concentrations in milligrams per liter except where noted)

CONSTITUENT	KNOXVILLE TENNESSEE (BETSON, 1978)	GOTTINGTEN GERMANY (RUPPERT, 1975)	NORTH CAROLINA AND VIRGINIA (WEIBUL, 1969)	MELROSE FLORIDA (BREZONIK, 1969)	CENTRAL FLORIDA (WANIELISTA, 1976)	CENTRAL FLORIDA (YOUSEF ET AL., 1985)
Specific conductance (μS/cm)			12	10–30	8–34	18
pH	5.1			5.3–6.8	4.7–6.4	5.2
Cl^-	4.0	0.9	0.1–1.1			
SO_4^{2-}	7.1	8.8	1.1–3.2	1.74		2.7
Na^+	1.5	0.8	0.3–1.1	0.8		2.7
K^+	2.6	1.0		0.29–1.85		2.2
Ca^+	3.8	3.0	0.2–1.2	0.13–0.21		1.6
Mg^{2+}	0.74	0.7		0.06–0.35		0.2
Organic N	2.5			0.32	0.01–0.63[a]	0.27
NH_3-N	0.41			0.208		0.11
NO_2-N				0.005		
NO_3-N	0.47		0.045–0.225	0.209	0.02–0.26	0.32[b]
OP-P				0.009	0.01–0.08	
TP-P	0.36	0.36		0.011	0.04–0.2	0.02
TOC					2.2	
BOD_5					1.1	
COD	65					
Cadmium						0.0025
Lead						0.038
Copper						0.066
Zinc						0.0082

[a]Total Organic N + NH_3-N
[b]NO_2-N + NO_3-N

Nitrogen and phosphorus are considered pollutants because they are nutrients which stimulate the growth of algae and other aquatic plants. This increase in aquatic activity is called *eutrophication*. The increased respiration due to nutrients and corresponding algae growth can lead to depressed dissolved oxygen levels in water and subsequent fish kills. The eutrophication of lakes is also an aesthetic problem because of odor, discoloration of the water, and the formation of algae scum on the water surface. Nutrient levels are typically monitored by measuring total nitrogen, total Kjehdahl nitrogen (TKN), organic nitrogen, nitrate, ammonia, total phosphorus, total organic carbon, and orthophosphorus. Eutrophication in water bodies can be monitored using algamass, chlorophyll a, turbidity, and color.

Organics directly affect the dissolved oxygen level in water. Dissolved oxygen (DO) is vital to all aquatic organisms which do not breathe air. Oxygen demand is typically measured as biochemical oxygen demand (BOD), chemical oxygen demand

(COD), and total organic carbon (TOC). Oxygen demand is primarily caused by organic material in the water which decays. As the matter decays the bacteria utilize dissolved oxygen in the water, causing DO levels to decrease.

The most common metals in stormwater runoff are zinc (Zn), copper (Cu), and lead (Pb). A specific problem caused by metals is toxicity. Oftentimes the metal concentrations may be less than toxic levels; however, certain metals may *bioaccumulate* causing problems on the upper end of the food chain.

A major source of pollution in rainfall is a depressed pH due to the attenuation of ions in the atmosphere. This phenomenon is called *acid rain.* Acid rain is found near heavily industrialized regions and is a side effect of air pollution emissions. Unpolluted rain is naturally acidic, mainly because CO_2 from the atmosphere dissolves in sufficient quantity to form carbonic acid, H_2CO_3. The equilibrium pH for pure rainwater is taken to be 5.6.

The pollutants contributing to acid rain are primarily sulfur and nitrogen compounds (Elliott et al., 1984). Chemical reactions in the atmosphere convert SO_2, NO_x, and volatile organic compounds (VOCs) to acidic compounds and associated oxidants (e.g., sulfuric acid, nitric acid, ozone, and hydrogen peroxide). Acid rain is considered an air pollution problem, and attempts to prevent this type of pollution are typically centered around reduction of the air pollution sources.

3.4
MEASUREMENT OF PRECIPITATION

Precipitation is measured as the vertical depth of water (or water equivalent in the case of snow) that would accumulate on a flat level surface if all the precipitation remained where it had fallen. The units for reporting the depth are inches and hundredths of an inch in the customary U.S. units and millimeters and tenths of a millimeter in the SI system.

3.4.1   Rainfall Gages

There are at least three types of gages commonly in use to record depth: tipping bucket, weighing, and float. Gages with standard 8-in. diameter collectors have been used but are, in general, not recording types. The depth of water has to be manually read from a storage reservoir.

The tipping-bucket gage (shown in Figure 3.5) works on the principle that water accumulated in the collector is funneled into a two-compartment bucket. Each bucket is designed to collect the equivalent of 0.01 in. or 0.1 mm of water over either an 8-in. or 10-in. diameter collector. Once one of the buckets is filled, the filled bucket will tip and empty its contents. The bucket on the other side is now in position to collect water from the funnel. When the bucket tips, an electrical signal for each 0.01 in. or 0.1 mm of precipitation is sent to a recording unit. As the buckets alternately fill and tip, a momentary closure of an electrical switch (mercury) is completed, causing the electrical signal. This type of gage is used primarily for the measurement of rainfall, and not for ice or snow measurements. The snow or ice must melt to be recorded.

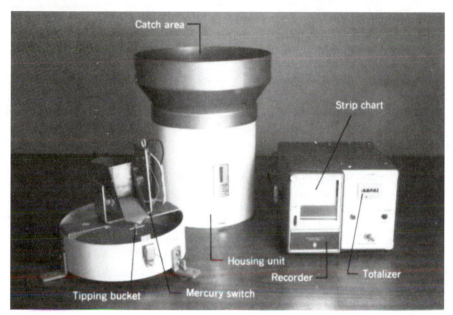

FIGURE 3.5 Tipping bucket housing unit mercury switch recorder strip chart totalizer.

The weighing-type gage measures the weight of rain or snow that accumulates in a bucket. The bucket sits on a scale that is calibrated to read an equivalent depth of water for a weight of precipitation. In remote areas, weighing-type gages have been known to operate for up to three months.

A float can be used to record the depth of water. The float is placed in the collector area or in a special reservoir of mercury or oil. As the depth of rainfall increases, the float rises and records the changes in depth with time. This type of gage can be easily damaged by freezing conditions.

Errors in measuring precipitation are usually small but tend to result in lower-than-expected readings. Instrument errors, however, must be protected against. Reduction of the collector area by damage to the collector, or covering of part of the collector, is a source of error. If the collector area is reduced, the depth of water recorded will be reduced. Other errors with the instruments are related to the weighing mechanism calibration, recording drive friction problems, and electrical current failures. All gages and recorders should be calibrated and serviced at least once per year.

☐ EXAMPLE PROBLEM 3.1

Part of a rain gage collector is covered during a storm event by debris. The debris reflected rain from the collector. Upon examination of the collector, it was found that 30% of the collector area was covered during rainfall. If the total amount of rain recorded was 0.51 in., what would be an estimate of the actual amount assuming a standard 8-in. diameter collector?

Solution

The collector diameter is 8 in., thus its area is $(\pi(8)^2/4) = 50.3$ in.^{2} The volume of rainfall recorded was 0.51 in. from a 35.2-in.^{2} area. The total volume of rainfall that should have been recorded as proportional to the total collector area of 50.3 in.^{2} and is

$$(35.2 \text{ in.}^2)/(0.51 \text{ in.}) = (50.3 \text{ in.}^2)/[X(\text{in.})]$$

$$\text{Actual estimate} = 0.73 \text{ in.} \quad \square$$

The positioning of the gage is also important in order to reduce errors in collecting precipitation. Obstacles that block precipitation from a collector must be avoided, and the collector must be positioned in a vertical plane. If a gage is inclined toward the wind, a greater amount of precipitation will be collected. Conversely, an incline away from the wind will reduce the collected volume.

Wind speed at the collector will reduce the estimate of precipitation because air is reflected upwards at the collector and precipitation is diverted. The error is greatest for light, or less dense precipitation, such as mist and snow. At a wind velocity of 20 mph (8.9 mps) Larson and Peck (1974) estimated about a 20% reduction in the volume of rainfall collected. For snow measurements, the reduction in volume was about 70%. The higher the gage is off the ground, the greater the error because of increased wind velocities. Trees, fences, and buildings have been used as windbreaks. Their height should be no higher than about one to two times the distance from the gage to the windbreak.

3.4.2 Networks of Rainfall Gages

The areal, or spatial, distribution of precipitation is related to meteorological and topographical factors. The number of precipitation gage stations per unit area (precipitation gage density) is generally less for flat regions than for mountainous regions in temperate climates. The use of daily, hourly, or more frequently measured precipitation data from a site is generally only relevant for that site. Using data from another site may produce errors. If data from one site are to be used at another site, less error is introduced if monthly or yearly precipitation data are used rather than daily or hourly data.

Comparing the areal distribution of total rainfall resulting from convective and cyclonic (frontal) storms, the convective storms tend to produce greater variability. For example, the two types of storms were compared for a precipitation network of six gages over an area of 600 km^2 (10 × 60 km) in central Florida. The convective storm had an average rainfall of 30.7 mm (1.21 in.) and a range of 12.5 to 63.5 mm, while the cold front produced an average of 21.6 mm (0.85 in.) and a range of 15.5 to 28 mm (Wanielista, 1977).

In St. Louis, Missouri, a network of 225 recording gages was distributed over a 2100 mi^2 urban area (Huff, 1974). Evidence available from these records indicates that the urban area was modifying rainfall intensities. It is generally believed that the climate in La Porte, Indiana, has changed primarily because of the large industrial complex in the Chicago area (Masters, 1974). Over 40 years, a 30 to 40% increase in precipitation has been noted.

Annual rainfall volumes for 1977 are compared in Table 3.5 to long-term averages over 30 years for various locations in Great Britain. These yearly data are important for regional water needs determination. Of equal importance is the yearly residual rainfall defined by Equation 3.6.

TABLE 3.5 Annual Areal Rainfall in 1977 Compared with the 1941–1970 Long-Term Annual Average by Water Authority Areas in England and Wales, and by River Purification board Areas in Scotland and in Northern Ireland

	1977 RAINFALL IN MM	LONG-TERM ANNUAL AVERAGE IN MM	1977 RAINFALL AS PERCENTAGE OF LONG-TERM AVERAGE
United Kingdom	1101	1090	101
England and Wales	925	912	101
Water authority areas			
North West	1206	1217	99
Northumbrian	874	879	96
Severn-Trent	839	773	109
Yorkshire	856	833	103
Anglian	583	611	95
Thames	744	704	106
Southern	979	794	100
Wessex	934	869	107
South West	1179	1194	99
Welsh	1403	1334	105
Scotland	1457	1431	102
Western Isles, Orkney and Shetland Islands Area	1316	1296	102
River purification board areas			
Highland	1642	1722	95
North East	1022	1023	100
Tay	1299	1255	104
Forth	1167	1117	104
Clyde	1802	1665	108
Tweed	1023	1003	102
Solway	1542	1425	108
Northern Ireland	1026	1095	94

Source: United Kingdom Meteorological Office, 1982. Surface Water: UK 1974–1976. Wallingford, U.K.

$$RR = RAIN - EVAP \qquad (3.6)$$

where

RR = residual rainfall (cm or in.)
$RAIN$ = long-term average rainfall (cm or in.)
$EVAP$ = long-term average evaporation (cm or in.)

The residual rainfall would be that which has the potential to recharge groundwaters and surface water reservoirs. In southern Wales, United Kingdom, yearly rainfall varies from 90 to 250 cm with residual rainfall varying from 50 to 200 cm.

Any network of precipitation gages should be planned to consider the intended use of the data and the economic impact. In more developed and densely populated regions that depend on the availability of scarce waters, the precipitation network operated by government groups is usually dense. For example, in the United Kingdom and Hawaii, there are about five to six stations per 100 mi^2, while in Alaska there is less than one per 2000 mi^2. In flat regions of temperate climate, one station per 200 mi^2 is recommended. In temperate mountainous regions, however, one station per 40 mi^2 is recommended. In polar regions, only one gage per 600 mi^2 is recommended (World Meteorological Organization, 1974).

The networks operated by the government of a country is believed to be responsive to large-scale, large-area projects. Specific related precipitation from convective storms need a greater density of gages. If a government network of gages is supplemented by other gages operated by private groups, the area density of gages is higher than expected and more useful for specific storm related data.

3.4.3   Radar Measurement of Rainfall

Radar can detect any type of hydrometeors in the atmosphere. A radar pulse for electromagnetic energy is used to determine the reflection of the hydrometeors. The reflection appears on a power display and is termed an echo. In general, the intensity (brightness) of the echo is a measure of the precipitation intensity. The distance from the radar site to the precipitation area can be measured by the time between emission of the radar pulse and receipt of the echo. The radar has an antenna that has to be orientated in the direction of the reflection. Thus, the spatial coordinates of the precipitation can be determined from the direction and distance measurements.

There can be a significant amount of ground clutter (interference from trees and buildings) when measuring hydrometeors in the atmosphere. To minimize the interference, a radar beam is directed upward at a slight angle. However, the height of the beam increases and accuracy decreases. Other factors affecting radar measurements are wind drift of particles, type of storm, distance from the radar to the storm, and the classification of the precipitation. However, radar can cover a complete area rather than relying on a single point measurement by a gage on the ground. Possibly, the joint use of radar and gage can lead to more accurate estimates of rainfall.

3.4.4   Missing Data

Precipitation measuring stations sometimes fail in providing a continuous record of precipitation. Instruments do malfunction and back-up systems may not always provide accurate data. A tipping-bucket gage may not function for a short period of time and the back-up volume gage may not provide time-related data. For a nonautomatic recording gage, an individual may fail to record the data or miss a visit to the site. Thus, there are generally missing data, the values of which must be estimated. There are two commonly used procedures for estimating daily precipitation depths, however, none are in common use for estimating hourly data (Paulhus and Kohler, 1952). The two procedures for estimating daily totals rely on the data from three adjacent stations. The locations of the adjacent stations are such that they are close to and approximately evenly spaced around the site with the missing data. Both procedures use the average annual precipitation (arithmetic average) at the three sites.

If the average annual precipitation at each of the three adjacent stations differs from the average at the missing data station by less than 10%, the following formula is used to estimate the missing daily data:

$$\overline{P}_X = \frac{(P_A + P_B + P_C)}{3} \tag{3.7}$$

where

$\overline{P}_X$ = estimated daily precipitation volume at the missing data site, X (depth)

P_A, P_B, P_C = estimated daily precipitation volume at the adjacent stations, A, B, and C (depth)

A simple arithmetic averaging of the data is used.

If the difference between the average annual precipitation at any of the adjacent stations and the missing data station is greater than 10%, a normal-ratio method is used. Normal is used as it refers to the arithmetic average. The method consists of weighting each adjacent station daily value by a ratio of the normal annual precipitation values and then average the numbers, or

$$\overline{P}_X = \tfrac{1}{3}[(N_X/N_A)P_A + (N_X/N_B)P_B + (N_X/N_C)P_C] \qquad (3.8)$$

where

N_X = average annual precipitation at the missing data site X (cm)

N_i = average annual precipitation at the adjacent sites (cm)

3.5

INTERPRETATION AND QUANTIFICATION OF PRECIPITATION DATA

To size water transport and storage systems, quantitative data for rainfall events must be provided. In some areas, these data are specified by regulations, however, it may be advantageous to update these data and certainly it is important to understand how these data were developed. These data can be defined in terms of:

1. Intensity (rate of rainfall), usually an average value for a given duration,
2. Duration of storm,
3. Time distribution of rainfall over the duration of storm,
4. Return period, and
5. Associated depth of rain.

All of these measures are required to adequately define a rainfall storm event.

3.5.1 Rainfall Intensity

Intensity, or depth of rainfall per unit time, is commonly reported in the units of millimeters per hour or (inches per hour). Weather stations utilizing gages that provide continuous records of rainfall can be used to obtain intensity data. These data are typically reported in either tabular form or graphical form (hyetograph).

Another way of reporting intensity data is the use of different time intervals. Figure 3.6 illustrates a hyetograph using a 15-min time interval for 6 in. of rain over 6 hours. It corresponds to a rainfall volume occurring once every 25 years for a specific region.

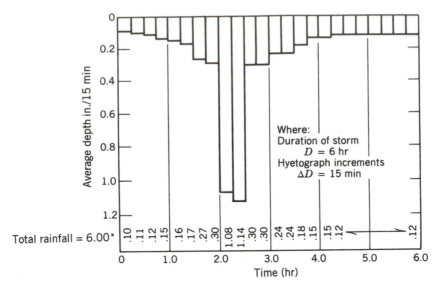

FIGURE 3.6   Hyetograph of 25-year design rainfall.

3.5.2 Cumulative Rainfall Diagram

A cumulative rainfall diagram that is a plot of cumulative rainfall versus time also is useful in runoff studies. At any given time during a storm, the intensity is the slope of the cumulative rainfall curve at that point in time. A graph of the cumulative rainfall diagram is shown in Figure 3.7. It can be used to determine the cumulative rainfall at any point during the duration of the storm event. The data of Figure 3.7 are obtained from Figure 3.6.

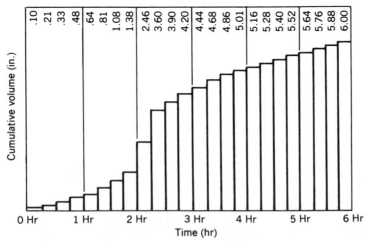

FIGURE 3.7   Cumulative rainfall diagram using the data of Figure 3.6.

TABLE 3.6 Storm Rainfall Variability (Maximum Events)

LOCATION	APPROXIMATE DEPTH		AVERAGE INTENSITIES		DURATION
	mm	in.	mm/hr	in./hr	
Funkiko, Formosa	1670	65.8	35	1.37	2 days
Cherrapunji, India	3330	131.1	20	0.78	7 days
Taylor, Texas, USA	585	23.0	24	0.96	24 hr
Hatteras, NC, USA	140	5.5	140	5.50	1 hr
Pensacola, FL, USA	60	2.4	720	28.80	5 min
Unionville, MD, USA	31	1.2	1860	72	1 min

❑ **EXAMPLE PROBLEM 3.2**

The maximum time it takes water to drain from a watershed under heavy rainfall is 45 min. Determine the maximum intensity (in./hr) in a 45-min. increment, assuming the cumulative rainfall diagram of Figure 3.7 is representative of the area where the watershed is located.

Solution

One must select from the diagram the maximum intensity associated with a 45-min period. Since Figure 3.7 was derived from Figure 3.6, either one can be used. Note, if one were to consider this intensity as being used to generate runoff rates or volumes, the watershed storage and infiltration capacity has to be considered. The maximum intensity of rainfall does not necessarily generate the maximum runoff condition. The maximum 45-min volume of rainfall from Figure 3.6 is 2.52 in. or 3.36 in./hr. ❑

Some indication of the variability of maximum rainfall storm volumes and intensities recorded at a location is shown in Table 3.6. As the storm duration decreases the average intensity increases.

3.5.3 Duration of Precipitation

The duration of a storm is the time from the beginning of rainfall to the point where the mass curve becomes horizontal indicating no further accumulation of precipitation within a certain time after the rain stops. In Figure 3.6, the storm duration is simply the width (time base) of the hyetograph. There may be frequent short time periods of no rainfall in a storm. Using simple empirical probability calculations, one can determine what the probability is for rain starting after a time interval of no rainfall. This time interval will vary with the type of storm (convective, cyclonic, etc.). But, in general, if the probability of rain starting after a specified time period of no rainfall (say 5 hr) is very small relative to other time periods (e.g., 0.001 vs 0.08), then the start of rainfall after the specific time (5 hr) would indicate a new storm.

Storm durations are usually reported over a range of a few minutes through hours, and days up to about 5 days. A 5-day storm may have time intervals of no rainfall greater than 5 to 10 hr. However, the use of daily durations up to about 3 or possibly

5 days is based more on total rainfall volume over that time period and time periods of 4 or 5 hr of no rain are not considered. Relative to longer duration storms with lower average intensities, higher intensities for short durations on fast responding watersheds where storm durations are about equal to travel time produce higher flow rates.

The choice of storm duration and thus intensities depends on the use of the rainfall data. The peak flow rates from an area require estimates of intensities associated with short time intervals while volume storage estimates for a watershed require long duration storms.

3.5.4 Dimensionless Cumulative Rainfall Diagrams

Since maximum rainfall volumes vary for different regions or for different locations in an area, a dimensionless volume axis for the cumulative rainfall diagram would be helpful for the time distribution of any rainfall volume. Also, duration of a storm may not always equal the duration used to construct the cumulative rainfall diagram. Therefore, the time distribution of the rainfall is normally given in a dimensionless cumulative rainfall diagram (also called a dimensionless mass diagram). This is a plot of the fraction of rainfall from a given volume of total rainfall as a function of the fraction of time for a given duration of a storm (Figure 3.8). The diagrams are generally specific to a region. Procedures for developing these dimensional diagrams can generally be classified as (Pilgrim and Cordery, 1975):

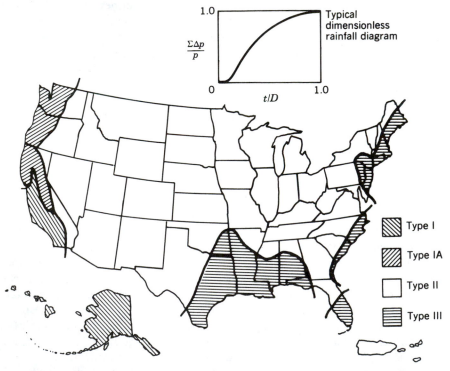

FIGURE 3.8 Approximate areas for NRCS (SCS) dimensionless rainfall diagrams and a typical plot (*Source:* United States Department of Agriculture Soil Conservation Service, 1986. *Technical Release #55,* Urban Hydrology, U.S. Department of Agriculture, SCS, Washington DC.).

1. Based on a large storm (usually one that produced a particular volume for an approximate duration)
2. Statistical averaging at each hour (usually many storms of similar duration)
3. Regression analysis (uses either same duration or mixed duration data)
4. The maximum intensities for selected durations

For a specified storm duration, some form a statistical analysis (regression or other curve fitting analysis) tends to produce the most realistic results. Usually, the larger volume storms are used.

Probably the most widely used diagrams among hydrologists were developed by the Soil Conservation Service (SCS) more recently known as the Natural Resources Conservation Service (NRCS) and were named Type I, Type IA, Type II, and Type III. Type II is widely used in the United States because of its geographic coverage. The approximate areas for these distributions and an example diagram are shown in Figure 3.8. The numerical values associated with Types II and III rainfalls are shown in Appendix C. Other dimensionless rainfall diagrams have been quantified by local and regional water management agencies. One such is the SCS Type II Florida Modified. These sources should be consulted because of the local nature and variability of storms. The Army Corps of Engineers also developed a dimensionless cumulative diagram and the numerical values are given in Appendix C. These values correspond to half hour intervals.

Given a known volume of rain for a given duration storm, the volume of rain at any time can be determined from the dimensionless cumulative rainfall diagram hyetograph. This time variation with watershed conditions directly determine the corresponding volume and flow rates of the surface runoff. High intensity rainfall at the beginning of a storm may result in a rapid rise in the runoff followed by a long recession of the flow. Conversely, if the more intense rainfall occurs toward the end of the storm, there will be a slow rise in runoff followed by a rapidly falling recession.

3.6

AVERAGE WATERSHED PRECIPITATION

Precipitation levels are variable over large geographical areas, such as the United States or Europe. Air mass movements, topography, and water/land locations are a few more important reasons for differences. In addition, specific locations have seasonal and yearly variations. All these factors complicate the scientific and engineering processes of design and operation of water resources systems. An indication of the geographical variability of mean average yearly rainfall is illustrated in Figure 3.9. These are only approximate values for a particular region, but geographical variability of yearly rainfall is clearly seen.

For smaller geographical areas, such as towns, cities, and water basins, rainfall volumes per storm event are most likely variable over the area. To illustrate this, consider raincells that are imaginary lines of equal precipitation (isohyets) over an area (Figure 3.10). Raincells for the St. Louis area have been reported in the literature (Huff, 1974). There are, in common use, three methods for estimating average precipitation for an area: isohyetal, Thiessen, and arithmetic average.

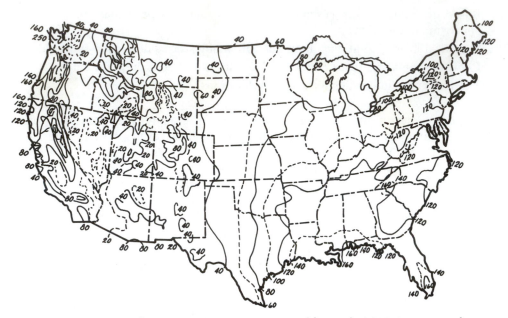

FIGURE 3.9 Mean annual precipitation (centimeters, converted from inches) (U.S. Department of Commerce, 1961).

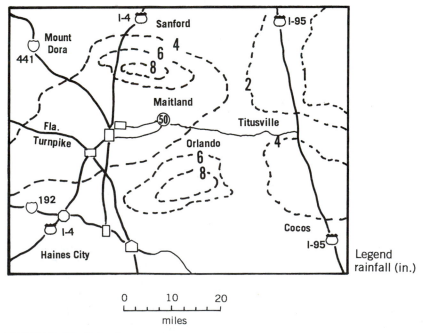

FIGURE 3.10 Raincells.

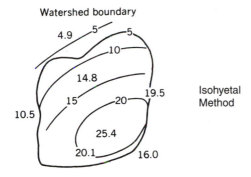

Watershed boundary

Isohyetal
Method

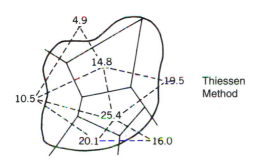

Thiessen
Method

FIGURE 3.11 Average precipitation (station location identified by the decimal point ex 4.9).

3.6.1 Isohyetal Method

The calculation of isohyets (i.e. contours of equal rainfall depth) is a reliable method of estimating average precipitation for a watershed, but the average is difficult to reproduce by another investigator because of the subjective nature (knowledge of storm morphology) of drawing isohyets. However, if precipitation values between precipitation station locations are determined by linear interpolation, the differences in average values should be reduced. The isohyetal calculations are well adapted for visual display (Figures 3.10 and 3.11). The area between each isohyet within the watershed is determined, and an average precipitation value is calculated. The isohyetal average is calculated as

$$\overline{P} = \sum_{i=1}^{n} W_i P_i \tag{3.9}$$

where
$\overline{P}$ = isohyetal average precipitation (mm)
P_i = isohyetal cell average precipitation (mm)
$W_i = A_i/A$; A_i − area of cell (km^2)
A = total area (km^2)
n = total number of cells

It should be noted that A_i is the area of the cell within the topographical drainage boundaries.

3.6.2 Thiessen Method

Another method for calculating average precipitation is the Thiessen method. The Thiessen method adjusts for the nonuniform location of gaging stations by attempting to determine the area of influence. The ratio of the area of influence of a station in the topographic basin to the total area in the topographic basin is a weighting factor to be applied for the calculation of the average value. If the procedure for calculating the weighting factors is followed, investigators are better able to duplicate the results of other investigators. The procedure is illustrated in Figure 3.11 and essentially involves connecting each precipitation station with straight lines, constructing perpendicular bisectors of the connecting lines and forming polygons with these bisectors. The area of the polygon is determined, and a weighted area is calculated using

$$W_i = A_p/A \tag{3.10}$$

where

W_i = weighted area, dimensionless
A_p = area of the polygon within the topographic basin (km^2)
A = total area (km^2)

The average precipitation using the Thiessen method is

$$\overline{P} = \sum_{i=1}^{n} W_i P_i \tag{3.11}$$

where

$\overline{P}$ = average precipitation (mm)
P_i = gage precipitation for polygon i
n = total number of polygons

3.6.3 Arithmetic Average Method

The arithmetic average method (Figure 3.11) uses only those gaging stations within the topographic basin and is calculated using

$$\overline{P} = \sum_{i=1}^{n} P_i/n \tag{3.12}$$

where

$\overline{P}$ = average precipitation depth (mm or in.)
P_i = precipitation depth at gage (i) within the topographic basin, (mm or in.)
n = total number of gaging stations within the topographic basin

◻ **EXAMPLE PROBLEM 3.3**

For the following watershed, estimate using three methods the average precipitation. The watershed is shown in Figure 3.11.

Solution

a. Isohyetal method

$$\bar{P} = \sum_i W_i P_i$$

ISOHYET (mm)	WEIGHTED AREA (W_i)	AVERAGE PRECIPITATION (P_i)	$W_i P_i$ (mm)
>20	0.30	22.7*	6.8
10	0.58	15.0	8.7
5	0.12	7.5	0.9
			$\bar{P} = 16.4$ mm

*(25.4 + 20)/2 = 22.7.

b. Thiessen method

$$\bar{P} = \sum_i W_i P_i$$

OBSERVED PRECIPITATION (P_i)	WEIGHTED AREA (W_i)	$W_i P_i$ (mm)
14.8	0.29	4.3
19.5	0.16	3.1
4.9	0.07	0.3
10.5	0.19	2.0
25.4	0.15	3.8
20.1	0.12	2.4
16.0	0.02	0.3
		$\bar{P} = 16.2$ mm

c. Arithmetic average (add those within the watershed)

$$\bar{P} = \sum_{i=1}^{n} P_i/n$$

$$\bar{P} = \frac{14.8 + 25.4 + 20.1}{3} = 20.1 \text{ mm} \quad ◻$$

TABLE 3.7   Relation of Areal Reduction Factor with Duration (D) and Area (A) for the U.K.

DURATION D	WATERSHED AREA (km²)									
	1	5	10	30	100	300	1000	3000	10,000	30,000
5 min	0.90	0.82	0.76	0.65	0.51	0.38	—	—	—	—
10 min	0.93	0.87	0.83	0.73	0.59	0.47	0.32	—	—	—
15 min	0.94	0.89	0.85	0.77	0.64	0.53	0.39	0.29	—	—
30 min	0.95	0.91	0.89	0.82	0.72	0.62	0.51	0.41	0.31	—
60 min	0.96	0.93	0.91	0.86	0.79	0.71	0.62	0.53	0.44	0.35
2 hr	0.97	0.95	0.93	0.90	0.84	0.79	0.73	0.65	0.55	0.47
3 hr	0.97	0.96	0.94	0.91	0.87	0.83	0.78	0.71	0.62	0.54
6 hr	0.98	0.97	0.96	0.93	0.90	0.87	0.83	0.79	0.73	0.67
24 hr	0.99	0.98	0.97	0.96	0.94	0.92	0.89	0.86	0.83	0.80
48 hr	—	0.99	0.98	0.97	0.96	0.94	0.91	0.88	0.86	0.82

Source: Flood Studies Report, 1975.

3.6.4   Areal Reduction Factors

As the area increases, the average precipitation for that area decreases. Using selected data from the U.K. Meteorological Office (1982), Table 3.7 was developed to illustrate the decrease (areal reduction) with increasing areas for various storm durations. The areal reduction factor is calculated as a ratio of area average precipitation to maximum precipitation measured at a point within the area.

3.7

EXTRAPOLATION OF POINT MEASUREMENTS TO WATERSHEDS

The point measure of rainfall depth and intensity from a gage is of value for estimating volume and runoff for larger areas. To accomplish this, the depth and intensity measured at a point must be considered as constant over an area or two or more point measures must be averaged. In this section, equations are presented to convert point measures to area measures. In the next section, methods to estimate average values are presented.

3.7.1   Depth and Watershed Volumes

The depth of rainfall or equivalent rainfall (if snow measures) recorded on a gage can be related to a watershed area if one can assume that the point estimate is reasonable and constant for the watershed area. Because of the great area variability of precipitation, the point estimate is usually used for small areas (few acres to square mile). To convert from depth to volume, one simply follows

$$V = PA(3630) \tag{3.13}$$

where
   V = volume of rainfall (CF)
   P = rainfall (in.)
   A = watershed area (acres)
   3630 = conversion factor (43,560 ft^2/acre divided by 12 in./foot)

3.7.2   Intensity and Watershed Discharge

The intensity measure at a gage also can be useful for scaling up to a watershed. From a basic mass per unit time balance of inputs and outputs, the rate of precipitation onto a watershed—if held constant—would equal the runoff from the watershed. The critical assumption is one of time to achieve steady-state conditions. If the intensity remains constant over the time it would take for the total contributing area to drain the rainfall excess to the output site, then precipitation intensity must equal runoff or

$$\text{Runoff (outflow)} = \text{precipitation intensity}$$

and to balance units, intensity must be multiplied by area, or

$$Q = iCA(1.008) \tag{3.14}$$

where
   Q = runoff rate (CFS)
   i = precipitation intensity (in./hr)
   CA = contributing area (acres)
   1.008 = conversion factor (CFS-hr/acre-in.)

It is important to note the contributing area is the area from which precipitation will result in runoff. An impervious area (pavement, roofs, etc.) with a discharge point is an example of a contributing area. The constant (1.008) may be dropped from the equation. Equation 3.14 is called the rational equation, primarily because it is a rational belief that outflow rate equals inflow rate. See also the discussion in Section 6.4.1.

3.8

INTENSITY-DURATION-FREQUENCY CURVES

Structures designed to control stormwater volumes and flows need a quantitative criteria to determine their size. The volume or flowrate which must be stored or conveyed by the system can be mathematically related to the precipitation; therefore, a predictive measure of the precipitation is required for stormwater system design. Two important stormwater parameters, intensity and duration, can be statistically related to a frequency of occurrence. The graphical representation of this relationship is the intensity-duration-frequency (IDF) curve. The IDF curve is a plot of average rainfall intensity versus rainfall duration for various frequency of occurrences (or return periods), as shown in Figure 3.12. This information can also be presented as an isohyetal map of intensity over an area for a given return period and duration as

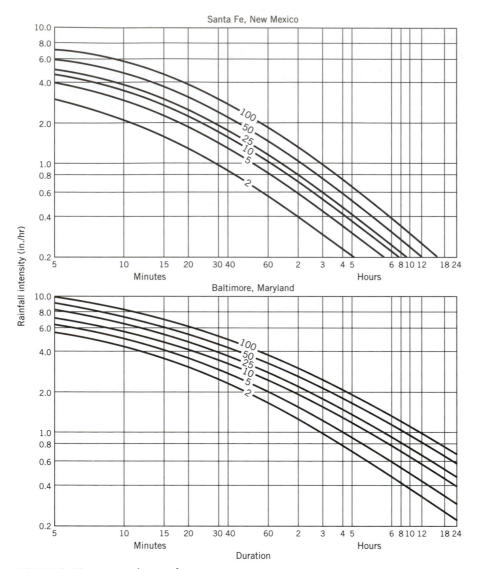

FIGURE 3.12   Intensity-duration frequency curves.

shown in Figure 3.13. These data are used for the design and operation of closed or open conduits, reservoirs, groundwater pumps, pollution control structures, assimilative capacity studies, and so on.

3.8.1   Precipitation Database

Even though the IDF curve is used for design based on rainfall volumes, the curve is ideally developed for calculating flow rates based on average intensity. The curve is typically developed from empirical data maintained by NOAA. These data are available as recordings taken either every 15 minutes or every hour. Using hourly data and a 24-hr duration, maximum average daily intensity can be determined by finding the 24 contiguous data points which have the greatest value. This value does not take

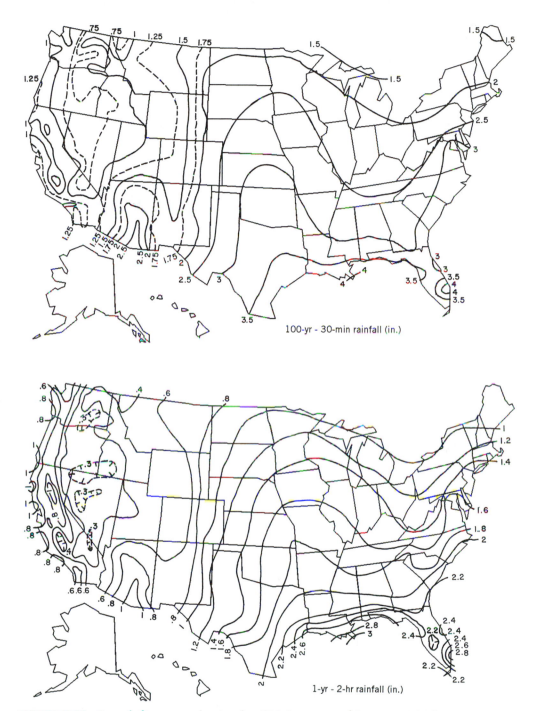

FIGURE 3.13 Example frequency—duration data (U.S. Department of Commerce, 1961).

TABLE 3.8 Example Hourly Precipitation Data

DATE	0100	0200	0300	0400	0500	0600	0700	0800	0900	1000	1100	1200	1300	1400	1500	1600	1700	1800	1900	2000	2100	2200	2300	2400	TOTAL
01/01/1986	0.00	.	.	.	.	.	.	.	.	.	.	.	.	.	.	.	.	.	.	.	.	.	.	.	0.00
01/04/1986	.	.	.	.	.	.	.	—	0.90	.	.	.	.	.	.	.	.	.	.	.	.	.	.	.	0.90
01/05/1986	.	.	.	.	.	.	.	.	.	.	.	.	.	0.30	.	.	.	.	.	.	.	.	.	.	0.30
01/08/1986	.	.	.	.	.	.	.	.	.	.	0.10	.	.	.	.	.	.	.	.	.	.	.	.	.	0.10
01/09/1986	.	.	.	.	.	.	.	.	.	.	.	.	.	.	.	.	.	.	.	.	.	.	.	0.20	0.20
01/10/1986	0.20	0.20	.	.	.	.	.	.	.	.	.	0.10	.	.	.	.	.	.	.	.	.	.	.	.	0.50
01/18/1986	.	.	.	.	.	.	.	.	.	.	.	.	.	.	.	.	.	.	0.10	.	.	.	.	.	0.10
01/26/1986	.	.	.	.	.	.	.	.	.	.	.	.	.	.	0.10	.	.	.	.	.	.	.	.	.	0.10
01/27/1986	.	.	.	.	.	.	.	.	.	.	.	.	0.10	.	.	.	.	.	.	.	.	.	.	.	0.10
02/01/1986	0.00	.	.	.	.	.	.	.	.	.	.	.	.	.	.	.	.	.	.	.	.	.	.	.	0.00
02/10/1986	.	.	.	.	.	.	.	.	.	.	.	.	.	.	.	.	0.10	.	.	.	.	.	.	.	0.10
02/18/1986	.	.	.	0.10	.	.	0.10	.	.	.	.	.	.	.	.	.	.	.	.	.	.	.	.	.	0.20
02/19/1986	.	.	.	.	.	0.10	.	.	.	.	.	.	.	.	.	.	.	.	.	.	.	.	.	.	0.10
02/23/1986	.	.	.	.	.	.	.	.	.	.	.	.	.	.	0.10	.	.	0.10	.	.	.	.	.	.	0.20
02/28/1986	.	.	.	.	.	.	.	.	.	.	.	.	.	.	0.10	.	.	.	.	.	.	.	.	.	0.10
03/01/1986	0.00	.	.	0.10	.	.	.	.	.	.	.	.	.	.	.	.	.	.	.	.	.	.	.	.	0.10
03/04/1986	.	.	.	.	.	.	.	.	.	.	.	.	.	0.10	.	.	.	.	.	.	.	.	.	.	0.10
03/10/1986	.	.	.	.	.	.	.	0.40	1.90	0.40	.	0.30	0.10	0.10	.	0.10	.	.	0.20	0.20	.	.	.	.	3.70
03/14/1986	.	.	.	.	.	.	.	.	.	.	.	.	0.40	0.30	.	.	.	.	.	.	.	.	.	.	0.70
03/15/1986	.	.	.	.	.	.	.	.	.	.	.	.	.	.	.	0.10	.	0.10	.	.	.	.	.	.	0.20
03/19/1986	.	.	.	.	.	.	.	.	.	.	.	.	.	.	.	.	.	.	0.40	.	.	.	.	.	0.40

Date																		Total
03/20/1986																		0.10
03/21/1986																		0.10
03/26/1986															0.10			0.60
03/27/1986								0.10			0.30	0.10	0.10	0.10	0.10			0.10
04/01/1986	0.00																	0.00
04/05/1986																		0.00
04/13/1986			0.10		0.10													0.10
04/21/1986											0.10	0.10						0.10
05/01/1986	0.00																	0.00
05/05/1986										0.10		0.10						0.10
05/09/1986					0.10	0.10	0.20				0.30	0.10	0.10	0.10				0.50
05/20/1986					0.10	0.20	0.10					0.10						0.50
06/01/1986	0.00					0.10					0.10							0.10
06/03/1986									0.20									0.30
06/07/1986								0.60	0.10									0.70
06/08/1986								0.10	0.20	0.30	0.30		0.10					0.70
06/09/1986	0.10							0.10	0.10	0.10								0.30
06/11/1986										0.20	0.20					0.10		0.30
06/13/1986							0.10	1.60	0.30	0.10	0.10							2.10
06/14/1986									0.80	0.20	0.20	0.20	0.10					1.30
06/15/1986							0.40											0.50
06/16/1986										0.10	0.10	0.20	0.10					0.10
06/17/1986										0.70	0.70	0.20	0.10	0.10	0.10		0.10	1.30

into account any precipitation which may have occurred before the 24-hr period and therefore may be unacceptable for use in volume based design. To develop an empirical frequency distribution, maximum 24-hr volumes and average intensity must be determined on an annual basis using many years of data or from a partial series basis using a cutoff minimum volume value. Once a sufficient number of years have been analyzed, the empirical frequency data are then compared to a theoretical distribution and return period predictions are made. These return period predictions for each duration analyzed are then plotted typically on a log–log scale, as shown in Figure 3.13. The Soil Conservation Service Technical Paper Number 40 (1961) has IDF curves of rainfall for the United States.

❑ **EXAMPLE PROBLEM 3.4**
_____

Using the hourly rainfall data in Table 3.8, determine the maximum 1-, 2-, 6-, and 12-hr duration volumes. Note that days with no rainfall are omitted.

Solution

By inspection of the table, we can determine the maximum intensities and the time and date they occurred.

maximum 1 hr = 1.9 in. (3/10/86 at 0900)

maximum 2 hr = 2.3 in. (3/10/86 at 0800–0900 and at 0900–1000)

maximum 6 hr = 3.1 in. (3/10/86 at 0800–1300)

maximum 12 hr = 3.5 in. (3/10/86 at 0800–1900)   ❑

3.8.2 Best Fit Curves

When using computers, calculators, or hand calculations, it may be more convenient to use equations rather than graphs. The observed data, in general, can usually fit an equation of the following form:

$$i = \frac{at^m}{(b + D)^n} \tag{3.15}$$

where

i = average rainfall (cm/hr or in./hr) for a fixed duration
t = frequency of occurrence (return period) (yr)
D = duration of storm (min or hr)
a, b, m, n = coefficients and exponents varying from one region to another

The common form of Equation 3.15 used for hydrologic analysis is one that fixes the frequency of occurrence, thus we eliminate t and m from the equation and assume the exponent n to equal unity, resulting in

$$i = \frac{a}{b + D} \tag{3.16}$$

TABLE 3.9 Rainfall Quantities and Intensities

RAINFALL DURATION (D)	QUANTITY FOR GIVEN FREQUENCIES						
	1	2	5	10	25	50	100
5 min	0.56[a] (6.72)[b]	0.64 (7.68)	0.78 (9.36)	0.87 (10.44)	0.98 (11.76)	1.11 (13.30)	1.18 (14.2)
10 min	0.86 (5.16)	0.99 (5.94)	1.20 (7.20)	1.34 (8.04)	1.51 (9.06)	1.71 (10.30)	1.82 (10.9)
15 min	1.09 (4.26)	1.25 (5.00)	1.51 (6.04)	1.69 (6.76)	1.91 (7.64)	2.16 (8.64)	2.30 (9.2)
30 min	1.51 (3.02)	1.73 (3.46)	2.10 (4.20)	2.50 (4.80)	2.65 (5.30)	3.00 (6.00)	3.20 (6.4)
60 min	1.95 (1.95)	2.25 (2.25)	2.70 (2.70)	3.10 (3.10)	3.40 (3.40)	3.80 (3.80)	4.20 (4.20)
2 hr	2.30 (1.15)	2.70 (1.35)	3.30 (1.65)	3.80 (1.90)	4.30 (2.15)	4.80 (2.40)	5.30 (2.65)
3 hr	2.50 (0.833)	2.90 (0.967)	3.70 (1.23)	4.30 (1.43)	4.80 (1.60)	5.30 (1.77)	5.90 (1.97)
6 hr	2.90 (0.483)	3.40 (0.567)	4.40 (0.733)	5.20 (0.867)	5.80 (0.967)	6.50 (1.08)	7.20 (1.20)
12 hr	3.30 (0.275)	4.00 (0.333)	5.20 (0.433)	6.20 (0.517)	7.00 (0.583)	7.80 (0.650)	8.70 (0.72)
24 hr	3.90 (0.163)	4.70 (0.196)	6.20 (0.258)	7.20 (0.300)	8.40 (0.350)	9.30 (0.388)	10.50 (0.43)

[a]Denotes inches of rainfall.
[b]Parentheses denote intensity (in./hr).

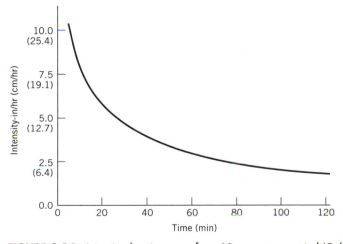

FIGURE 3.14 Intensity-duration curve for a 10-year return period (Golding, 1977).

TABLE 3.10   Best-Fit Intensity/Duration Curves

FREQUENCY (yr)	EQUATION	CORRELATION COEFFICIENT
1	$i^b = \dfrac{167.08}{23.210 + D^c}$	$r^2 = .999$
2	$i = \dfrac{195.29}{23.885 + D}$	$r^2 = .999$
5	$i = \dfrac{247.36}{26.185 + D}$	$r^2 = .996$
10	$i = \dfrac{289.07}{27.736 + D}$	$r^2 = .996$
25	$i = \dfrac{324.23}{27.969 + D}$	$r^2 = .996$
50	$i = \dfrac{357.76}{26.864 + D}$	$r^2 = .996$
100	$i = \dfrac{401.78}{28.864 + D}$	$r^2 = .995$

[a]Good only for durations less than 180 min.
[b]i = rainfall intensity (in./hr).
[c]D = duration (min).

Standard statistical bivariate regression procedures are then used to solve for the coefficients a and b. Additional examples of these equations in graphical form are found in Appendix C.

□   **EXAMPLE PROBLEM 3.5**

To illustrate the development of the above equation, rainfall storms of various return periods (frequencies) and durations are shown in Table 3.9. These data are obtained from local climatological reports. A graph relating intensity to duration of storm is illustrated in Figure 3.14 for a frequency of occurrence of 10 years.

Solution

Develop "best" fit equations using the least squares criteria. By our using bivariate regression techniques, equations are developed relating intensities and duration. These equations are listed in Table 3.10 (Golding, 1977).   □

3.9

SNOW

Snow can cause significant problems both by accumulation and by melting thus creating potential floods. Deep accumulations can cause transportation and structural failures.
The density of snowpack varies with the time since initial accumulation and other

meteorological conditions. Newly fallen snow may have as little as 5% water while long accumulated snowpack (close to ice) may have over 80% water.

Snow depth is measured by rain gages, snow stakes, and density determinations. Wind affects the accuracy of the measurement devices, thus shields are frequently used to protect gages from wind. Snow stakes are calibrated and driven into the ground. The height of the stake must be longer than most expected accumulations, or additional stakes must be added during the snow accumulation periods. Generally, many stakes are used for an area because of local variations in snow depth caused by wind. The stakes outline a snow course and are usually spaced out 15 to 30 m (50–100 ft) apart.

Stream gaging stations are useful for the measurement of water equivalents and density of the snowpack. By measuring the runoff volume, the equivalent depth of water over the watershed can be estimated. By knowing the average depth of snow, the density can be calculated.

3.9.1 Physical Nature and Thermal Properties

Snowflakes are crystals with a dendritic structure that form a low density snowpack immediately after falling. However, over time the density increases. The density increases because of many factors:

1. Surface heat exchange at the snow–air interface due to radiation, convection, and condensation.
2. A conduction process at the ground surface.
3. Compaction due to its own weight.
4. Percolation of liquid (from melt or rain) through the snowpack.
5. Movement of air at the snow–air interface.
6. Air and snow temperature variations.

Of all these factors, the ones that have been primarily related to snowflake density are near-surface air temperature and wind speed. As near-surface air temperature increases, the snowflake density increases. The U.S. Army Corps of Engineers (1956) illustrated this increase of snowflake density as air temperature increases to follow a nonlinear relationship. An average density value for calculating average water equivalents is 0.10 g/cm^3. Also, as wind velocity increases, density increases.

Average snowpack density increases with time, and a settlement of depth of snowpack can be seen. On a daily basis, snowfall depth is recorded; however, the cumulative snowfall is greater than the cumulative snowpack depth over time (except on the very first day, the snowpack depth may equal the snowfall depth). Density of the snowpack will increase with depth into a snowpack. For one location, the U.S. Army Corps of Engineers (1956) reported a variation from 0.08 g/cm^3 at the surface to about 0.35 g/cm^3 at about 8 to 12 ft below the surface. The average density was about 0.27 g/cm^3 for this particular snowpack. At any particular time and depth into snowpack, ice may form and thus density may vary considerably. Ice forms when snow melts at the surface, percolates into the snow, and then freezes.

The depth of snowmelt (d_m) can be calculated if representative values for depth of snowpack and density of snowpack are available with the density of water at 0°C equal to 1 g/cm^3.

$$d_m = \rho_s d_s / \rho_m \tag{3.17}$$

where

d_m = depth of melt (cm)
ρ_m, ρ_s = melt and snowpack density (g/cm³)
d_s = depth of snowpack (cm)

The depth of snowpack water (d_w) in a depth of snowpack (d_s) can be estimated by an indirect measure of average snowpack temperature (T_s) in degrees Celsius below zero (a positive value) and the snowpack density (ρ_s). The depth of snowpack water can be derived by knowing the heat capacity (H_c) of the snowpack, which depends on the snowpack specific heat, its density, and average snowpack column temperature. Investigations by the U.S. Army Corps of Engineers (1956) and knowledge that specific heat has a constant value of 0.5 calories per gram per degree Celsius for a range of snowpack density of 0.05 to 0.90, lead to Equation 3.18. The heat required per unit area to raise the temperature of the snowpack to 0°C is

$$H_c = \int_0^{d_s} \rho_s c_s T_s \, dz \tag{3.18}$$

where

H_c = heat capacity (calories/cm²)
c_s = snowpack specific heat (calories/gram °C)
T_s = snowpack temperature (°C)
z = depth for integration (cm)

Upon integration and assuming average values with depth

$$H_c \cong \rho_s c_s T_s d_s \tag{3.19}$$

Before significant melt can occur, an additional latent heat of fusion must be supplied and is written as

$$\tag{3.20}$$
$$H_c = L_f \rho_w d_w$$

where

H_c = heat content (cal/cm²)
L_f = latent heat of fusion = 80 calories per gram
ρ_w = density of liquid water = 1 g/cm³
d_w = water depth equivalent of the cold snowpack (cm)

For water, the latent heat of fusion is a quantity of heat required to convert 1 g of ice to water. Equating Equations 3.19 and 3.20:

$$d_w = (\rho_s c_s T_s d_s)/(\rho_w L_f) = (\rho_s d_s T_s)/160 \tag{3.21}$$

The depth of water of the cold snowpack compared to the depth of snowmelt is usually less for the same snowpack density and snowpack depth.

The collection of accurate snowpack density data are now always possible for large areas. Furthermore, terrain features, wind velocities, and near surface air temperatures must be used to determine the rate of melt. Equations 3.17 and 3.21 provide estimates of potential quantities, but the rate of melt (depth per day) is not available. To accomplish this, energy budget methods are useful.

❑ **EXAMPLE PROBLEM 3.6**
_____

Data on newly fallen snow density and air temperature are available from the U.S. Army Corps of Engineers (1956). Using a computer program, develop a best-fit linear and power equation relating density as a dependent (Y) variable to surface air temperature as an independent (X) variable. The partial data listing is:

DENSITY (g/cm³)	AIR TEMPERATURE (°F)	DENSITY (g/cm³)	AIR TEMPERATURE (°F)
0.04	14	0.09	22
0.06	17	0.10	23
0.08	18	0.09	24
0.08	19	0.11	25
0.09	20	0.12	30
0.08	20.5	0.13	31
0.08	21	0.18	32

Solution

Using the linear regression option of your computer program or solving by hand computation a linear regression model (see Chapter 2), the equations are

$$\text{linear:} \qquad \rho = -0.036 + 0.0058T \qquad 14 \le T \le 32$$

$$\text{power:} \qquad \rho = 0.0013T^{1.41} \qquad 14 \le T \le 32$$

where
ρ = density of snowfall (g/cm³)
T = degrees Fahrenheit (°F) ❑

3.9.2 Energy Budget

There exist at least six heat effects on a snowpack, which accounts for a change in heat stored in a pack during a time interval (Δt). In equation form, these are (units of calories):

$$\Delta H = H_S + H_L + H_C + H_{CS} + H_G + H_P \tag{3.22}$$

where
ΔH = heat storage change
H_S = short-wave solar radiation

H_L = net long-wave radiation exchange
H_C = convective from the air
H_{CS} = condensation or sublimation
H_G = conduction from the ground
H_P = advection from precipitation

If heat is added to the snowpack column, it is a positive value, removed heat has a negative value. When ΔH exceeds the cold content H_C, melt will occur. Since it requires 80 cal/g for each square centimeter or 80 cal/cm (density of water is 1 g/cm³) the depth of melt in centimeters is

$$d_m = \Delta H/80 \tag{3.23}$$

where ΔH = energy change in calories.

The calculation of each heat effect requires extensive time related data. Thus, the Corps of Engineers (1956) developed other statistical measures to estimate the rate as a function of daily meteorological data. They completed extensive studies in the western portion of the United States. These studies are detailed sufficiently to separate rainy and rain-free periods of melt. During rainfall, heat transfer by convection and condensation is important. During rain-free periods, radiation becomes significant. During rainfall conditions, estimation equations in the U.S. system of units are:

For partial forest cover (<60% of the area)

$$\text{SNM} = 0.09 + (0.029 + 0.0084kU + 0.007i)(T_a - 32) \tag{3.24}$$

and for heavily forested cover (>60% of the area)

$$\text{SNM} = 0.05 + (0.074 + 0.007i)(T_a - 32) \tag{3.25}$$

where
SNM = daily snowmelt (in./day)
   k = watershed constant (0.3 for dense areas to 1.0 for clear plains)
   U = average wind velocity at 50 feet above snow (mph)
   i = average rainfall intensity (in./day)
   T_a = temperature of saturated air measured 10 ft above the snow (°F)

During rain-free conditions, a variety of equations exist as a function of percent forest area. The equation for heavy forested areas is

$$\text{SNM} = 0.074(0.53T_a' + 0.47T_d') \tag{3.26}$$

where
   T_a' = temperature difference between that at 10 ft and the snow surface (°F)
   T_d' = temperature difference between the 10-ft dewpoint and snow surface temperature (°F)

3.10
SUMMARY

- Precipitation forms when nuclei are present, onto which water attaches and grows to over 0.5 mm in diameter.
- Hydrometeor is the general term for the various forms of rain, snow, and other classifications of precipitation.
- A hyetograph is a plot of intensity versus time.
- A dimensionless mass diagram is a plot of the fraction of rainfall from a given volume of total rainfall as a function of the fraction of time for a given storm duration.
- Precipitation is measured both as depth and intensity at a point on the ground.
- Depth of precipitation can be converted to a watershed volume by assuming the depth is constant over the watershed area.
- The intensity of runoff is related to the intensity of rainfall. The rainfall intensity must remain constant over the time it would take the total contributing area to drain the rainfall excess.
- The network of gages provides precipitation data for point on the ground. For a wider regional estimate, averaging techniques must be used. Commonly used ones are the arithmetic average, isohyetal, and Thiessen.
- Daily snowmelt can be estimated using an energy-budget method and for specific conditions, Equations 3.24, 3.25, and 3.26 can be used.
- Intensity–duration–frequency curves reflect the rainfall history of an area. By specifying the risk (frequency), an intensity can be estimated for a specific watershed.
- Rainfall excess can be assumed to be directly related to precipitation, resulting in a coefficient relating both. This coefficient is called the runoff coefficient and can be defined as the ratio of rainfall excess to precipitation. In later chapters, the runoff coefficient will be used in another formula for estimating runoff.

3.11
PROBLEMS

3.11.1 Hand Solutions

1. A rain gage is located in a 2.5-acre impervious watershed with no initial abstraction. The gage records 1.0 in. of rainfall in one hour. The maximum intensity was 2.4 in. of rainfall/hour for 10 min. Assume that 10 min is a time during which all parts of the watershed can contribute to a discharge point. What is the volume of rainfall in cubic feet and the maximum runoff rate?

2. Develop a hyetograph (graphical presentation) for the following rainfall data. Plot the intensity (inches/hour) for 30 minute intervals.

TIME (min)	CUMULATIVE RAIN (in.)
30	0.04
60	0.38
90	1.07
120	1.44
150	1.62
180	1.70

3. What is the average rainfall intensity for a 1-hr and 6-hr rain event with a return period of 100 years for Orange County, Florida, area? Use the intensity–duration–frequency (IDF) Curves of Appendix C.

4. What is the total rainfall volume for the storms of Problem 3?

5. If 55 mm of rain is recorded for a 6-hr storm by one rain gage for a watershed area of 10 km^2, but the runoff from the watershed indicates only 45 mm of rain has fallen on the entire area, what is the areal reduction factor? How does this compare to the results reported in the U.K. (Table 3.7)?

6. Using Figures 3.6 and 3.7, determine the average rainfall intensities in inches/hour for the following time periods.
   a. From 3:00 hour to the 4:00 hour
   b. For the first half of the 6-hr storm
   c. The maximum in any one hour
   d. The maximum in any 15-min period
   e. The last half hour

7. Obtain hourly rainfall data from a local climatological report and develop a hyetograph and a cumulative rainfall curve.

8. For the cumulative rainfall curve of Figure 3.15, develop a 1-hr listing of rainfall intensities for a storm of 24-hr duration. The maximum storm volume is for a 100-yr storm event. To obtain the volume, use any intensity–duration–frequency (IDF) curve or the ones in Appendix C. State your references.

9. Using the NRCS (SCS), Type II dimensionless cumulative rainfall curve of Appendix C, develop cumulative rainfall depths on an hourly basis for a total rainfall of 203 mm (8 in.). Do the same for the Corps of Engineers standard distribution. Compare results.

10. Develop for each half-hour increment the cumulative rainfall for a 24-hr storm using the NRCS (SCS) Type III curve of Appendix C and a 254-mm (10 in.) rainfall.

11. Describe the drainage area that is defined by the USGS Station #1 in Figure 3.16 by drawing on the map the appropriate boundaries and compute the area of the drainage basin in square miles.

    Now, from the precipitation stations, identified by (4.00 in.) calculate the average precipitation by three methods: (1) arithmetic mean, (2) Thiessen, and (3) isohyetal.

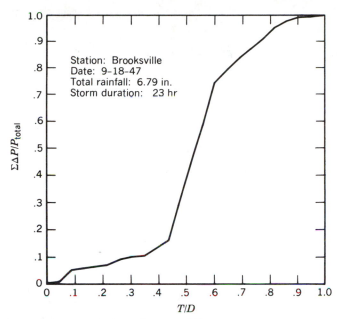

FIGURE 3.15 Actual cumulative rainfall curve.

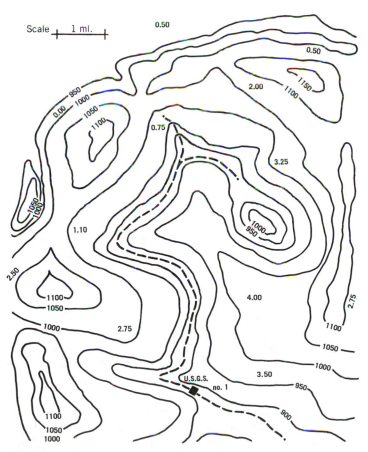

FIGURE 3.16 USGS No. 1 drainage area.

12. Plot the intensity duration data of this problem and fit by "graphical" (eye) technique the best line.

DURATION (min)	INTENSITY (in./hr)
10	4
15	3.2
20	2.7
30	1.9
60	1.2
120	0.8
180	0.6

13. Using two dimensionless cumulative rainfall diagrams as applied to a rural water-shed with sandy type soils (high infiltration), which hyetograph would produce the maximum peak discharge if soil saturation occurs at about half of the rain-fall volume?

14. Develop a dimensionless rainfall diagram using the following incremental precipi-tation data from a 24-hr storm.

TIME (hr)	ΔP(in) (in 4-hr increments)
0	0
4	1.00
8	2.00
12	3.00
16	2.00
20	1.00
24	1.00

Using one of the IDF curves in Appendix C and a 24-hr–50-yr event, estimate the rainfall volume during the first 12 hr of the 24-hr storm using the dimensionless rainfall diagram just developed.

15. Using the IDF curve for Santa Fe, New Mexico (Figure 3.12), answer the follow-ing questions.
   a. What is the rainfall intensity for a 50-yr storm event with a duration of 60 min?
   b. If the duration were 5 hr for the 50-yr storm event, what is the total rainfall?
   c. If the watershed response were only 30 min, what is the maximum 30-min intensity using the 50-yr storm event? Compare this to a storm with a frequency of once every 2 yr.
   d. If the time of concentration for an impervious 20-acre watershed were 30 min, what is an estimate of the peak discharge for the 50-yr storm event? Express your answer in CFS units. What assumptions did you make?

e. What is an estimate of rainfall excess for the same storm as used in part (d)? Express your answer in cubic feet.

16. Using Figure 3.12 and a 50-yr storm event over 24-hr provide a listing using 1-hr intervals of a dimensionless cumulative rainfall diagram. Assume the maximum intensity occurs in hour 12, the next in hour 11, the next in hour 13, and so forth.

17. Use an IDF curve of your choice, possibly for your region, and provide a 1-hr listing of a dimensionless cumulative rainfall diagram. Also state your assumptions for the ranking of each value.

18. Snow is sampled for density and depth over a watershed approximately 4 mi² in area. The average density of water in the 27-in. snowpack was estimated to be 40%. What is the expected volume of runoff in acre-feet from this area if no additional evaporation occurs and all the snow melting goes to rainfall excess?

19. Using the U.S. Army snowmelt equation for a partial forest covered a 160 acre area with a watershed constant of 0.5, average wind speed of 10 mph (50 ft above snow), temperature of 40°F, and an average rainfall of 0.4 in./day, what is the expected daily snowmelt in acre-feet/day?

20. For the following 1-hr hyetograph measured at Baltimore, Maryland:
    a. What is the average intensity in cm/hr?
    b. What is the volume of rainfall in m³ and liters if the watershed is 4000 m²?
    c. What is the approximate return period?

TIME (min)	INTENSITY (cm/hr)
0–10	2.0
10–20	6.0
20–30	12.0
30–40	8.0
40–50	6.0
50–60	3.0

21. Using the NRCS (SCS) nondimensional Type II rainfall distribution and a 10-in., 24-hr storm event, list the cumulative and interval rainfall volumes for every 2-hr time interval.

22. A 12-hr duration, one in 25-yr storm, occurs in Orange County, Florida. What is the average intensity of rainfall for the first 3 hr? Use the NRCS (SCS) Type II rainfall distribution. State all assumptions.

23. What are differences between cyclonic and tropical cyclone weather systems?

24. What is the dew-point temperature? How is it related to relative humidity?

25. Calculate the average minimum daily temperature for the month of March using the data of Table 3.3.

26. Pick any four days from Table 3.3 and show how the value of average temperature was calculated.

27. Given the following 24 hourly values of temperature, calculate the average using two methods.

HOUR	TEMP., °C	HOUR	TEMP., °C	HOUR	TEMP., °C
1 am	20	9	15	5	20
2	19	10	16	6	20
3	17	11	17	7	19
4	16	12	18	8	18
5	14	1 pm	19	9	17
6	13	2	20	10	17
7	14	3	21	11	16
8	15	4	21	MN	16

28. For the total monthly rainfall of Table 3.3, calculate the monthly rainfall excess assuming the runoff coefficient is .6.

29. If the runoff coefficient of the previous problem changed to .8 for daily rainfall greater than .5 in., what is the daily and monthly sum for rainfall excess?

3.11.2 Computer-Assisted Problems

1. Using SMADA, create two rainfall data files using the rainfall data of this chapter and Appendix C or other files for your local area.

2. Develop, using the REGRESS program, a mathematical equation to estimate the dimensionless cumulative rainfall curve for the following data:

TIME (hr)	P/P_{total}	TIME (hr)	P/P_{total}
0	0	13	0.453
2	0.01	14	0.585
3	0.05	15	0.74
4	0.065	16	0.79
5	0.07	17	0.835
6	0.075	18	0.87
7	0.09	19	0.91
8	0.10	20	0.95
9	0.11	21	0.975
10	0.135	22	0.985
11	0.17	23	0.99
12	0.29	24	1.00

3. Using REGRESS, estimate the "best" equation for one of the cumulative rainfall diagrams you have used in the past. What is the next best equation? How did you determine the best? For purposes of this problem, if a polynomial equation fit is used, do not exceed a 3 degree equation.

4. Develop your own computer program to calculate the average of any type of precipitation data (i.e., rainfall depth, snow equivalent depth, or intensities). The program must relate the units of the problem to the program user.

5. Using REGRESS estimate a "best" fit line for the following rainfall intensity and duration data. Postulate at least three different relationships.

DURATION (min)	INTENSITY (in./hr)
10	4.0
15	3.2
20	2.7
30	1.9
60	1.2
120	0.8
180	0.6

6. For Example Problem 3.6, execute the computer program (regress) to obtain the solution. Also present a graphical display of both equations and estimates of density for a temperature of 28°F.

TABLE 3.11 Volume Predictions (in inches) for Given Durations and Return Periods Using Log Pearson Type III Distribution

RETURN PERIOD	1 HOUR	2 HOUR	4 HOUR	6 HOUR	10 HOUR	12 HOUR	24 HOUR
200	7.41	9.23	11.08	10.70	11.03	11.79	12.32
100	6.65	8.19	9.82	9.92	10.22	10.83	11.45
50	5.93	7.21	8.63	8.76	9.38	9.85	10.54
25	5.22	6.28	7.50	7.79	8.50	8.85	9.60
10	4.30	5.10	6.07	6.50	7.26	7.48	8.27
5	3.60	4.22	5.02	5.48	6.21	6.37	7.15
3	3.06	3.57	4.22	4.68	5.34	5.46	6.21
2	2.59	3.01	3.54	3.96	4.52	4.63	5.34

TABLE 3.12 Intensity Predictions (in inches/hr) for Given Durations and Return Periods Using Log Pearson Type III Distribution

RETURN PERIOD	1 HOUR	2 HOUR	4 HOUR	6 HOUR	10 HOUR	12 HOUR	24 HOUR
200	7.41	4.62	2.77	1.78	1.10	0.98	0.51
100	6.65	4.10	2.46	1.62	1.02	0.90	0.48
50	5.93	3.61	2.16	1.46	0.94	0.82	0.44
25	5.22	3.14	1.88	1.30	0.85	0.74	0.40
10	4.30	2.55	1.52	1.08	0.73	0.62	0.34
5	3.60	2.11	1.26	0.91	0.62	0.53	0.30
3	3.06	1.79	1.06	0.78	0.53	0.46	0.26
2	2.59	1.51	0.89	0.66	0.45	0.39	0.22

TABLE 3.13 Curve Fit Coefficients for IDF Curve Developed Apalachicola, Florida Data

RETURN PERIOD	a	b	R^2
200	12.55	0.833	0.9877
100	11.88	1.063	0.9619
50	10.95	1.193	0.9466
25	10.02	1.357	0.9230
10	8.57	1.529	0.8963
5	7.62	1.877	0.8253
3	6.63	2.000	0.8041
2	5.61	2.011	0.8011

3.11.3 Case Studies-Development of an Intensity-Duration-Frequency (IDF) Curve

Using the data from the case study in Chapter 2, we will develop an IDF curve for Apalachicola, Florida. By comparing the fits of each distribution using all seven sets (durations 1 hr to 24 hr) of data it was shown that the 2 parameter Log Normal and the Log Pearson Type III fit the data sufficiently to use in the generation of the IDF curves. The Log Pearson Type III distribution will be used to generate these curves.

Using the DISTRIB program, it is possible to determine a prediction for a number of return periods for each duration. These predictions are shown in Table 3.11. By dividing each of the volumes in Table 3.11 by the respective duration, we can develop a set of rainfall intensities, as shown in Table 3.12. These intensity data are then fit

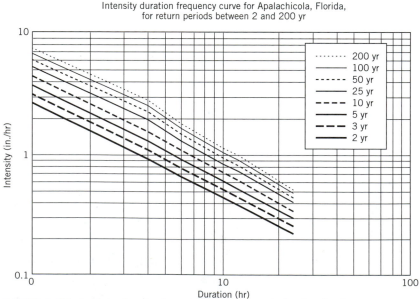

FIGURE 3.17 Intensity-duration-frequency-curve for Apalachicola, Florida.

to the curve fit form;

$$i = \frac{a}{b + D}$$

The values for the coefficients a, and b and the respective correlation coefficient R^2 are shown in Table 3.13. These intensities are plotted on a log-log scale to give the IDF curve, as shown in Figure 3.17.

3.12

REFERENCES

Betson, Roger. 1978. "Bulk Precipitation and Streamflow Quality Relationships in Urban Areas," *Water Resources Research,* **14(6):** 1165–1169.

Brezonik, P. L., et al. 1969. *Eutrophication Factors in North Central Florida Lakes,* Bulletin Series No. 134, University of Florida, Gainesville, FL.

Chow, V. T. 1964. *Handbook of Applied Hydrology.* McGraw-Hill, New York.

Golding, B. L. 1977. "Volusia County 208 Comprehensive Rainfall Analysis," Howard, Needles, Tammen, and Bergendoff, Orlando, FL, February.

Hansen, J., Johnson, D., Lacis, A., Lebedeff, S., Lee, P., Rind, D., and Russell, G. 1981. "Climate Impact of Increasing Atmospheric Carbon Dioxide," *Science* **213:** 4511.

Huff, F. A. 1974. "The Distribution of Heavy Rainfall in a Major Urban Area." *Proceedings of the National Symposium on Urban Rainfall and Runoff and Sediment Control,* D. T. Y. Kao, Ed., University of Kentucky, Lexington, July, pp. 53–59.

Larson, L. W. and Peck, E. L. 1974. "Accuracy of Measurements for Hydrologic Modeling," *Water Resources Research,* **10(4),** pp. 857–863, August.

Masters, G. M. 1974. *Introduction to Environmental Science and Technology.* Wiley, New York, p. 218.

Meteorological Office. 1982. Surface Water: United Kingdom 1974–76, HMSO Water Data Unit, Wallingford, U.K.

National Environmental Research Council. 1975. U.K. Institute of Hydrology Flood Studies Report, Volume II, Wallingford, U.K.

Paulhus, J. L. H. and Kohler, M. A. 1952. "Interpretation of Missing Precipitation Records," *Monthly Weather Review,* **80,** pp. 129–133, August.

Pilgrim, D. H. and Cordery, I. 1975. "Rainfall Temporal Patterns for Design Floods," *Journal of the Hydraulics Division, ASCE,* NYI, January, pp. 81–95.

Ruppert, H. 1975. "Geotechnical Investigations on Atmospheric Precipitation in a Medium-Sized City," *Water, Air and Soil Pollutants,* **4:** 447–763.

U.S. Army Corps of Engineers. 1956. *Snow Hydrology,* North Pacific Division, Portland, Oregon, June 30.

U.S. Department of Commerce. 1961. "Rainfall Frequency Atlas of the United States," Tech. Paper #40, U.S. Department of Commerce, Washington, DC, May.

Wanielista, M. P. 1976. *Nonpoint Source Effects,* Report ESEI-76-1, University of Central Florida, Orlando, FL, p. IV-8.

Wanielista, M. P. 1977. *Orlando Area 208 Study,* Report to Black Crow & Eidesness and the East Central Florida Regional Planning Council," (January 31, 1977). Winter Park, FL.

World Meteorological Organization. 1974. *Guide to Hydrometeorological Practices,* 3rd Ed. WMO Tech. Paper No. 82, pp. 3.8–3.10, Geneva.

Yousef, Y. A., et al. 1985. *Consequential Species of Heavy Metals,* BMR-85-286 (FL-ER-29-85), Florida Department of Transportation, Tallahassee, FL.

4

EVAPOTRANSPIRATION

Evapotranspiration is the sum total of water returned to the atmosphere from surface and ground (soil) water, ice, snow, and vegetation. Evapotranspiration is the sum of evaporation and transpiration. Evaporation is water vapor from all but vegetation. Water movement through a plant that is lost to the atmosphere is called transpiration. Transpiration is related to the type of plants and the quantity of sunlight. Transpiration rates are similar to evaporation rates when the stomata of plants are open, and appear to be controlled by the diameter of the stomata openings. When the stomata are closed, transpiration rates continue but at a very slow rate (Daubernmire, 1959).

4.1

EVAPORATION FROM WATER

Evaporation is understood to be a cooling process because heat is removed from the surface where evaporation has taken place. Energy must be available for the vaporization process and are chiefly solar and advective. Advective winds carry heat into a watershed from other heated surfaces. In addition, vapor pressures at the surface and the overlying air must be different to allow for evaporation. If the overlying air is saturated, evaporation rates will be reduced to near zero. Also, on some lakes vapor blankets form; thus, evaporation is near zero during these times.

There are three general methods commonly in use for measuring evaporation, which are mainly indirect methods: (1) measurements from evaporation pans, such as a class A pan and the British Pan, (2) water budgets, and (3) correlations with climatic data.

4.1.1 Evaporation Pans

To estimate evaporation, the class A evaporation pan is the most widely used method in the United States. The pan is an unpainted, galvanized iron 4-ft (122 cm) diameter circular container. It is usually filled to a depth of 20 cm and refilled when the depth has fallen to $\leq$ 18 cm. The water surface is measured daily with a hook gage. Precipitation is measured by the standard rain gage. A class A evaporation station would also include an anemometer, mounted 6 in. above the pan rim.

TABLE 4.1 Selected Pan Evaporation Data[a] in (mm/yr)

BRITISH PAN		CLASS A PAN	
AREA	YEARLY AVERAGES	AREA	YEARLY AVERAGES
Brecon, Wales	508	Bartlett Dam, Arizona	3089
Cardiff, Wales	652	Lincoln, Nebraska	1290[b]
Bath, England	625	Seattle, Washington	810[b]
Birmingham, England	490	Vicksburg, Mississippi	1311
London, England	650	Newark, California	1478
Belfast, N. Ireland	550	West Palm, Florida	1546
Yorkshire, England	400	Vero Beach, Florida	1580
Edinburgh, Scotland	590	Hoaeae, Hawaii	1589
Wick, Scotland	500	Norris, Tennessee	1059

[a]Approximate values based on Meteorological Data in the Country (Wallingford, 1977, and U.S. Weather Bureau, 1958).
[b]Ice cover on pan, thus inoperative during part of the year.

Pan evaporation is used to estimate lake evaporation. The lake evaporation (E_L) is usually calculated for yearly time periods using a pan coefficient (p_c) or

$$E_L = p_c E_p \tag{4.1}$$

The pan coefficient on an annual basis has been reported to vary between 0.65 and 0.82 (Kohler et al., 1955). For short time periods, the coefficient has been reported for well-watered grass to vary between 0.35 and 0.85 (Shih et al., 1983).

Some selected values of pan evaporation data are given in Table 4.1. These yearly averages are shown to vary within a country and certainly the method of measurement also would be variable. Hot, dry areas like Arizona have high readings. These evaporation rates also vary with the time of the year, the greatest usually being during the periods of intense sunlight (solar energy) and least during cold cloud-covered days. In the northern hemisphere, evaporation potential is greatest in the summer, while the winter months have the greatest evaporation potential for the southern hemisphere.

TABLE 4.2 Average Evaporation Rates from Water Surfaces (cm/mo and in./mo)

LOCATION	JANUARY	APRIL	JULY	OCTOBER	ANNUAL
Columbia, SC	4.0 (1.6)	11 (4.4)	16 (6.5)	11 (4.4)	130 (51)
Eastport, ME	2.0 (0.8)	3 (1.1)	5 (2.0)	4 (1.6)	40 (16)
Galveston, TX	2.3 (0.9)	6.6 (2.6)	16 (6.3)	11 (4.4)	109 (43)
Miami, FL	7.6 (3.0)	13 (5.0)	13 (5.3)	10 (4.0)	127 (50)
Oklahoma City, OK	4.0 (1.6)	12 (4.7)	26 (10.2)	16 (16.3)	167 (66)
Salt Lake City, UT	2.0 (0.8)	8.9 (3.5)	27 (10.6)	10 (4.0)	140 (55)
Yuma, Az	9.9 (3.9)	20 (8.0)	34 (13.4)	20 (8.0)	254 (100)

TABLE 4.3   South Florida Average Annual Evaporation

STATION	ANNUAL EVAPORATION, mm (in.)
Vero Beach	1240 (49)
Belle Glade	1120 (44)
Hialeah	1195 (47)
Loxahatchee	1175 (46)
Okeechobee	1060 (42)
Tamiami Trail	1165 (46)
Miami Metro area	1270 (50)

This variability is shown in Table 4.2. The data are very useful to indicate regional rates. However, if evaporation is necessary for long-term studies of local regions then more specific estimates are necessary. In an area of about 200 mi^2, local yearly rates may vary among each location as shown in Table 4.3. However, note that the variability is less than a comparison among regional or country yearly rates (Table 4.1). Monthly variability for specific local areas are shown in Table 4.4. This variability must be expected and can be roughly explained by solar energy changes reflected in temperature and other weather condition changes.

The National Oceanic and Atmospheric Administration (NOAA) in the United States reports daily data on pan evaporation, wind movement above the pan and temperature of the water at the surface in the pan. Using these data, one can estimate empirical relationships for pan evaporation or predict surface water (lake) evaporation.

TABLE 4.4   Monthly Evaporation Losses

MONTH	SOUTHEAST COAST, FL		SOUTHWEST COAST, FL	
	in.	mm	in.	mm
January	2.2	55	2.03	52
February	2.9	74	2.67	68
March	4.3	110	3.77	96
April	5.2	123	4.63	118
May	5.7	144	5.64	143
June	5.3	134	4.84	123
July	5.3	135	4.82	122
August	4.9	124	4.28	109
September	4.4	110	3.62	92
October	3.8	96	3.40	86
November	2.7	68	2.42	61
December	2.1	53	1.88	48
Annual total	48.5	1234	44.00	1118

Source: Partly from Boyd, 1986.

4.1.2. Water Budget

Another estimate depends on an accurate water budget in which evaporation is the only unknown variable. As an example, assume a lake that has accurate measures on inflow and outflow. Using the surface inventory equations of earlier chapters.

$$\text{Change in storage} = \text{inputs} - \text{outputs}$$

$$\Delta S = P + R + BI - BO - T - E - O \tag{4.2}$$

where

ΔS = change in reservoir storage (mm)
P = precipitation (mm)
R = surface water inflow (mm)
BI = groundwater inflow (mm)
BO = groundwater outflow (mm)
T = transpiration (mm)
E = evaporation (mm)
O = surface water releases (mm)

Assuming a reservoir with little vegetation and lined to prevent groundwater additions or depletions, evaporation can be measured as accurately as precipitation and surface water discharge measurements using:

$$E = P + R - O + \Delta S \tag{4.3}$$

In some areas, the water budget has been used successfully to estimate lake evaporation. Perhaps, the best known study was done on Lake Hefner, Oklahoma (U.S. Geological Survey, 1952, 1954).

4.1.3 Correlations to Climatic Data

Empirical formulas have been developed to relate either pan or actual lake evaporation to atmospheric measures. The form of the equations are similar and in general are related to vapor pressure and wind speed.

$$E = f(\Delta e, U) \tag{4.4}$$

where

Δe = changes in vapor pressure from the water to the air
U = wind speed

Equations of the form of Equation 4.4 are generally referred to as mass-transfer equations, (Critchfield, 1983; Rosenberg, Brad, and Verma, 1983). A typical equation developed in connection with the Lake Hefner study is as follows:

$$E_L = 0.00241(e_o - e_{a8})U_8 \tag{4.5}$$

where

E_L = evaporation rate in inches per day

e_o = saturation vapor pressure at the water surface in inches of mercury

e_{a8} = vapor pressure in air over the lake at an elevation of 8 m, in inches of mercury

U_8 = wind speed over the lake at an elevation of 8 m, in miles per day

The correlation was further defined by Kohler et al. (1955) and others as:

$$E_p = (e_0 - e_a)^n(m + bU) \tag{4.6}$$

where

E_p = daily pan evaporation (in./day)

e_0 = saturation vapor pressure at water surface temperature (in. of mercury)

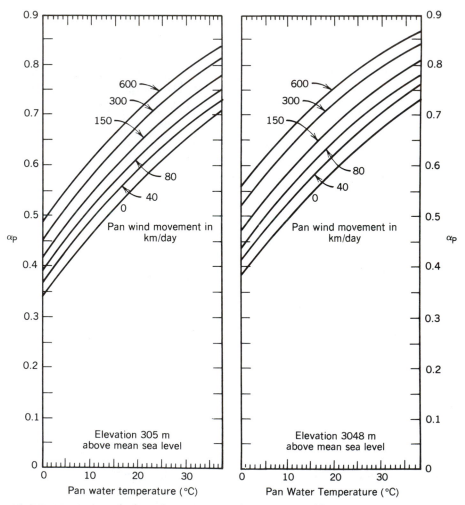

FIGURE 4.1   Portion of advected energy (into a class A pan) used for evaporation. (from Kohler, 1955).

e_a = atmospheric vapor pressure at air temperature (in. of mercury)

U = wind movement (mpd) − 6 in. above pan rim

n, m, and b = constants

Using data from many sites, Kohler et al. (1955) estimated these constants to be 0.88, 0.37, and 0.0041, respectively, using English system units for wind speed (mpd), vapor pressure (in. of mercury) (see Vapor Pressure table in the Appendix), and evaporation (in./day). Additional correlations were performed to estimate pan evaporation for those areas not serviced by a class A pan station.

Other important climatic variables are mean daily air and water temperature, wind movement (advective energy), and solar radiation (solar energy). Not all the advective energy is used for evaporation. That portion used, designated by α_p, can be estimated using the data from Kohler et al. (1955), as shown in Figure 4.1 for class A pans.

To estimate lake evaporation, a general formula can be used with differences in temperature, mean daily air speed and elevation above sea level. Kohler developed such a formula, which was reduced to a coaxial graph and is shown in Figure 4.2.

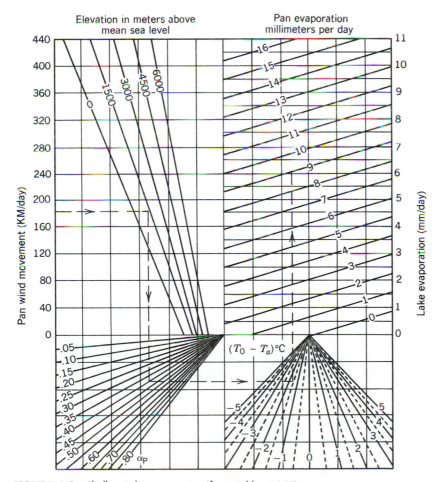

FIGURE 4.2 Shallow Lake evaporation (from Kohler, 1955).

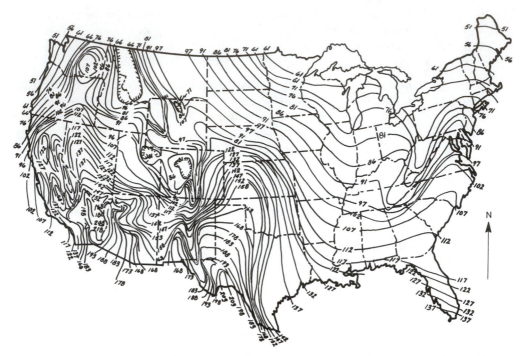

FIGURE 4.3 Average annual lake evaporation data (in centimeters) (period 1946–1955) (from Kohler, 1959).

Knowing the elevation above mean sea level, wind speed, and differences in water (T_0) and air (T_a) temperatures, estimates can be made for α_p, then given pan evaporation, lake evaporation is obtained.

For generalized values of average annual lake evaporation in the United States, Kohler et al. (1959) presented data, which have been converted to the metric system of units and are shown in Figure 4.3. These evaporation rates are for shallow lakes that are not severely polluted. Changes in evaporation from lakes due to depth and water quality are minor. Thus, the evaporation data in Figure 4.2 are applicable as a first estimate and, if significant, more detailed studies at the site may be necessary. Sometimes, these correlation equations require data that are costly to obtain, thus their use may be limited.

4.2

TRANSPIRATION AND EVAPOTRANSPIRATION

Factors affecting transpiration are similar to those affecting evaporation plus plant physiological factors, such as leaf structure, plant diseases and behavior of stomata. Soil moisture content is also important and perhaps one of the more important limiting factors.

Transpiration for small plant areas can be determined by a closed container in which humidity changes are measured. The soil can be sealed to prevent evaporation from soil. These experiments are performed onsite or by use of a phytometer, which is a container with a particular plant rooted in it. Precise determinations of transpiration are difficult, and extrapolations to other areas can be misleading. Exact environmental and physiological conditions should be reported when measuring transpiration. Water budgets are valuable but again require estimates of other variables, and, thus, the transpiration estimates are frequently only as accurate as the measurements of the other variables.

If total monthly consumptive use (evapotranspiration) can be measured for a vegetative area, and the monthly evaporation is known, the transpiration rate per time period can be estimated by using

$$T = ET - E \tag{4.7}$$

where

T = transpiration rate (mm/time)
ET = evapotranspiration rate (mm/time)
E = evaporation rate (mm/time)

Estimates for evapotranspiration are made by measuring losses of water using soil sampling tubes and lysimeters (pervious bottom tubes). Field measures are in general very costly or difficult; thus, empirical equations have been developed using generally available climatic data. Some typical equations are:

1. Thornthwaite's (1944):

$$ET = 1.6 \left(\frac{10t}{TE} \right)^{a} \tag{4.8}$$

where

$a = 0.49239 + 0.01792 TE$
ET = monthly evapotranspiration (cm)
t = mean monthly temperature (°C)
TE = Thornthwaite's temperature efficiency

$$\text{index} = \sum_{i=1}^{12} (t_i/5)^{1.514}$$

[This equation usually has to be adjusted for time of year (month) and latitude.]

2. Blaney and Criddle (1950):

$$ET = kpt/100 \tag{4.9}$$

where

k = consumptive use coefficient
p = percent of daytime hours per year in the study month
t = mean monthly temperature (°F)
ET = monthly evapotranspiration (in.)

Values for k and p are found in Tables 4.5 and 4.6. When a range of k values is presented, the lower values are for coastal areas with the higher values for arid lands. Also shown in Table 4.5 are ranges of water transpired during the growing season per equal weight of dry matter grown.

The above equations estimate potential evapotranspiration. It is that quantity of water vapor that is not constrained by inadequate moisture supply. When soil moisture is a constraint, potential evapotranspiration is never attained. The water budget method is one way of estimating evapotranspiration rather than potential evapotranspiration. The water budget method is more appropriate for monthly, seasonal or yearly time intervals. The calculation of starting and ending watershed storage can reduce the error of estimation, but over a longer period of time the difference in storage can be assumed equal to zero.

❏   **EXAMPLE PROBLEM 4.1**

Assume the following situations for a small watershed in northern Indiana. The six-month seasonal precipitation is 70 cm, runoff is 20 cm, and the change in groundwater storage is 15 cm. What are the monthly evapotranspiration rates, assuming no initial abstraction?

Solution

$$\Delta S = P - Q - ET$$

or

$$15 = 70 - 20 - ET: ET = 35 \text{ cm/6 mo}$$

TABLE 4.5   Seasonal Potential Consumptive-Use Coefficients, k

Crop	LENGTH OF GROWING SEASON OR PERIOD	k	GRAM OF WATER PER GRAM DRY MATTER GROWN
Alfalfa	Between frosts	0.80–0.85	700–1000
Beans	3 months	0.60–0.70	350–600
Corn	4 months	0.75–0.85	250–350
Cotton	7 months	0.65–0.75	500–700
Orchard, Citrus	7 months	0.50–0.65	300–600
Walnuts	Between frosts	0.70	250–400
Deciduous	Between frosts	0.60–0.70	300–700
Pasture, grass	Between frosts	0.60–0.75	300–600
Ladino clover	Between frosts	0.80–0.85	300–800
Potatoes	$3\frac{1}{2}$ months	0.65–0.75	300–600
Rice	3–5 months	1.00–1.20	609–900
Sugar beets	6 momths	0.65–0.75	300–500
Tomatoes	4 months	0.70	500–800
Vegetables, small	3 months	0.60	400–800

Source: Criddle, 1958; Blaney, 1959.

TABLE 4.6 Daytime Hours Percentages, p

LATITUDE (DEG)	JAN.	FEB.	MAR.	APR.	MAY	JUNE	JULY	AUG.	SEPT.	OCT.	NOV.	DEC.
North												
60	4.67	5.65	8.08	9.65	11.74	12.39	12.31	10.70	8.57	6.98	5.04	4.22
50	5.98	6.30	8.24	9.24	10.68	10.91	10.99	10.00	8.46	7.45	6.10	5.65
40	6.76	6.72	8.33	8.95	10.02	10.08	10.22	9.54	8.39	7.75	6.72	6.52
35	7.05	6.88	8.35	8.83	9.76	9.77	9.93	9.37	8.36	7.87	6.97	6.86
30	7.30	7.03	8.38	8.72	9.53	9.49	9.67	9.22	8.33	7.99	7.19	7.15
25	7.53	7.14	8.39	8.61	9.33	9.23	9.45	9.09	8.32	8.09	7.40	7.42
20	7.74	7.25	8.41	8.52	9.15	9.00	9.25	8.96	8.30	8.18	7.58	7.66
15	7.94	7.36	8.43	8.44	8.98	8.80	9.05	8.83	8.28	8.26	7.75	7.88
10	8.13	7.47	8.45	8.37	8.81	8.60	8.86	8.71	8.25	8.34	7.91	8.10
0	8.50	7.66	8.49	8.21	8.50	8.22	8.50	8.49	8.21	8.50	8.22	8.50
South												
10	8.86	7.87	8.53	8.09	8.18	7.86	8.14	8.27	8.17	8.62	8.53	8.88
20	9.24	8.09	8.57	7.94	7.85	7.43	7.76	8.03	8.13	8.76	8.87	9.33
30	9.70	8.33	8.62	7.73	7.45	6.96	7.31	7.76	8.07	8.97	9.24	9.85
40	10.27	8.63	8.67	7.49	6.97	6.37	6.76	7.41	8.02	9.21	9.71	10.49

Source: From Criddle, 1959.

TABLE 4.7   North Florida Evapotranspiration Data (Latitude 28°N, Longitude 80°W)

MONTH	EVAPOTRANSPIRATION (cm/mo)	MONTH	EVAPOTRANSPIRATION (cm/mo)
January	3.05	July	12.95
February	4.88	August	11.40
March	6.86	September	9.15
April	10.29	October	6.60
May	12.20	November	4.55
June	12.95	December	3.00

or

$$ET = 5.83 \text{ cm/mo} \quad \square$$

With the above example problem, it is assumed that evapotranspiration does not vary from month to month. This is a poor assumption because over a 6-month period, vegetation and climate changes are most probable. Using data from a north Florida watershed, the following evapotranspiration data (Table 4.7) were calculated assuming storage changes were negligible. Average monthly data were used.

TABLE 4.8   Ratio Between Evapotranspiration from Well-Watered Grass and Evaporation from Class A Pan

	CASE 1: PAN SURROUNDED BY SHORT GREEN CROP				CASE 2: PAN SURROUNDED BY DRY-SURFACE GROUND			
	UPWIND FETCH OF GREEN	RELATIVE HUMIDITY PERCENT			UPWIND FETCH OF DRY	RELATIVE HUMIDITY PERCENT		
WIND (km/day)	CROP (m)	LOW 20–40	MED 40–70	HIGH > 70	FALLOW (m)	LOW 20–40	MED 40–70	HIGH > 70
Light	0	0.55	0.65	0.75	0	0.7	0.8	0.85
	10	0.65	0.75	0.85	10	0.6	0.7	0.8
< 170 km/day	100	0.7	0.8	0.85	100	0.55	0.65	0.75
	1000	0.75	0.85	0.85	1000	0.5	0.6	0.7
Moderate	0	0.5	0.6	0.65	0	0.65	0.75	0.8
	10	0.6	0.7	0.75	10	0.55	0.65	0.7
170–425 km/day	100	0.65	0.75	0.8	100	0.5	0.6	0.65
	1000	0.7	0.8	0.8	1000	0.45	0.55	0.6
Strong	0	0.45	0.5	0.6	0	0.6	0.65	0.7
	10	0.55	0.6	0.65	10	0.5	0.55	0.65
425–700 km/day	100	0.6	0.65	0.7	100	0.45	0.5	0.6
	1000	0.65	0.7	0.75	1000	0.4	0.45	0.55
Very strong	0	0.4	0.45	0.5	0	0.5	0.6	0.65
	10	0.45	0.55	0.6	10	0.45	0.5	0.55
> 700 km/day	100	0.5	0.6	0.65	100	0.4	0.45	0.5
	1000	0.55	0.6	0.65	1000	0.35	0.4	0.45

Source: From Doorenbos and Pruitt (1974).

Evapotranspiration data are applicable to a specific place with certain climatic and vegetative conditions. Investigators must examine the water resources data for a region to determine available data or to develop new data. Another indirect method for estimating evapotranspiration is to use the evaporation from a class A pan and convert the pan evaporation to evapotranspiration using results from other studies or developing ratios of evapotranspiration using a controlled mass balance and evaporation from a class A pan.

Doorenbos and Pruitt (1974) developed ratios of evapotranspiration and evaporation from a class A pan (Table 4.8). Their coefficients apply to conditions of initially dry soil. These results indicate that ET from grasses can vary from 0.35 to 0.85 of that from a class A pan. The range depends on the surrounding soil/plant conditions, wind speed, relative humidity, and upwind fetch.

4.3

EVAPORATION FROM SNOW

The depth of evaporation (E_s) from snow surfaces can be estimated using a form of Dalton's law, when atmospheric vapor pressure is less than snowpack surface vapor pressure.

$$E_s = k_e \overline{u}_b (e_s - e_a)(z_a z_b)^{1/6}(\Delta t) \tag{4.10}$$

where

$\quad E_s$ = depth of evaporation (in.)
$\quad \overline{u}_b$ = average wind speed in miles per hour at elevation B above the snowpack
$\quad e_a$ = atmospheric vapor pressure in millibars (mbar) at elevation A (note $e_a < e_s$)
$\quad e_s$ = saturated vapor pressure in millibars at the snowpack
z_a, z_b = height above snowpack (ft) for e_a and $\overline{u}_b$

and for Δt in days, the constant (k_e) is estimated as (U.S. Army Corps of Engineers, 1956):

$$k_e = 0.00635 \text{ in. ft}^{1/3} \text{ hr per day mbar mile}$$

The vapor pressure of the surface film of melting snow is 6.11 mbar. For the same wind conditions, evaporation from snow is about one-fourth that from water at 30°C and dewpoint at 15°C. This and other data lead to the assumption that maximum evaporation from a snow surface is about 0.2 in. of water per day.

The direct transformation of ice to a vapor or the reverse is called sublimation. The heat released during sublimation is about 680 cal/cm³. The snowpack depth (Δd_e) should be reduced by an amount indirectly proportioned to the density of the snowpack or

$$\Delta d_e = (\rho_w/\rho_s)(E_s) \tag{4.11}$$

where

ρ_w = density of water = 1 g/cm^3
ρ_s = snowpack density (g/cm^3)
Δd_e = snowpack depth change due to evaporation (in.)

where E_s units are inches.

4.4
SUMMARY

- Evaporation and transpiration estimates are necessary for long-term water budgets, especially those used for irrigation and reservoir capacity studies. The usual time periods are weeks to months.

- Experimental estimates of evaporation are made using evaporimeters or an evaporation pan. These estimates are higher than what can be expected from an open water body. Adjustment factors have been developed to convert "instrument" values to true lake evaporation.

- Evaporation rates vary with the weather conditions (see Tables 4.1 and 4.2). Cloud cover, wind speed, and temperature are three of the meteorological parameters commonly used to estimate evaporation.

- To estimate evapotranspiration, there are two equations in general use— Thornthwaite and Blaney and Criddle. Otherwise, a mass balance to estimate evapotranspiration can be used.

- For the same wind conditions, evaporation from snow is about one-fourth that from water at 30°C and dewpoint at 15°C.

4.5
PROBLEMS

4.5.1 Hand Problems

1. How much water is needed in acre-feet for 100 ac of corn growing at North latitude °27.5 during the month of May if precipitation is assumed at zero and 150% of the water must be provided? Means monthly temperature is 67°F.

2. A River Basin Regulatory Person must release water from a reservoir to satisfy a downstream need of 48,000 m^3/day during the month of July. The average daily class A pan evaporation is 5 mm and the pan coefficient is 0.70. Estimate how much water must be released from the reservoir to satisfy the 48,000 m^3/day need if the average river width is 61 m and the distance down the center of the river from the reservoir to point of need is 78 km. Express your answer in terms of both m^3 and ac-ft. Neglect or assume that net infiltration into and out of the river from groundwater sources is negligible and there is no transpiration.

3. Estimate the mean monthly evaporation rate for the O'Hare International Airport

watershed. Local climatological data supplied by NOAA for the Airport are shown for months in the year 1974.

MONTH	PRECIPITATION (in.)	OUTLET WATER (in.)
March	2.40	1.42
April	4.27	3.05
May	5.09	3.60
June	4.69	3.50
July	2.96	1.65

The water leaving the watershed is estimated from a U.S. Geologic Survey gaging station and is also shown. Comment on your answers by explaining why the results are low relative to the annual average values given in Figure 4.3. What assumptions have you made? Assume storage changes are minimal and no transpiration or infiltration.

4. a. Compute the weekly evaporation from a class A pan if the precipitation and water added to bring the level of water in the pan to a fixed level are as follows.

WEEK	1	2	3	4
Rainfall (in.)	0.00	1.04	1.84	0.42
Water added (in.)	0.90	0.04	−0.70[a]	0.93

[a]Water taken from pan.

  b. If the pan evaporation coefficient for this period of time is 0.8 and an adjacent lake has a surface area of 150 ac, what is the evaporation from the lake expressed in inches and cubic feet for the 4-week period of time?
  c. How can evaporation volume of this lake be reduced?

5. a. Using empirical formula (4.6) for evaporation with a water temperature of 60°F, an air temperature of 80°F and a wind speed of 10 mph, what is the daily pan evaporation? Relative humidity is 40%.
  b. If this is considered the average daily evaporation, what region of the United States is this characteristic of (use Figure 4.3 as comparison)?

6. Estimate the maximum evapotranspiration of corn grown at 27° north latitude over April and May. The average monthly temperatures are 70°F and 75°F, respectively. How much water must be added (irrigated) if rainfall for these 2 months is 3 in. and we lose 2 in. to infiltration and initial abstraction?

7. One month of rainfall and pan evaporation data have been collected and a correlation to a mass balance for lake evaporation is attempted to determine the pan coefficient. The lake evaporation was 6.4 in. and the pan evaporation was 9.3. What is the pan coefficient?

8. It is desired to build a small reservoir (average surface area @ 600 acres) in Southwest Florida. Use a water budget approach to determine the reservoir storage volume needed to assure a constant regulated discharge of 5.0 cfs. Typical low water period inflows in acrefeet are:

MONTH	INFLOW (ac-ft)	MONTH	INFLOW (ac-ft)
J	150	J	360
F	170	A	275
M	175	S	255
A	240	O	200
M	360	N	170
J	425	D	160

Use Table 4.4 for evaporation losses.

9. A conversion of land from tomato farming to citrus is planned. Is it true that less water will be used to supplement rainfall based on a growing season of one year? Explain and support your answers with calculations and referenced assumptions.

4.5.2. Case Studies: Data Sources for Evapotranspiration Estimation

There are many methods to estimate actual and potential evapotranspiration rates for various regions of the country. Many of these methods rely on the availability of existing data for this estimation. The world wide web has greatly simplified the collection of these data and made it possible to perform estimates which previously required long hours of data searching. Included in Table 4.9 are suggested world wide web links for data collection and estimation of evapotranspiration.

TABLE 4.9   Suggested World Wide Web Links for Further Investigation of Evapotranspiration

WEB ADDRESS	DESCRIPTION
http://jei.umd.edu/jei/penman.html	This is an on-line calculator for using the Penman equation for use in calculating evapotranspiration.
http://weather.nmsu.edu/bulletin.htm	Contains Blaney-Criddle crop coefficients and other useful information, including necessary meteorological data for Los Cruces, New Mexico
http://cando.dwr.co.gov/manuals/cu/html/tech_over/ref_method.html	This link contains an overview of the Penman-Monteith equation for evapotranspiration estimation.
http://met-www.cit.cornell.edu/nrcc_database.html	Excellent database resource for climatic data required in evapotranspiration calculation.
http://twri.tamu.edu/~twri/twripubs/New Waves/v8n2/research-5.html	Overview of real-time estimation of evapotranspiration conducted at Texas A&M University
http://www.moenet.com.au/~keebs/etfig.html	Example of evapotranspiration calculations performed on various crops

4.6

REFERENCES

Beaver, R. D. 1977. "Infiltration in Stormwater Detention/Percolation Basin Design," Res. Report, College of Engineering, Florida Technological University, Orlando, FL.

Beaver, R. D., Hartman, J. P., and Wanielista, M. P. 1977. "Infiltration and Stormwater Retention/Detention Ponds," Stormwater Retention/Detention Basin Seminar, Y. A. Yousef, Ed., Florida Technological University, Orlando, FL.

Blaney, H. F. 1959. "Monthly Consumptive Use Requirements for Irrigated Crops," Proceedings of the American Society of Civil Engineers Journal, Irrigation and Drainage Division **85:** 1–12, March.

Blaney, H. F., and Criddle, W. D. 1950, "Determining Water Requirements in Irrigated Areas from Climatological and Irrigated Data," SCS, TP-96, August.

Boyd, C. E. 1986. "Influence of Evaporation Excess on Water Requirements for Fish Farming," *Conference on Climate and Water Management, A Critical Era,* American Meteorological Society, Asheville, NC, August.

Chow, V. T. 1964. *Handbook of Applied Hydrology.* McGraw-Hill Book Co., New York, p. 12.7.

Criddle, W. D. 1959. "Methods of Computing Consumptive Use of Water," Proceedings of the American Society of Civil Engineers Journal, Irrigation and Drainage Division **84:** 1–27, January.

Critchfield, Howard J. 1983. *General Climatology* (4th Ed.). Prentice-Hall, Inc., Englewood Cliffs, NJ.

Dawkins, E. and Associates. 1977. "North Pinellas 201 Facility Plan," City of Orlando, FL.

Daubernmire, R. F. 1959. *Plants and Environment, 2nd Ed.* New York, Wiley.

Doorenbos, J., and Pruitt, W. O. 1974. *Guidelines for Prediction of Crop Water Requirements.* Foreign Agricultural Organization, Rome, Italy, Irrigation and Drainage Paper No. 25.

Green, W. H. and Ampt, G. A. 1911. "Studies on Soil Physics I, The Flow of Air and Water Through Soils," *Journal of Agricultural Science,* **4,** 1–24.

Jones, Frank E. 1992. *Evaporation of Water.* Lewis Publishers, Chelsea, MI.

Kohler, M. A., Nordenson, T. J., and Fox, W. E. 1955. "Evaporation from Pans and Lakes," Research Paper No. 38, U.S. Weather Bureau, Washington, DC.

Kohler, M. A., Nordenson, T. J., and Baker, D. R. 1959. "Evaporation Maps for the United States," U.S. Weather Bureau Technical Paper #37, Washington, DC.

Overton, D. E. and Meadows, M. E. 1976. *Stormwater Modeling.* New York, Academic Press.

Rosenberg, N. J., Brad, B. L., and Verma, S. B. 1983. *Microclimate: The Biological Environment.* John Wiley & Sons, New York.

Schomaker, C. E. 1966. "The Effect of Forest and Pasture on the Disposition of Precipitation," *Marine Farm Research,* July.

Seminole County, Florida. 1975. Soil Conservation Service, U.S. Department of Agriculture. Sanford, Florida, pp. 73–125.

Skaggs, R. W. and Khaleel, R. 1982. "Infiltration," in *Hydrologic Modeling of Small Watersheds.* C. T. Haan, Ed. American Society of Agricultural Engineers, St. Joseph, MI, 121–166.

Thornthwaite, C. W. et al. 1944. "Report of the Committee on Transpiration and Evaporation, 1943–44," *Transactions of the American Geophysics Union,* **25,** Part V, 683–693.

Todd, D. K. 1980. *Ground Water Hydrology.* Wiley, New York.

U.S. Department of Agriculture. 1972. Soil Conservation Service. National Engineering Handbook, Section 4, Washington, DC. Note that the Soil Conservation Service is now the Natural Resources Conservation Service.

note: the Soil Conservation Service is now the National Resources Conservation Service.

U.S. Department of Agriculture. 1951. *Soil Survey Manual #18,* Washington, DC.

U.S. Department of Agriculture. 1986. Soil Conservation Service. "Urban Hydrology for Small Watersheds," Technical Release No. 55, Washington, DC.

note: the Soil Conservation Service is now the National Resources Conservation Service.

U.S. Geological Survey. 1952. Quadrangle Size Map (Eastern U.S.) USGS Distribution Section, 1200 South Eads Street, Arlington, VA 22202, or (Western U.S.) Distribution Section, Federal Center, Denver, CO 80225.

U.S. Geological Survey. 1954. "Water-Loss Investigations: Vol. 1-Lake Hefner Studies," Paper No. 269 (Reprint of USGS Circular No. 229, 1952).

U.S. Geological Survey. 1954. "Water-Loss Investigations: Lake Hefner Studies, Base Data Report," Paper No. 270.

U.S. Army Corps of Engineers. 1956. *Snow Hydrology,* North Pacific Division, Portland, Oregon, June 30.

U.S. Weather Bureau. 1958 and 1980. Technical Paper 13, U.S. Government Printing Office, Washington, DC, with updates.

Wallingford. 1977. *Surface Water: United Kingdom, 1974–76,* Her Majesties Surface Water Office, Water Data Unit, 1982. Wallingford, U.K.

Walton, W. D. 1970. *Groundwater Resource Evaluation.* McGraw-Hill, New York.

WATERSHED CHARACTERISTICS AND INFILTRATION

A runoff hydrograph is defined as an expression for surface water discharge over time. It is the expression of the *watershed characteristics* that invariably govern the relationship between rainfall and the resulting runoff or runoff hydrograph. These watershed characteristics include area, shape, drainage patterns, land use, land and channel properties, land and drainage slopes, and the infiltration capacity of the soil. An accurate representation of these characteristics is essential to the accurate estimate of runoff from the watershed.

5.1
GENERAL CHARACTERISTICS OF WATERSHEDS

Watershed characteristics can be grouped into two categories; topographical and infil-tration. Topographical information includes areas, slopes, depression areas, and any stream patterns on the watershed. Infiltration characteristics will determine the amount of precipitation which is stored below the surface of the soil. A watershed is a very dynamic and complex system. The simplest model of watershed runoff is the rational equation introduced in Chapter 3. The historical development of this equation is included in Chapter 6.

$$Q = CiA \tag{5.1}$$

In this equation the watershed is modeled with two watershed characteristics: the rational coefficient (C), and the watershed area (A). The effects of infiltration and depression storage are incorporated into the value of C. This equation provides an estimate of the rate of runoff.

More complex methods for the determination of runoff are available. They require a more detailed mathematical description of the watershed characteristics. These characteristics will be discussed in this chapter.

5.1.1   Area and Impervious Area

Watersheds can be characterized by certain descriptive parameters. These parameters will affect the shape and magnitude of any resulting hydrograph. The total amount

135

of runoff resulting from a precipitation event on a watershed can be estimated using a mass balance.

$$R = P - F - A \tag{5.2}$$

where
P = total volume of precipitation
R = total volume of rainfall excess
F = total volume of infiltration
A = total volume of abstraction (surface storage)

All of these parameters can be represented in depth units (inches, centimeters, millimeters). This unit system makes this equation independent of the watershed area. In practice it may be desired to present these volumes in more traditional volume units of cubic feet, liters, or acre-feet. In this case the volumes will be dependent on the *watershed area*. The watershed area is the total surface area of the drainage basin in question. This area can be subdivided into two areas: the *pervious area* and the *impervious area*. The pervious area allows for soil infiltration where the impervious area does not. If a watershed were 100% impervious then the infiltration term of the mass balance would always be zero.

Precipitation which falls on an impervious area will be stored, or will flow directly to the watershed outlet, or will flow onto pervious watershed areas where it may infiltrate. The portion of the impervious area which is sloped so that precipitation flows directly to the watershed outlet is called *directly connected impervious area* (*DCIA*). That portion of the impervious area which allows for flow onto the pervious region is said to be *not directly connected*. In hydrograph generation these regions may be routed separately or they be combined by taking a weighted average of the areal characteristics of the watershed.

5.1.2 Drainage System and Land Cover

Other watershed characteristics are details of the watershed soils and land cover. After initial abstraction, water flows over land to a natural or man-made drainage system. The conduit slope, hydraulic roughness, channel storage or length, impervious area, infiltration volume, and watershed shape affect the hydrograph shape from the watershed, known as the discharge hydrograph (Figure 5.1).

As shown in Figure 5.1, discharge hydrograph shapes and peak flow rates vary greatly from one location to another. The slope of the drainage affects the time it takes for water to flow to a discharge point. The greater the slope, the less the time of travel relative to a less steep slope. If the rainfall excess were the same for both slope conditions, the peak would occur at less time from the start of rainfall excess and be larger for steeper slope conditions. The same result would occur if the roughness of the transport system affected the travel time. Watershed storage may result from natural depression areas, infiltration, and man-made storage. For similar rainfalls on watersheds of homogeneous (same) land use, the effect of storage on hydrograph shapes is shown in Figure 5.1e. The rising limbs of hydrographs for directly drained watersheds is steeper than that for watersheds with storage. For larger storage systems,

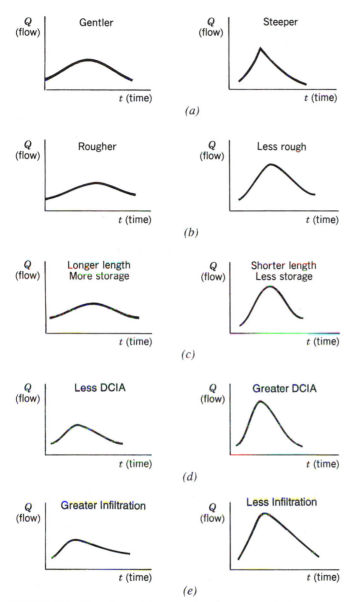

FIGURE 5.1   The effects of drainage characteristics on discharge hydrographs. (*a*) Slope. (*b*) Roughness. (*c*) Storage—length of time. (*d*) Directly connected impervious area (DCIA). (*e*) Infiltration volume.

the hydrograph starts well after the start of rainfall. The recession limb may be controlled partially by groundwater conditions, but if both watersheds have similar stream characteristics (bed depth, slope, area), the falling limbs will be similar. The area under the hydrographs (rainfall excess) will, however, be different, since the watershed with storage will produce a lower rainfall excess. Since infiltration can be considered watershed storage, hydrograph shapes should also reflect similar shapes as those for surface storage.

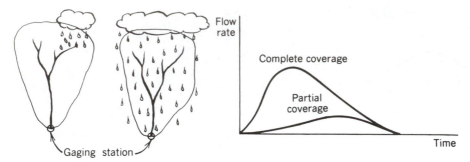

FIGURE 5.2   Area coverage effects on hydrographs.

5.1.3   Area Storm Coverage

The area coverage of a storm on a watershed also affects the hydrograph shape. For a watershed with some directly connected impervious areas, a comparison of hydrograph shapes resulting from a localized storm and a storm over the entire watershed is shown in Figure 5.2. The location of the localized storm will affect the time of occurrence of the peak discharge. A rainfall near the outlet will result in a peak near the start of the storm and rapid passage of the streamflow. Rainfall in remote portions of the watershed will result in the runoff at the outlet being spread out over a longer time period. The peak will occur later in time and be lower than the peak resulting from localized rainfall near the outlet. Storm movement away from or towards a watershed also affects the time of occurrence of the peak discharge (Figure 5.3). Storms moving toward the gaging station generally produce a greater peak relative to movement in the opposite direction.

5.1.4   Stream Order

To obtain a more accurate hydrograph shape, a watershed can be divided into individual streams and separate hydrographs computed for each. Then the separate hydrographs are routed using methods presented in later chapters. Stream order is a numbering system for the surface drainage segments which can assist in the identification of hydrographs. The smallest conduit or the only one for a watershed is designated order 1 (Figure 5.4). When two first-order conduits join, a conduit of order 2 is formed. Two conduits of the same order must join to increase the order of the new conduit.

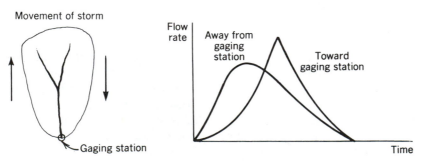

FIGURE 5.3   Storm direction effects on hydrographs.

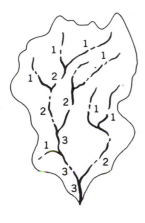
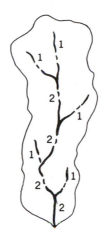

FIGURE 5.4   Stream orders.

The order number is dimensionless; therefore, it is possible to compare corresponding order numbers from two dissimilar watersheds.

The drainage area of a watershed or stream order will be that area contributing surface flow to the drainage conduit. The drainage area of a second-order basin consists of the sum of the drainage areas leading into the second-order conduit plus contributing areas (A_0) along the conduit. Figure 5.5 illustrates the areas and can be written as

$$A_u = \sum_{i=1}^{n} A_{1_i} + \cdots + \sum_{i=1}^{n} A_{i-1_i} + \sum_{i=1}^{n} A_{0_i} \tag{5.3}$$

where

A_i = drainage area of the i stream order
A_1 = drainage area of the stream order 1
A_0 = drainage area along the stream
$\;n$ = total number of contributary areas

Drainage area is used as input data in many mathematical models, and existing relationships to other hydrological data have been developed, for instance,

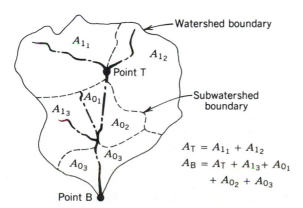

$$A_T = A_{1_1} + A_{1_2}$$
$$A_B = A_T + A_{1_3} + A_{0_1}$$
$$+ A_{0_2} + A_{0_3}$$

FIGURE 5.5   Subwatersheds.

1. Drainage area versus order for a single large watershed:

$$\text{Log } A_u = KU \tag{5.4}$$

where

A_u = watershed area
K = proportionality constant
U = stream order, dimensionless

2. Stream length versus watershed area for geographically similar areas:

$$L = k_1 A^{k_2} \tag{5.5}$$

where

L = conduit length
A = watershed area
k_1, k_2 = constants (consistent units)

3. Discharge versus watershed area for geographically similar areas:

$$Q = K_1 A^{K_2} \tag{5.6}$$

where

Q = discharge at specified return period
A = watershed area
K_1, K_2 = constants (consistent units)

Empirical relationship three (Equation 5.6) can be found in basic hydrology texts. The exponent K_2 has been found to usually vary between 0.5 and 1.0 if discharge is expressed as ft³/sec and drainage area as square miles.

The length of flow of a conduit can be scaled directly from development maps, from U.S. Geological Survey quadrangle sheets, or measured directly on the ground by field crews. The length of flow combined with slope, cross sections, and roughness characteristics of the conduit are used to determine velocities of flow, flow rates, depth of flow, and time of travel.

In general, a plot of slope vs. distance in a watershed will reveal higher gradients at the head of the watershed than at the end. Slope is a dimensionless number and is the vertical distance (drop) divided by the horizontal distance. In areas of steep slopes, erosion is higher than areas of flat slopes because velocity and discharges are higher.

5.2

TIME OF CONCENTRATION

The time of concentration is the longest travel time it takes a particle of water to reach a discharge point in a watershed. There are three common ways that waters are transported: overland flow, pipe flow (storm sewer), and channel flow, including gutter flow. Each method has a separate formula for estimating time of concentration.

5.2.1 Izzard's Formula

The time of concentration for overland flow can be calculated using many different formulas. As the size of the watershed decreases, overland flow becomes dominant for the calculation of time of concentration. Izzard (1944) conducted experiments on pavements and turf. He developed a dimensionless hydrograph for surface flow laminar regions. When using Izzard's formula, well-defined channels should not be evident. The time of concentration is the same as time to equilibrium developed by Izzard and the maximum runoff value of flow is calculated by using

$$t_c = \frac{41 \, KL^{1/3}}{i^{2/3}} \qquad (\text{for } i \times L < 500) \tag{5.7}$$

where

t_c = time of concentration (min)
L = overland flow distance (ft)
i = rainfall intensity (in./hr)
$K = \dfrac{0.0007i + c_r}{S^{1/3}}$

and

For $iL < 500$ in.-ft/hr.

S = slope (ft/ft)
c_r = retardance coefficient, given as

Very smooth asphalt	0.007
Tar and sand pavement	0.0075
Crushed-slate roof	0.0082
Concrete	0.012
Tar and gravel pavement	0.017
Closely clipped sod	0.046
Dense bluegrass	0.060

5.2.2 Kerby's Equation

Kerby (1959) also developed an equation for overland flow:

$$t_c = c(Lns^{-0.5})^{0.467} \quad \text{for} \quad L < 365 \text{ m (1000 ft)} \tag{5.8}$$

where

t_c = time of concentration (min)
L = length of flow (ft) (generally less than 1000 ft)
s = slope (ft/ft)
c = 0.83 (when using feet) or 1.44 (when using meters)
n = retardance roughness coefficient

Smooth pavements	0.02
Poor grass, bare sod	0.30
Average grass	0.40
Dense grass	0.80

5.2.3 Kirpich's Equation

Kirpich (1940) developed an equation that can be used for rural areas to estimate t_c. The Kirpich equation (5.9) is based on data reported by Ramser (1927) for six small agricultural watersheds near Jackson, Tennessee. The slope of these watersheds was steep with well-drained soils. Timber cover ranged from zero to 56%, and watershed areas ranged from 1.2 to 112 acres.

$$t_c = 0.0078(L^{0.77}/S^{0.385}) \tag{5.9}$$

where

t_c = time of concentration (min)
L = length of travel (ft)
S = slope (ft/ft)

5.2.4 Kinematic Wave

The kinematic wave equation (Ragan, 1971; Fleming, 1975) can be used to estimate time of concentration when there exists a kinematic wave (velocity not changing with distance but changing at a point). The time of concentration equation for these conditions is

$$t_c = \frac{0.93[L^{0.6}N^{0.6}]}{i^{0.4}S^{0.3}} \tag{5.10}$$

where

t_c = time of concentration (min)
L = overland flow length (ft)
N = Manning's roughness coefficient for overland flow (see Table 5.1)
i = rainfall intensity (in./hr)
S = average slope of overland flow path (ft/ft)

The length of the overland flow segment generally should be limited to 300 ft. Manning's N values of Table 5.1 were determined specifically for overland flow conditions. Equation 5.10 generally involves a cumbersome trial and error process using the following steps.

1. Assume a trial value of rainfall intensity (i).
2. Find the overland travel time (t_c), using Equation 5.10.
3. Find the actual rainfall intensity for a storm duration of t_c from the appropriate intensity–duration–frequency (IDF) curve for your area. Also, record the intensity for t_c.
4. Compare rainfall intensities, if they are not the same, select a new trial rainfall intensity and repeat step 1.

TABLE 5.1   Overland Flow Manning's *N* Values[a]

	RECOMMENDED VALUE	RANGE OF VALUES
Concrete	0.011	0.01–0.013
Asphalt	0.012	0.01–0.015
Bare sand	0.010	0.010–0.016
Graveled surface	0.012	0.012–0.030
Bare clay-loam (eroded)	0.012	0.012–0.033
Fallow (no residue)	0.05	0.006–0.16
Plow	0.06	0.02–0.10
Range (natural)	0.13	0.01–0.32
Range (clipped)	0.08	0.02–0.24
Grass (bluegrass sod)	0.45	0.39–0.63
Short grass prairie	0.15	0.10–0.20
Dense grass	0.24	0.17–0.30
Bermuda grass	0.41	0.30–0.48
Woods	0.45	— —

Note: These values were determined specifically for overland flow conditions and are not appropriate for conventional open channel flow calculations.
[a] Values are from Engman (1983), with additions from the Florida Department of Transportation Drainage Manual (1986).

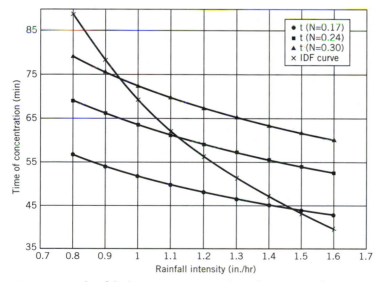

FIGURE 5.6   Plot of the kinematic wave equation with IDF curve solution to Example Problem 5.1.

☐ **EXAMPLE PROBLEM 5.1**

Using the kinematic wave equation and a spreadsheet, estimate and give a range of values for the time of concentration for an overland flow length of 150 ft with an average slope of 0.001 ft/ft. The flow is through dense grass. Use the IDF curve developed in Example Problem 2.14.

Solution

Since the recommended value for n for dense grass (from Table 5.1) is 0.24, and the range of values is 0.17 to 0.30, we will use all three of these values to estimate time of concentration. Using a spreadsheet, a table can be developed as shown. Cell B4 contains the equation

$$+0.93 * \$B\$1^{\wedge}0.6 * B2^{\wedge}0.6/(\$A4^{\wedge}0.4 * \$D\$1^{\wedge}0.3)$$

	A	B	C	D
1	L	150	S	0.001
2	N	0.17	0.24	0.3
3	i(in/hr)	t(N = 0.17)	t(N = 0.24)	t(N = 0.30)
4	0.7	59.48	73.15	83.63
5	0.8	56.39	69.35	79.28
6	0.9	53.79	66.16	75.63
7	1	51.57	63.43	72.51
8	1.1	49.64	61.05	69.80
9	1.2	47.94	58.96	67.41
10	1.3	46.43	57.11	65.29
11	1.4	45.08	55.44	63.38
12	1.5	43.85	53.93	61.66
13	1.6	42.73	52.56	60.08

By plotting the IDF equation from Example Problem 2.14 with t_c vs. i data from the spreadsheet, a solution can be determined for each of the individual n values used. The solutions are for $n = 0.17$, $i = 1.47$ in./hr, and $t_c = 44$ min; for $n = 0.24$, $i = 1.12$ in./hr, and $t_c = 61$ min; and for $n = 0.30$, $i = 0.94$ in/hr, and $t_c = 74$ min. Our range of values is therefore 44 min–74 min with an estimate of 61 min.

5.2.5 Natural Resources Conservation Service Equation (formerly the Soil Conservation Service)

The Soil Conservation Service (USDA, 1975) used two techniques, which are essentially hydraulic wave equations. The simpler of the two estimation equations relates time of concentration to watershed lag, or

$$t_c = 1.67t_L \tag{5.11}$$

where

t_L = watershed lag time in hours (from the center of mass of rainfall excess to the time of peak runoff)

and

$$t_L = L^{0.8} \frac{(S' + 1)^{0.7}}{1900w_s^{0.5}} \tag{5.12}$$

where

L = watershed hydraulic length (ft)
S' = potential watershed storage (in.) defined by Equation 5.25
w_s = average watershed slope (percentage value)

Perhaps the most frequently used table to aid in calculating overflow velocities was presented by the Soil Conservation Service (SCS) (1975) and is reproduced as Figure 5.7. This alternate SCS method requires an estimate of overland slope and a description of the cover crop or land use. Thus, some engineering judgment must be exercised. Generally, estimates are made for each relatively constant slope area. Large areas of constantly varied slopes and ditches should be divided into smaller homogeneous areas with regard to slope and cover type.

5.2.6 Bransby Williams Equation

A common equation used for the calculation of time of concentration with some historical significance is the Bransby Williams (1922) equation

$$t_c = 21.3L \frac{1}{A^{0.1}S^{0.2}} \tag{5.13}$$

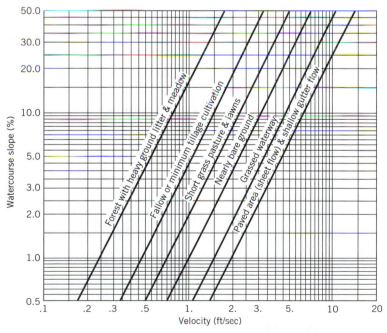

FIGURE 5.7 The average velocities for estimating travel time for overland flow (SCS method). (*Source:* SCS, 1975.)

where

L = length of channel from divide to outlet in miles
A = watershed area in square miles
S = slope of a linear profile having the same area under it as the actual profile of the main stream in ft/ft.

5.2.7 Federal Aviation Agency Equation

The Federal Aviation Agency (FAA, 1970) developed an equation from airfield drainage data which uses the runoff coefficient as used by the rational method.

$$t_c = \frac{1.8(1.1 - C)L^{0.50}}{S^{0.33}} \qquad (5.14)$$

where

C = rational Coefficient
L = maximum length of overland flow in feet
S = slope in percent of longest overland flow path

5.2.8 Manning's Equation

In storm sewer gutters and open channels, Manning's equation (Chow, 1959) to calculate average velocities is frequently used:

$$v = (1.486/n)(R^{2/3})(S^{1/2}) \qquad (5.15)$$

where

v = velocity (ft/sec)
R = hydraulic radius, ft = $D/4$ for pipes flowing full
S = slope (ft/ft)
n = roughness coefficient (see Table 5.2)

For gutters, an n value of 0.021 is recommended for a belted or broomed finish on a concrete pavement and 0.018 for a smooth, trowel-finished concrete gutter. A length of gutter to establish uniform flow depth may be in the order of 50 ft or more. Appendix F details additional calculation steps for computing time of concentration in storm sewer gutters.

❑ EXAMPLE PROBLEM 5.2

Consider a watershed shown in Figure 5.8. Compute the time of concentration for the basin from points A through D.

Solution

From A → B

From Figure 5.7, v = 2.2 ft/sec,

t_c = length/velocity = 1000 ft/2.2 ft/sec = 455 sec

TABLE 5.2 Values of the Roughness Coefficient, n

TYPE OF CHANNEL AND DESCRIPTION	MINIMUM	NORMAL	MAXIMUM
A. Closed conduits flowing partly full			
a. Brass, smooth	0.009	0.010	0.013
b. Steel			
1. Lockbar and welded	0.010	0.012	0.014
2. Riveted and spiral	0.013	0.016	0.017
c. Cast iron			
1. Coated	0.010	0.013	0.014
2. Uncoated	0.011	0.014	0.016
d. Wrought iron			
1. Black	0.012	0.014	0.015
2. Galvanized	0.013	0.016	0.017
e. Corrugated metal			
1. 6 by 1 in. corrugations	0.020	0.022	0.025
2. 6 by 2 in. corrugations	0.030	0.032	0.035
3. Smooth wall spiral aluminum	0.010	0.012	0.014
f. Concrete			
1. Culvert, straight	0.010	0.012	0.013
2. Culvert with bends	0.011	0.013	0.014
3. Sewer with manholes, inlet, etc., straight	0.013	0.015	0.017
g. Sanitary sewers	0.012	0.013	0.016

B. Channel conditions $n = (n_0 + n_1 + n_2 + n_3)m$		Values	
a. Material involved	Earth		0.020
	Rock cut	n_0	0.025
	Fine gravel		0.024
	Coarse gravel		0.028
b. Degree of irregularity	Smooth		0.000
	Minor	n_1	0.005
	Moderate		0.010
	Severe		0.020
c. Relative effect of obstruction	Negligible		0.000
	Minor	n_2	0.010–0.015
	Appreciable		0.020–0.030
	Severe		0.040–0.060
d. Vegetation	Low		0.005–0.010
	Medium	n_3	0.010–0.025
	High		0.025–0.050
	Very high		0.050–0.100
e. Degree of meandering	Minor		1.000
	Appreciable	m	1.150
	Severe		1.300

Source: U.S. Department of Transportation, 1985, and W.L. Cowan, 1956.

Solution

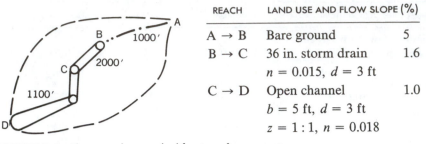

REACH	LAND USE AND FLOW SLOPE (%)	
A → B	Bare ground	5
B → C	36 in. storm drain	1.6
	$n = 0.015$, $d = 3$ ft	
C → D	Open channel	1.0
	$b = 5$ ft, $d = 3$ ft	
	$z = 1:1$, $n = 0.018$	

FIGURE 5.8   The example watershed for time of concentration.

From B → C

$$v = (1.486/0.015)(3/4)^{2/3}(0.016)^{1/2} = 10 \text{ ft/sec}$$

(usually pipe flow velocity is between 3 and 10 ft/sec)

$$t_c = \text{length of B–C/velocity} = 2000 \text{ ft}/(10 \text{ ft/sec}) = 200 \text{ sec}$$

From C → D

$$v = (1.486/0.018)(1.78)^{2/3}(0.01)^{1/2} = 12.1 \text{ ft/sec}$$

$$t_c = \text{length of C–D/velocity} = 1100/12.1 = 91 \text{ sec}$$

Total $t_c = 455 + 200 + 91 = 746$ sec   ❑

The Manning equation is the most widely used to estimate velocity and time of concentration for sewer pipe, open channels, and gutters. Overland flow time of concentration requires more judgment and the comparison of results from a few equations. The time of concentration is an important description of a watershed because of its vital role in defining the shape of a runoff hydrograph.

5.3

INFILTRATION

Infiltration is the entry of waters into the ground. The rate and quantity of water which infiltrates is a function of soil type, soil moisture, soil permeability, ground cover, drainage conditions, depth of water table, and intensity and volume of precipitation. The soil type helps identify the size and number of capillaries through which water may flow into the ground, while moisture content helps identify capillary potential and relative conductivity. For soils with a low moisture content, capillary potential is high and conductivity is low. Soil moisture will increase soil conductivity. Depth of water table affects the potential amount of water which can infiltrate into the soil. Higher water tables mean that the potential infiltration volume may be limited. Soil type with its water conditions and intensity with volume of precipitation affect the amount of water from precipitation which actually infiltrates into the soil.

5.3.1 Soil and Hydrologic Classification

Soil types and degree of soil saturation are the major determining factors in infiltration. Infiltration has been related to soil texture by Rawls et al. (1982). Minimum infiltration rates are shown in Table 5.3 and in general are related to texture class, water capacity, and the Natural Resources Conservation Service (NRCS) hydrologic soil groups (Table 5.4).

Sand, silt, clay, and decaying materials are the primary particles in soil. The U.S. Department of Agriculture defines soil in terms of percentage sand, silt, and clay as shown in Figure 5.9 (U.S. Department of Agriculture, 1951). Soil is generally classified by five factors; climate, slope, biological activity, parent material, and age. These factors determine the soil drainage characteristics. Using more than 3,000 specifically named soil types, the NRCS divided each into one of four hydrologic groups (Table 5.4).

Some sands, such as Leon Fine, are given a dual classification as A/D. The first letter applies to the drained condition while the second letter applies to the undrained natural condition. The drained condition can only occur if a closely spaced underdrain system or open ditch drainage system is installed to lower the naturally high groundwater. This is expensive and is generally not performed except in urban areas where a water problem exists. Urbanization, in general, will lower the groundwater table because of open ditches, the construction of underground utilities which cut the organic pan layers, and construction in general which results in remolded soils. An extensive classification of many soils has been completed by the NRCS (U.S. Department of Agriculture, 1975). A partial listing is shown in Appendix H.

TABLE 5.3 Hydrologic Soil Properties Classified by Soil Texture

TEXTURE CLASS	EFFECTIVE WATER CAPACITY (in./in)	MINIMUM INFILTRATION RATE (f_c) (in./hr)	SCS HYDROLOGIC SOIL GROUPING[a]
Sand	0.35	8.27	A
Loamy sand	0.31	2.41	A
Sandy loam	0.25	1.02	B
Loam	0.19	0.52	B
Silt loam	0.17	0.27	C
Sandy clay loam	0.14	0.17	C
Clay loam	0.14	0.09	D
Silty clay loam	0.11	0.06	D
Sandy clay	0.09	0.05	D
Silty clay	0.09	0.04	D
Clay	0.08	0.02	D

Source: Rawls et al., 1982.
[a]Specific named soil types may have a different SCS soil classification than the general one of this column for a texture class.

TABLE 5.4 SCS Hydrologic Soil Groups

SOIL GROUP[a]	DESCRIPTION
A	Lowest runoff potential. Includes deep sands with very little silt and clay; also, deep, rapidly permeable gravel.
B	Moderately low runoff potential. Mostly sandy soils less deep and less aggregated than A, but the group as a whole has above average infiltration after thorough wetting.
C	Moderately high runoff potential. Comprises shallow soils and soils containing considerable clay and colloids, though less than those of group D. The group has below average infiltration after saturation.
D	Highest runoff potential. Includes mostly clays of high swelling percentage, but the group also includes some shallow soils with nearly impermeable subhorizons near the surface.

Source: U.S. Department of Agriculture, Soil Survey Manual #18, Washington DC, 1951.
[a]A mixed designation, (i.e., B/D) refers to drained/undrained natural situation.

5.3.2 Green-Ampt Method

A number of theoretically based infiltration models have evolved over the years. Many of these models are based on the work performed by Green-Ampt in the early 1900s with the application of groundwater methods to infiltration.

The Green-Ampt (1911) method of infiltration estimation is based on Darcy's law.

$$f(t) = K\frac{\Delta h}{\Delta z} \tag{5.16}$$

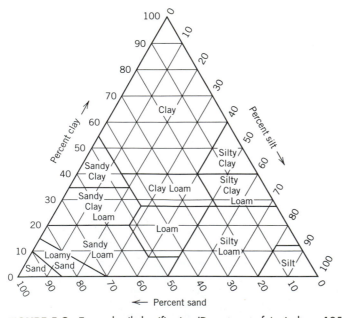

FIGURE 5.9 Textural soil classification (Department of Agriculture, 1951).

by making the assumptions:

1. The soil surface is covered by a pool of water whose depth can be neglected,
2. There is a distinctly definable wetting front in the soil which can be viewed as a plane separating a uniformly wetted infiltrated zone from a totally dry uninfiltrated zone,
3. Once the soil is wetted, the water content in the wetted zone does not change as infiltration continues (hydraulic conductivity K in this zone is constant), and
4. There is a negative constant pressure just above the wetting front.

In this case capillary suction (Ψ) plus depth of percolating water add to give the pressure head (Δh) for flow in the vertical direction. In terms of the depth of percolating water (H) this equation becomes

$$f(t) = K \left(\frac{H + \Psi}{H} \right) \tag{5.17}$$

where $f(t)$ = infiltration rate (L/T).

The volume of this water is the product of the difference in the initial soil moisture content and the final soil moisture content times the depth of percolating water.

$$F(t) = H(\theta_s - \theta_i) = \eta H \tag{5.18}$$

where

$F(t)$ = cumulative infiltration volume (L)
θ_s = saturated soil water content (fraction of total volume)
θ_i = initial soil water content (fraction of total volume)
η = the fillable pore space or effective porosity ($\theta_s - \theta_i$)
H = depth of percolating water

Substitution of $f(t) = dF(t)/dt$ results in the following equation;

$$f(t) = \eta \frac{dH}{dt} \tag{5.19}$$

Setting Equation 5.17 equal to Equation 5.19 gives the form

$$\eta \frac{dH}{dt} = K \left(\frac{H + \Psi}{H} \right) \tag{5.20}$$

which can then be integrated with initial conditions $H = 0$ at time $= 0$ and combined with Equation 5.18 to give the final form of the Green-Ampt equation.

$$Kt = F(t) - \eta \Psi \ln \left[\frac{\eta \Psi + F(t)}{\eta \Psi} \right] \tag{5.21}$$

where

K = hydraulic conductivity of the soil (L/T)
Ψ = capillary suction of the soil at the wetting front (L)

TABLE 5.5 Green-Ampt Parameter Estimates

SOIL TYPE	TOTAL POROSITY	EFFECTIVE POROSITY	CAPILLARY SUCTION (in.)	HYDRAULIC CONDUCTIVITY (in./hr)
Sand	0.437	0.417	1.95	4.135
Loamy sand	0.437	0.401	2.41	1.205
Sandy loam	0.453	0.412	4.33	0.510
Loam	0.463	0.434	3.50	0.260
Silty loam	0.501	0.486	6.57	0.135
Sandy clay loam	0.398	0.330	8.60	0.085
Clay loam	0.464	0.390	8.22	0.046
Silty clay loam	0.471	0.432	10.75	0.033
Sandy clay	0.430	0.321	9.41	0.025
Silty clay	0.479	0.423	11.50	0.018
Clay	0.475	0.385	12.45	0.012

From Rawls and Brakensiek, 1985; Rawls et al., 1983.

η = effective soil porosity (fraction)
t = time (T)
$F(t)$ = cumulative infiltration volume at time t (L)

This equation shows that infiltration rate is a function of total volume infiltrated at time t. The solution for $F(t)$ is solved by trial and error for any time t. Table 5.5 shows some estimates for the parameters used in the Green-Ampt equation.

□ **EXAMPLE PROBLEM 5.3**

Estimate capillary soil suction using the Green-Ampt equation if it takes 2 hr for an infiltration wetting front to travel 1 ft below the ground and $\eta = 0.4$. Soil hydraulic conductivity is measured as 1 in./hr.

Solution

Using the Green-Ampt equation and substituting results in

$$(1 \text{ in./hr})(2 \text{ hr}) = (0.4)(12 \text{ in.}) - 0.4\Psi \ln \left[\frac{(0.4)\Psi + (12 \text{ in.})(0.4)}{0.4\Psi} \right]$$

using the "solve for" function on a spreadsheet, or by trial and error, on a hand calculator the solution $\Psi = 7$ in. can be found.   □

5.3.3 Modifications to the Green-Ampt Equation

A number of modifications have been made to improve the accuracy of the Green-Ampt method over the years. Bouwer (1966) proposes that the hydraulic conductivity K be substituted with the hydraulic conductivity of the wetted zone, not the fully saturated conductivity. He suggests using a conductivity equal to $\frac{1}{2}$ the fully saturated conductivity in the equation.

The Green-Ampt equation can be easily modified to account for a hydraulic head above the surface. This modification may be desired if depth of surface water is significant compared to the capillary suction. Bouwer (1969) and Onstad et al. (1973)

suggested modifications in capillary suction head to account for changes in that parameter with soil wetness. Brakensiek (1970) extended the model to allow for multiple soil layers. Skaggs and Khaleel (1982) outlined estimation procedures for the parameters in the Green-Ampt equation. Verification of the assumptions made in the derivation of Green-Ampt are made in Todd (1980).

5.3.4 NRCS—Curve Number Method

There are many interrelated factors that influence infiltration volumes and rainfall excess. In general terms, these are climatic and watershed related. Infiltration and thus rainfall excess will vary during a storm event. One empirical description for infiltration and rainfall excess is the curve number method. At the start of precipitation, the intensity of rainfall is usually less than the rate at which water is stored. As depression storage becomes filled, and the soil and vegetative cover becomes saturated, rainfall excess increases. When soil, depression area, and vegetation storage approach ultimate saturation, storage will approach a potential saturation value (S') and infiltration rate approaches zero. Then the rainfall excess rate will equal the precipitation rate. Rainfall excess (R) and watershed storage (S) are derived from precipitation and the soil type. A possible relationship over time is shown in Figure 5.10 and rainfall excess (R) is expressed as

$$R = P - S \qquad (5.22)$$

where

R = rainfall excess
P = rainfall volume
S = storage volume on and within the soil (initial abstraction plus infiltration)

At saturation, the rate of rainfall excess is equal to the intensity of precipitation. A proportional relationship can be developed as

$$\frac{S}{S'} = \frac{R}{P} \qquad (5.23)$$

where

S = storage at any time (mm, in.)
S' = storage at saturation (mm, in.)
R = rainfall excess at any time (mm, in.)
P = precipitation at any time (mm, in.)

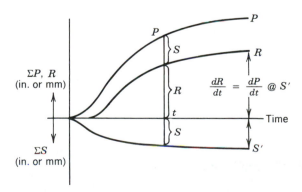

FIGURE 5.10 Time variability of hydrologic events.

Since $S = P - R$, substituting into Equation 5.23 yields

$$\frac{(P - R)}{S'} = \frac{R}{P}$$

or

$$R = \frac{P^2}{(P + S')} \qquad \text{for } I_A = 0 \tag{5.24}$$

Additional work done by the NRCS and reported in various publications (Kent, 1973) identified an empirical relationship between the initial abstraction and storage and, thus, developed an equation where the initial abstraction was assumed equal to $0.2S'$. However, abstraction values for urban areas were shown to be less if the soil types were A or B, and, in fact, Golding (1986) recommends values of $0.075S'$ and $0.10S'$ for A and B type urban soils, respectively. Using more than 3,000 soil types divided into four hydrologic groups, the NRCS developed runoff curve numbers (CN) to estimate S' in Equation 5.24. The maximum storage of water is estimated in millimeters and inches using the following:

Metric (mm)   English (in.)

$$S' = (25{,}400/CN) - 254 \quad \text{and} \quad S' = (1000/CN) - 10 \tag{5.25}$$

and rainfall excess using

$$R = (P - 0.2S')^2/(P + 0.8S') \qquad \text{if } P > 0.2S'$$

and $\qquad\qquad R = 0 \quad \text{if } P \leq 0.2S'.$ $\qquad\qquad\qquad\qquad\qquad$ (5.26)

A table for S' and the starting values of rainfall excess with time is shown in Table 5.6.

Runoff curve numbers can be estimated if the soil classification and the cover crop (land use) are known. In Tables 5.7 and 5.8, runoff curve numbers (CN) are shown for areas with the same ground cover and soils.

The NRCS has established three antecedent moisture conditions for use with CN:

CONDITION	DESCRIPTION
1	A condition of drainage basin soils where the soils are dry but not to wilting point
2	The average case
3	When heavy rainfall or light rainfall with low temps have occurred producing high runoff potential

To adjust the curve number (CN) for wet (condition 3) and dry (condition 1) moisture soils, if condition 2 is available, Table 5.9 can be used.

TABLE 5.6 Saturation Values and Start of Rainfall Excess for Initial Abstraction of 20% for Given Curve Number[a]

CN FOR CONDITION	s' VALUES[a] (in.)	EXCESS CURVE STARTS WHERE $P =$ (in.)	CN FOR CONDITION	s' VALUES[a] (in.)	EXCESS CURVE STARTS WHERE $P =$ (in.)	CN FOR CONDITION	s' VALUES[a] (in.)	EXCESS CURVE STARTS WHERE $P =$ (in.)
100	0	0	76	3.16	0.63	52	9.23	1.85
99	0.101	0.02	75	3.33	0.67	51	9.61	1.92
98	0.204	0.04	74	3.51	0.70	50	10.0	2.00
97	0.309	0.06	73	3.70	0.74	49	10.4	2.08
96	0.417	0.08	72	3.89	0.78	48	10.8	2.16
95	0.526	0.11	71	4.08	0.82	47	11.3	2.26
94	0.638	0.13	70	4.28	0.86	46	11.7	2.34
93	0.753	0.15	69	4.49	0.90	45	12.2	2.44
92	0.870	0.17	68	4.70	0.94	44	12.7	2.54
91	0.989	0.20	67	4.92	0.98	43	13.2	2.64
90	1.11	0.22	66	5.15	1.03	42	13.8	2.76
89	1.24	0.25	65	5.38	1.08	41	14.4	2.88
88	1.36	0.27	64	5.62	1.12	40	15.0	3.00
87	1.49	0.30	63	5.87	1.17	39	15.6	3.12
86	1.63	0.33	62	6.13	1.23	38	16.3	3.26
85	1.76	0.35	61	6.39	1.28	37	17.0	3.40
84	1.90	0.38	60	6.67	1.33	36	17.8	3.56
83	2.05	0.41	59	6.95	1.39	35	18.6	3.72
82	2.20	0.44	58	7.24	1.45	34	19.4	3.88
81	2.34	0.47	57	7.54	1.51	33	20.3	4.06
80	2.50	0.50	56	7.86	1.57	32	21.2	4.24
79	2.66	0.53	55	8.18	1.64	31	22.2	4.44
78	2.82	0.56	54	8.52	1.70	30	23.3	4.66
77	2.99	0.60	53	8.87	1.77	0	infinity	infinity

[a] Watershed area is considered to have the same ground cover and soils. If directly connected impervious areas (discharge point hydraulically connected to watershed) exist, then calculate excess for the directly connected area separately from the remaining area. Rainfall excess from the directly connected areas will appear earlier in a storm event than excess from soils not directly connected. The curve number for the areas and initial abstraction are different.

Miller and Veissman (1972) modified the NRCS curve number procedure for urban drainage basins. The residential land use in Table 5.7 assumes an average percent impervious area. If this percentage were changed, the curve number would be changed. Runoff curve numbers for the unpaved (pervious) portions of urban basins are shown in Table 5.10.

For an urbanized area, if the unpaved (pervious) area had a grass cover on 80% of that area and was in class B soil, the CN would be 61 (Table 5.10). If the grass cover were considered poor (grass on 40% of the area), the CN would be 79 in class B soil. This is contrasted to a CN of 86 on bare ground cover in class B soil.

Some watersheds contain subareas with different infiltration characteristics. These subwatersheds can be routed separately or combined by the use of a *composite curve*

TABLE 5.7 Curve Numbers for Urban Land Uses[a]

COVER DESCRIPTION		CURVE NUMBERS FOR HYDROLOGIC SOIL GROUP			
COVER TYPE AND HYDROLOGIC CONDITION	AVERAGE % IMPERVIOUS AREA[b]	A	B	C	D
Fully developed urban areas (vegetation established)					
Open space (lawns, parks, golf courses, cemeteries, etc.)[c]					
Poor condition (grass cover < 50%)		68	79	86	89
Fair condition (grass cover 50 to 75%)		49	69	79	84
Good condition (grass cover > 75%)		39	61	74	80
Impervious areas:					
Paved parking lots, roof, driveways, etc. (excluding right-of-way)[d]		98	98	98	98
Streets and roads:					
Paved; curbs and storm sewers (excluding right-of-way)		98	98	98	98
Paved: open ditches (including right-of-way)		83	89	92	93
Gravel (including right-of-way)		76	85	89	91
Dirt (including right-of-way)		72	82	87	89
Western desert urban areas:					
Natural desert landscaping (pervious areas only)		63	77	85	88
Artificial desert landscaping (impervious weed barrier, desert shrub with 1–2-in. sand or gravel mulch and basin borders)		96	96	96	96
Urban districts:					
Commercial and business	85	89	92	94	95
Industrial	72	81	88	91	93
Residential districts by average lot size:					
$\frac{1}{8}$ acre or less (town houses)	65	77	85	90	92
$\frac{1}{4}$ acre	38	61	75	83	87
$\frac{1}{3}$ acre	30	57	72	81	86
$\frac{1}{2}$ acre	25	54	70	80	85
1 acre	20	51	68	79	84
2 acres	12	46	65	77	82
Developing urban areas					
Newly graded areas (pervious areas only, no vegetation)		77	86	91	94
Idle lands (*CN*s are determined using cover types similar to those in Table 5.8).					

Source: Reproduced from U.S. Department of Agriculture,–SCS (1986).
[a]Average runoff condition, Antecendent Moisture Condition (AMC) II, and Ia = $0.2S'$.
[b]The average percent impervious area shown was used to develop the composite *CN*s. Other assumptions are as follows: impervious areas are directly connected to the drainage system, impervious areas have a *CN* of 98, and pervious areas are considered equivalent to open space in good hydrologic condition.
[c]*CN*s shown are equivalent to those of pasture. Composite *CN*s may be computed for other combinations of open space cover type.
[d]In some warmer climates, a curve number of 95 may be used.

TABLE 5.8 Runoff Curve Numbers for Hydrologic Soil-Cover Complexes
(Antecedent Moisture Condition II)

	COVER		HYDROLOGIC SOIL GROUP			
LAND USE	TREATMENT OR PRACTICE	HYDROLOGIC CONDITION	A	B	C	D
Fallow	Straight row	—	77	86	91	94
Row crops	Straight row	Poor	72	81	88	91
	Straight row	Good	67	78	85	89
	Contoured	Poor	70	79	84	88
	Contoured	Good	65	75	82	86
	Contoured and terraced	Poor	66	74	80	82
	Contoured and terraced	Good	62	71	78	81
Small grain	Straight row	Poor	65	76	84	88
		Good	63	75	83	87
	Contoured	Poor	63	74	82	85
		Good	61	73	81	84
	Contoured and terraced	Poor	61	72	79	82
		Good	59	70	78	81
Close-seeded	Straight row	Poor	66	77	85	89
legumes[a]	Straight row	Good	58	72	81	85
or	Contoured	Poor	64	75	83	85
rotation	Contoured	Good	55	69	78	83
meadow	Contoured and terraced	Poor	63	73	80	83
	Contoured and terraced	Good	51	67	76	80
Pasture or range		Poor	68	79	86	89
		Fair	49	69	79	84
		Good	39	61	74	80
	Contoured	Poor	47	67	81	88
	Contoured	Fair	25	59	75	83
	Contoured	Good	6	35	70	79
Meadow		Good	30	58	71	78
Woods		Poor	45	66	77	83
		Fair	36	60	73	79
		Good	25	55	70	77
Farmsteads		—	59	74	82	86
Roads (dirt)[b]		—	72	82	87	89
(hard surface)[b]		—	74	84	90	92

Source: U.S. Department of Agriculature *National Engineering Handbook*, Soil Conservation
Service U.S. Department of Agriculture Section 4, Chapter 9, Hydrologic Soil Cover Complexes,
1972. Washington, DC.
[a] Close drilled or broadcast.
[b] Including right-of-way.

TABLE 5.9 *CN* Adjustments

CN CONDITION 2	CORRESPONDING *CN*	
	CONDITION 1 (DRY)	CONDITION 3 (WET)
100	100	100
95	87	98
90	78	96
85	70	94
80	63	91
75	57	88
70	51	85
65	45	82
60	40	78
55	35	74
50	31	70
45	26	65
40	22	60
35	18	55
30	15	50

Source: U.S. Department of Agriculture, 1972.

TABLE 5.10 Runoff Curve Numbers (*CN*): Pervious Areas—Urban Watersheds (Moisture Condition 2)

LAND USE	HYDROLOGIC SOIL CLASS			
	A	B	C	D
Bare ground	77	86	91	94
Gardens or row crop	72	81	88	91
Good grass (cover on greater than 75% of the pervious area)	39	61	74	80
Fair grass (cover on 50–75% of the pervious area)	49	69	79	84
Poor grass (cover on less than 50% of the pervious area)	68	79	86	89
Fair woods	36	60	73	79

Source: U.S. Department of Agriculture, 1972.

number. A composite curve number is defined as an areal weighted average of the curve numbers for the watershed subregions.

$$CCN = \frac{\Sigma\, CN_i A_i}{\Sigma\, A_i} \tag{5.27}$$

where
CCN = the composite curve number
CN_i = curve number for region i
A_i = watershed area for region i (acres)

Directly connected impervious areas should not be included with calculations of a composite curve number especially for low volume rainfalls. However, a composite curve number can be used with large volume rainfalls and on near completely impervious areas.

□ EXAMPLE PROBLEM 5.4

Consider a watershed for which the specification of curve numbers (CN's), is desired and 75% is directly connected impervious and 25% is pervious (equally divided among poor and good grass cover) and in class C soil. Also calculate the composite CN.

Solution

Using Tables 5.7 and 5.10:

Calculations for Each Land Use		Calculation for
COVER	CN	COMPOSITE CN
Impervious (directly connected)	98 (Table 5.7)	98(0.75) + 80(0.25) = 94
Pervious (good grass) on 50%	74 (Table 5.10)	
Pervious (poor grass) on 50%	86 (Table 5.10)	

Average of pervious area = .50(74) + .50(86) = 80.   □

Calculations for maximum soil storage and curve numbers were completed for the Kissimmee River Basin in Florida (Huber et al., 1976): Results illustrated the validity of Table 5.7 and some changes due to local groundwater conditions.

If the appropriate curve number is known, the rainfall excess can be calculated. In fact, the NRCS has developed a chart (Table 5.11) and a plot (Figure 5.11) of these relationships. The investigator must determine the land use and soil type. Then, assuming a soil moisture condition, he must calculate a curve number. Using Table 5.11 or Figure 5.11, read directly the volume of runoff, given a volume of rainfall.

TABLE 5.11 Rainfall Excess (in.) Using Curve Numbers (CN)

RAINFALL (in.)	CURVE NUMBER (CN)[a]								
	60	65	70	75	80	85	90	95	98
1.0	0	0	0	0.03	0.08	0.17	0.32	0.56	0.79
1.2	0	0	0.30	0.07	0.15	0.28	0.46	0.74	0.99
1.4	0	0.02	0.06	0.13	0.24	0.39	0.61	0.92	1.18
1.6	0.01	0.05	0.11	0.20	0.34	0.52	0.76	1.11	1.38
1.8	0.03	0.09	0.17	0.29	0.44	0.65	0.93	1.29	1.58
2.0	0.06	0.14	0.24	0.38	0.56	0.80	1.09	1.48	1.77
2.5	0.17	0.30	0.46	0.65	0.89	1.18	1.53	1.96	2.27
3.0	0.33	0.51	0.72	0.96	1.25	1.59	1.98	2.45	2.78
4.0	0.76	1.03	1.33	1.67	2.04	2.46	2.92	3.43	3.77
5.0	1.30	1.65	2.04	2.45	2.89	3.37	3.88	4.42	4.76
6.0	1.92	2.35	2.80	3.28	3.78	4.31	4.85	5.41	5.76
7.0	2.60	3.10	3.62	4.15	4.69	5.26	5.82	6.41	6.76
8.0	3.33	3.90	4.47	5.04	5.62	6.22	6.81	7.40	7.76
9.0	4.10	4.72	5.34	5.95	6.57	7.19	7.79	8.40	8.76
10.0	4.90	5.57	6.23	6.88	7.52	8.16	8.78	9.40	9.76
11.0	5.72	6.44	7.13	7.82	8.48	9.14	9.77	10.39	10.76
12.0	6.56	7.32	8.05	8.76	9.45	10.12	10.76	11.39	11.76

Source: Reproduced from U.S. Department of Agriculture, SCS, 1986.
[a]To obtain rainfall excess for *CN*s and other rainfall amounts not shown in this table, use an arithmetic interpolation.

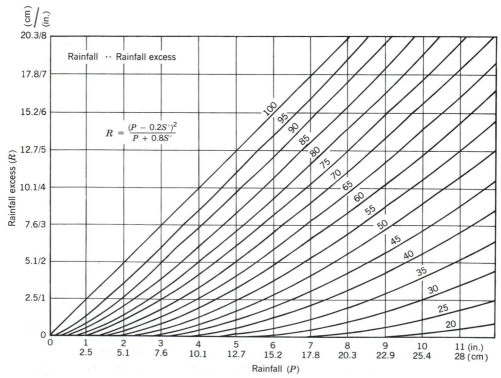

FIGURE 5.11 Rainfall excess—*CN* curves.

☐ *EXAMPLE PROBLEM 5.5*

Estimate infiltration volumes (inches) and rainfall excess (inches and CFS) from a 10-acre area on an hourly basis using the *NRCS/CN* method and the following data:

$$CN = 90$$

$$P = 2.5 \text{ in.}$$

$$D = 4 \text{ hr}$$

Solution

$$S' = \frac{1000}{CN} - 10 = 1.11''$$

GIVEN: ⟶		CALCULATED: ⟶			
TIME (hr)	P (in.)	R^a (in.)	r(t) (in./hr)	(CFS)[b]	ΣF^c (in.)
0	0	0	0	0	0
1	0.7	0.14	0.14	1.5	0.56
2	1.6	0.76	0.62	6.2	0.84
3	2.1	1.18	0.42	4.2	0.92
4	2.5	1.53	0.35	3.5	0.97

[a]Calculated from $R = \dfrac{(P - 0.2S')^2}{(P + 0.8S')}$.

[b]$r(t)(\text{CFS}) = \dfrac{(\text{in./hr})(\text{acres})(43{,}560 \text{ ft}^2/\text{ac})}{12 \text{ in./ft} \times 3600 \text{ sec/hr}}$

$$= (\text{in./hr})(\text{acres})(1.008).$$

[c]$\Sigma F = P - R.$ ☐

5.3.5 Horton Equations

Permeabilities and rates of soil infiltration, will fluctuate with time and location. Laboratory and field experiments are performed to determine rates. Since permeabilities can vary over a range from 10^{-7} cm/sec for sandy clays to 10^{-2} cm/sec for loose sands, the designer is confronted with many decisions. Laboratory testing using constant or falling head parameters usually are of limited value since there is usually too much soil disturbance and the laboratory boundary conditions and gradients are often different from those in the field. Field testing using borehole or percolation tests is more reliable than laboratory testing for small (<2000 m²) areas, but care must be taken to ensure representative testing locations.

For large percolation/recharge basins or areas, considerably more geotechnical surveying and analysis are necessary. Factors involved in such works are discussed by Walton (1970). A fairly reliable percolation test for use in small retention/detention

TABLE 5.12   Double-Ring[a] Infiltrometer Results for Horton Infiltration Parameters

SITE	TEST NO.	INITIAL WATER DEPTH IN DRUM (in.)	f_0 (in./hr)	f_c (in./hr)	K (hr^{-1})	USDA[b] PERMEABILITY (in./hr)
Wimbledon Park	1	2.0	20	6.3	49.1	10 → 20 +
(Lakeland-Blanton)	2	3.7	32	7.8	36.9	in./hr
	3	5.9	60	13	48.8	
	4	8.3	29	13	25.1	
	5	10.2	40	16	16.6	
	6	11.0	55	19.4	15.4	
Cross Creek	1	3.1	1.25	0.19	8.0	<10 in./hr
(Lakeland-Blanton	2	3.5	0.84	0.26	5.3	
with some organics)	3	3.5	0.65	0.08	2.3	
	4	3.6	1.57	0.19	4.1	
	5	3.6	2.6	0.42	8.4	
	6	5.0	1.2	0.26	0.8	
	7	5.9	1.05	0.13	6.0	
Lake Nan	1	3.9	3.4	0.57	4.0	5 → 10
(Blanton-Pomello-Plummer)	2	5.9	2.3	0.73	6.0	in./hr
	3	9.1	2.4	0.56	8.4	
	4	9.8	5.1	0.83	6.8	

Source: Beaver, 1977.
[a]ASTM D3385-76 Procedure.
[b]From Seminole County, Florida Soil Survey Supplement, Soil Conservation Service, U.S. Department of Agriculture, September 1975.

pond design is the double-ring infiltrometer (Chow, 1964). A double-ring infiltrometer is simply a 55-gal drum as an outer ring with a 10 in. or 12 in. inner ring. Example results of these percolation tests are shown in Table 5.12.

Beaver (1977) found using an infiltrometer, that infiltration can be represented by Horton's equation (Horton, 1939, 1940). This method gives an expression for time-varying infiltration. The Horton equation is shown as Equation 5.2B and drawn as shown in Figure 5.12. Also, in Figure 5.12, the rate of precipitation is compared to the rate of infiltration. The volume of infiltration is the area under the infiltration curve and the volume of rainfall is the area under the rainfall intensity curve.

$$f(t) = f_c + (f_0 - f_c)e^{-Kt} \qquad (5.28)$$

where

$f(t)$ = infiltration rate as a function of time cm/hr (in./hr) or other consistent ones
  f_c = final, or ultimate, infiltration rate—for a hydraulic gradient of unity, this is analogous to the soil permeability
  f_0 = initial infiltration rate
  K = recession constant (hr^{-1}) or other consistent units
  t = time-units compatible with K

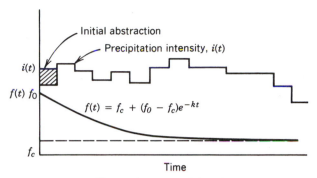

FIGURE 5.12 Infiltration (exponential decay).

The total volume of infiltrate using Horton's equation is determined by integrating the area under the curve, or

$$F = \int_0^t f(t)\, dt = f_c t + \frac{(f_0 - f_c)}{K}(1 - e^{-Kt}) \tag{5.29}$$

where F = total infiltration volume, cm (in.) or other consistent units.

During a rainfall event, the application of Equation 5.29 has to be adjusted because rainfall intensity may be lower than the rate of water infiltration potential. Also, another empirical formulation by Holtan et al. (1975) or the theoretical equations of Green and Ampt (1911) also may provide a more accurate fit to the field observed data.

Figures 5.13 and 5.14 illustrate the results of one specific infiltrometer test on a pond floor at Cross Creek in Maitland, Florida (Beaver 1977). The Cross Creek basin has an approximate 280-m³ (10,000-ft³) volume capacity, and the site covers less than 0.25 ac. The bottom soils would be classified as Lakeland–Blanton fine sands using USDA criteria. There were slight organics in the surface soils. Field samples indicate a porosity of 33 to 35%, a coefficient of uniformity of 2 to 3 and an effective grain size (D_{10}) of 0.10 to 0.13 mm. Figure 5.13 is plotted from field infiltrometer tests while Figure 5.14 is derived by differentiating the infiltration curve of Figure 5.13.

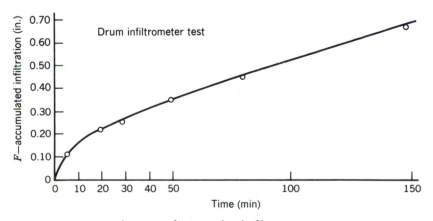

FIGURE 5.13 Example site-specific accumulated infiltration.

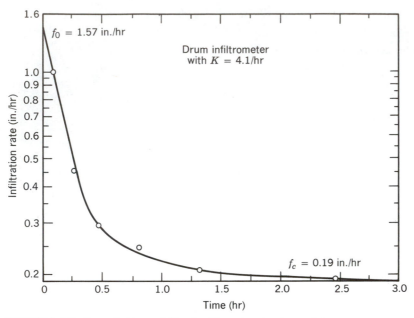

FIGURE 5.14   Example Horton infiltration illustrating the shape of the curve.

5.3.6 Water Budget

If infiltration is the only unknown in a water budget and the variables can be easily measured, then a water budget would produce accurate results. But it is usually difficult to measure depression area and interception storage. Water held on the ground forming ponds or water films is referred to as water in depression storage. This water may eventually evaporate and infiltrate depending on ground cover. For small impervious watersheds, 1.0 to 3.0 mm of depression storage is possible. This storage generally decreases as the slope of the watershed increases. For pervious areas, depression storage is greater, ranging from 2.5 mm (0.1 in.) for clay to 5 mm (0.2 in.) for sandy soils (Hicks, 1944).

Intercepted water is that which adheres to the surface of plants. In urban areas with 10% foliage, Overton and Meadows (1976) estimated that approximately 2.5 mm (0.1 in.) of water is intercepted during the first hour of a storm. For more dense foliage areas, Schomaker (1966) estimated 24% of annual precipitation was intercepted by trees. In areas where the precipitation volume is light (<2–3 mm) and there is considerable ground cover, the intercepted water can be significant and potential evaporation high. Thus, precipitation volumes reaching the ground may be near zero.

The sum of depression and interception storage is initial abstraction, so named because the initial precipitation will not result in runoff if depression and interception storage are not saturated. Initial abstraction can be measured by knowing the volume of rainfall, runoff, and infiltration. Generally, initial abstraction is estimated from field observations included with infiltration estimates or directly measured from other known quantities. The measurement of runoff and rainfall on near totally impervious areas (parking lots) results in estimating initial abstraction at 1 mm (0.04 in.) (Wanielista and Shannon, 1977). For dense vegetative areas or flat urban areas, initial abstraction as high as 3 or 4 mm (0.12–0.16 in.) was used (Wanielista and Shannon, 1977).

5.4

DEVELOPMENT OF TRANSIENT RAINFALL EXCESS

Rainfall excess rate, also referred to as the instantaneous hydrograph or excess, is a representation of the unrouted flow from a watershed. It can be expressed as the result of the mass balance

$$R = P - F - A \quad \ldots \text{ for a time period } \Delta T \tag{5.30}$$

where

R = cumulative rainfall excess (in.)
P = cumulative precipitation (in.)
F = cumulative infiltration (in.)
A = cumulative additional watershed abstraction (in.)

Precipitation is typically given in units of volume (inches) during incremental time periods (Δt). An instantaneous or excess hydrograph is a plot of rainfall excess rate (dR/dt) versus time. This instantaneous hydrograph reflects the results of abstraction and infiltration on a watershed, but does not reflect the transient runoff from the hydrograph.

5.4.1 Rainfall, Rainfall Excess, and Runoff

Where a runoff hydrograph represents the flow versus time from a watershed, the excess or instantaneous hydrograph represents the flow occurring on the watershed at any point in time which is destined to become runoff. The techniques by which the excess hydrograph becomes a runoff hydrograph are discussed in Chapter 6. Neglecting abstraction the rainfall excess rate can be expressed

$$\frac{dR}{dt} = \frac{dP}{dt} - \frac{dF}{dt} \tag{5.31}$$

since precipitation rate is also rainfall intensity (i) and infiltration rate is expressed as f, this equation becomes (modified to prevent negative values)

$$\frac{dR}{dt} = i - f \quad \text{for } i > f$$

$$\frac{dR}{dt} = 0 \quad \text{for } i \leq f \tag{5.32}$$

Since rainfall values are typically given as a volume over a time period, or a cumulative total volume since start of rain, rainfall intensity can be calculated as

$$i = \frac{P_{t+\Delta t} - P_t}{\Delta t} \tag{5.33}$$

Infiltration rate can likewise be calculated as

$$f = \frac{F(t + \Delta t) - F(t)}{\Delta t} \tag{5.34}$$

EXAMPLE PROBLEM 5.6

For the following 15-min precipitation accumulations from a 2-hr storm, estimate the infiltration rate and excess rate in inches per hour using the NRCS curve number method for infiltration. The CN for the watershed is 80. A spreadsheet should be used to develop the solution.

	A	B
1		Cumulative
2	Time	precipitation
3	(mins)	(in.)
4	15	.1
5	30	.3
6	45	.6
7	60	.9
8	75	1.1
9	90	1.2
10	105	1.3
11	120	1.4

Solution

Using the equation $S' = (1000/CN) - 10$ and the relationships for F, f, P, R, and i, the following spreadsheet (Table 5.13) can be developed.   ❑

5.4.2 Separate versus Combined Routing

A common method to produce an excess hydrograph from watersheds containing both pervious and impervious regions is to calculate excess from both the impervious and pervious regions separately. Directly connected impervious area excess is routed directly to the outlet of the watershed allowing for no infiltration. Non-directly connected impervious area excess is routed to the pervious area and is added to the amount of pervious precipitation which is available for infiltration (see Figure 5.15).

For each rainfall step the amount of excess R can be calculated by determining

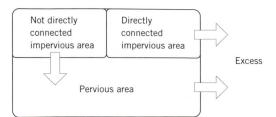

FIGURE 5.15 Representation of separate routing.

TABLE 5.13 Spreadsheet for Example Problem 5.6

	A	B	C	D	E	F	G	H
1	S'	2.5						
2	Time	Cumulative	Incremental	Precipitation	Cumulative	Excess	Cumulative	Infiltration
3	minutes	Precipitation	Precipitation	Rate (in./hr)	Excess	Rate (in./hr)	Infiltration	Rate (in./hr)
4	0	0	0	0	0		0.00	
5	15	0.1	0.1	0.4000	0.00	0.0000	0.10	0.4000
6	30	0.3	0.2	0.8000	0.00	0.0000	0.30	0.8000
7	45	0.6	0.3	1.2000	0.00	0.0154	0.60	1.1846
8	60	0.9	0.3	1.2000	0.06	0.2053	0.84	0.9947
9	75	1.1	0.2	0.8000	0.12	0.2438	0.98	0.5562
10	90	1.2	0.1	0.4000	0.15	0.1480	1.05	0.2520
11	105	1.3	0.1	0.4000	0.19	0.1633	1.11	0.2367
12	120	1.4	0.1	0.4000	0.24	0.1772	1.16	0.2228

The cell equations apply:

B1: 1000/80 - 10   C5: B5 - B4   D5: 60*C5/(A5 - A4)

E5: @IF(B5 > 0.2*B1, ((B5 - 0.2*B1)^2/(B5 + 0.8*B1)),0)   G5: B5 - E5   H5: D5 - F5

F5: (E5 - E4)/(A5 - A4)*60

the excess from both of the impervious regions using the curve number of impervious area. The excess from the nondirectly connected impervious area is then added to the precipitation on the pervious area.

$$P_P = P + \frac{A_{\mathrm{NDCIA}}}{A_P} R_{\mathrm{NDCIA}} \tag{5.35}$$

where

$\quad P_P$ = precipitation on pervious area (in.)
A_{NDCIA} = area of nondirectly connected impervious (acres)
$\quad A_P$ = pervious area (acres)
R_{NDCIA} = excess from nondirectly connected impervious (in.)
$\quad P$ = precipitation (in.)

This precipitation is then used to find the infiltration and the rainfall excess from the pervious region. This excess is averaged with the excess from the directly connected impervious area using an areal weighted average.

$$R = R_P \frac{A_P}{A_P + A_{\mathrm{DCIA}}} + R_{\mathrm{DCIA}} \frac{A_{\mathrm{DCIA}}}{A_P + A_{\mathrm{DCIA}}} = \frac{R_P A_P + R_{\mathrm{DCIA}} A_{\mathrm{DCIA}}}{A_T - A_{\mathrm{NDCIA}}} \tag{5.36}$$

where

$\quad R$ = Watershed rainfall excess (in.)
R_{DCIA} = excess from directly connected impervious area (in.)
A_{DCIA} = area of directly connected impervious (acres)
A_{NDCIA} = area of nondirectly connected impervious (acres)
$\quad A_T$ = total watershed area (acres)
$\quad R_P$ = excess from pervious region (in.)

❑ **EXAMPLE PROBLEM 5.7**
_____

A 150-acre watershed with a pervious curve number of 70 has 50 acres of impervious land on it (curve number 98). Half of this land is directly connected and the other half flows onto the pervious region. What is the rainfall excess if 2 in. of precipitation occurs on this watershed.

Solution

The rainfall excess from the impervious region can be calculated using a curve number of 98,

$$S' = \frac{1000}{CN} - 10 = \frac{1000}{98} - 10 = 0.2$$

and

$$R_I = \frac{(P - 0.2S')^2}{(P + 0.8S')} = \frac{(2 - (0.2)(0.2))^2}{(2 + (0.8)(0.2))} = \frac{3.8416}{2.16} = 1.78 \text{ in.}$$

Since there are 25 acres [(0.50)(50 acres)] not directly connected impervious, the effective depth of precipitation on the pervious region is modified to

$$P_P = 2 + 1.78 \left(\frac{25 \text{ acres}}{50 \text{ acres}} \right) = 2.89 \text{ in.}$$

Using a curve number of 70 for this region $S' = 4.286$, excess from pervious region is calculated as

$$R_P = \frac{(2.89 - (0.2)(4.286))^2}{(2.89 + (0.8)(4.286))} = \frac{4.132}{6.319} = 0.65 \text{ in.}$$

This is then added with the directly connected impervious area excess to give

$$R = \frac{R_P A_P + R_{\text{DCIA}} A_{\text{DCIA}}}{A_T - A_{\text{NDCIA}}} = \frac{(0.43)(100) + (1.78)(25)}{150 - 25} = 0.88 \text{ in.}$$

The same watershed using a composite curve number of 84 gives an excess of 0.74 in. Separate routing gives a more accurate estimate of the rainfall excess than a composite curve number because it more closely approximates the physical scenario of the watershed. ❑

5.5

WATER QUALITY

Even though the primary aim of stormwater control is to attenuate the quantity of water flowing from a watershed to prevent adverse downstream effects, the importance of controlling stormwater quality should not be overlooked. Water quality of watershed runoff varies from one watershed to another; however, it is possible to estimate the amount of pollutant, or *pollutant load,* if certain watershed characteristics are known. The pollutant load is defined as the mass of a given pollutant which flows from the watershed over a given period of time per acre of watershed. It is often expressed in units kg/(ha-yr) or pounds/(acre-year). This pollutant load is important in that it adversely affects the water quality of downstream water bodies.

5.5.1   Concentrations

Dissolved or suspended constituents in water are quantitatively expressed in terms of concentration, typically in mg/l. For various watershed landuses concentrations of a number of pollutants can be estimated in stormwater runoff. Pollutants such as metals will usually have higher concentrations in runoff from commercial and industrial areas. Nutrients, such as phosphorus and nitrogen show high concentrations in runoff from agricultural landuses and from watersheds such as golf courses where fertilizers are used. Residential watersheds show higher concentrations of most pollutants with greater population density.

Numerous studies have been performed to estimate the typical concentrations from a number of characteristic watershed types. These studies show a large variance in concentration for the same general landuse.

5.5.2 Event Mean Concentration

The event mean concentration (EMC) is the average concentration which occurs in the runoff from a specific rainfall event.

$$EMC = \frac{M_P}{V_R} \qquad (5.37)$$

where

EMC = event mean concentration
  M_P = total mass of specific pollutant (mg)
  V_R = total volume of runoff (l)

In practice it is very difficult to collect the entire volume of runoff from a rainfall event and measure the mass of pollutant present in that volume. To measure *EMC*, flow samples are taken at regular intervals during the event from the runoff. The runoff is also gaged to give an estimate of flowrate throughout the event. The samples are taken to the lab and concentration for each sample is determined. A flowrate weighted average of these concentrations can then be calculated.

$$EMC = \frac{\Sigma\, C_i Q_i}{\Sigma\, Q_i} \qquad (5.38)$$

where

C_i = concentration of sample i
Q_i = flow rate of runoff when sample i was taken

5.5.3 Loading Rates

There exists a great variability in the mass load of any given pollutant from one storm to the next and from one location to another. To get an accurate estimate of the pollutant load it is recommended that many storms be sampled or that events are sampled over a long stretch of time (i.e., years). Once the pollutant mass is known, it can be expressed in terms of a *loading rate* with units of kg/ha-yr or lb/acre-yr. The loading rate can be calculated for a given time period of collection (years) by Equation 5.39.

$$LR = M_{\text{pollutant}}/A_{\text{watershed}} \qquad (5.39)$$

where

LR = loading rate
$M_{\text{pollutant}}$ = total mass of pollutant over time period of collection
$A_{\text{watershed}}$ = area of the watershed from which pollutant was collected

This loading rate can be used to estimate pollutant loads from watersheds which are similar in landuse, size, and slope conditions to the watershed tested.

❑ EXAMPLE PROBLEM 5.8

Using the Pollutant Loading PLOAD software which comes with your textbook, estimate the annual pollutant load of total nitrogen (TN) if the watershed contains 2 acres of pasture and $\frac{1}{2}$ acre of highway. The annual precipitation is 60 in.

TABLE 5.14 Results for Example Problem 5.8

```
                    Pollutant Loading Analysis
     Watershed Number      : 1
     Watershed Area(acres) : 2
     Rational Coefficient  : .2
     Land Use              : Pasture
     Rainfall for Analysis : 60
```

Pollutant Name	Watershed Loading	
	kg/year	lb/year
BOD	12.348	27.227
SS	338.829	747.118
TN	6.125	13.505
TP	0.494	1.089
LEAD	0.099	0.218
COPPER	0.025	0.054
ZINC	0.074	0.163

```
     Watershed Number      : 2
     Watershed Area(ac)    : .5
     Rational Coefficient  : .6
     Land Use              : Highway
     Rainfall for Analysis : 60
```

Pollutant Name	Watershed Loading	
	kg/year	lb/year
BOD	21.486	47.376
SS	244.490	539.101
TN	3.408	7.515
TP	0.667	1.470
LEAD	0.130	0.286
COPPER	0.019	0.041
ZINC	0.111	0.245

```
               Analysis of Entire Watershed
     Total Watershed Area (ac) : 2.5
     Effective Area (ac)       : .7
     Rainfall for Analysis     : 60
```

Pollutant Name	Total Loading	
	kg/year	lb/year
BOD	33.834	74.603
SS	583.320	1286.219
TN	9.533	21.020
TP	1.161	2.559
LEAD	0.228	0.504
COPPER	0.043	0.095
ZINC	0.185	0.408

Solution

Start the Pollutant Loading program and open the default landuse.luf file. Select Calculate, Pollutant Load from the pull-down menus. Enter a rainfall of 60 in. and check the annual button. Next scroll the land uses and enter 2 acres when reaching the landuse "Pasture." Increment the watershed number to 2 by using the spin button, scroll through the land uses, and enter 0.5 acres when reaching the landuse of highway. When finished, the results will be printed in a text viewer and will appear like Table 5.14.   ❏

5.6
SUMMARY

A watershed is an area of land in which water incident upon the area flows overland to a common outlet. Watersheds can be characterized by their area, shape, drainage pattern, landuse, and other properties. The characteristics are used to determine the shape a hydrograph will take in response to a rainfall event on the watershed or to determine the volume or peak flowrate of water from the watershed.

- For use in hydrology a number of watershed characteristics have been defined, such as the time of concentration. Landuse, length of overland flow, and watershed slope are commonly used to determine the time of concentration. In determining the time of concentration a number of different formula exist. The engineer should use more than one equation when estimating the time of concentration of a watershed and apply engineering judgment to estimate the correct value.
- Water quality is an important consideration of watershed runoff. Water quality can be highly variable; however, estimates can be made of the volume and concentration of various pollutants if watershed characteristics such as land use are known.
- Infiltration is the movement of water into the ground. Percolation relates to groundwater and refers to the movement of water in the ground. For the prediction of rainfall excess, infiltration and evapotranspiration rates are necessary for a specified time period.
- The NRCS–CN method of estimating infiltration requires an estimate of the land use and the hydrologic soil types to determine an empirically derived curve number that is used to estimate storage. If the initial abstraction portion of saturated storage is assumed to be equal to 20%, then Equation 5.26 can be used to estimate rainfall excess. If initial abstraction is assumed to be zero, then Equation 5.24 can be used. The common formula is Equation 5.26: $R = (P - 0.2S')^2/(P + 0.8S')$.
- Table 5.7 is used frequently and it should be emphasized that the curve number is a composite value.
- Table 5.11 can be used to aid in calculating rainfall excess using the NRCS–CN

method. However, the formulas are easy to use and should be used and then checked against values obtained from tables and figures.

- Infiltration rates and volumes can be estimated using data from a double-ring infiltrometer. Site conditions do vary requiring a number of tests depending on soil types and groundwater levels. An exponential equation usually can be used to reproduce the experimental data.

5.7
PROBLEMS

5.7.1 Hand Problems

1. Estimate the volume of water that will infiltrate into a soil before surface saturation occurs using the Green–Ampt equation. The following data are known: (1) saturated moisture content is 0.25, (2) the initial moisture content is zero, (3) the average rainfall rate is 2 in./hr, (4) the average capillary suction head is 4 in., and (5) $K_s = 1$ in./hr. What is the maximum soil storage volume if the soil is homogeneous to a water table depth of 5 ft?

2. Assuming an average soil moisture condition on hydrologic soils group C, calculate the runoff volume (inches) for a 100-ac suburban development with the following land use if rainfall is 4 in. You must use the NRCS–CN method.

LAND USE	PERCENTAGE OF LAND
$\frac{1}{4}$-ac residential lots	40
$\frac{1}{8}$-ac condominiums	20
Commercial area with curbs	25
Open space, gras cover = 85%	15

What is the area of retention (no outlet) or percolation basin in square feet and acres to store the runoff water (rainfall excess) if the maximum depth of storage is 5 ft, excluding debris storage and freeboard? Assume a rectangular pond with vertical sides.

3. Consider a proposed development in class B soil of 1000 ac of which 400 ac will be impervious, 50 ac will be water surfaces (lakes, detention and canals), 200 ac will be open space in good condition, and the remaining acres will be in fair grass cover on 50 to 75% of the land area. What CN number (weighted) would you use? Discuss results if the water surface were not contributary to the drainage system.

4. For the experimental setups of Figure 5.16, determine equations to predict infiltration rates and exfiltration rates. The double-ring infiltrometer is used for infiltration. Exfiltration is to be performed on a previously constructed pipe underground. The recorded data and observations for discussion and analysis are shown.

Exfiltration Data				Infiltration Data		
ELAPSED TIME (min)	INCREMENTAL TIME (min)	METER READING (gal)	INCREMENT VOL. (gal)	ELAPSED TIME	VOLUME H_2O ADDED (in.^{3})^a	VOLUME H_2O ADDED (LITERS)
1	1	218.5	1.7	17 sec	76.7	1.25
2	1	220.1	1.6	1 min	82.9	1.35
3	1	221.5	1.4	2 min	41.5	0.68
4	1	223.1	1.6	3 min	16.5	0.27
5	1	224.5	1.4	4 min	8.3	0.136
6	1	225.9	1.4	5 min	8.3	0.136

Exfiltration Pipe Data: Length—4.0 ft Diameter—1.75 in. Total volume—115.5 in^3.

^{a}To maintain a constant head: Head—2.0 in. (5.08 cm) Inside pipe dia.—10 in. (25.4 cm) Inside pipe area—78.5 in^2 (506.7 cm^2).

5. Using the relationship of Figure 5.17 with D_{10} equal to 0.5, 1, and 2 mm, respectively, with the data of Problem 1, provide a sensitivity analysis plot for volume as a function of D_{10}.

6. A hydrologist designing a stormwater drainage system requires an infiltration experiment for a new pond area to evaluate the infiltration characteristics of the clay loam soil. A ring infiltrometer test was made on the soil. The results of the

Experimental setup

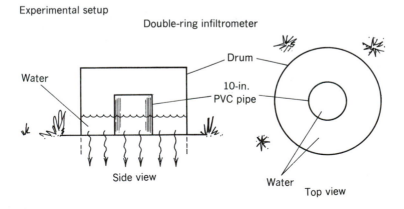

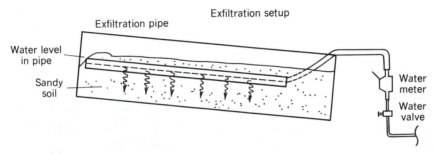

FIGURE 5.16   Double-ring infiltrometer and exfiltration experimental apparatus.

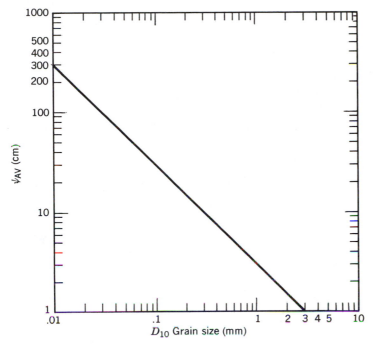

FIGURE 5.17 Suction potential versus grain size for well-graded cohesionless soils (after Weaver, R. J. Research Report 69-2, 1971. New York State Department of Transportation).

test are given in the data table below. The inside diameter of infiltrometer is 35 cm and the area is 962 cm².

a. Determine the infiltration capacity for the time intervals in the experiment.
b. What is the initial infiltration, f_0, in Horton's equation?
c. What is the ultimate infiltration, f_c, in Horton's equation?

ELAPSED TIME (min) (1)	Δt (hr) (2) $\Delta(1) \div 60$	VOLUME OF WATER ADDED SINCE START (cc) (3)	f (cm) (4) (3) $\div$ AREA	ΔF (cm) (5) $\Delta(4)$	$f(t)$ (cm/hr) (6) (5) $\div$ (2)
0		0			
2		300			
5		650			
10		1190			
20		1950			
30		2500			
60		3350			
90		3900			
150		4600			

7. For Example Problem 5.5, change the precipitation to 0, 1.4, 3.2, 4.2, and 5.0 in. over 20 ac of area and calculate the instantaneous rainfall excess in CFS.

8. What are the precipitation increments in 30-min intervals and the cumulative precipitation resulting from a 4.2-in. storm using the NRCS Rainfall Distribution Type II? Assume a 6-hr-duration storm. Now, what shapes do the rainfall excess hydrographs have (draw cumulative rainfall excess graphs) for the rainfall distribution if the *CN* values were 60 and 80?

9. a. Use the NRCS–*CN* method to estimate rainfall excess for an area before and after development. The rainfall is estimated at 6 in. The hydrologic soil type is A. The predevelopment condition is open space with grass on about 50% of the area. The postdevelopment condition is residential with a 38% impervious cover.

   b. If the watershed were 25 ac, how many acre-feet of ponding is necessary to store all the rainfall excess from the postdevelopment?

10. The runoff coefficient and the curve number are important parameters for the design of stormwater holding ponds. A water management district specifies the 25-yr return period storm and a duration of 6 hr for all rainfall data. During this storm event, the volume of rainfall excess for a 100-ac urban watershed was measured at 725,000 ft³. For this area, what is the runoff coefficient and curve number? Assume (from field measurements) that there is no initial abstraction.

11. For a 50-ac watershed, estimate the volume (millimeters and acre-feet) of infiltration in 3 hr from a rainfall with maximum intensity of 50 mm/hr if the initial infiltration rate is 100 mm/hr and the final rate is 40 mm/hr. The recession constant is 30/hr. What assumptions did you make to validate your answer?

12. A 300-rural-acre watershed with initial abstraction is known to have a curve number of 61. If the precipitation on the watershed is as follows, what is the runoff at each time increment in cubic feet per second using the NRCS-*CN* rainfall excess procedure?

TIME (min)	ΔP (in.)
0	0
5	0.07
10	0.90
15	0.32
20	0.04

13. A land use change is proposed for a 48-acre vegetated area in NRCS Hydrologic Soil Type A with 98% of the area covered with dense grass. A developer has plans to directly connect with concrete 24 of the 48 acres. The water management district requires that the runoff from 3 in. of rainfall must be stored on an adjacent flood area. What size pond (acre-feet) is calculated: a) using a composite curve number for the total area, and b) separating the areas into directly connected and pervious areas (assumes the impervious area directly discharges to the pond)? Explain the differences in the two calculations.

14. What is an estimate of the maximum rainfall excess (cubic feet) for a 300-acre watershed in the Orlando, Florida, area which has a hydrologic soil type A with

a fair grass cover and $\frac{1}{4}$ acre lot sizes if the design must correspond to a one in 5-yr return frequency? The watershed time of concentration is 2 hr. State and reference all your assumptions. Also, compare this result for a storm in your area.

15. What is an estimate of the infiltration volume (acre-feet and inches) for a 1-acre proposed pond area after 4 hr of infiltration with an assumed initial infiltration rate of 10 in./hr, a final rate of 2.4 in./hr, and a recession limb of 1.5/hr? For these conditions, what is the texture class and the NRCS Hydrologic Grouping for the soil? If the rainfall on the watershed were 3 in. and the volume of infiltrate after 4 hr was the volume of rainfall excess, what is the runoff coefficient?

16. A 50-acre residential area with $\frac{1}{2}$-acre lot sizes in NRCS Hydrologic Type A is being planned on an open space that has good condition grass. The design storm event for the area is 8.64 in. What is the pre- and postcondition rainfall excess volume in acre-feet and inches using the NRCS-CN method applied only to the pervious area plus the runoff from the impervious area assuming all the impervious area is directly connected. Assume the precondition has no impervious area.

17. A double-ring infiltrometer is used to estimate a Horton type equation that can be used to predict infiltration rates and volume for a proposed retention pond. The initial and final rates are 10 and 2 in./hr, respectively. The recession rate is 1.5/hr. What is the estimated rate of infiltration after one-half hour of infiltration? Also, what is the volume of infiltration (cubic feet, liters, and inches) for a 1-ac pond infiltration area after 2 hours of infiltration?

18. A watershed with an average slope of 0.001 ft/ft has a land cover of unmowed pasture. The area of the watershed is 2 acres. The watershed drains into an open trapezoidal grass-lined channel with a base width of 2 ft and side slopes of 4:1. The furthest point in the watershed is 300 ft from the channel. The channel is 100 ft in length and empties into a wet detention pond. Give an estimate of the time of concentration of the watershed and a possible range of time of concentration. Justify your numbers.

19. A rural forest with ground litter is being converted to a single-family residential area. Overland flow is expected in the predevelopment condition rural area for a distance of 2500 ft and a slope of 1% (1 ft/100 ft). The longest time of travel for the postdevelopment condition is the sum of 300 ft of dense grass lawn followed by a storm sewer designed to transport at minimum velocity for a distance of 1800 ft. What is the time of concentration (in minutes) for the predevelopment and the postdevelopment conditions.

5.7.2 Computer-Assisted Problems

1. Using the SMADA computer program and a rainfall volume and time distribution for a 25-yr–2-hr storm of your choice (for your area) develop the infiltration volume (watershed storage) for the following watershed condition using the NRCS–CN procedure to estimate rainfall excess and infiltration for the pervious area. Use the Santa Barbara Urban Hydrograph Generation routine to get a printout.

$$\text{Area} = 200 \text{ ac}$$

$$t_c = 120 \text{ min}$$

$$\% \text{ Impervious} = 40$$

$$\% \text{ DCIA} = 70$$

$$CN = 60 \text{ (pervious area)}$$

2. Using the Santa Barbara Hydrograph Generation Procedure, provide a plot of infiltration volume (inches) versus percent impervious values for percent impervious values of 20, 40, 60, 80, and 90 using the same watershed conditions as in Problem 1 (except variable percent imperviousness). Also, construct a plot of peak discharge versus percent imperviousness. Note the time of concentration is held constant. Comment on your results.

3. On the graph of Problem 2, also change the pervious area curve number to 85 and plot the resulting peak discharge and infiltration volumes as a function of percent imperviousness. Comment on your results.

4. It is more reasonable to expect the time of concentration to decrease as the percent impervious area which is directly connected increases. Develop the peak discharge versus the percent impervious relationship (graphical plot) using the data of Problem 1 except for the following changes:

t_c (min)	% IMPERVIOUS
150	20
120	40
90	60
70	80
60	90

5.7.3 Case Studies

Using a 25-yr return period and a 6-hr storm from the site in Apalachicola, Florida, develop a rainfall excess hyetograph and a plot of infiltration rate using a Natural Resources Conservation Service (NRCS) Curve Number (CN) of 70. You should use a spreadsheet to perform all repetitive calculations.

Solution

From the case study in Chapter 3 we can determine that the 25-yr, 6-hr storm has an average intensity of 1.3 in./hr. Therefore, the total volume of rainfall for this storm is (6 hr)(1.3 in./hr) = 7.8 in. We will develop the rainfall excess hyetograph on $\frac{1}{2}$-hr increments and since the average rainfall intensity is 1.3 in./hr, the rainfall per $\frac{1}{2}$-hr increment will be assumed at 0.65 in.

Using the NRCS method of estimating infiltration, rainfall excess can be determined using the two following equations.

$$S' = \frac{1000}{CN} - 10$$

$$R = \frac{(P - 0.2S')^2}{P + 0.8S'} \quad \text{for} \quad P > 0.2S'$$

$$R = 0 \qquad\qquad \text{for} \quad P \le 0.2S'$$

TABLE 5.15 Calculation of Rainfall Excess Using a Spreadsheet

	A	B	C	D	E	F	G
1	CN	S'					
2	70	4.29					
3			Cumulative Volume			Rate	
4	Time	Precipitation	Excess	Infiltration	Precipitation	Excess	Infiltration
5	(hr)	(in.)	(in.)	(in.)	(in/hr)	(in/hr)	(in/hr)
6	0.00	0.00	0.00	0.00	0	0	0
7	0.50	0.65	0.00	0.65	1.30	0.00	1.30
8	1.00	1.30	0.04	1.26	1.30	0.08	1.22
9	1.50	1.95	0.22	1.73	1.30	0.36	0.94
10	2.00	2.60	0.50	2.10	1.30	0.56	0.74
11	2.50	3.25	0.86	2.39	1.30	0.71	0.59
12	3.00	3.90	1.26	2.64	1.30	0.81	0.49
13	3.50	4.55	1.71	2.84	1.30	0.89	0.41
14	4.00	5.20	2.19	3.01	1.30	0.95	0.35
15	4.50	5.85	2.69	3.16	1.30	1.00	0.30
16	5.00	6.50	3.21	3.29	1.30	1.04	0.26
17	5.50	7.15	3.74	3.41	1.30	1.07	0.23
18	6.00	7.80	4.29	3.51	1.30	1.10	0.20

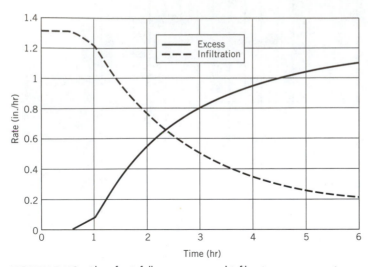

FIGURE 5.18   Plot of rainfall excess rate and infiltration rate versus time.

where

  R = cumulative rainfall excess
  P = cumulative precipitation
 CN = NRCS curve number

Using the equations in spreadsheet form we construct the following spreadsheet as shown in Table 5.15.

```
Cell B2    1000/A2-10
Cell C6    @IF(B6<=.2*$B$2,0,(B6-.2*$B$2)^2/(B6+.8*$B$2))
```

The equation in cell C6 is copied to the entire C column. By using B2 instead of B2 in the equation we ensure that the fixed address representing S′ is used for all calculations of rainfall excess.

```
Cell D6    B6-C6
```

The calculation of infiltration as the difference between precipitation and excess $(I = P - R)$. This calculation is also copied to all cells in column C.

```
Cell E7    (B7-B6)/(A7-A6)    Precipitation rate
Cell F7    (C7-C6)/(A7-A6)    Excess rate
Cell G7    (D7-D6)/(A7-A6)    Infiltration rate
```

Infiltration rate and excess rate can be plotted against time in Figure 5.18.

5.8

REFERENCES

Barsby Williams, G. 1922. "Flood Discharge and the Dimensions of Spillways in India," *The Engineer* (London), **121:** 321–322, September.

Beaver, R. D. 1977. "Infiltration in Stormwater Detention/Percolation Basin Design," Res. Report, College of Engineering. Florida Technological University, Orlando, Fl.

Beaver, R. D., Hartman, J. P., and Wanielista, M. P. 1977. "Infiltration and Stormwater Retention/Detention Ponds," Stormwater Retention/Detention Basin Seminar, Y. A. Yousef, Ed., Florida Technological University, Orlando, FL.

Bouwer, H. 1969. "Planning and Interpreting Soil Permeability Measurements," *Journal Irrigation Drainage Engineering,* **95 (IR3):** 391–402.

Brakensiek, D. L. 1970. Infiltration of Water into Non-uniform Soil, discussion. *Journal of the Irrigation and Drainage Division,* Proceedings of the American Society of Civil Engineers. New York, NY.

Cowan, W. L. 1956. "Estimating Hydraulic Roughness Coefficients," *Agricultural Engineering,* **37(7):** 473–475, July.

Chow, V. T., 1959. *Open-Channel Hydraulics.* McGraw-Hill, New York, pp. 108–114.

Federal Aviation Authority (FAA). 1970. Advisory circular on airport drainage. Report A/C 150-5320-58, U.S. Department of Transportation. Washington, D.C.

Fleming, G. 1975. *Computer Simulation Techniques in Hydrology.* Elsevier, New York.

Golding, B. C. 1986. DABRO-Drainage Basin Runoff Model—A Computer Program, Hieldebrand Software, 8992 Islesworth Court, Orlando, FL 32819.

Green, W. H. and Ampt, G. A. 1911. "Studies on Soil Physics I, The Flow of Air and Water Through Soils," *Journal of Agricultural Science,* **4,** 1–24.

Hicks, W. I. 1944. "A Method of Computing Urban Runoff," *Transactions of the American Society of Civil Engineers,* **125:** 1217–1253.

Horton, R. E. 1939. "An Approach Toward a Physical Interpretation of Infiltration Capacity," *Transactions of the American Geophysics Union,* **20:** 693–711.

Horton, R. E. 1940. "An Approach Toward a Physical Interpretation of Infiltration Capacity," *Proceedings of Soil Science Society of America,* **5:** 399–317.

Huber, W. C., Heaney, J. P., Bedient, P. B., and Bowden, J. P. 1976. *Environmental Resources Management Studies in the Kissimmee River Basin,* University of Florida, Gainesville, May.

Izzard, C. F. 1944. "The Surface–Profile of Overland Flow." *Transactions of the American Geophysics Union,* **25:** 959–968.

Kent, K. M. 1973. "A Method for Estimating Volume and Rate of Runoff in Small Watersheds," U.S. Department of Agriculture, Soil Conservation Service, TP-149, April.

Kerby, W. S. 1959. "Time of Concentration for Overland Flow," *Civil Engineering,* **29(3):** 174.

Kirpich, Z. P. 1940. "Time of Concentration of Small Agricultural Watersheds," *ASCE Civil Engineering,* **10(6):** 362, June.

Miller, C. R. and Veisman, W. 1972. "Runoff Volumes from Small Urban Watersheds," *Water Resources Res.* **8(2):** April.

Onstad, C. A., Olson, T. C., and Stone, L. R. 1973. "An infiltration model tested with monolith moisture measurements," *Soil Science* **116(1):** 13–17.

Overton, D. E., and Meadows, M. E. 1976. *Stormwater Modeling.* Academic Press, New York.

Ragan, R. M. 1971. "A Nomograph Based on Kinematic Wave Theory for Determining Time of Concentration for Overland Flow," Report #44, College Park, MD, University of Maryland, December.

Rawls, W. J., Brakensiek, D. L, and Saxton, K. E. 1982, "Estimation of Soil Properties," *Transactions of the American Society of Agricultural Engineers,* **25(5):** 1316–1320.

Rawls, W. J., and Brakensiek, D. L., 1983. "A Procedure to Predict Green Ampt Infiltration Parameters," *Advanced Infiltration, America Society of Agricultural Engineering,* 102–112.

Rawls, W. J., and Brakensiek, D. L. 1985. "Prediction of Soil Water Properties for Hydrologic Modeling," *Watershed Management in the Eighties,* ASCE, 293–299.

Schomaker, C. E. 1966. "The Effect of Forest and Pasture on the Disposition of Precipitation," *Marine Farm Research,* July.

Skaggs, R. W., and Khaleel, R. 1982. "Infiltration," in *Hydrologic Modeling of Small Watersheds.* C. T. Haan, Ed. American Society of Agricultural Engineers, St. Joseph, MI, 121–166.

Todd, D. K. 1980. *Ground Water Hydrology.* Wiley, New York.

U.S. Department of Agriculture. 1951. *Soil Survey Manual #18,* Washington, DC.

U.S. Department of Agriculture. 1972. Soil Conservation Service. *National Engineering Handbook,* Section 4, Washington, DC.

U.S. Department of Agriculture, Soil Conservation Service (USDA-SCS). 1975. *National Engineering Handbook,* Section 4, Washington, DC.

U.S. Department of Agriculture. 1986. Soil Conservation Service. "Urban Hydrology for Small Watersheds," Technical Release No. 55, Washington, DC.

U.S. Geological Survey. 1952. Quadrangle Size Map (Eastern United States.) USGS Distribution Section, 1200 South Eads Street, Arlington, VA 22202, or (Western United States.) Distribution Section, Federal Center, Denver, CO 80225.

U.S. Department of Transportation. 1985. *Hydraulic Design of Highway Culverts,* Report No. FHWA-IP-85-150 Federal Highway Administration, McLean, VA, September, p. 34.

Walton, W. C. 1970. *Groundwater Resource Evaluation.* McGraw-Hill, New York.

Wanielista, M. P., and Shannon, E. 1977. *An Evaluation of Best Management Practices for Stormwater,* East Central Florida Regional Planning Council, July, Winter Park, FL.

6

HYDROGRAPHS

A hydrograph is a graph of flow rate versus time. It is also referenced as a listing of flow rate data versus time. It is one of the more useful concepts of hydrology and is used frequently in stormwater management. The hydrograph forms much of the ideas and concepts for later chapters of this book.

This chapter develops mathematical descriptions for a hydrograph and explains how various shapes result from watershed conditions. A streamflow hydrograph is composed of both surface runoff and groundwater that has infiltrated into a stream. Groundwater flow is composed of interflow (fast responding from groundwater areas close to the stream) and base flow (relatively constant over longer time periods). Runoff hydrograph computation procedures are developed first using discrete flow values for specific times and then a continuous time, variable rainfall excess procedure.

6.1

HYDROGRAPH PROPERTIES/SHAPES

A hydrograph is typically a plot but can be a listing of flow rates versus time for a specific conduit. It consists of both surface (overland) and groundwater flows that infiltrate from the ground to the surface conduit. Groundwater inflow that is relatively persistent over time is called base flow. Groundwater flows for shorter periods of time are sometimes called interflow. Many shapes for hydrographs are possible, one such is shown in Figure 6.1a. The groundwater flows are caused by many factors in the soil, some of which can be distinguished in analyzing the hydrograph recession curve (tail of the hydrograph). Thus ground and surface flow rates can be estimated from streamflow rates.

A typical surface runoff hydrograph is shown in Figure 6.1b. The hydrograph consists of three general parts: (1) rising limb or concentration curve, (2) crest segment or peak discharge, and (3) recession curve or falling limb. Note that the rainfall excess or effective rainfall volume is the same as the volume of runoff (area under the runoff hydrograph). From this effective rainfall and the watershed characteristics, a runoff hydrograph will have at least the following properties:

1. *Lag Time (L).* The time interval from the center of mass of the rainfall excess to the peak of the resulting hydrograph.

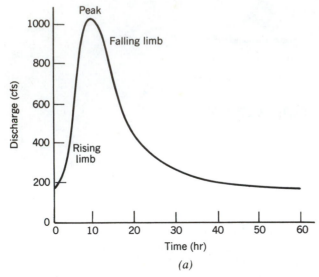

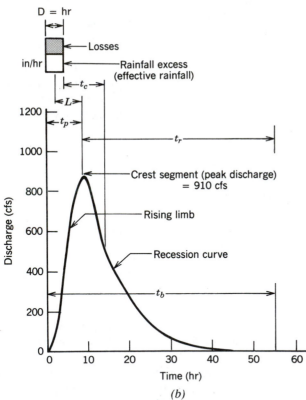

FIGURE 6.1 Hydrograph properties. (*a*) Streamflow (surface and groundwater) hydrograph. (*b*) Surface runoff.

2. *Time to Peak (t_p).*   The time interval from the start of rainfall excess to the peak of the resulting hydrograph.
3. *Time of Concentration (t_c).*   The time interval from the end of rainfall excess to the inflection point (change of slope) on the recession curve. (As considered in the SCS analysis.) Also, the longest time for water to flow to a discharge point from any point in the watershed. (As typically used in the rational method.)
4. *Recession Time (t_r).*   Time from the peak to the end of surface runoff.
5. *Time Base (t_b).*   Time from the beginning to the end of surface runoff.

6.1.1 Hydrograph Records

Since most of the printed records report flow rates on a daily basis, most hydrographs for large watersheds are reported on a daily basis. However, if the rainfall excess passes from the watershed in less than a day, the hydrograph must reflect a time scale in hours or possibly minutes. One must be aware of the time frame over which flow rates occur and analyze accordingly.

The mean (arithmetic average) daily discharge is a representation of flow from midnight to midnight. A reasonable expectation of hydrograph variability for small and large watersheds is shown in Figure 6.2. Once the flow rates have been measured at time intervals which reflect hydrograph variability, a mathematical analysis can be done to describe the hydrograph shape and aid in predicting peak flow rates and volume discharge.

6.1.2 Stream Types by Streamflow Hydrograph Analysis

Hydrographs for a stream define the relative contributions of both surface and groundwaters. The magnitude of the hydrograph recession time will aid in determining the long term water yield of a stream. Annual hydrographs that have rarely zero flow volume (water yield) and relatively long recession times are characteristic of perennial streams, an example of which is shown in Figure 6.3. Perennial streams have dependable water yield potentials during the year. Water yield is maintained primarily because stored groundwater is released very slowly. The groundwater storage is always at a sufficiently high level to maintain the groundwater table above the bottom of the stream.

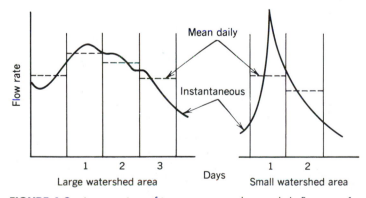

FIGURE 6.2 A comparison of instantaneous and mean daily flow rates for large and small watersheds.

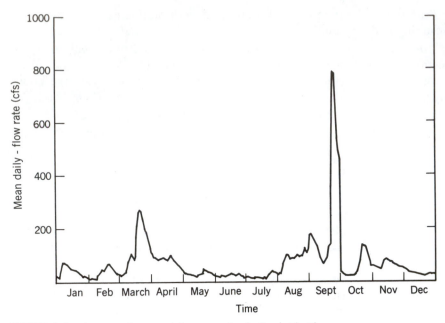

FIGURE 6.3 An example perennial stream—Reedy Creek, Florida.

In contrast, intermittent streams have limited groundwater storage and release stored waters at a faster rate. The recession times are generally shorter than perennial streams, and stream flows fall to zero during extended dry period (Figure 6.4). Base flow or interflow exists only during and shortly after heavy rainfall periods, and the water yield is based primarily on surface runoff.

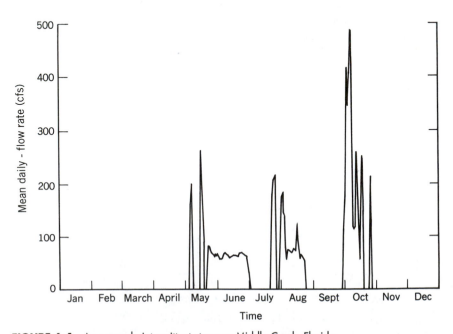

FIGURE 6.4 An example intermittent stream—Middle Creek, Florida.

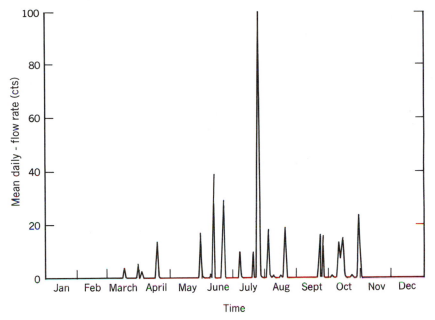

FIGURE 6.5 An example ephemeral stream—Colorado Springs, Colorado. (From Schuly, 1980).

There are some streams that have no interflow and base flow because the soils forming the side walls and bed of the channel are relatively impervious materials. In other cases, the water table is always lower than the stream bottom and runoff waters actually recharge the groundwater from the stream bed. These streams are labeled as ephemeral ones with a typical annual hydrograph shown in Figure 6.5. Note that most likely there is a single recession constant that is higher in magnitude than the other two types of streams. This type of stream is not dependable for water yield.

6.2

MATHEMATICAL DESCRIPTION OF HYDROGRAPH SHAPE

One method of describing a hydrograph shape is to assume that watershed outflow rate is a linear function of storage. This method has been referred to as a single linear reservoir model (Pedersen et al., 1980). It appears to be appropriate for small watersheds and rainfall events of short duration.

6.2.1 Mathematical Description of the Rising Limb

For the rising limb of a hydrograph, a mathematical description can be developed using a mass balance equation with generation equal to zero:

$$\text{Input} - \text{output} \pm \text{generation} = \text{accumulation}$$

$$r - Q = dS/dt \tag{6.1}$$

where
   r = rainfall excess rate (ft³/sec)
   Q = outflow rate (ft³/sec)
   S = storage for rainfall excess (ft³)
   t = time frame for analysis (sec)

Assuming an approximate linear relationship exists as shown in Figure 6.6 between outflow rate and storage

$$Q = KS \tag{6.2}$$

where
   S = storage volume (ft³)
   K = storage coefficient (sec⁻¹)
   Q = outflow rate (ft³/sec)

and in differential form:

$$dQ = K\, dS \tag{6.3}$$

Substituting for dS from Equation 6.3 into Equation 6.1 results in

$$r - Q = dQ/(K\, dt) \tag{6.4}$$

Integration of Equation 6.4 when $t = 0$ and $Q = 0$ results in

$$Q(t) = r(1 - e^{-Kt}) \tag{6.5}$$

Equation 6.5 is valid only for the rising limb of a hydrograph developed using a linear relationship between storage and outflow rate. As long as rainfall excess and stream flow data are available, the appropriate K (storage coefficient) can be estimated to "fit" the rising limb of the hydrograph. It is possible to use piecewise linear approximations of the storage–discharge relationship and thus have different K factors for different storage ranges, where K may vary with the watershed storage conditions. In this case, the runoff rate–storage relationship will be nonlinear and may be difficult to estimate. Other procedures for curve fitting including nonlinear responses are discussed in Chapter 2.

$$Q = KS \tag{6.6}$$

storage relationship.

FIGURE 6.6 A linear storage relationship.

6.2.2 Hydrograph Groundwater Flow Separation (Recession Limb)

The recession limb of a streamflow hydrograph has been analyzed because it results from both surface runoff and groundwater flows. By subtracting groundwater flow rates from the stream hydrograph, an estimate of the runoff hydrograph and effective rainfall for the watershed and rainfall conditions is possible. Since groundwater flows are determined by both long-term and short-term groundwater storage, the groundwater portion of the recession limb is most likely a combination of groundwater flows resulting from many previous rainfall events.

Since both surface and groundwater flows will be directly related to surface and groundwater storage, assume the following differential relationship:

$$dQ = K' \, dS \qquad (6.6)$$

where

S = storage volume (L^3)
K' = storage coefficient, most likely a composite of 2 or 3 coefficients (t^{-1})
Q = outflow rate (L^3/t)

The mass balance with input equal to zero and no generation is

$$\text{Input} - \text{output} \pm \text{generation} = \text{accumulation}$$

$$Q = dS/dt \qquad (6.7)$$

Substituting for dS from Equation 6.6 into 6.7 and integrating with $Q = Q_p$ at $t_r = t - t_p$, where t_r is time from the peak (recession time), t_p is the time to peak, t is the time from start of rainfall excess, and Q_p is peak flow rate, yielding

$$Q_{t_r} = Q_p e^{-K't_r} \qquad (6.8)$$

Equation 6.8 is analogous to a storage tank (watershed) losing its water at an exponential rate as shown in Figure 6.7.

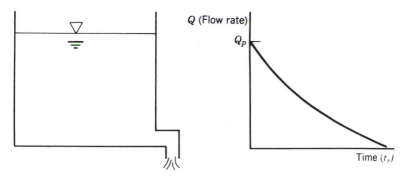

FIGURE 6.7   The recession curve analogy to a storage tank.

The storage coefficient (recession coefficient) is rarely constant over the entire recession curve. Groundwater flows into a stream depend on the soil types at different locations. Still other flows depend on the elevation relationship (head) of the groundwater table to the stream surface over a very long period of time (base flow). Thus, it is more accurate to represent the recession curve by a composite exponential, or

$$Q = Q_i e^{-K_i' t_r} \tag{6.9}$$

where

Q_i = intercept for any segment "i" at $t_r = 0$ (cfs)
K_i' = slope for any segment i (1/time) (storage coefficient)

The Q_i's are the flow rate coefficients. Both the flow rate and storage coefficients can be estimated knowing the exponential form of Equations 6.8 and 6.9. Thus, by plotting the logarithm values of Q/Q_p versus time, lines of constant slope can be determined that result in estimates of the storage coefficients and the flow rate coefficients. This estimation procedure, which yields storage coefficients, inflection points, and estimates of base flow, is called the logarithm recession method.

❏ **EXAMPLE PROBLEM 6.1**

Estimate equations for the recession limb based on the following streamflow data. Assume that there are three types of flow that affect the recession limb: surface runoff, interflow, and base flow. The peak discharge occurred at 12:00 noon and was 200 cfs.

t_r (hr)	FLOW (cfs)	t_r (hr)	FLOW (cfs)
0 (noon)	200	10	20
1	142	11	19
2	96	12	18
3	60	13	17
4	42	14	16.5
5	35	15	16
6	31	16	15
7	28	17	14.5
8	25	18	14
9	22		

Solution

The data are plotted as shown in Figure 6.8 following the form of Equation 6.9. Thus the data plotted are

t_r	Q/Q_p	t_r	Q/Q_p
0	1.00	10	0.10
1	0.71	11	0.095
2	0.48	12	0.90
3	0.30	13	0.085
4	0.21	14	0.083
5	0.18	15	0.08
6	0.155	16	0.075
7	0.14	17	0.073
8	0.125	18	0.07
9	0.11		

The storage coefficients are the slopes from Figure 6.8. Note that the direct calculation yields a coefficient to the base "10," thus to convert to the base e logarithm, one must multiply by 2.3(ln(10)). The estimate of the slope is simplified if one considers a complete log cycle for the change in ordinate value; thus, each slope is calculated using

$$K' = [\log(1) - \log(0.1)]/\Delta t_r$$

or

$$K' = 1/\Delta t_r \qquad \text{base 10} \tag{6.10}$$

For a slope change (in this case $t_r \geq 10$ hr), the base flow rate coefficient is estimated from Figure 6.8 using the intercept of the exponential lines where the intercept ($t_r = 0$), which is 0.155.

$$Q_b = 0.155(200) = 31 \text{ cfs}$$

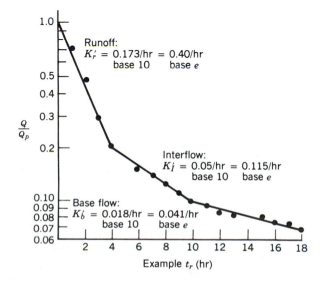

FIGURE 6.8 Example Problem 6.1, recession curves.

For $10 > t_r > 4$ hr, the interflow rate coefficient is

$$Q_I = 0.32(200) = 64 \text{ cfs}$$

The final equation forms are:

$$t_r \geq 10 \text{ hr} \qquad Q = 31e^{-0.041t_r}$$
$$4 \text{ hr} \leq t_r < 10 \text{ hr} \qquad Q = 64e^{-0.115t_r}$$
$$0 \leq t_r < 4 \text{ hr} \qquad Q = 200e^{-0.40t_r}$$

To illustrate the final curve fit by the equations for the original data, see Figure 6.9.   ❏

Other methods for separating base flows from stream hydrographs exist. The fixed time method assumes that surface runoff always ends after a fixed time interval.

$$\tau = (DA)^n \tag{6.11}$$

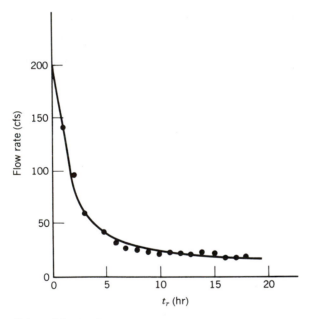

Selected Comparisons

t_r	STREAM FLOW	EQUATIONS
0	200	200
3	40	60
5	35	36
10	20	21
18	14	15

FIGURE 6.9 Example Problem 6.1, a hydrograph and equation comparison. The · denotes field data; the − denotes equation form.

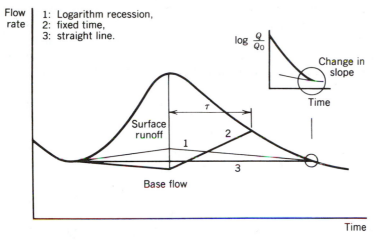

FIGURE 6.10   Base flow separation techniques.

where

τ = time from the peak to the end of the runoff hydrograph (days)

DA = drainage area (square miles)

n = recession constant

Frequently the recession constant has a value of about 0.2 but must be derived for a particular location. The derivation is frequently dependent on the inflection point from a semilogarithmic plot of flows. The straight line and logarithm method also depend on inflection points. A comparison of methods is shown in Figure 6.10. The logarithm recession method is based on theory and is preferred. However the straight line and fixed time methods require less calculations and can be used when validated for a region. Other base flow separation methods illustrate reasonable separation but no standard procedures exist.

6.3
UNIT HYDROGRAPH FROM STREAMFLOW DATA

The area under the hydrograph represents the volume of runoff from a watershed. The area under a unit hydrograph represents the volume of runoff equivalent to a unit depth (1 in.) over the entire watershed. It is a description of the many watershed characteristics that affect the runoff process. The shape of the unit hydrograph expresses a set of characteristics of the land and soil that are assumed to be repeated for similar duration storms. If the duration changes, the time base of the hydrograph changes.

The method for determining unit hydrographs from streamflow data evolved from the work of Sherman (1932). Others (Clark, 1945) expanded on this work. It was originally assumed that the rainfall excess occurs over a fixed hydrograph time base, and consequently, for a given watershed, the hydrograph shape, time to peak, and recession time were constant. For a specific watershed, the unit hydrograph resulting from a given quantity of rainfall excess can be used to generate another hydrograph from a different quantity of rainfall excess if the storm durations are the same. For longer storm durations, the rainfall excess can be divided into smaller time periods, each of duration equal to that used for the unit hydrograph. This assumes that the unit hydrograph does not change with watershed conditions. If the soil becomes saturated or channel velocities increase with increasing cumulative rainfall excess, the shape of the unit hydrograph may change.

A unit hydrograph may be derived from streamflow data or by assuming a specific shape based on watershed conditions. Using streamflow records, the base flow (and other groundwater flows, if present) are subtracted from the streamflow. The resulting hydrograph is the runoff hydrograph. Thus, the runoff hydrograph for the watershed can be specified. Next, the area under the runoff hydrograph (rainfall excess) is calculated. The rainfall excess is in cubic feet if the streamflow is measured in cubic feet per second and the base of the hydrograph is converted to seconds. The volume of runoff is divided by the watershed area obtaining a depth of rainfall excess. Frequently, it is called runoff depth. Since the duration of rainfall excess has been assumed to be constant and the base of the hydrograph is constant, the volume of runoff resulting from various rainfall excesses is proportional to the corresponding discharge (ordinate) values. Thus, the proportionality factor of a given runoff depth to the unit hydrograph runoff depth (1 in.) can be used to convert discharge values in any hydrograph to those resulting from a unit hydrograph. Steps for the calculation procedure are illustrated in Figure 6.11.

The following guidelines should be considered for developing the unit hydrograph if the procedures of Figure 6.11 are followed:

1. Independence in storm event and streamflow (one runoff event does not affect another).

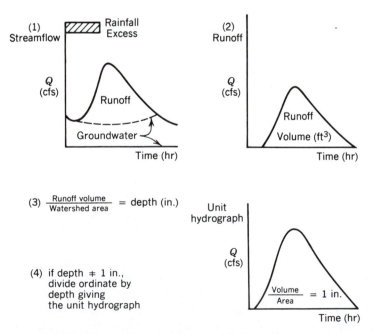

FIGURE 6.11   The steps for calculating a unit hydrograph.

2. For most situations, direct runoff should be greater than one-half inch (this improves the accuracy of the unit hydrograph).

3. Record many storms and average the ordinate values.

Since the assumption that two storms would have the same duration and area distribution is rarely true, there arose a need to improve and generalize the unit hydrograph method. Early studies (Clark, 1945) introduced the concept of instantaneous unit graphs: that rainfall intensities of short duration (minutes) produce instantaneous hydrographs. In the lagging procedure (Morgan and Hullinhors, 1939), each of these instantaneous hydrographs is added to produce a final hydrograph. Hydrographs from long duration storms are assumed to consist of several instantaneous unit hydrographs. For example, a 2-hr unit hydrograph is created from two 1-hr unit hydrographs, and two 4-hr unit hydrographs form an 8-hr hydrograph, and so on. Also, a combination unit hydrograph can be formed by adding the ordinates of the two smaller duration storms and then dividing by 2 (see Figure 6.12). Note that the lagging procedure uses storms of equal duration.

6.3.1   A Convolution Computation Procedure Using Discrete Time Data

Rarely is there a constant rainfall excess over a single time increment. Usually, the rainfall excess varies with time. Consider a unit hydrograph method that divides a rainfall into successive shorter time events, each of constant rainfall excess and equal times. Each rainfall excess value is multiplied by the unit hydrograph to obtain the resulting ordinate values displaced in time by the start of each rainfall excess. This multiplication is defined in mathematical terms as a convolution. Then, the total runoff hydrograph is the superposition of each hydrograph as initiated by its rainfall excess.

   The basic premise of the unit hydrograph method is that the individual hydrographs obtained by multiplying the ordinates of the unit hydrograph by the various successive rainfall excess increments, when properly arranged with respect to time, can be added to give the total runoff hydrograph. Examine Figure 6.13, which illustrates a design hydrograph resulting from three successive 15-min rainfall excess increments (0.6, 1.0, and 0.4 in.). First, the individual hydrographs of runoff from each rainfall excess increment were computed by multiplying the ordinates of the 15-min unit hydrograph by the respective rainfall excess increment and plotted as shown in Figure 6.13. Note that each of the individual hydrographs start at the same time as the incremental rainfall excess producing it. The total storm runoff (design) hydrograph is then obtained

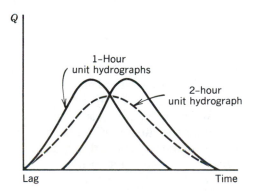

FIGURE 6.12   The example lagging procedure.

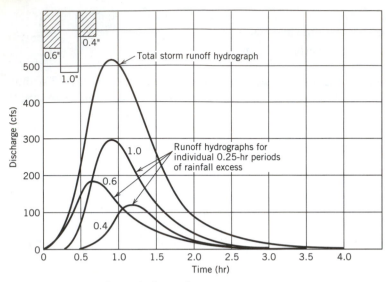

FIGURE 6.13 Combining hydrographs.

by adding the ordinates of each of the individual incremental hydrographs. This procedure is very analogous to the influence line procedure in structural analysis, and the unit hydrograph could be considered as sort of an influence line of the contributary drainage watersheds. One should immediately recognize that the above procedure of computing the resulting, individual incremental hydrographs and then summing the ordinates is very time consuming if there are many rainfall excess increments involved. For example, if there are 15-min increments of a 6-hr rainfall, one may have to compute and add up the ordinates of as many as 24 individual hydrographs, each displaced by 15 min.

However, there is a relatively quick procedure, developed by the U.S. Bureau of Reclamation (BR) (1952), for applying rainfall excess increments to a unit hydrograph that can be used to compute the design hydrograph by the unit hydrograph method. This procedure is easily adopted to computation by a pocket calculator with a memory.

☐ ***EXAMPLE PROBLEM 6.2***

Develop a composite hydrograph using the rainfall excess from a 25-yr, 6-hr rainfall.

Solution

The first step is to compute a unit hydrograph for the particular watershed. The ordinate values of the unit hydrograph are for an actual watershed and are shown in Table 6.1, column 2. To compute a 25-yr frequency design hydrograph, one applies to this unit hydrograph the rainfall excess increments resulting from a 25-yr frequency, 6-hr design rainfall. There are 23 increments of rainfall excess. Theoretically, no rainfall excess occurs during the first 15 min and very little up to hour 1.0, such that, for all practical purposes, rainfall prior to hour 1.0 can be ignored.

TABLE 6.1   Computation of Composite Flood Hydrographs (BR Procedure) Initial Matrix

			MADE BY	DATE	JOB NO
		0.23	CHECKED BY	DATE	SHEET NO
RAINFALL EXCESS		0.23	25-YR FREQUENCY — 6-HR DESIGN RAINFALL		
IN REVERSE ORDER		0.29	OCP PHASE 6 — DRAINAGE BASIN (270.6 AC.), CN = 95		
		0.29			
		1.12			
		1.03			

TIME (hr)	ORDINATES OF 15-MINUTE UNIT U (ft^3/sec)		DESIGN FLOOD HYDROGRAPH (ft^3/sec)	TIME (hr)	ORDINATES OF 15-MINUTE UNIT U (ft^3/sec)	RAINFALL-EXCESS (in.)	DESIGN FLOOD HYDROGRAPH (ft^3/sec)
		0.28					
		0.20					
		0.15					
0	0	x 0.13	= [0]	8.00			
0.25	68			8.25			
0.50	246			8.50			
0.75	293			8.75			
1.00	202			9.00			
1.25	121			9.25			
1.50	72			9.50			
1.75	41			9.75			
2.00	26			10.00			
2.25	15			10.25			
2.50	9			10.50			
2.75	6			10.75			
3.00	3			11.00			
3.25	1			11.25			
3.50	0			11.50			
3.75				11.75			
4.00				12.00			
4.25				12.25			
4.50				12.50			
4.75				12.75			
5.00				13.00			
5.25				13.25			
5.50				13.50			
5.75				13.75			

To demonstrate the BR procedure for applying these 20 rainfall excess increments to the unit hydrograph set up a standard form (Table 6.1). In the first column is the time in successive 0.25-hr increments. Second, write the rainfall excess increments in reverse order on a smaller piece of paper, place the smaller piece of paper at the top of the third column as shown in Table 6.1, and then start multiplying the

TABLE 6.2   Computation of Composite Flood Hydrographs (BR Procedure) First Calculation

MADE BY	DATE	JOB NO
CHECKED BY	DATE	SHEET NO

RAINFALL EXCESS — IN REVERSE ORDER

0.18
0.23
0.23
0.29
0.29
1.12
1.03
0.28
0.20
0.15
0.13

25-YR FREQUENCY — 6-HR DESIGN RAINFALL

OCP PHASE 6 — DRAINAGE BASIN (270.6 AC.), CN = 95

TIME (hr)	ORDINATES OF 15-MINUTE UNIT U (ft^3/sec)		DESIGN FLOOD HYDROGRAPH (ft^3/sec)	TIME (hr)	ORDINATES OF 15-MINUTE UNIT U (ft^3/sec)	RAINFALL-EXCESS (in.)	DESIGN FLOOD HYDROGRAPH (ft^3/sec)
0	0	x 0.15	= $\boxed{0}$	8.00			
0.25	68	x 0.13	= $\boxed{9}$	8.25			
0.50	246			8.50			
0.75	293			8.75			
1.00	202			9.00			
1.25	121			9.25			
1.50	72			9.50			
1.75	41			9.75			
2.00	26			10.00			
2.25	15			10.25			
2.50	9			10.50			
2.75	6			10.75			
3.00	3			11.00			
3.25	1			11.25			
3.50	0			11.50			
3.75				11.75			
4.00				12.00			
4.25				12.25			
4.50				12.50			
4.75				12.75			
5.00				13.00			
5.25				13.25			
5.50				13.50			
5.75				13.75			
6.00				14.00			

unit hydrograph ordinates by the rainfall excess increments and summing up the total as you slide the small piece of paper down. Starting the small piece of paper (slide) on the first line, as shown in Table 6.1, only multiply the first rainfall excess increment by 0, the first unit hydrograph ordinate. Moving the slide down an additional line, as shown in Table 6.2, multiply and sum up the first three lines

TABLE 6.3 Computation of Composite Flood Hydrographs (BR Procedure) Intermediate Calculation

					MADE BY	DATE	JOB NO
		0.15					
		0.18			CHECKED BY	DATE	SHEET NO
RAINFALL EXCESS		0.23		25-YR FREQUENCY — 6-HR DESIGN RAINFALL			
IN REVERSE ORDER		0.23		OCP PHASE 6 — DRAINAGE BASIN (270.6 AC.), $CN = 95$			
		0.29					
		0.29					

TIME (hr)	ORDINATES OF 15-MINUTE UNIT U (ft^3/sec)		DESIGN FLOOD HYDROGRAPH (ft^3/sec)	TIME (hr)	ORDINATES OF 15-MINUTE UNIT U (ft^3/sec)	RAINFALL-EXCESS (in.)	DESIGN FLOOD HYDROGRAPH (ft^3/sec)
		1.12					
		1.03					
		0.28					
0	0	x 0.20	= 0	8.00			
0.25	68	x 0.15	= 9	8.25			
0.50	246	x 0.13	= 42	8.50			
0.75	293			8.75			
1.00	202			9.00			
1.25	121			9.25			
1.50	72			9.50			
1.75	41			9.75			
2.00	26			10.00			
2.25	15			10.25			
2.50	9			10.50			
2.75	6			10.75			
3.00	3			11.00			
3.25	1			11.25			
3.50	0			11.50			
3.75				11.75			
4.00				12.00			
4.25				12.25			
4.50				12.50			
4.75				12.75			
5.00				13.00			
5.25				13.25			
5.50				13.50			
5.75				13.75			
6.00				14.00			
6.25				14.25			

TABLE 6.4 Computation of Composite Flood Hydrographs (BR Procedure) Final Hydrograph

TIME (hr)	ORDINATES OF 15-MINUTE UNIT U (ft^3/sec)	RAINFALL-EXCESS (in.)	DESIGN FLOOD HYDROGRAPH (ft^3/sec)	TIME (hr)	ORDINATES OF 15-MINUTE UNIT U (ft^3/sec)	RAINFALL-EXCESS (in.)	DESIGN FLOOD HYDROGRAPH (ft^3/sec)
0	0		0				
0.25	68		9				
0.50	246		42				
0.75	293		89				
1.00	202		138				
1.25	121		244				
1.50	72		479				
1.75	41		694				
2.00	26		685				
2.25	15		557				
2.50	9		446				
2.75	6		365				
3.00	3		304				
3.25	1		256				
3.50	0		218				
3.75			189				
4.00			167				
4.25			152				
4.50			143				
4.75			138				
5.00			135				
5.25			126				
5.50			95				
5.75			60				
6.00			35				
6.25			21				
6.50			12				
6.75			7				
7.00			4				
7.25			2				
7.50			1				
			0				

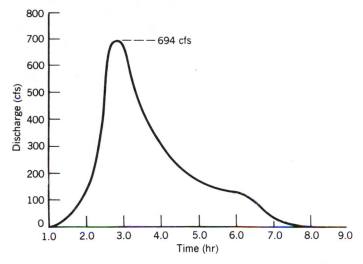

FIGURE 6.14 A design hydrograph, 25-yr frequency, 6-hr rainfall, unit hydrograph method.

as follows:

$$Q_1 = R_1 U_1 = 0.13 \times 0 = 0$$

$$Q_2 = R_1 U_2 + R_2 U_1 = (0.13 \times 68) + (0.15 \times 0) = 9$$

which is the second value of the design hydrograph at time 0.25 hr. The slide is then moved down another line as shown in Table 6.3 with calculations as follows:

$$Q_3 = R_1 U_3 + R_2 U_2 + R_3 U_1$$

$$Q_3 = (0.13 \times 246) + (0.15 \times 68) + (0.20 \times 0) = 42$$

with the result being placed in the third line of the design hydrograph (at time 0.50 hr). This process of summation and addition is continued until the computation of the design hydrograph is completed, as shown in Table 6.4. The design hydrograph as computed by this method is shown in Figure 6.14. ❑

The computation of a design hydrograph by the unit hydrograph method procedure requires much work if done manually. However, the various procedures outlined in this chapter can be programmed for computer or calculator computation. The unit hydrograph method has the advantage of acceptability in that it is universally understood and used in design. As previously mentioned, it does require a design rainfall with determination of rainfall excess and a unit hydrograph shape.

6.3.2 Matrix Procedure

A matrix is a set of numbers arranged in a rectangular array. From the calculations using the BR method, it was shown that a composite hydrograph is the result of multiplying rainfall excess by the unit hydrograph. It is a discrete convolution method. The rainfall excess is one rectangular array and the unit hydrograph is another. A streamflow record is always a combination of flows (composite) resulting from many rainfall events or increments of a single event; thus, one will rarely, if ever, have a hydrograph resulting from 1 in. of rainfall excess in one time increment. The challenge is to decompose a streamflow hydrograph into flow resulting from individual rainfall events and recognize that each of these flows results from a rainfall applied to the unit hydrograph. The unit hydrograph can then be identified using matrix methods. Thus, matrix and vector forms of the data are required.

Define the unit hydrograph as a vector of flows, that is, a column vector U, which has j unit hydrograph values. Also, define the rainfall excess R of i periods of time. Thus, the resulting hydrograph can be expressed in general matrix form as

$$Q = R \times U \tag{6.12}$$

where

$$R = \begin{bmatrix} R_1 & 0 & 0 & \cdots & 0 \\ R_2 & R_1 & 0 & \cdots & 0 \\ R_3 & R_2 & R_1 & \cdots & 0 \\ R_i & R_{i-1} & & & R_1 \end{bmatrix} \qquad U = \begin{bmatrix} U_1 \\ U_2 \\ U_3 \\ U_j \end{bmatrix} \qquad Q = \begin{bmatrix} Q_1 \\ Q_2 \\ Q_3 \\ Q_k \end{bmatrix}$$

and the computations are expanded as

$$Q_1 = R_1 U_1$$
$$Q_2 = R_2 U_1 + R_1 U_2$$
$$Q_3 = R_3 U_1 + R_2 U_2 + R_1 U_3$$

etc.

where
the hydrograph has k values, $k = i + j - 1$
and i = number of rainfalls excess values
j = number of unit hydrograph values

6.3.3 Computation Procedure for Deriving a Unit Hydrograph

The most common streamflow hydrographs result from widely varying rainfall excesses. A constant rainfall excess usually cannot be assumed over the duration of the storm, and it is difficult to separate the streamflow hydrographs resulting from a particular rainfall excess. However, it is possible using matrix methods, to solve for the unit hydrograph. Rewriting Equation 6.12, one obtains

$$R^{-1} \times Q = R^{-1} \times R \times U$$

or

$$U = R^{-1} \times Q \tag{6.13}$$

Thus, knowing that the resulting streamflow (**Q**) vector and the rainfall excess inverse exists as a square matrix with a nonvanishing determinant, the unit hydrograph can be found. Since the rainfall excess is not a square matrix, a transformation must be done. Then the resulting equation is

$$\boldsymbol{R}^T \times \boldsymbol{Q} = \boldsymbol{R}^T \times \boldsymbol{R} \times \boldsymbol{U}$$

and

$$\boldsymbol{U} = (\boldsymbol{R}^T \times \boldsymbol{R})^{-1} \boldsymbol{R}^T \times \boldsymbol{Q} \qquad (6.14)$$

where $\boldsymbol{R}^T$ is the transpose of the matrix $\boldsymbol{R}$. It is found by interchanging the rows for columns or in general notation:

$$\boldsymbol{R} = \begin{bmatrix} R_1 & 0 \\ R_2 & R_1 \\ R_3 & R_2 \\ 0 & R_3 \end{bmatrix} \quad \text{and} \quad \boldsymbol{R}^T = \begin{bmatrix} R_1 & R_2 & R_3 & 0 \\ 0 & R_1 & R_2 & R_3 \end{bmatrix} \qquad (6.15)$$

☐ **EXAMPLE PROBLEM 6.3**

Derive a unit hydrograph from streamflow data resulting from a storm of variable rainfall excess. Assume that the base flow record also is available. Runoff is not available.

Streamflow and Baseflow Data (cfs-hr)

TIME (hr)	EXCESS (in.)	STREAMFLOW (cfs)	BASEFLOW (cfs)	RUNOFF (cfs)
0	0	10	10	0
1	0.5	70	10	60
2	1.0	220	20	200
3	0	230	20	210
4	1.0	290	30	260
5	0	300	30	270
6		215	30	185
7		180	30	150
8		145	30	115
9		120	30	90
10		95	30	65
11		70	30	40
12		40	30	10
13		30	30	0

Solution

1. First calculate runoff as runoff = streamflow − baseflow.
2. Now arrange the values in a matrix form similar to Equation 6.12, or $Q = R \times U$ where $i = 4$, $k = 13$, and $j = k - i + 1 = 10$ nonzero values for the unit hydrograph.

Thus,

$Q_1 = R_1 U_1$

$60 = 0.5(U_1)$ and $U_1 = Q_1 R_1^{-1} = 60/0.5 = 120$ cfs

$Q_2 = R_1 U_2 + R_2 U_1$

$200 = 0.5(U_2) + 1.0(120)$ $U_2 = 160$ cfs

$Q_3 = R_1 U_3 + R_2 U_2 + R_3 U_1$

$210 = 0.5(U_3) + 1.0(160) + 0.0(120)$ $U_3 = 100$ cfs

$Q_4 = R_1 U_4 + R_2 U_3 + R_3 U_2 + R_4 U_1$

$260 = 0.5(U_4) + 1.0(100) + 0.0(160) + 1.0(120)$ $U_4 = 80$ cfs

$Q_5 = R_1 U_5 + R_2 U_4 + R_3 U_3 + R_4 U_2 + R_5 U_1$

$270 = 0.5(U_5) + 1.0(80) + 0.0(100) + 1.0(160) + 0.0(120)$ $U_5 = 60$ cfs

$Q_6 = R_1 U_6 + R_2 U_5 + R_3 U_4 + R_4 U_3$

$185 = 0.5(U_6) + 1.0(60) + 0.0(80) + 1.0(100)$ $U_6 = 50$ cfs

$Q_7 = R_1 U_7 + R_2 U_6 + R_3 U_5 + R_4 U_4$

$150 = 0.5(U_7) + 1.0(50) + 0.0(60) + 1.0(80)$ $U_7 = 40$ cfs

$Q_8 = R_1 U_8 + R_2 U_7 + R_3 U_6 + R_4 U_5$

$115 = 0.5(U_8) + 1.0(40) + 0.0(50) + 1.0(60)$ $U_8 = 30$ cfs

$Q_9 = R_1 U_9 + R_2 U_8 + R_4 U_6$

$90 = 0.5(U_9) + 1.0(30) + 1.0(50)$ $U_9 = 20$ cfs

$Q_{10} = R_1 U_{10} + R_2 U_9 + R_4 U_7$

$65 = 0.5(U_{10}) + 1.0(20) + 1.0(40)$ $U_{10} = 10$ cfs

$Q_{11} = R_1 U_{11} + R_2 U_{10} + R_4 U_8$

$40 = 0.5(U_{11}) + 1.0(10) + 1.0(30)$ $U_{11} = 0$ cfs

Note that previous computed values are used in later computations therefore roundoff errors tend to build. Care must be used in the computations. ❑

6.4

SYNTHETIC HYDROGRAPHS

A synthetic hydrograph is one developed with minimum use of streamflow data, a very common situation in practice. As an example, consider a pending land use change. Obviously, hydrographs must be synthesized for the anticipated postdevelopment conditions.

A runoff hydrograph is defined as an expression for surface water discharge over time. It is an expression of the watershed characteristics that invariably govern the relationship between rainfall and the resulting runoff. It represents the integrated effects of rainfall and watershed characteristics, such as area, shape, drainage patterns, land use, land and channel properties, land and drainage slopes, and the infiltration capacity of the soil. Thus, any synthetic hydrograph must be related to watershed and precipitation conditions. Some of the methods to achieve these relationships are presented in this chapter. The methods discussed are those in common use: rational, SCS hydrograph, unit-hydrograph, contributing area, Santa Barbara, and discrete unit-time hydrograph.

6.4.1   Rational Method

The rational method is one of the oldest and was originally used to only estimate the peak discharge. More recently, it has been used to develop a particular hydrograph shape.

If rainfall intensity remains constant over the time interval required to completely drain a watershed, then the runoff (intensity) would be equal to the rainfall intensity. From a mass balance relating rainfall intensity to runoff intensity both intensities can be equated and expressed in the following formula with suitable conversion factors.

$$Q = K'iCA \qquad (6.16)$$

where

Q = runoff (cfs)
i = rainfall intensity (in./hr)
CA = net effective area (ac)
K' = conversion factor = 1.008 (cfs-hr/ac-in.) or in the metric system = 0.00278 m³/s/ha-mm/hr)

The assumptions for use of the formula requires a delineation of the contributing area and intensity remains constant over the time period required to drain the area (time of concentration). The contributing area can be related to the characteristics of the watershed that contribute runoff. For impervious areas that are hydraulically connected (water flow is continuous), runoff and rainfall excess must come from this area. However, other areas may contribute during heavy or additional rainfall conditions making the contributing area larger. the impervious area that contributes runoff is frequently called the directly connected impervious area (DCIA). In a residential area, roofs on homes may not directly drain to another impervious surface or a sewer, thus the roofs would not be directly (hydraulically) connected.

For watersheds that have long travel times, it is almost impossible to have a constant intensity over that time period. This limits the use of Equation 6.16 to short time of travel watersheds. Examples of such watersheds are paved areas with curb, gutter, and sewers. Thus, the contributing area, rainfall intensity, and hydraulic characteristics are constant or fixed. As presented by Mulvaney (1851) and used by Kuichling (1889), Equation 6.16 can be restated as the rational formula:

$$Q_p = CiA \qquad (6.17)$$

where

Q_p = peak discharge (cfs)
C = runoff coefficient (dimensionless)
i = rainfall intensity (in./hr)
A = watershed area (ac)

The conversion factor (1.008) was dropped from Equation 6.17. The form of Equation 6.17 is called the rational formula primarily because the units of the quantities are approximately numerically consistent. The runoff coefficient can be simply related to the contributing area by dividing the contributing area by the total area. For a contributing area that is well defined, the runoff coefficient also is well defined. However, if there is no well-defined contributing area, then the runoff coefficient must be determined from published data (Table 6.5) or from runoff studies that relate the volume of runoff to the volume of precipitation.

TABLE 6.5   Runoff Coefficients C Recurrence Interval ≤ 10 years[a]

DESCRIPTION OF AREA	RUNOFF COEFFICIENTS	CHARACTER OF SURFACE	RUNOFF COEFFICIENTS
Business		Pavement	
Downtown	0.70 to 0.95	Asphalt or concrete	0.70 to 0.95
Neighborhood	0.50 to 0.70	Brick	0.70 to 0.85
Residential		Roofs	0.70 to 0.95
Single family	0.30 to 0.50	Lawns, sandy soil	
Multiunits, detached	0.40 to 0.60	Flat, 2%	0.05 to 0.10
Multiunits, attached	0.60 to 0.75	Average, 2–7%	0.10 to 0.15
Residential, suburban	0.25 to 0.40	Steep, 7% or more	0.15 to 0.20
Apartment	0.50 to 0.70	Lawns, heavy soil	
Industrial		Flat, 2%	0.13 to 0.17
Light	0.50 to 0.80	Average, 2–7%	0.18 to 0.22
Heavy	0.60 to 0.90	Steep, 7% or more	0.25 to 0.35
Parks, cemeteries	0.10 to 0.25		
Railroad yard	0.20 to 0.35		
Unimproved	0.10 to 0.30		

Source: From *Design and Construction of Sanitary and Storm Sewers. ASCE Manual of Practice No. 37,* 1970. Revised by D. Earl Jones, Jr.
[a]For 25- to 100-yr recurrence intervals, multiply coefficient by 1.1 and 1.25, respectively, and the product cannot exceed 1.0.

The basic assumptions for using the rational formula are

1. The rainfall intensity must be constant for a time interval at least equal to the time of concentration (typically taken to be the average intensity over the time period).
2. The runoff is a maximum when the rainfall intensity lasts as long as the time of concentration.
3. The runoff coefficient is constant during the storm.
4. The watershed area does not change during the storm.

These assumptions appear reasonable for well-defined watersheds with a short time of concentration (typically assumed less than 20 min). A rainfall intensity associated with the time of concentration and frequency of occurrence can be obtained from intensity–duration–frequency curves and the runoff coefficient from standard published results (Table 6.5). The calculations are simple and the numerical data easily obtained.

As demonstrated by Williams (1950), Pagan (1972), Mitchi (1974) and others, the peak flow, Q_p, as computed by the rational method, $Q_p = CiA$, is actually the peak of a triangular hydrograph. Consider a rainfall of constant intensity i uniformly distributed over a particular drainage basin and of a duration, D, as shown on the hyetograph of Figure 6.15. The volume of runoff (rainfall excess), V_1, the lower (clear) portion of the hyetograph, is equal to $CiDA$, where C is the runoff coefficient in the rational method as normally used—a dimensionless ratio of the total volume of runoff to the total volume of rainfall; i is the intensity of rainfall in inches/hour; D is the rainfall duration in hours, and A is the area of the drainage basin in acres. The units of runoff are acre-in./hr or ft^3/sec (1.008 ac-in./hr = 1 ft^3/sec; however, ignore the 1.008) which, from previous discussion, are recognized as a volume. The upper portion of the hyetograph, again a volume, is equal to the fraction of rainfall remaining onsite $(1 - C)$ times rainfall volume (iAD). From Figure 6.15 the volume, V_2, of the triangular hydrograph is equal to $t_c Q_p$, where t_c is the time of concentration of the drainage area. Note that in this case we have assumed that the time to peak, t_p, previously defined as the time from the start of rainfall excess to the peak of the hydrograph, is equal to the time of concentration, t_c. Since the volume of runoff, V_1, as represented by the lower portion of the hyetograph, must equal the volume of the triangular hydrograph, V_2, and making the assumption that the duration of rainfall D is equal to the time of concentration, t_c, of the basin, the rational method equation is obtained. Thus, the peak flow, Q_p, as computed by the rational method, is shown to be the same as the height of an isosceles triangle with base equal to $2t_c$. Minor modifications to include curve lines can be made as long as the peak and rainfall excess remain the same. An exponential curve approximation also can be made.

From an examination of the isosceles triangular hydrograph shown in Figure 6.15, the shape is different from the typical hydrograph shown in Figure 6.1. For instance, the recession curve portion of this hydrograph is too short. However, the rational method hydrograph shape is approximately that shown in Figure 6.16. Curved rising and falling limbs are also possible as long as the volume of runoff remains equal to rainfall excess.

Rainfall intensity that continues past the time of concentration produces a constant runoff rate beyond the time of concentration. As before, consider the rainfall is uniformly distributed over the watershed and of equal intensity, i, throughout its

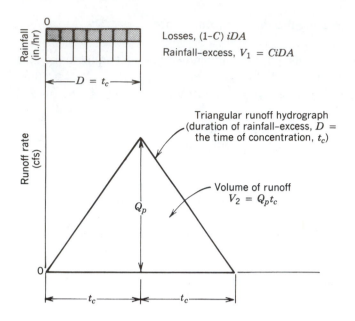

Assume: Contributing area varies linearly with time.

Since:
$$V_1 = V_2$$
$$CiDA = Q_p t_c$$

But:
$$t_c = D$$

Therefore:
$$Q_p = CiA$$

where
$$Q_p = \text{ft}^3 / \text{sec}$$
$$i = \text{in.} / \text{hr}$$
$$A = \text{ac}$$

Figure 6.15 A rational method hydrograph with derivation.

duration, D. Consider a situation in which the duration of the rainfall, D, is greater than the time of concentration, t_c, as shown in Figure 6.17. Again the volume of runoff, V_1, the lower portion of the hyetograph, is equal to $CiAD$. The volume, V_2, of the trapezoidal hydrograph is equal to $Q_p D$.

The rational method for calculating hydrograph peak runoff is used frequently by designers. Essentially, the users must estimate a runoff coefficient, rainfall intensity and watershed area. Using this runoff coefficient, the slope of the land, and the overland flow distance, overland time of travel can be computed using Figure 6.17 and the C factor of Table 6.5. Note the range of values for the runoff coefficient. Also, remember from previous chapters that the runoff coefficient can vary with soil moisture conditions and the period of time and volume of rainfall. Conservative designs producing higher runoff peaks require a higher value for the runoff coefficient. Also, the time of concentration must be consistent with the use of the data.

The coefficients in Table 6.5 are applicable for storms of 2- to 10-yr return frequen-

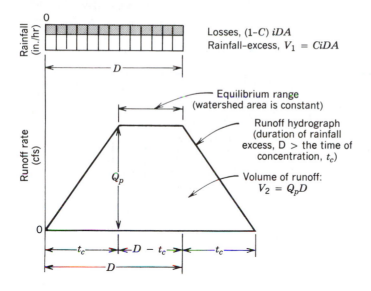

Figure 6.16 A rainfall hyetograph $D > t_c$ with resulting hydrograph and derivation.

cies and were originally developed when many streets were uncurbed and drainage was conveyed in roadside swales. For recurrence intervals longer than 10 yr, the indicated runoff coefficients should be increased, assuming that nearly all of the rainfall in excess of that expected from the 10-yr recurrence interval rainfall will become runoff and would be accommodated by an increased runoff coefficient.

The rational formula has been used to estimate the peak discharges from areas with a relatively low time of concentration, that is, a few minutes up to about 20 min. Thus, watershed areas are generally homogeneous in hydrologic processes (same runoff areas, similar soil storage, etc.) and usually less than 50 to 100 ac in size. The larger areas generally have steeper drainage slopes. The lower time of concentration is more consistent with the assumption of constant rainfall intensity, or in other words, it is not probable to have constant rainfall intensity over longer periods of time. When designing for a peak discharge, the duration for the rainfall interval should approximate the time of concentration. The longer the time of concentration, the longer the rainfall interval and thus, there is less of a chance of constant rainfall intensity.

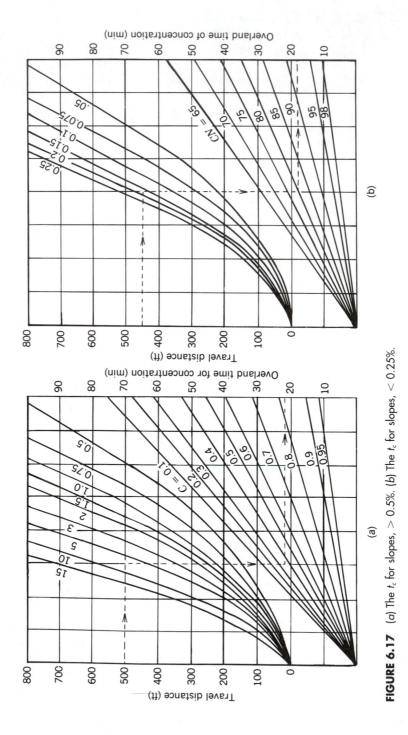

FIGURE 6.17 (a) The t_c for slopes, > 0.5%. (b) The t_c for slopes, < 0.25%.

210

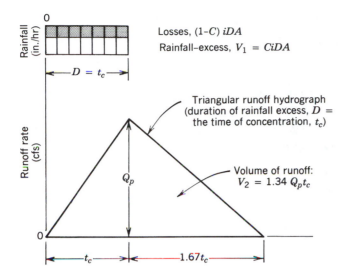

FIGURE 6.18 A SCS (NRCS) typical hydrograph.

6.4.2 The SCS (NRCS) Hydrograph

The SCS (NRCS) hydrograph shape has a lesser peak and a longer recession time relative to the rational hydrograph. If the recession curve were longer to make the triangular hydrograph more nearly represent a more true-to-life situation, say of a length equal to $1.67t_c$, Figure 6.18 would result. Again setting V_1 equal to V_2, the peak Q_p of this hydrograph is computed to be equal to $0.75CiA$. This fact should immediately indicate that the rational method is a conservative design procedure because of existing hydrograph shapes. Since

$$V_1 = V_2$$

$$CiDA = 1.34Q_pt_c$$

But

$$t_c = D$$

Therefore,

$$Q_p = 0.75CiA* \qquad (6.18)$$

where
Q_p = ft^3/sec
i = in./hr
A = ac

Also consider a case in which the duration of the rainfall, D, is less than the time to peak, t_p, as shown in Figure 6.19. Again, the volume of runoff, V_1, the lower portion of the hyetograph, is equal to $CiAD$. The volume, V_2, of the triangular runoff hydrograph is equal to $\frac{1}{2}Q_p(t_p + t_r)$.

---

*If A = square miles, then Equation 6.18 is rewritten as $Q_p = 484\,CiA$.

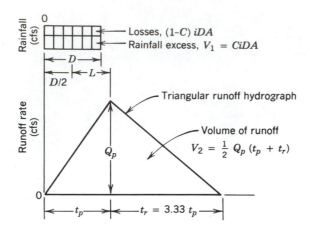

FIGURE 6.19   The rainfall hyetograph $D < t_p$.

Since

$$V_1 = V_2$$

$$CiAD = \tfrac{1}{2}Q_p(t_p + t_r)$$

However,

$$t_r = 3.33t_p$$

Therefore,

$$Q_p = \frac{2CiAD}{4.33t_p}$$

However,

$$t_p = D/2 + L$$

Therefore,

$$Q_p = \frac{0.46CiAD}{D/2 + L} \tag{6.19}$$

Thus, from Equation 6.19, it is noted that the peak discharge is less than that calculated by Equation 6.18. The difference results because the recession limb time is 3.33 times longer than the time to peak, thus the hydrograph peak would have to be attenuated if rainfall excess were the same (volume of runoff is the same in Figures 6.18 and 6.19).

The specific shape of the hydrograph is important. Reasonable estimates of the recession time and the time to peak must be made. Some guidance is available from the SCS (USDA, 1975) and other researchers (Capice, 1984). However, the guidelines are general and rely on the judgment of the hydrologist, engineer, planner, and others. Given the watershed area (A) and a hydrograph, which is a triangular shape, guidance for the selection of K is given in Table 6.6. Equations 6.20 and 6.21 can be used to calculate the falling limb time and peak discharge.

$$t_f = xt_p \qquad (6.20)$$

and
$$Q_p = KCiA \qquad (6.21)$$

where
K = peak attenuation factor
A = watershed, square miles or acres
i = average intensity (in./hr)
x = falling limb factor, dimensionless
x = $(1291/K) - 1$ (with A in square miles)
= $(2/K) - 1$ (with A in acres)
t_f = falling limb time (same units as t_p)

Again, the user of the factors in Table 6.6 should attempt with field data to justify the choice with the field-derived hydrographs for similar type areas. Table 6.6 assumes a hydrograph shape is available. However, for a nongaged area, time of concentration and composite curve number can be estimated. From the curve number (see Chapter 5), an initial abstraction can be calculated ($I_a = 0.2S'$: $S' = [1000/CN] - 10$) and using Figure 6.20 with a total rainfall (P), the peak attenuation factor is obtained.

Note: The peak discharge is calculated by using $Q_p = KAR$

where
Q_p = peak discharge (cfs)
R = rainfall excess (in.)
A = watershed area (mi^2)

The peak rate can be reduced if the watershed has lakes and swamps through which it flows (USDA-SCS, 1986). For a lake area of 1%, the peak discharge can be reduced by the factor 0.87 and for 3%, the factor is 0.75.

TABLE 6.6 Triangular Shaped Hydrograph Attenuation Factors

| | FALLING | PEAK ATTENUATION FACTOR $(K)^a$ | |
GENERAL DESCRIPTION	LIMB FACTOR (x)	A = SQUARE MILES	A = ACRES
Rational formula	1	645	1.00
Urban, steep slopes	1.25	575	0.89
Typical SCS	1.67	484	0.75
Mixed urban/rural	2.25	400	0.62
Rural, rolling hills	3.33	300	0.46
Rural, slight slopes	5.5	200	0.31
Rural, very flat	12.0	100	0.16

[a] Includes 1.008 conversion factor.

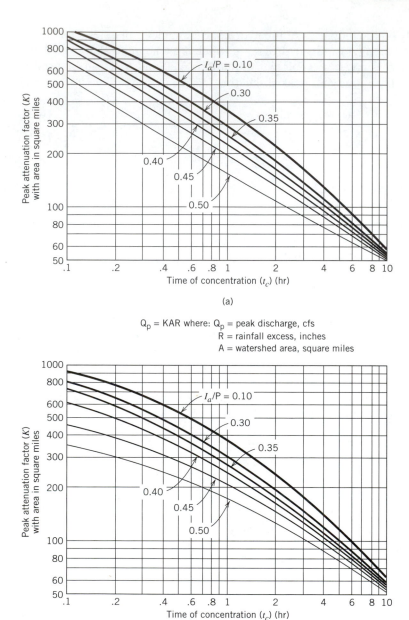

$Q_p = KAR$ where: Q_p = peak discharge, cfs
R = rainfall excess, inches
A = watershed area, square miles

FIGURE 6.20 The peak discharge factor for the updated SCS method. (*a*) For SCS type II rainfall distribution. (*b*) For SCS type III rainfall distribution. (Reproduced from U.S. Department of Agriculture-SCS, 1986.)

6.4.3 SCS (NRCS) Unit Hydrograph

The rational formula hydrograph shape, the SCS hydrograph shape with any peak attenuation factor, or any shape obtained from streamflow records can be used to develop a unit hydrograph. Once the unit hydrograph is specified, rainfall excess increments can be applied for each duration and a composite hydrograph obtained.

In this chapter, the unit hydrograph was defined and developed from streamflow data. It was shown that the unit hydrograph for a watershed is the runoff hydrograph resulting from a unit (1 in.) of rainfall excess from the watershed area during a specified rainfall period of time. The specified period of time is an interval that is brief enough so that natural fluctuations of rainfall intensity during that interval will not materially affect the shape of that hydrograph. Also, one can develop a synthetic unit hydrograph given watershed, rainfall excess, and hydrograph shape information, but no stream-flow data.

The derivation and application of the unit hydrograph are based on the following assumptions:

1. For a particular point in a watershed, identical rainfalls with same antecedent conditions produce identical hydrographs.
2. For a particular point in a watershed, the time base of all hydrographs from rainfalls of the same duration with the same antecedent conditions are equal. A graphic example is shown in Figure 6.21a.
3. For a particular point in a drainage basin, the ordinates of all hydrographs from rainfalls of the same duration with the same antecedent conditions are proportional to the volume of the rainfall excess. Therefore, if the ordinates of each hydrograph are divided by the volume of rainfall excess that produced it, the resulting unit graphs will be identical in shape.
4. The ordinates of several partial hydrographs obtained by multiplying the unit hydrograph by successive rainfall excess amounts of unit duration may be added to obtain the total hydrograph of runoff. A graphic example is shown in Figue 6.21b.

The basic premise of the unit hydrograph, therefore, is that individual hydrographs resulting from the successive increments of rainfall excess that occur throughout a storm period will be proportional in discharge throughout their length, and that when properly arranged with respect to time, the ordinates of the individual hydrographs

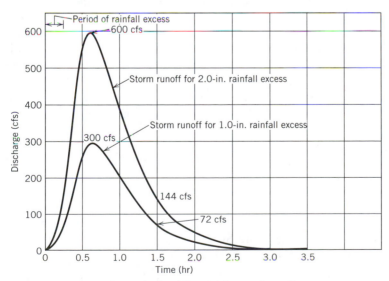

FIGURE 6.21a Hydrographs for different volumes of runoff.

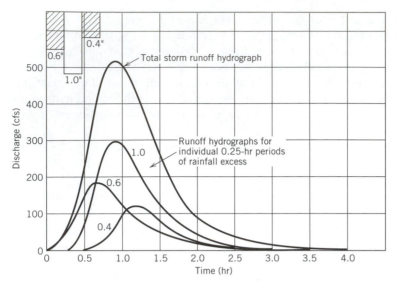

FIGURE 6.21*b* Combining hydrographs.

can be added to give ordinates representing the total storm discharge. Constructing a hydrograph of storm discharge resulting from the 3 to 15-min increments of rainfall excess as shown in Figure 6.21*b* will illustrate calculation procedures. First, the individual hydrogaphs resulting from each 15-min increment of rainfall excess are computed by multiplying the ordinates of the 15-min unit hydrograph shown in Figure 6.21*a* by the increments of rainfall excess in inches, each individual hydrogaph beginning at the same time as its respective rainfall excess. The hydrograph of total storm discharge is then obtained by summing the ordinates of the individual hydrographs as is shown in Figure 6.21*b*. To express total discharge, the ordinates of the storm discharge hydrograph must be increased by the amount of the estimated groundwater discharge.

The principles as mentioned above are not rigorously true for all channels. Channel storage varies with the stage, so that the unit hydrographs of large flows will differ from those of small flows. However, it has been found by experience that the unit hydrograph method gives results sufficiently accurate for most practical problems if reasonable judgment is used in its application. A procedure commonly in use is the synthetic unit hydrograph procedure developed by the SCS of the U.S. Department of Agriculture (USDA-SCS, 1975). In this procedure, the peak discharge, q_p, of the synthetic unit hydrograph is computed by the following equation:

$$q_p = \frac{(484AR)}{t_p} \qquad (6.22)$$

where
q_p = the peak discharge (ft^3/sec)
A = the area of the drainage basin (mi^2)
R = rainfall excess (in.) (1.0 in for unit hydrogaphs)
t_p = the time to peak (hr)
484 = peak attenuation factor, K

In the SCS procedure, the time to peak, t_p, in hours is computed by the following equation:

$$t_p = (D/2) + L \qquad (6.23)$$

where
D = the duration of the rainfall excess
L = the lag time (hr) (time between the centroid of the rainfall excess and the peak of the unit hydrograph)

These are the same terms as previously defined and shown on Figure 6.1.

For the purposes of computing t_p, the lag time, L, is assumed equal to $0.6t_c$, where t_c is the time of concentration of the drainage basin in hours. Substituting into Equation 6.23

$$t_p = (D/2) + 0.6t_c \qquad (6.24)$$

Substituting for peak flow (Equation 6.22), the peak of the SCS unit hydrograph can be calculated using the following equation:

$$q_p = \frac{484AR}{(D/2) + 0.6t_c} \qquad (6.25)$$

In this procedure, the time of concentration of the drainage basin is computed using the conventional procedures of computing and then summarizing overland and channel flow time.

Dimensionless hydrograph ratios developed by the SCS from a study of many storms for use with Equation 6.25 are shown in Table 6.7. These dimensionless ratios for both a curvilinear and a triangular unit hydrograph give the relationship between q_p (the peak discharge) and any other Q and t_p and any other time, t. Actually, for plotting the triangular unit hydrograph only three points are necessary.

In Figure 6.18, Equation 6.26 was derived with $t_p = (D/2) + L$.

$$Q_p = \frac{0.75CiAD}{(D/2) + L} \qquad (6.26)$$

If we multiply by 640 to allow the drainage area A in acres to be given in square miles and multiply again by 1.008, with $CiAD = AR$ and $L = 0.6t_c$, Equation 6.26 is converted to

$$Q_p = \frac{484AR}{(D/2) + 0.6t_c} \qquad (6.27)$$

Equation 6.27 can be recognized as the equation for peak flow of the NRCS (SCS) synthetic unit hydrograph given as Equation 6.26. Therefore, in Figure 6.18 the Soil Conservation Service's synthetic unit hydrograph equation has been derived.

If rainfall excess can be considered to be estimated by approximately the same rate (e.g., 0.1 in./hr, 2 mm/hr) over the duration of a storm event, then Equation 6.25

TABLE 6.7   Ratios for Dimensionless Hydrographs for $K = 484$

TIME	CURVILINEAR HYDROGRAPH		TRIANGULAR HYDROGRAPH[a]	
$(t/t_p)^b$	DISCHARGE $(q/q_p)^b$	MASS (Q_a/Q)	DISCHARGE[b] (q/q_p)	MASS (Q_a/Q)
0	0	0	0	0
0.1	0.015	0.001	0.1	0.004
0.2	0.075	0.006	0.2	0.015
0.3	0.16	0.018	0.3	0.034
0.4	0.28	0.037	0.4	0.060
0.5	0.43	0.068	0.5	0.094
0.6	0.60	0.110	0.6	0.135
0.7	0.77	0.163	0.7	0.184
0.8	0.89	0.223	0.8	0.240
1.0	1.00	0.375	1.0	0.375
1.1	0.98	0.450	0.94	0.448
1.2	0.92	0.517	0.88	0.516
1.3	0.84	0.577	0.82	0.579
1.4	0.75	0.634	0.76	0.639
1.5	0.65	0.683	0.70	0.694
1.6	0.57	0.727	0.64	0.744
1.8	0.43	0.796	0.52	0.831
2.0	0.32	0.848	0.40	0.900
2.2	0.24	0.888	0.28	0.951
2.4	0.18	0.916	0.16	0.984
2.6	0.13	0.938	0.04	0.999
2.8	0.098	0.954	0[c]	1[c]
3.5	0.036	0.984		
4.0	0.018	0.993		
4.5	0.009	0.997		
5.0	0.004	0.999		
Infinity	0	1		

Reproduced from USDA-SCS, 1986.
[a]To draw the triangular hydrograph, points only at $(t/t_p) = 0$, 1 and 2.67 are needed.
[b]Given q_p and t_p.
[c]At $(t/t_p) = 2.67$.

can be used to estimate the peak for the whole storm and the dimensionless hydrograph shapes of Table 6.7 can be used. It is important to note, however, that the attenuation factor (K, Table 6.6) for the dimensionless hydrographs of Table 6.7 is 484. Dimensionless hydrographs can be calculated for other peak attenuation factors. This has been done for any attenuation factor and has been incorporated into the stormwater management computer program, Stormwater Management and Design Aid (SMADA). In this computer program, the user is asked to specify any factor and unit hydrograph method.

If the rainfall excess were to vary with each rainfall increment throughout the storm event (typical design hyetograph), then the unit hydrograph shape can be applied to each rainfall excess increment and distributed with time. This is the more common application of the unit hydrograph theory. The method can assimilate different intensity—duration rainfall characteristics. This usually results in a more accurate simulation of runoff.

An example problem to illustrate the very basic calculations of the unit hydrograph procedure using the typical SCS hydrograph shape ($K = 484$) is developed for one rainfall time increment.

❑ **EXAMPLE PROBLEM 6.4**

For an actual drainage basin with data shown in Table 6.8, compute a unit hydrograph using the SCS procedure.

Solution

The first parameter to determine is the duration of the unit hydrograph which, to produce reasonable peak flow estimates, should be in the range of one-fifth to one-eighth of the time of concentration. However, a longer time, up to one-half of the lag time, can be used for planning or preliminary calculations. The lag time for a time of concentration of 55 min (0.92 hr) is

$$L = 0.6t_c = 0.6 \times 0.92 = 0.55 \text{ hr}$$

Assuming a duration of one-half the lag time, we compute the duration of our unit hydrograph as follows:

$$D = 0.50 \times 0.55 = 0.28 \text{ hr}$$

Round off to $D = 0.25$ hr for computation convenience.

TABLE 6.8 OCP-Phase Drainage Basin Physical and Topographical Characteristics

Drainage area, ac	270.6
Percent impervious (land area)	75
Percent pervious (land area)	25
Pervious area Makeup, %	
Good grass cover	12.5
Poor grass cover	12.5
Time of concentration, min	55
Hydrologic soil group (SCS classification)	C

From Golding, 1974.

The time to peak, t_p, and the peak discharge, q_p, of the 15-min (0.25 hr) unit hydrograph are then computed as follows by substituting in Equations 6.22 and 6.23. Assume that $t_c = 0.92$ hr,

$$t_p = (D/2) + 0.6t_c = (0.25/2) + 0.6(0.92) = 0.68 \text{ hr}$$

$$q_p = \frac{484AR}{t_p} = \frac{(484)(0.42)(1)}{0.68} = 298.94 \text{ ft}^3/\text{sec}$$

As the drainage area is actually slightly greater than 0.42 mi^2, use $q_p = 300$ ft^3/sec.

The curvilinear unit hydrograph for 15-min intervals is computed by multiplying t_p and q_p by the dimensionless hydrograph ratios given in the first two columns of Table 6.7. This computation of the 15-min curvilinear unit hydrograph is shown in Table 6.9 and has been plotted on Figure 6.22. If we want to draw a triangular 15-min unit hydrograph to compute a design hydrograph, only the basic information must be known, the peak q_p ($= 300.17$ ft^3/sec) occurs at a time t_p ($= 0.68$ hr) from the start of the hydrograph, and the base of the triangular hydrograph is equal to $2.67t_p$ ($= 1.82$ hr). The 15-min triangular unit hydrograph is also shown in Figure 6.22. Frequently, the "tail" of the hydrograph is not important, thus the triangular shape gives adequate results for the peak estimate and flows immediately adjacent to the peak.

To use the 15-min curvilinear unit hydrograph, or for that matter the triangular shape, determine the discharge, q, at time equal to 0.25-hr intervals (clock hour readings), which can be done by scaling the ordinates of the plotted 15-min unit hydrograph on Figure 6.22. The discharges at clock intervals of 0.25 hr for the 15-min curvilinear unit hydrograph have been scaled and area also given in Table 6.9. ❏

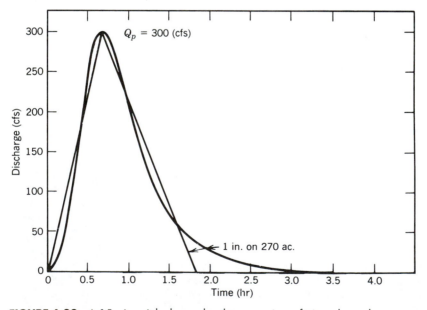

FIGURE 6.22 A 15-min unit hydrograph—the comparison of triangular and curvilinear dimensionless hydrographs.

TABLE 6.9 Calculation of Unit Hydrograph SCS Procedure
($t_p = 0.68$, $q_p = 300$)

COMPUTATION OF UNIT HYDROGRAPH				CLOCK HOUR READINGS[a]	
TIME RATIO (t/t_p)	DISCHARGE RATIO (q/q_p)	TIME (T) (hr)	DISCHARGE (q) (ft^3/sec)	TIME (hr)	DISCHARGE (ft^3/sec)
0	0	0	0	0	0
0.1	0.015	0.07	5	0.25	73
0.2	0.075	0.14	23	0.50	246
0.3	0.16	0.20	48	0.75	293
0.4	0.28	0.27	84	1.00	202
0.5	0.43	0.34	129	1.25	121
0.6	0.60	0.41	180	1.50	72
0.7	0.77	0.48	231	1.75	41
0.8	0.89	0.54	267	2.00	26
0.9	0.97	0.61	291	2.25	15
1.0	1.00	0.68	300	2.50	9
1.1	0.98	0.75	294	2.75	6
1.2	0.92	0.82	276	3.00	3
1.3	0.84	0.88	252	3.25	1
1.4	0.75	0.95	225	3.50	0
1.5	0.65	1.02	195		
1.6	0.57	1.09	171		
1.8	0.43	1.22	129		
2.0	0.32	1.36	96		
2.2	0.24	1.50	72		
2.4	0.18	1.63	54		
2.6	0.13	1.77	39		
2.8	0.098	1.90	29		
3.0	0.075	2.04	23		
3.5	0.036	2.38	11		
4.0	0.018	2.72	5		
4.5	0.009	3.06	3		
5.0	0.004	3.40	1		

[a] Time from start of rainfall, rainfall excess assumed to start immediately after start of rainfall.

6.4.4 Clark Unit Hydrograph

The Clark Unit Hydrograph procedure is a two-step procedure for the development of a unit hydrograph. The first step of the procedure is the development of a time area (TA) curve based on watershed characteristics. This curve is then routed through a linear reservoir to produce the final unit hydrograph.

The TA curve relates time to the fraction of the total watershed area, which contributes to runoff. A *TA* curve can be developed by determining this contribution for time intervals between 0 and the total time of concentration of the watershed. These time intervals are drawn onto a plan drawing of the watershed, and the total contributing area of each of these isochrones is determined. A set of isochrones drawn on a watershed is shown in Figure 6.23. A number of methods for determination of this *TA* curve have been developed. HEC-1 and SMADA assume a generic shape to the watershed. The total time of concentration of the watershed is broken into a number of intervals and for each of these intervals the ratio of this time to the total time of concentration is calculated.

$$T_i = t_i/t_c \qquad (6.28)$$

where

T_i = ratio of time to total time of concentration
t_c = total watershed time of concentration
t_i = time step in question

The cumulative TA curve may be developed from equations (6.29) and (6.30) (Hydrologic Engineering Center, 1987), if the true time area curve is not available,

$$TA_i = 1.414T_i^{1.5} \qquad (0 <= T_i < 0.5) \qquad (6.29)$$

$$1 - TA_i = 1.414(1 - T_i)^{1.5} \qquad (0.5 < T_i < 1.0) \qquad (6.30)$$

where

TA_i = cumulative value of time area curve
T_i = ratio of time to total time of concentration

Once the *TA* curve is developed, the Clark unit hydrograph is generated by routing this *TA* curve through a linear reservoir with a routing parameter R. The routing parameter R is used to calculate a routing coefficient c using the following equation:

$$c = \frac{2\Delta t}{2R + \Delta t} \qquad (6.31)$$

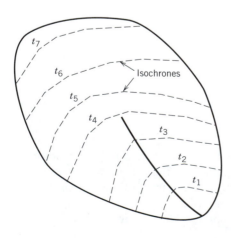

FIGURE 6.23   Example of a watershed with isochrones drawn.

where

c = linear routing coefficient (unitless)
R = Clark storage coefficient (time units)
Δt = time step of analysis (time units)

The Clark storage coefficient may be evaluated from information at the point of inflection on the recession limb of a measured hydrograph:

$$R = -\frac{Q}{dQ/dt} \tag{6.32}$$

or estimated as the travel time from basin divide to basin outlet. Hall (1984) has identified seven definitions in the literature. The routing coefficient and the TA curve are used to find the instantaneous unit hydrograph (IUH) using the linear routing equation;

$$IUH_i = c\overline{TA}_i + (1 - c)IUH_{i-1} \tag{6.33}$$

where

IUH_i = increment of the instantaneous unit hydrograph
c = linear routing coefficient
$\overline{TA}_i$ = average time area ordinate at step i. $\overline{TA}_i = 0.5(TA_i + TA_{i-1})$

The final unit hydrograph can then be generated by averaging two instantaneous unit hydrographs which are the time step Δt apart.

$$UH_i = 0.5(IUH_i + IUH_{i-1}) \tag{6.34}$$

where

IUH_i = increment of the instantaneous unit hydrograph
UH_i = increment of the unit hydrograph

❑ EXAMPLE PROBLEM 6.5

A 15-mi^2 watershed in the western part of the United States has a time of concentration of 6 hr. If the Clark storage coefficient is estimated to be 8 hr, calculate the unit hydrograph.

Solution

The first step is to develop a TA curve for the watershed. Equations 6.29 and 6.30 will be used to calculate the TA curve. A time step of 1 hr will also be used (see Table 6.10).

The TA curve is then routed through a linear reservoir using a routing coefficient calculated using Equation 6.31

$$c = \frac{2\Delta t}{2R + \Delta t} = \frac{2(1)}{2(8) + 1} = 0.1176$$

TABLE 6.10   Development of a TA Curve for Example Problem 6.5

TIME	TIME/t_c	CUMULATIVE TA	INCREMENTAL TA
0.0	0.000	0.000	0.000*
1.0	0.143	0.076	0.076*
2.0	0.286	0.216	0.140*
3.0	0.429	0.397	0.181*
4.0	0.571	0.603	0.207**
5.0	0.714	0.784	0.181**
6.0	0.857	0.924	0.140**
7.0	1.000	1.000	0.076**

*Equation 6.29 used
**Equation 6.30 used

After routing the *TA* curve through the linear reservoir, the instantaneous unit hydrograph shown in Figure 6.24 is developed. The final unit hydrograph is developed by averaging the instantaneous unit hydrograph with itself offset by one time step. The final results are shown in Table 6.11.   ❏

6.4.5   Santa Barbara Urban Hydrograph

The Santa Barbara Urban Hydrograph method (SBUH method) was developed by Mr. James M. Stubchaer of the Santa Barbara County (California) Flood Control and Water Conservation District and was first presented at the National Symposium on Urban Hydrology and Sediment Control held at the University of Kentucky (1975). One is immediately impressed by the SBUH method's ease of application as it computes a hydrograph directly without going through an intermediate process as the unit hydrograph method does. A comparison with other methods currently in use indicates the SBUH method's apparent accuracy and ease of application.

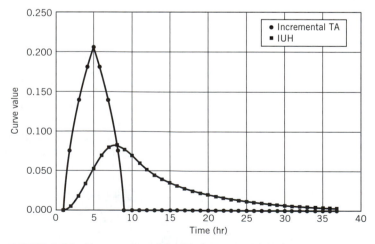

FIGURE 6.24   Linear routing of the *TA* Curve to produce the *IUH*.

TABLE 6.11 Final Results of the Generation of a Unit Hydrograph by the Clark Method for Example Problem 6.5

TIME STEP	INCREMENTAL TA	IUH	OFFSET IUH	FINAL UH
1	0.000	0.0000	0.0045	0.0022
2	0.076	0.0045	0.0167	0.0106
3	0.140	0.0167	0.0335	0.0251
4	0.181	0.0335	0.0524	0.0430
5	0.207	0.0524	0.0690	0.0607
6	0.181	0.0690	0.0797	0.0743
7	0.140	0.0797	0.0830	0.0814
8	0.076	0.0830	0.0778	0.0804
9	0.000	0.0778	0.0686	0.0732
10	0.000	0.0686	0.0605	0.0646
11	0.000	0.0605	0.0534	0.0570
12	0.000	0.0534	0.0471	0.0503
13	0.000	0.0471	0.0416	0.0444
14	0.000	0.0416	0.0367	0.0392
15	0.000	0.0367	0.0324	0.0345
16	0.000	0.0324	0.0286	0.0305
17	0.000	0.0286	0.0252	0.0269
18	0.000	0.0252	0.0223	0.0237
19	0.000	0.0223	0.0196	0.0209
20	0.000	0.0196	0.0173	0.0185
21	0.000	0.0173	0.0153	0.0163
22	0.000	0.0153	0.0135	0.0144
23	0.000	0.0135	0.0119	0.0127
24	0.000	0.0119	0.0105	0.0112
25	0.000	0.0105	0.0093	0.0099
26	0.000	0.0093	0.0082	0.0087
27	0.000	0.0082	0.0072	0.0077
28	0.000	0.0072	0.0064	0.0068
29	0.000	0.0064	0.0056	0.0060
30	0.000	0.0056	0.0050	0.0053
31	0.000	0.0050	0.0044	0.0047
32	0.000	0.0044	0.0039	0.0041
33	0.000	0.0039	0.0034	0.0036
34	0.000	0.0034	0.0030	0.0032
35	0.000	0.0030	0.0027	0.0028
36	0.000	0.0027	0.0023	0.0025
37	0.000	0.0023		

Although the SBUH method was originally programmed for a desk-size computer in the BASIC computer language, the computations can be done manually as will be demonstrated here. It is also easily handled by a programmable pocket-type calculator. With the SBUH method, a unit hydrograph shape can be specified by choice of the calculation time interval (Δt) and a time of concentration.

The SBUH method is similar to the contributing area procedure for hydrograph computation in which subwatershed hydrographs in a watershed are developed and then routed to determine an outflow hydrograph from the watershed. However, in the SBUH method, the final design (outflow) hydrograph is obtained by routing the instantaneous hydrograph for each time period (obtained by multiplying the various incremental rainfall excesses by the watershed area in acres) through an imaginary linear reservoir with a routing constant dependent on the time of concentration of the watershed. Therefore, the difficult and time-consuming process of preparing an effective area time of concentration curve for the watershed is eliminated.

A step-by-step description of the SBUH method is given as follows:

1. Runoff depths for each time period are calculated using the following equations.

   Impervious area runoff, $R(I) = d'P(\Delta t)$(depth)   (6.35)

   Pervious area runoff, $R(P) = (1 - d')[P(\Delta t) - F(\Delta t)]$(depth)   (6.36)

   Total runoff depth, $R(\Delta t) = R(I) + R(P)$(depth)   (6.37)

   where
   $P(\Delta t)$ = rainfall depth during time increment Δt (depth)
   $F(\Delta t)$ = infiltration during time increment Δt (depth)
      d' = directly connected impervious portion of drainage basin (fraction)
      Δt = incremental time period (hr) (i.e., 0.25, 0.50, etc.)

2. The instantaneous hydrograph is then computed by multiplying the total runoff depth, $R(\Delta t)$ for each time period, Δt, by the watershed area, A, and dividing by the time increment, Δt, and converted to m³/s or ft³/s.

$$I(\Delta t) = \frac{R(\Delta t)A}{\Delta t} \qquad \text{(m}^3\text{/s or ft}^3\text{/s)} \tag{6.38}$$

   (Note that as in the Rational method, the conversion factor 1.008 is dropped when using inches and acres and the total watershed area is used rather than a contributing area at each time step.)

3. The final design (outflow) hydrograph, $Q(\Delta t)$, is then obtained by routing the instantaneous hydrograph, $I(\Delta t)$, through an imaginary reservoir with a time delay calculated using the time of concentration, t_c, of the watershed. This flood routing is done by use of the following equation to estimate routed flow Q.

$$Q(2) = Q(1) + K_r[I(1) + I(2) - 2Q(1)] \tag{6.39}$$

   where
$$K_r = \frac{\Delta t}{(2t_c + \Delta t)} \tag{6.40}$$
      I = inflow to imaginary reservoir
      t_c = time of concentration

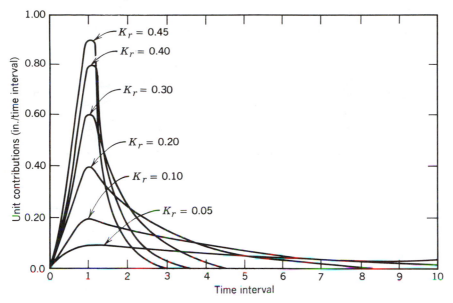

FIGURE 6.25   Normalized Santa Barbara unit hydrographs. (From Walsh and Wanielista, 1982.)

The shape of the discharge hydrograph is determined by the value of K_r (the routing constant). In Figure 6.25, the shapes of unit hydrograph are shown for various K_r values.

In the SBUH method, the impervious portion of the watershed is considered to be the directly (hydraulically) connected portion of the impervious area as previously discussed. All of the rain that falls on this impervious portion of the basin is considered rainfall excess. The equation for impervious area runoff can be modified for depression storage or evaporation if applicable by the following equation:

$$R(I) = d'[P(\Delta t) - D(\Delta t)]EVP \qquad (6.41)$$

where

$R(I)$ = impervious area runoff (depth)
   d' = impervious portion of the drainage basin (fraction)
$P(\Delta t)$ = rainfall depth during time increment (Δt) (depth)
$D(\Delta t)$ = depression storage during time increment (Δt) (depth)
EVP = portion of rainfall excess which is evaporated before runoff (fraction)

❏ EXAMPLE PROBLEM 6.6

To illustrate the SBUH method, and for purposes of comparison with other hydrograph methods, a design hydrograph by the SBUH method is required. A computer program (SMADA) included with this book is used for the 270.6-ac watershed basin as shown in Table 6.12. The results are also shown in Table 6.12 following exactly the step-by-step computation procedure for the SBUH method listed previously.

The rainfall increments given in column 3 are taken directly from the 25-yr frequency, 6-hr rainfall hyetograph. Because the pervious and impervious areas are considered separately in the SBUH method, it is easier to specify a curve number

TABLE 6.12   Design Hydrograph

Given: Area = 270.6 ac   Compute: 1. Routing Constant (K_r)

$d' = 0.75$

$\Delta t = 15$ min $= 0.25$ hr

$t_c = 55$ min $= 0.92$ hr

$$K_r = \frac{\Delta t}{2t_c + \Delta t} = \frac{0.25}{2(0.92) + 0.25} = 0.12$$

$CN = 54$ (pervious area)

2. Design hydrograph (via computer output)

Note: Rainfall excess for pervious area calculated assuming initial abstraction saturated, thus, $R = P^2/(P + S')$

TIME INCREMENT	TIME (min)	RAINFALL DEPTH (in.)	INFILTRATION % INIT ABST (in.)	RUNOFF DEPTH (in.)	INSTANT HYDROGRAPH (cfs)	WATERSHED HYDROGRAPH (cfs)
1	15	0.10	0.025	0.075	81.49	9.78
2	30	0.11	0.027	0.083	90.35	28.05
3	45	0.12	0.028	0.092	99.38	44.09
4	60	0.15	0.034	0.116	125.37	60.48
5	75	0.16	0.035	0.125	135.06	77.21
6	90	0.17	0.036	0.134	144.94	92.28
7	105	0.27	0.055	0.215	233.04	115.49
8	120	0.30	0.057	0.243	262.72	147.26
9	135	1.08	0.180	0.900	973.84	260.31
10	150	1.14	0.155	0.985	1065.68	442.58
11	165	0.32	0.039	0.281	304.68	500.80
12	180	0.28	0.032	0.248	268.32	449.37
13	195	0.24	0.026	0.214	231.18	401.46
14	210	0.24	0.025	0.215	232.22	360.72
15	225	0.18	0.018	0.162	174.82	322.99
16	240	0.16	0.016	0.144	155.84	285.15
17	255	0.14	0.014	0.126	136.69	251.82
18	270	0.13	0.012	0.118	127.20	223.05
19	295	0.13	0.012	0.118	127.45	200.07
20	300	0.12	0.011	0.109	117.86	181.49
21	315	0.12	0.011	0.109	118.07	166.25
22	330	0.12	0.011	0.109	118.27	154.71
23	345	0.11	0.010	0.100	108.58	144.80
24	360	0.11	0.010	0.100	108.74	136.13
25	375	0.00	0.000	0.000	0.00	116.50
26	390	0.00	0.000	0.000	0.00	88.54
27	405	0.00	0.000	0.000	0.00	67.29
28	420	0.00	0.000	0.000	0.00	51.14
29	435	0.00	0.000	0.000	0.00	38.87

(*CN*) for the pervious area and estimate infiltration increments for the pervious area only. Using the methods previously outlined, a *CN* of 54 is used to compute the infiltration for the pervious area.

The design hydrograph as computed by the SBUH method is plotted on Figure 7.22. Note that the runoff hydrograph shape is different from that obtained by the rational formula (peak at 6 hr and equal to $(0.8)(1)(270) = 216$ cfs with $c = 0.8$, $i = 1$ in./hr. and area = 270 ac). ❑

The basic equations for the SBUH Method are simple enough for computerization on even the smallest programmable pocket-type calculator, which is a big advantage of the SBUH method. Another advantage of the SBUH method is that it does not have the tendency to overcompute the peak of the hydrograph. However, as in the unit hydrograph method, it does require the promulgation of a design rainfall, the determination of rainfall excess, and a hydrograph shape.

6.4.6 Continuous Convolution Method

Until this discussion, unit hydrograph generation procedures depended on using the same hydrograph shape and constant rainfall excess for equal time intervals. Now, using basic mass balances to develop a continuous functional form, rainfall excess and routing can be considered variable with time. In the analysis of streamflow data, three different situations can exist: (1) the streamflow hydrograph shapes do not change drastically with changing flows, thus the watershed acts as a linear time invariant system; (2) the streamflow hydrograph shapes change with time during a rainfall event; and (3) the past input elements of rainfall excess may affect the present hydrograph shape such that the sum of separate hydrographs do not directly add to form the total. The watershed then acts as a nonlinear time variant system. For situation 1, synthetic hydrograph procedures are appropriate and are based on the concept of the unit hydrograph from this chapter. Situation 2 is solved by this convolution procedure which allows changing rainfall excess and routing functions over a continuous time interval. Situation 3 is not presented in this text.

In the historical development of unit hydrograph procedures, it was assumed that the watershed produces a series of rainfall excesses that can be multiplied by the same hydrograph shape and added together to form the final hydrograph. As long as the watershed continues to produce the same hydrograph shape, the unit hydrograph procedure is reasonable and acceptable. In watersheds that experience changing groundwater and surface storage conditions and whose drainage channels are such that velocities change drastically with flow rates, a given hydrograph shape will be valid only over a range of rainfall excess values.

For example, consider a watershed with highly permeable soils, low water tables, and very little relief resulting in channels with low velocity gradients (similar to coastal regions). When rainfall occurs, infiltration will recharge the superficial water table causing it to rise. As rain continues, the ground will become saturated and runoff occurs at a higher intensity. At the beginning of rainfall, the hydrograph shape is relatively flat because it is primarily composed of groundwaters infiltrating into the drainage system. As the groundwater table rises toward the ground surface, the hydrograph shape will become steeper or flow rates will change faster with time. The change in flow rates appear gradually. Once the infiltrated waters enter a channel, velocities may change. Thus, the watershed storage conditions and drainage geometry can produce different hydrograph shapes.

Recall from unit hydrograph theory that the product of the unit hydrograph and rainfall excess is the estimate of the watershed hydrograph. Given any hydrograph shape and a rainfall excess, the product of the two results in a hydrograph for the given rainfall excess (watershed) and channel response (velocity of flow). Snyder et al. (1970) labeled the channel response or routing function as a state function because it describes the current translation state of the watershed drainage system. The rainfall excess is characteristic of the watershed soil and initial abstraction conditions, thus it is labeled as a characteristic function. A mathematical procedure which is basically the integration of the product of two functions is called convolution. The general form of the convolution integral is

$$Q_t = \int_0^t r(\tau)g(t-\tau)\,d\tau \tag{6.42}$$

where

$\quad Q_t$ = hydrograph flow rate (ft^3/sec)
$\quad r(\tau)$ = rainfall excess rate or characteristic function as a function of a time parameter (τ) for integration (ft^3/sec)
$g(t-\tau)$ = hydrograph or routing function offset in real time by τ

Convolution integrals can be solved by either discrete processes or by functional forms. Both methods are available for computer programming solution. The discrete forms allows one to specify any shape hydrograph without the necessity of a mathematical equation form, but fixed time intervals are used. The function forms however have the advantage of more efficient computer programming, variable time intervals, and numerical stability.

6.4.6A   Convolution with Constant Rainfall Excess

A constant characteristic function represents a constant rate of rainfall excess. This is reasonable for short time periods and permits a detailed explanation of the convolution method including solution details. Rainfall excess that is constant over a time period can result from an impervious area with constant intensity rainfall. Thus, the characteristic function is a constant denoted by the parameter a. The state function will be assumed to be exponential because the exponential function can be used to estimate the shape of stream hydrographs. The resulting functional form derived from a mass balance (see Appendix G) results in the convolution form:

$$Q_t = \int_0^t a(ke^{-k(t-\tau)})\,d\tau \tag{6.43}$$

where

a = constant rainfall excess, flow rate units (ft^3/sec)
k = routing coefficient (1/t)

The value of the routing coefficient (k) will determine the particular form of the state (hydrograph routing) function and must be determined. Various forms are illustrated in Figure 6.26. The routing function reflects the drainage characteristics such as geometry,

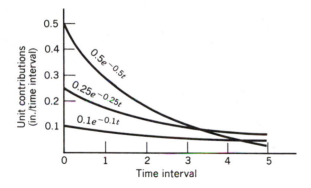

FIGURE 6.26 A routing function form.

slope, channel transections, etc. The routing function with high recession ($0.5e^{-0.5t}$) reflects a rapidly responsive drainage watershed. A sluggish drainage watershed with an attenuated peak can be represented by a routing function with a low recession initially ($0.1e^{-0.1t}$). The choice of the routing function (k value) and the characteristic function (a) determine the shape of the hydrograph.

Assuming end of rainfall occurs at time (D), the integration of Equation 6.43 produces

$$Q_t = a(1 - e^{-kt}) \qquad 0 \le t \le D \tag{6.44}$$

and

$$Q_t = ae^{-kt}(e^{kD} - 1) \qquad t > D \tag{6.45}$$

There are two unknowns, a and k. However, from rainfall and streamflow data these unknowns can be determined. The peak is defined when time equals storm duration:

$$Q_p = a(1 - e^{-kD}) \tag{6.46}$$

Also, the area under the hydrograph must reflect the rainfall excess over the watershed, or

$$R = \int_0^D a(1 - e^{-kt})\, dt + \int_D^t ae^{-kt}(e^{kD} - 1)\, dt \tag{6.47}$$

with time in units of hours and k as per hour values, Equation 6.47 integrates to Equation 6.48 in units of cubic feet.

$$R = (60 \text{ s/m} \times 60 \text{ m/h})[aD + (a/k)e^{-kt}(1 - e^{kD})] \tag{6.48}$$

and if $t = \infty$, Equation 6.48 reduces to

$$R = 3600(aD) \tag{6.49}$$

where
a = peaking parameter (ft^3/sec)
D = duration of rainfall excess (hr)

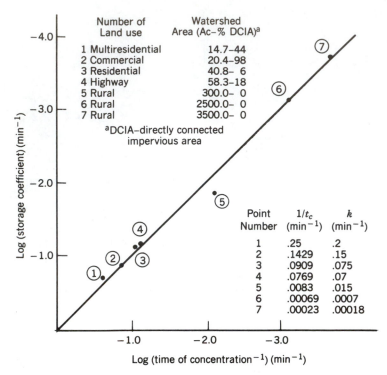

FIGURE 6.27 The relationship between the storage coefficient k and $1/$time of concentration for seven watersheds. The coefficient k was chosen as the one that best fits the runoff hydrograph whereas time of concentration was estimated from the watershed characteristics.

Note that Equation 6.49 is the same result as that shown using the rational formula assumption of an isosceles triangle. There results two equations and two unknowns: namely, a and k in Equation 6.46 and 6.48, and only one unknown in Equation 6.49.

The convolution parameters a and k generally have to be estimated from runoff or streamflow data by using a curve fit procedure. Curve fit procedures were discussed in Chapter 2. However, in many situations, streamflow data are not available, and the parameters a and k are known from other sources. Also, the routing coefficient k is the inverse of the time it takes to drain the watershed—this relationship is shown in Figure 6.27. By knowing the time of concentration, the routing coefficient can be determined.

☐ **EXAMPLE PROBLEM 6.7**

Continuing with the above assumption of a constant rainfall excess and the hydrograph routing function shape assumed as an exponential, develop a hydrograph for a 10-acre area with a runoff coefficient of 0.5, and rainfall of 4 in. over 3 hr. The time to peak is assumed at 3 hr, which can be verified by stream measures.

Solution

The assumption of any hydrograph procedure can be used, but for simplicity, assume that the runoff given by the Rational formula is used:

$$Q_p = iCA(1.008) = 0.5(4/3)(10)(1.008) = 6.72 \text{ cfs}$$

and for rainfall excess (in cubic feet):

$$R = CPA \text{ (conversion factors)}$$
$$= 0.5(4 \text{ in.})(10 \text{ ac})(1/12 \text{ in./ft})(43,560 \text{ ft}^2/\text{ac})$$
$$= 72,600 \text{ ft}^3$$

First, solve Equations 6.46 and 6.48 by trial and error. Assume that $k = 0.50$ and solve for a from Equation 6.46, assuming peak discharge is 6.72 cfs.

$$a = Q_p/(1 - e^{-kD}) = 8.65 \text{ cfs}$$

Does Equation 6.47 have a value of 72,600 ft³? Solving for R with $t = 7$ hr and $D = 3$ hr yields $R = 88,000$ ft³, thus the value is greater than 72,600 ft³. The trial-and-error solution is very time consuming but, with the aid of a computer program and a reasonable hydrograph ending time, the equations can be solved, and the comparisons are shown in Figure 6.28. The rate of increase of the rising limb can be reduced if the contributing area was made a function of time. Thus, a more reasonable shape would result. ❑

6.4.6B Convolution and Variable Rainfall Excess

A more reasonable distribution of rainfall excess is one that increases with time and decreases after the period of maximum rainfall intensity. Consider a straight line mathematical expression for the variable rainfall excess as shown in Figure 6.29. The equations for rainfall excess are

$$r(\tau) = d + b\tau \text{ for increasing excess}$$

and

$$r(\tau) = f - g\tau \text{ for decreasing excess} \tag{6.50}$$

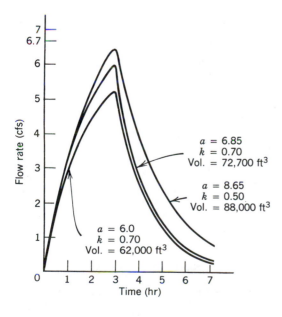

FIGURE 6.28 A hydrograph shape with rectangular rainfall excess. Note that 7 hr was considered the maximum hydrograph base time because of the small (10 ac) homogeneous area. Volumes were calculated over a 7-hr period by using Equation 6.48, and peak discharges were calculated by using Equation 6.46.

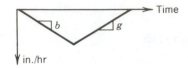

in./hr   **FIGURE 6.29**   Rainfall excess.

the convolution integral for the increasing excess is

$$Q_t = k \int_0^t (d + b\tau)e^{-k(t-\tau)} \, d\tau \tag{6.51}$$

It is important to note that this equation is defined when rainfall excess starts. Thus, the function for different rainfall excesses may lag the other in time. The parameter (τ) is used rather than real "clock" time. Integration of Equation 6.51 by separation yields

$$Q_t = (d + b\tau)(1 - e^{-k\tau}) + \frac{b}{k}[(k\tau + 1)e^{-k\tau} - 1]$$

$$0 \le \tau \le D' \qquad \text{where } D' = \text{end of excess function} \tag{6.52}$$

and

$$Q_t = (d + b\tau)(e^{kD'} - 1)e^{-k\tau} - \frac{b}{k}[(k(\tau - D') + 1)e^{-k(\tau-D')} - (k\tau + 1)e^{-k\tau}]$$

$$\text{for } \tau \ge D' \tag{6.53}$$

Note here that D' is defined as the end of the particular continuous rainfall excess function. If the rate of excess changes, a new function for rainfall excess has to be defined of length (D'). For decreasing excess, (f) is substituted for (d), and ($-g$) is substituted for b. The equations are mathematically clear, but the constant k must be determined to "fit" the physical situation or streamflow data.

The use of a triangular shaped distribution of rainfall excess produces a more realistic hydrograph shape. The hydrograph is developed from the rainfall excess expressed in flow units. Conversion from inches per hour to a flow rate establishes an instantaneous rainfall excess in flow rate units (i.e., cfs).

☐   **EXAMPLE PROBLEM 6.8**

Determine the constants in the equations for rainfall excess assuming triangular shaped rainfall excess distributions. The watershed area is 10 ac, the runoff coefficient is 0.5, the storm duration is 3 hr, the rainfall excess is maximum (peaks) at $1\frac{1}{2}$ hr, and the total rainfall volume is 4 in. Also, draw the rainfall excess diagram in cfs units.

Solution

The triangular shaped rainfall excess equations are

Increasing rainfall excess (in./hr): $d + b\tau$ $0 \le t \le \frac{3}{2}$ hr

Falling rainfall excess (in./hr): $f - g\tau$ $\frac{3}{2} < t \le 3$ hr

First, transpose time axis is to use Equation 6.50:

$$d + b\tau \qquad 0 \le \tau \le \tfrac{3}{2}\,\text{hr}$$

$$f - g\tau \qquad 0 \le \tau \le \tfrac{3}{2}\,\text{hr}$$

The rainfall excess diagram is

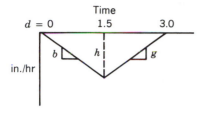

The volume of rainfall excess is = 0.5(4) = 2 in. Thus, the volume of the rainfall excess diagram is calculated by the area of a triangle; and the peak rainfall excess follows:

$$\text{Volume} = \text{in.} = \tfrac{1}{2}(b)(h) = \tfrac{1}{2}(\text{hr})(\text{in./hr})$$

$$2\,\text{in.} = \tfrac{1}{2}(3\,\text{hr})(h) \ \text{or} \ h = \tfrac{4}{3}\,\text{in./hr}$$

Determine the d and b coefficients for unit watershed area:

$$0 \le \tau \le \tfrac{3}{2} \qquad d + b\tau = 0 \qquad @ \qquad \tau = 0 \ \ d = 0$$

$$d + b\tau = \tfrac{4}{3} \qquad @ \qquad \tau = \tfrac{3}{2}$$

and
$$0 + b(\tfrac{3}{2}) = \tfrac{4}{3} \qquad \text{or} \qquad b = \tfrac{8}{9}\,\text{in./hr}^2$$

Determine the f and g coefficients for unit watershed area:

$$0 \le \tau \le \tfrac{3}{2} \qquad f - g(0) = \tfrac{4}{3} \qquad @ \qquad \tau = 0 \qquad \text{or} \qquad f = \tfrac{4}{3}$$

$$f - g(\tfrac{3}{2}) = 0 \qquad @ \qquad \tau = \tfrac{3}{2}$$

and
$$\tfrac{4}{3} - g(\tfrac{3}{2}) = 0 \qquad \text{or} \qquad g = \tfrac{8}{9}\,\text{in./hr}^2$$

Now, translate the rainfall excess in units of inches/hour to units of cfs. For area in acres: (in./hr × acres × 1.008 = cfs):

$$\text{Increasing excess} \ (0 + \tfrac{8}{9}\tau)10.08 = 0 + 9.0\tau\,(\text{cfs}) \qquad 0 \le \tau \le \tfrac{3}{2}$$

$$\text{Falling excess} \quad (\tfrac{4}{3} - \tfrac{8}{9}\tau)10.08 = 13.5 - 9.0\,\tau\,(\text{cfs}) \qquad 0 \le \tau \le \tfrac{3}{2}$$

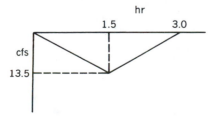

☐ **EXAMPLE PROBLEM 6.9**

Assuming a hydrograph routing factor $k = 0.7$, and the rainfall excess of Example Problem 6.8, develop a graphical representation of the hydrograph. Compare the results by using a k of 0.5 and plot.

Solution

Calculations are done by using Equations 6.52 and 6.53. The results are shown in Table 6.13 and Figure 6.30.   ☐

TABLE 6.13   Example Calculations—Example Problem 6.9

$k = 0.7$

| | | FIRST EXCESS $(0 + 9\tau)$ | | | SECOND EXCESS $(13.5 - 9\tau)$ | | |
| | | | | | | | |
t	τ	$\tau \leq 3/2$	$\tau > 3/2$	τ	$\tau \leq 3/2$	$\tau > 3/2$	TOTAL
0	0	0	0	—	0	0	0
1	1	2.53[a]	0	—	0	0	2.53
1.5	1.5	5.14	0	0	0	0	5.14
2.0	2.0	—	3.62[b]	0.5	3.28[a]	0	6.91
2.5	2.5	—	2.55	1.0	4.27	0	6.82
3.0	3.0	—	1.80	1.5	3.63	0	5.43
3.5	3.5	—	1.27	2.0	—	2.56[b]	3.83
4.0	4.0	—	0.89	2.5	—	1.80	2.69
5.0	5.0	—	0.44	3.5	—	0.90	1.34
6.0	6.0	—	0.22	4.5	—	0.45	0.67
7.0	7.0	—	0.11	5.5	—	0.22	0.33

[a]Use Equation 6.52.
[b]Use Equation 6.53.

$k = 0.5$

| | | FIRST EXCESS $(0 + 9\tau)$ | | | SECOND EXCESS $(13.5 - 9\tau)$ | | |
| | | | | | | | |
t	τ	$\tau \leq 3/2$	$\tau > 3/2$	τ	$\tau \leq 3/2$	$\tau > 3/2$	TOTAL
0	0	0	0	—	—	—	0
1.0	1.0	1.92	—	—	—	—	1.92
1.5	1.5	4.00	—	0	—	—	4.00
2.0	2.0	—	3.12	0.5	2.47	—	5.59
2.5	2.5	—	2.43	1.0	3.39	—	5.82
3.0	3.0	—	1.89	1.5	3.12	—	5.01
3.5	3.5	—	1.47	2.0	—	2.43	3.90
4.0	4.0	—	1.15	2.5	—	1.89	3.04
5.0	5.0	—	0.70	3.5	—	1.15	1.85
6.0	6.0	—	0.42	4.5	—	0.70	1.12
7.0	7.0	—	0.26	5.5	—	0.42	0.68

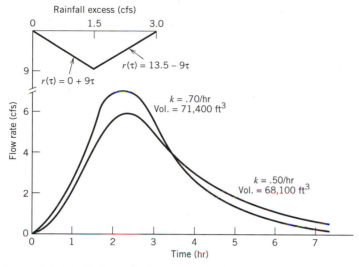

FIGURE 6.30 Hydrographs from triangular-shaped rainfall excess.

□ ___EXAMPLE PROBLEM 6.10___

The following rainfall excess is divided into five separated durations because the slope of the rainfall excess curve changes. It was developed from rainfall data and represents a typical rainfall hydrograph that has peak intensity near the middle of a storm. The rainfall excess also reflects a watershed with high initial storage.

TIME INTERVALS (hr)	RAINFALL EXCESS (cfs)	ROUTING COEFFICIENT k	COMMENTS
0–1	0		No excess
1–2	$0 + 8\tau$	0.3	Increasing
2–3	8	0.4	Constant excess
3–4	$8 - 2\tau$	0.5	Decreasing excess
4–5	$6 - 6\tau$	0.3	Decreasing excess

The rainfall excess diagram is shown in Figure 6.31:

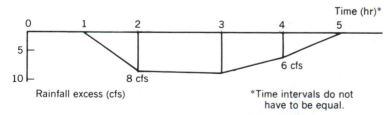

*Time intervals do not have to be equal.

FIGURE 6.31 A rainfall excess diagram—Example Problem 6.10.

TABLE 6.14 Example Problem 6.10: Hydrograph Tabulation

EXCESS: t	$0 + 8\tau$ $k = 0.3$	8τ $k = 0.4$	$8 - 2\tau$ $k = 0.5$	$6 - 6\tau$ $k = 0.3$	TOTAL (cfs)
1	0	0	0	0	0
2	1.088[a]	0	0	0	1.088
3	0.8064[b]	2.64[c]	0	0	3.446
4	0.597	1.77[d]	2.722	0	5.089
5	0.443	1.185	1.651	0.739	4.018
6	0.328	0.794	1.001	0.547	2.6754
7	0.243	0.532	0.607	0.405	1.787
8	0.170	0.357	0.368	0.300	1.205
9	0.133	0.239	0.223	0.223	0.818
10	0.099	0.160	0.135	0.165	0.559

Example Calculations: <u>using</u>

[a]$Q_2 = [0 + 8(1)](1 - e^{-0.3(1)}) + (8/0.3)[(0.3(1) + 1)e^{-0.3(1)} - 1]$ (6.52)

[b]$Q_3 = [0 + 8(2)](e^{(0.3)1} - 1)e^{-0.3(2)} - (8/0.3)[(0.3(1) + 1)e^{-0.3(1)} - (0.3(2) + 1)e^{-0.3(2)}]$ (6.53)

[c]$Q_3 = 8(1 - e^{-0.4(1)})$ (6.44)

[d]$Q_4 = 8(e^{0.4(1)} - 1)e^{-0.4(2)}$ (6.45)

Solution

Using the superposition of rainfall excess generated hydrographs and the convolution procedures for both constant and variable rainfall excess, Table 6.14 and Figure 6.32 are developed. ❑

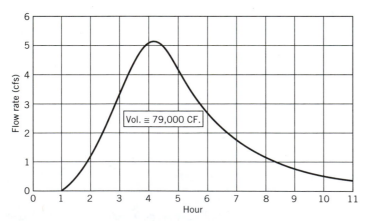

FIGURE 6.32 Example Problem 6.10, hydrograph.

6.4.7 Contributing Area Method

When estimating hydrograph shapes for large mixed land use and mixed soil conditions, different hydrograph shapes may result from the different soil and ground cover. If an average soil and ground cover were used for an entire watershed, the estimated hydrograph may not adequately represent the mixed soil and ground cover. It is more accurate to generate the hydrograph shape for each land use with the same soil and ground cover system. All contributing area hydrograph estimates are then added (linearly) with respect to time to produce a final hydrograph shape. The Contributing Area method must use some method of hydrograph generation; usually the equation $(Q = iCA)$ as a hydrograph procedure is used. As in the Rational Method itself, the Contributing Area method requires estimates for rainfall intensity, watershed areas, and a runoff coefficient (C) to translate precipitation to rainfall excess in addition to a more detailed description of the watershed. The Rational method for calculating hydrographs assumes the watershed to have similar soils and ground cover for estimating a hydrograph at the discharge or other decision point (inlets). The duration of the storm must be divided into short time increments (ΔD). This is the same concept as used for the hyetograph time intervals.

The concept of directly or hydraulically connected impervious areas is important when using the Contributing Area method. This method requires a knowledge of flows from one watershed to the next. Those impervious areas (i.e., pavements) that are directly connected to the outlet of the watershed either by overland flow by other impervious areas or by conduits are considered to be directly or hydraulically connected impervious areas.

Another important concept to consider is the proper selection of the boundary of the watershed for which the hydrograph is to be estimated. In general, the maximum watershed size is any one drained by a single principal drainage conduit or water course such as subbasins A, B, and C in Figure 6.33. This is partly because most synthetic hydrograph procedures were established by experiments on watersheds with single principal water courses. These watersheds are considered to be homogeneous. If, in Figure 6.33, watersheds A and B are considered as a single watershed, computation would be difficult because uniform rainfall would most likely produce a double-hump hydrograph and the rational method produces a single peak.

In addition to the same land use and soil conditions criteria, inlets can be used to define a subbasin. Generally, the more a major watershed is subdivided into smaller ones, the more accurate the final result. Of course, dividing a large watershed into many will create additional work, which may not be warranted and may not actually

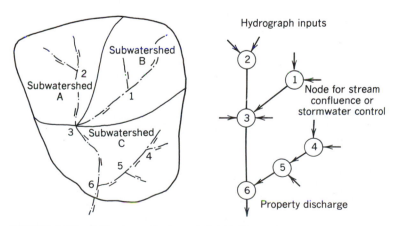

FIGURE 6.33   Subbasins and schematic (stick) diagrams.

increase the accuracy because the accuracy of a hydrograph is directly dependent on how correct the basic information is that is used in the design.

Examining Figure 6.33, one would compute separate hydrographs for subbasins A, B, and C and combine them for each time calculation as follows:

1. Combine the hydrograph for subbasins A and B by vertical addition of the flow hydrograph at each time.
2. Route (by flood-routing procedures) the combined hydrograph of subbasins A and B through the length of the receiving channel of subbasin C.
3. Add the hydrograph from subbasin C to the routed combined hydrograph from subbasins A and B.

If the principal water course draining subwatershed C entered the receiving channel at the upper end of C (node 3), combine all three hydrographs by vertical addition

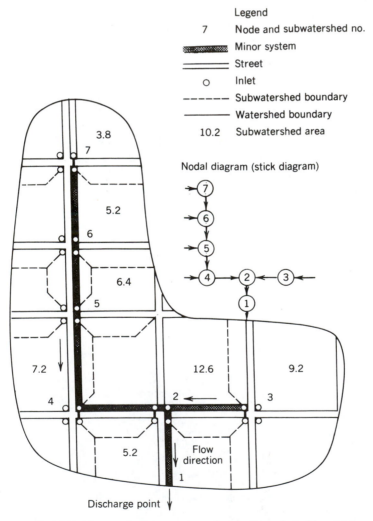

FIGURE 6.34 The contributing area watershed for Example Problem 6.11.

of flow and then route the combined hydrograph through the length of the receiving channel in subwatershed C.

Construction of a nodal (stick) diagram indicating flow direction would aid in describing the watersheds and keeping an inventory of calculations. The computations are not complex but strict computation procedures must be followed. Thus, an example problem is beneficial.

❑   **EXAMPLE PROBLEM 6.11**

Consider the watershed of Figure 6.34. The minor system is the drain pipes underground. Overland flow in the streets, parking lots, and lawns is called the major system. The total watershed is defined as an urban area. For better accuracy, the total area is subdivided into seven homogeneous contributing areas. The rational formula is used to estimate the runoff hydrograph shape. The time of concentration is estimated for each contributing area. This plus a list of contributing areas, the rational "C" for each area, and the equivalent impervious area (product of area and "C") are shown in Table 6.15.

TABLE 6.15   Area, Equivalent Impervious Area, and Time of Concentration for Example Problem 6.11

UP STREAM NODE / DOWN-STREAM NODE	SURFACE TYPE (RATIONAL C)	TOTAL CONTRIBUTING AREA (ac)	EQUIVALENT IMPERVIOUS AREA (C × AREA) (ac)	TIME OF CONCENTRATION TO DOWNSTREAM NODE 1 (min)
1/1	Commercial (0.85)	5.2	4.42	5.0
2/1	Commercial (0.85)	12.6	10.71	13.2
3/2	Residential (0.36)	9.2	3.31	23.8
4/2	Commercial 12% Residential 88% (0.88 × 0.36) +(0.12 × 0.85) = 0.42	7.2	3.02	20.5
5/4	Commercial 42% Residential 58% (0.58 × 0.36)] +(0.42 × 0.85) = 0.565	6.4	3.62	27.8
6/5	Residential (0.36)	5.2	1.87	34.4
7/6	Residential (0.36)	3.8	1.37	40.0
	Σ	49.6	28.32	

TABLE 6.16   Cumulative Time of Concentration Related to Effective Area

UPSTREAM NODE / DOWNSTREAM NODE	Σ t_c ALONG FLOW PATH	Σ EFFECTIVE AREA (ac)
1/1	5.0	4.42
2/1	13.2	15.13
3/2	23.8	21.46
4/2	20.5[a]	18.15[a]
5/4	27.8	25.08
6/5	34.4	26.95
7/6	40.0	28.32

[a]Flow path is from node 4 to 2, and area 4 contributes in time before area 3 contributes.

Next, it is necessary to estimate the time of concentration in the watershed to the point of discharge. In doing this, the schematic nodal diagram in Figure 6.34 is helpful. The time of concentration is estimated along the flow path. The maximum flow rate will occur when the concentration is the smallest value because rainfall intensity is larger for shorter duration storms. Thus, the shortest time of concentration is recorded in Table 6.16. By considering the time of concentration equal to the incremental duration (ΔD) of rainfall, the peak flows are estimated for that time increment. Thus, the peak flows can be calculated for each subwatershed and each rainfall intensity if the time of concentration is known for that subwatershed. Figure 6.35 illustrates the format for computing peak flows at each subwatershed. Note that the discharge hydrograph must be calculated for each hyetograph

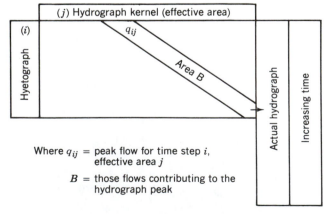

Where q_{ij} = peak flow for time step i, effective area j

B = those flows contributing to the hydrograph peak

FIGURE 6.35   The computation format—contributing area method.

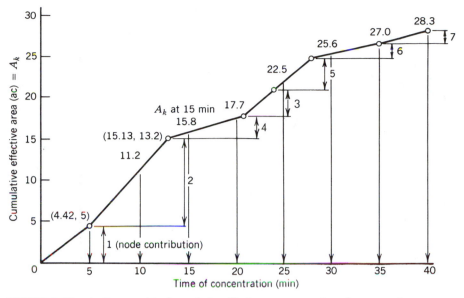

FIGURE 6.36 An S-curve plot of cumulative effective area versus time of concentration.

time step (ΔD) and the hydrograph estimates for each time step must therefore be summed. The summation is done on a diagonal because of the time it takes to reach the discharge point. The time increments (steps) for the hyetograph must be specified. The time steps must be at most 20% of the watershed time of concentration or at least five steps must be specified. The effective area is then estimated for each time step and obtained from the time of concentration estimate. If the hyetograph time steps were every 5 min, the first 5 min has an equivalent impervious area of 4.42 ac. This is the "effective area" for contributing runoff.

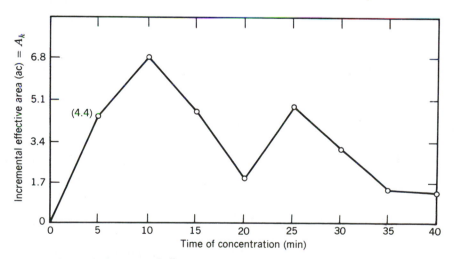

FIGURE 6.37 An incremental effective area—concentration curve ($\Delta t = 5$ min).

TABLE 6.17 Calculation Sheet for Storm 1 △

HYETOGRAPH		SUBWATERSHED HYDROGRAPH ESTIMATES FOR EACH TIME STEP								ROUTED HYDROGRAPH	
TIME (min)	PRECIP. (in./hr)	A_{40} 1.3	A_{35} 1.4	A_{30} 3.1	A_{25} 4.8	A_{20} 1.9	A_{15} 4.6	A_{10} 6.8[b]	A_5 4.4	DISCHARGE (cfs)	TIME (min)
5	0.6	0.78	0.84	1.86	2.88	1.14	2.76	4.08	2.64[a] →	2.64	5
10	0.3	0.39	0.42	0.93	1.44	0.57	1.38 ↗	2.04 ↗	1.32 →	5.40	10
15	0.2	0.26	0.28	0.62	0.96	0.38	0.92	1.36	0.88 →	5.68	15
20	0.1	0.13	0.14	0.31	0.48	0.19	0.46	0.68	0.44	4.32	20
25	0.1	0.13	0.14	0.31	0.48	0.19	0.46	0.68	0.44	5.49	25
30	0.1	0.13	0.14	0.31	0.48	0.19	0.46	0.68	0.44	5.26	30
										4.06	35
										2.95	40
										1.65	45
										1.19	50
										0.58	55
										0.27	60
										0.13	65
										0.00	70

[a] $Q = i(A)_k$ where $(A)_k = CA_k$ and k = hyetograph time step. $Q = 0.6(4.4) = 2.64$ cfs.
[b] $A_{10} = 11.2 - 4.4 = 6.8$ Ac
Note: CA is the incremental contributing area.

TABLE 6.18 Incremental Values for each 5-min Hyetograph Interval

HYETOGRAPH TIME (min)	CUMULATIVE EFFECTIVE AREA (ac)	INCREMENTAL EFFECTIVE AREA (ac)
0	0	0
5	4.4	4.4
10	11.2	6.8
15	15.8	4.6
20	17.7	1.9
25	22.5	4.8
30	25.6	3.1
35	27.0	1.4
40	28.3	1.3

Table 6.17 shows the calculations for a storm with higher intensity at the start of the storm and a decreasing intensity with time. The effective contributing areas were estimated from Figure 6.36. Since the hyetograph time step is 5 min, the effective area associated with each 5-min increment has to be determined. The cumulative area is estimated from Figure 6.36, the values of which are repeated in Table 6.18 to one-place accuracy since it was read from a graph. The incremental values of Table 6.18 are for each 5-min hyetograph interval. Figure 6.37 graphically represents the effective areas and illustrates the variable contribution of area as it affects runoff. If the incremental effective areas were about the same, the hyetograph would primarily determine the hydrograph shape. Note for the example problem that the effective areas of Figure 6.37 illustrate possibly two separate hydrograph peaks (there are two incremental effective area "humps.") Example calculations for the incremental hydrograph values at 10 min are

AREA	RATIONAL Q_p
1	$Q_p = (CA)(i) = (4.4)(0.3) = 1.32$ cfs
2	$Q_p = (6.8)(0.3) = 2.04$ cfs
etc.	

and the discharge hydrograph at 10 min is $4.08 + 1.32 = 5.40$ cfs. The flow from area 2 takes 10 min to reach the discharge point; thus, the incremental hydrograph value from the first 5 min of the storm is used.

Table 6.19 shows calculations for a different hyetograph. The two resulting hydrographs are shown in Figure 6.38 (Walsh and Wanielista, 1982). Note the double "hump" on the first hydrograph and the differences between the two hydrographs. Because the rain volume is the same for both storms (about 0.12 in.), one would expect the same peak discharge if the rational formula were applied to the same

TABLE 6.19 Calculation Sheet for Storm 2 (Denoted by ⊙ on Figure 6.38)

HYETOGRAPH		SUBWATERSHED HYDROGRAPH ESTIMATES FOR EACH TIME STEP								ROUTED HYDROGRAPH	
TIME (min)	PRECIP. (in./hr)	A_{40} 1.3	A_{35} 1.4	A_{30} 3.1	A_{25} 4.8	A_{20} 1.9	A_{15} 4.6	A_{10} 6.80	A_5 4.42	DISCHARGE (cfs)	TIME (min)
5	0.1	0.13	0.14	0.31	0.48	0.19	0.46	0.68	0.44	0.44	5
10	0.2	0.26	0.28	0.62	0.96	0.38	0.92	1.36	0.88	1.56	10
15	0.4	0.52	0.56	1.24	1.92	0.76	1.84	2.72	1.76	3.58	15
20	0.4	0.52	0.56	1.24	1.92	0.76	1.84	2.72	1.76	5.59	20
25	0.2	0.26	0.28	0.62	0.96	0.38	0.92	1.36	0.88	6.30	25
30	0.1	0.13	0.14	0.31	0.48	0.19	0.46	0.68	0.44	5.67	30
										5.04	35
										4.41	40
										3.21	45
										2.18	50
										1.11	55
										0.40	60
										0.13	65

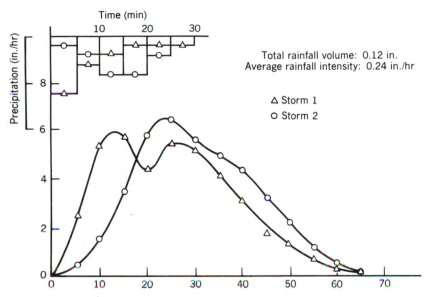

FIGURE 6.38 Hydrographs resulting from two different hyetographs.

area. Note, however, that the hyetographs are different. If a constant rainfall intensity were assumed (say 0.24 in./hr) then the rational formula estimate of the peak discharge is 0.24(28.3) = 6.8 cfs. ☐

6.5

SUMMARY

A hydrograph is a presentation of the time variable nature of flow rates. The presentation can be both in graphical or tabular form. Streamflow is a combination of surface flow rates (runoff) and groundwater flow rates. Hydrographs can be developed using synthetic methods. The development of a synthetic hydrograph requires specific knowledge of watershed characteristics and rainfall excess. With this information, peak discharge or hydrograph shape can be estimated.

- A streamflow hydrograph can be composed of both surface runoff and groundwater infiltrating into a stream.
- Streamflow hydrograph analysis on a yearly basis is helpful in determining the type and yield of a stream.
- A linear model for the rising limb of a hydrograph was developed. It resulted in Equation 6.5. Its development assumes a linear relationship among storage and flow rate. This is a reasonable assumption for a range of data for most streamflow hydrographs.
- Two commonly used methods for separating base flow from runoff in a stream-

flow hydrograph are fixed time and logarithm. The logarithm method and equations are presented and developed.

- Synthetic hydrograph procedures are useful for the generation of hydrographs when no streamflow data are available.

- Time of concentration is a necessary parameter for the generation of hydrographs using the procedures of this chapter. For gutter and sewer systems, the Manning equation is useful to estimate velocity and then time of concentrations. For overland flow, at least two other equations have been frequently used.

- Watershed characteristics are important for the determination of hydrograph shapes.

- Watershed area-time curves can be used with intensity data to generate hydrographs.

- The rational method is used extensively for peak discharge estimation for small watersheds with well-defined contributing areas. The rainfall intensity is assumed constant for a time period equal to the longest travel time for the watershed.

- For larger mixed land use areas, the rational method can be considered simplistic, and like some time-tested methods, it generally produces an overestimate of the peak discharge.

- The SCS typical hydrograph shape ($K = 484$) estimates a peak discharge that is about 75% of that estimated using the rational formula, assuming all other factors are equal.

- The peak attenuation factor (K) in the general hydrograph formula ($Q_p = KCiA$) can be changed to develop other hydrograph shapes.

- The Santa Barbara hydrograph method is based on a formula to route rainfall excess. Directly connected impervious areas can be considered. A hydrograph shape factor is inherent in its use.

- Given rainfall intensities and rainfall excess that are different for fixed time periods with a unit-hydrograph for each time period, the convolution procedure can be used to generate a hydrograph shape. A discrete unit–time hydrograph method is outlined in the text and is available in some computer programs like the one (SMADA) included with this book.

- Computer programs can be used to generate hydrographs using the rational, Santa Barbara, SCS, or unit hydrograph shape.

- The unit hydrograph is for one unit (inch or centimeter) of rainfall excess (runoff). Groundwater flow must be estimated and subtracted from streamflow to determine the runoff. The unit of rainfall excess is assumed to be spread out over the entire watershed.

- A computation procedure using rainfall excess and a unit hydrograph developed from streamflow data is presented to generate a composite hydrograph.

- Matrix methods also can be used to generate a composite hydrograph from a unit hydrograph and rainfall excess. Computer programs are available to do this.

- A unit hydrograph can be derived from a streamflow hydrograph using matrix methods.

- The mathematical equations for the general case of changing rainfall excess and hydrograph shapes were developed using a convolution procedure. Formulas were developed that can be used for any routing and rainfall excess condition.

6.6

PROBLEMS

6.6.1 Hand Problems

1. For a 50-ac single-family residential area and a rainfall intensity of 3 in./hr for 1 hr, develop a hydrograph using an assumed average runoff coefficient from Table 6.5 and assume the rational triangular hydrograph shape. What is the peak discharge if a maximum runoff coefficient is assumed?

2. Calculate the time of concentration for overland flow on sandy soil of 5% slope for a distance of 500 ft by using the SCS (NRCS) method. The soil has a high water table resulting in a potential watershed storage of only 0.5 in. Compare these figures by using the data in Figure 6.17. Discuss the results.

3. Explain in your own words the meaning of rainfall excess as it affects time of concentration.

4. Calculate the peak runoff from a residential area with similar watershed soil and surface characteristics. The area is 20 ac in size with 40% imperviousness. Use a rainfall intensity of 3 in./hr for 1 hr. Do the calculations by using the rational formula and the SCS typical hydrograph procedure. Compare results and discuss assumptions. The pervious area does not contribute runoff.

5. Assume that the hydrograph of Problem 4 had an attentuation factor of 520. What is the peak discharge? Now, assume that only 50% of the impervious area is directly connected and the soil is very permeable and can store 6 in. of water before saturation. The duration of rainfall is 1 hr. What is your estimate of peak discharge? Again, state your reasoning.

6. Using detailed calculations, estimate the time of concentration using Izzard's formula and the kinematic equation. There is 300 ft of overland flow, the rain intensity is for a short duration storm, the flow slope is 0.05, and the area is sodded.

7. By using longhand calculations and the SCS unit–time discrete hydrograph procedure, develop the resulting triangular hydrograph for one time step using the watershed data of Table 6.8, a hydrograph attenuation factor of 250, and a constant rainfall of 2 in. over 4 hr. Use $\Delta D = 10$ min = duration of one rainfall increment; the curve number method for infiltration is used to estimate rainfall excess from the pervious area.

8. Increase the attenuation factor to 350 with the same input assumptions of Problem 7 and discuss your results. Now change the hyetograph of 0.25 in. for the first ΔD, 0.50 in. for the second ΔD, 1 in. for the third ΔD, and 0.25 in. for the fourth ΔD. How do the resulting peak discharges compare.

9. Prove mathematically one of the attenuation factors (other than 484) of Table 6.6.

10. Calculate the storage needed for the rainfall excess from a 30-ac, 40% directly connected impervious area, and a 0.5-in. rainstorm. State your assumptions. Develop a hydrograph if the time of concentration is 1 hr and the rain lasted 1 hr. Prove the area under the hydrograph equals your pond storage (by example calculations). The runoff coefficient is 0.72.

11. Develop two instantaneous hydrographs for the watershed of Table 6.8. Use the hyetograph of Figure 6.39. The percentage of the impervious area that is directly

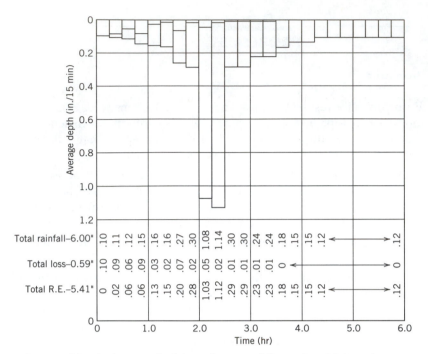

FIGURE 6.39 The hyetograph of 25-yr design rainfall curve No. 95. Note that rainfall excess values are adjusted to present a constantly increasing and then a decreasing rainfall excess. The actual computations produce no rainfall excess during the first hour.

connected is 60 and should be used in the analysis. One hydrograph is estimated without any stormwater management. The second hydrograph is estimated considering 1 in. of initial abstraction over the total area (stormwater management). Compare the resulting hydrograph shapes and comment on your results.

12. Use the assumptions of the rational method (triangular hydrograph, constant rainfall intensity) and compute the peak discharge for the data of Problem 11. Compare and discuss your results. Why is the rational method not a reasonable approximation for the peak in this case?

13. A county requires use of the 1 in 10-year storm frequency for detention basin design (use curves from Appendix C or other similar ones). What duration (1 or 6 hr) produces the largest detention basin, if the output hydrograph is 10 ft^3/sec, the watershed area is 50 ac, the rational coefficient is 0.8, the output hydrograph starts at 1 hr at a constant 10 ft^3/sec, and the time of concentration is 1 hr? Only compare using average intensities and the rational formula.

14. For Watershed No. 2 in Figure 6.40, and the data of Table 6.20, develop a discharge hydrograph by using the contributing area method. Compare your results to the peak discharge calculated using the rational formula. Use the hypothetical storm intensity of Figure 6.41 or make up your own of similar complexity. Compare your results to a hydrograph by using the data of Example Problem 6.11 (Watershed No. 1).

15. Compare the two hydrographs of Figure 6.38 with the hydrograph resulting from the use of the Rational Formula applied to the total watershed of Figure 6.34.

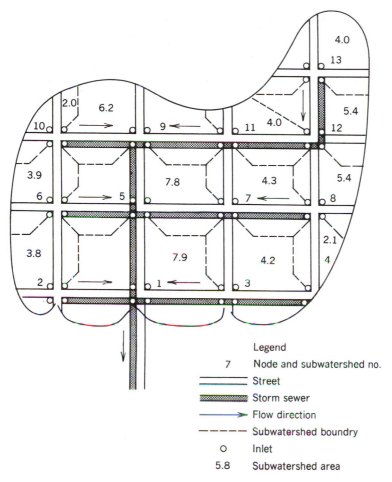

FIGURE 6.40 A plan view of watershed No. 2.

16. A pre- versus post-development hydrograph analysis for a 100-ac watershed must be completed. The precondition watershed has a runoff coefficient of 0.2 and a hydrograph shape with a peak attenuation factor of 0.31. The postcondition is most likely to have a hydrograph shape similar to the standard SCS–peak attenuation factor and a runoff coefficient of 0.4. If the time of concentration is one hour and the rainfall intensity is 5 in./hr, draw both hydrographs and estimate the peak discharge. Next, estimate the storage volume that approximates the condition of pre- versus postpeak discharge. Express storage in terms of cubic feet.

17. Using Example Problem 6.11, change the rainfall hyetograph to a constant intensity of 0.25 in./hr for 30 min and calculate the peak discharge using the other conditions of the example. Compare this to the peak discharge estimate using the rational formula.

18. Estimate the time and magnitude of a peak discharge from a 10-ac completely impervious smooth pavement area where the overland length is 1000 ft at a slope of 0.5 ft/100 ft. The geographic area is in Orange County, Florida and the design

TABLE 6.20   Data Sheet for Watershed 2 Unit Hydrograph Determination

UPSTREAM NODE/ DOWNSTREAM NODE	SURFACE TYPE (RATIONAL C)	TOTAL CONTRIBUTING AREA (ac)	EFFECTIVE CONTRIBUTING AREA (ac)	TIME OF CONCENTRATION TO DOWNSTREAM NODE (min)	Σt_c to DOWNSTREAM NODE (min)
1/1	Residential (0.36)	7.9	2.84	6.1	6.1
2/1	Residential (0.36)	3.8	1.37	5.2	11.3
3/1	Residential (0.36)	4.2	1.51	5.2	11.3
5/1	Residential (0.36)	7.8	2.81	6.1	12.2
4/3	Residential (0.36)	2.1	0.76	5.2	16.5
6/5	Residential (0.36)	3.9	1.40	5.2	17.4
7/5	Commercial 50% Residential 50% (0.61)	4.3	2.60	5.2	17.4
9/5	Residential (0.36)	6.2	2.23	6.1	18.3
10/9	Residential (0.36)	2.0	0.72	4.2	22.5
8/7	Commercial (0.85)	5.4	4.59	5.2	22.6
11/9	Residential 20% Commercial 80% (0.75)	4.0	3.01	5.2	23.5
12/11	Commercial (0.85)	5.4	4.59	5.2	28.7
13/12	Commercial (0.85)	4.0	3.4	6.1	34.8

return period (frequency) is 1 in 10 yr. Regulations require the use of Kerby's equation for calculation of time of concentration.

19. Provide flow estimates for a synthetic unit hydrograph at times equal to half the peak flow time, at the peak flow time, 1.5 times the peak flow time, twice the peak flow time, and 5 times the peak flow time using the SCS curvilinear hydrograph shape. The watershed is a 40-ac commercial area with a percent impervious directly connected area of 32 ac and a 15-min time of concentration. Assume a rainfall duration of half the lag time. Also, what is the peak discharge if the attenuation factor were 400?

20. A 3-ft diameter storm sewer transports water to a discharge point. We are interested in the hydrograph shape at the discharge point. The sewer is 2000 ft long at a slope of 0.001 and $n = 0.012$.

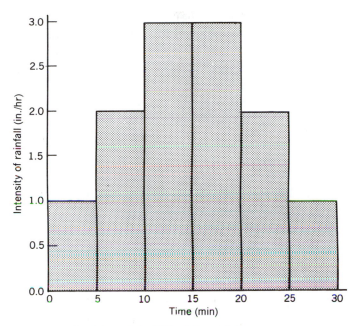

FIGURE 6.41 A hypothetical ½-hr, 1-in. rainfall event.

    a. If the sewer is flowing full at peak discharge, estimate the routing constant for the Santa Barbara method. The computation interval is 2 min.

    b. What is the peak discharge?

    c. If the resulting discharge hydrograph can be represented by a rational hydrograph shape, what is the volume discharge in cubic feet?

21. Determine the following from the hydrograph of Figure 6.42: L, t_p, t_r, t_B, and the volume of rainfall excess in cubic feet.

22. Comment on the peak discharge and rainfall excess changes resulting from the same volume and intensity of rainfall after converting a land from:

    a. Pervious to impervious

    b. Directly drained (hydraulic connection) to indirectly drained

    c. A high-water-table sandy soil to a low-water-table sandy soil

    d. No surface storage to surface storage through which runoff waters must travel

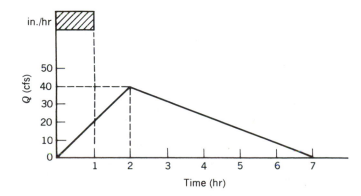

FIGURE 6.42 The hydrograph for Problem 1.

23. Derive Equation 6.5 from Equation 6.1 and Equation 6.8 from Equation 6.7.

24. Given the following recession limb of a streamflow hydrograph, develop the recession equations for flow rate.

TIME (hr)	FLOW (cfs)	TIME (hr)	FLOW (cfs)
0	300 (peak)	6	30
1	208	7	28
2	144	8	26
3	91	10	24
4	63	14	20
5	45		

25. Pick any streamflow record and separate the base flow using the straight line and fixed time methods if τ in the fixed time method is equal to $1.5T_p$.

26. Consider the following runoff hydrograph. If the watershed area is 100 acres, what is the rainfall excess expressed in inches under the runoff hydrograph?

TIME (hr)	RUNOFF (cfs)	TIME (hr)	RUNOFF (cfs)
0	0	6	45
1	70	7	30
2	160	8	20
3	110	9	12
4	80	10	5
5	60	11	0

27. Develop a unit hydrograph for Problem 26. Also plot the results.

28. What size (acres) of watershed produced the following unit hydrograph? The flow rates are average hourly data. What is precipitation volume if the runoff coefficient were 0.5?

TIME (hr)	RUNOFF (cfs)
0	0
1	90
2	50
3	30
4	20
5	10
6	0

29. Given the unit runoff hydrograph of Problem 28, develop a composite runoff hydrograph given the following rainfall excess values. Set up the problem and complete it as if you were using the matrix method.

TIME (hr)	R (in.)
1	0.2
2	1.0
3	0.5

30. Assume that the surface runoff hydrograph of Figure 6.1(b) is a unit hydrograph resulting from a 5-hr storm event. The peak discharge occurs at hour 9 and is 910 cfs. A 10-hr storm event is divided into two equal time intervals with 0.5 in. of rainfall excess in the first 5 hr and 1.0 in. of rainfall excess in the next 5 hr. What is the discharge at hours 9 and 14, assuming the unit hydrograph can be used?

31. Compute a design hydrograph resulting from the following rainfall excess increments using the following unit hydrograph, which is in 0.25-hr increments: 70, 182, 137, 68, 33, 16, 9, 5, 2, 1, 0 cfs @ 2.75 hr.

Time (hr)	0.25	0.50	0.75
Rainfall excess increment (in.)	0.50	1.25	0.75

32. For the convolution procedure and assuming a constant rainfall excess, derive Equation 6.44 from Equation 6.43 (show all work).

33. Assume that rainfall excess for an 8-hr rainfall on a 40-ac watershed can be represented by a first rainfall excess,

$$r(t) = 0 + 25t \qquad 0 \le t \le 4 \text{ hr}$$

and the next excess $r(t) = 100 - 25t \qquad 4 \le t \le 8$ hr

where $r(t)$ = rainfall excess in cfs. Develop the hydrograph assuming that the routing factor is 0.25. *Hint:* Convert real time t to τ such that $r(\tau) = 100 - 25\tau$ between 4 and 8 hr real time.

34. From Problem 33, what is the runoff coefficient if the total rainfall volume used for design were 10 inches?

35. Do the detailed calculations for Example Problem 6.9. Change the rainfall excess and the routing coefficient in the first time interval to be a constant 2 cfs and 0.30, respectively, and in the second interval to $2 + 8(\tau)$. Compare results and comment on the peak as related to initial abstraction.

36. For the watershed of Table 6.21 and Figure 6.40 with a rainfall of 2 in./hr for 30 min and a discharge of 6 cfs at 2 hr, estimate the parameters using convolution (determine k and a).

TABLE 6.21 Data Sheet for Watershed 2 (Figure 6.40): Unit Hydrograph Determination

UPSTREAM NODE/ DOWNSTREAM NODE	SURFACE TYPE (RATIONAL C)	TOTAL CONTRIBUTING AREA (ac)	EFFECTIVE CONTRIBUTING AREA (ac)	TIME OF CONCENTRATION TO DOWNSTREAM NODE (min)	Σ t_c TO DOWNSTREAM NODE (min)
1/1	Residential (0.36)	7.9	2.84	6.1	6.1
2/1	Residential (0.36)	3.8	1.37	5.2	11.3
3/1	Residential (0.36)	4.2	1.51	5.2	11.3
5/1	Residential (0.36)	7.8	2.81	6.1	12.2
4/3	Residential (0.36)	2.1	0.76	5.2	16.5
6/5	Residential (0.36)	3.9	1.40	5.2	17.4
7/5	Commercial 50% Residential 50% (0.61)	4.3	2.60	5.2	17.4
9/5	Residential (0.36)	6.2	2.23	6.1	18.3
10/9	Residential (0.36)	2.0	0.72	4.2	22.5
8/7	Commercial (0.85)	5.4	4.59	5.2	22.6
11/9	Residential 20% Commercial 80% (0.75)	4.0	3.01	5.2	23.5
12/11	Commercial (0.85)	5.4	4.59	5.2	28.7
13/12	Commercial (0.85)	4.0	3.4	1.3	30.0

37. For the following listing of a runoff hydrograph, determine an equation for the recession limb using an exponential decay function.

TIME (hr)	FLOW RATE (cfs)
6 (peak)	300
7	220
8	165
9	120
11	73
14	30
17	12

38. Obtain stream flow data for one or more years. Plot the data and describe the type of stream it is.

6.6.2 Computer Problems

1. Using the matrix method computer program (EZMAT) and the following data for rainfall excess and a unit hydrograph, develop a composite runoff hydrograph.

		UNIT HYDROGRAPH	
TIME (hr)	RAINFALL EXCESS (in.)	TIME (hr)	FLOW RATE (cfs)
0	0	0	0
0.25	0.13	0.25	68
0.50	0.15	0.50	246
0.75	0.20	0.75	293
1.00	0.28	1.00	202
1.25	1.03	1.25	121
1.50	1.12	1.50	72
1.75	0.29	1.75	41
2.00	0.29	2.00	26
2.25	0.23	2.25	15
2.50	0.23	2.50	9
2.75	0.18	2.75	6
3.00	0.15	3.00	3
3.25	0.15	3.25	1
3.50	0.12	3.50	0
3.75	0.12		
4.00	0.12		
4.25	0.12		
4.50	0.12		
4.75	0.12		
5.00	0.12		

2. Change the unit hydrograph shape to the following attenuated values and comment on the composite runoff hydrograph if the same rainfall excess were used.

TIME (hr)	ORDINATE (cfs)
0	0
0.25	68
0.50	200
0.75	250
1.00	240
1.25	172
1.50	72
1.75	41
2.00	26
2.25	15
2.50	9
2.75	6
3.00	3
3.25	1
3.50	0

3. If additional on-site retention by diversion of rainfall excess reduces the rainfall excess stated in Problem 1 to 0.00, 0.00, 0.00, 0.15, 1.03, 1.12, 0.29, 0.29, 0.23, and 0.23, what effect does this have on the peak discharge?

4. Using the regression program (REGRESS) with the data of Problem 2, estimate the parameters for a straight line equation and a polynomial of degree 3. Comment and provide a graphic display of the results.

5. For the watershed of Table 6.22, develop a hydrograph for a 25-yr, 24-hr storm using the Santa Barbara Urban Hydrograph procedure. At first, assume the impervious area is directly connected, then assuming only 50% directly connected impervious area. Note that the pervious curve number must be used if SMADA is being used. Comment on the changes in the hydrograph shape when all the impervious area is directly connected versus only partial connection.

6. Compare the resulting hydrographs of Problem 5 with a hydrograph generated using the SCS unit graph method with hydrograph attenuation of 350. Discuss your results.

TABLE 6.22   Hypothetical Watershed

Watershed area = 250 ac
Time of concentration = 60 min
% Impervious = 60
% $DCiA$ = 100
CN for pervious area = 80

7. Determine the hydrograph shape at a proposed river crossing of an interstate highway. The watershed is 300 ac with little impervious area, C type soils, and $\bar{t}_c = 3$ hr. State all your assumptions.

8. Increase by 50% the time of concentration in Problem 5 and compare peak discharges. Also, assume a 50% directly connected impervious area for the comparisons.

9. Using the SMADA computer program and a rainfall volume and time distribution for a 25 yr, 2 hr storm of your choice (for your area), develop a runoff hydrograph and the infiltration volume (watershed storage) for the following watershed condition (use Santa Barbara Hydrograph Routine).

$$\text{Area} = 200 \text{ ac}$$

$$t_c = 120 \text{ min}$$

$$\% \text{ Impervious} = 40$$

$$\% \text{ } DCiA = 70$$

$$CN = 60 \text{ (pervious area)}$$

10. Provide a plot of infiltration volume (inches) versus percent impervious values for percent impervious values of 20, 40, 60, 80, and 90 using the same watershed conditions as in Problem 5 (except variable percent imperviousness). Also, construct a plot of peak discharge versus percent imperviousness. Note that the time of concentration is held constant. Comment on your results.

11. On the graph of Problem 10, also change the pervious area curve number to 85 and plot the resulting peak discharge and infiltration volumes as a function of percent imperviousness. Comment on your results.

12. For some watersheds, it is more reasonable to expect the time of concentration to decrease as the percent impervious area (which is directly connected) increases. Develop a peak discharge versus the percent impervious relationship (graphical plot) by using the data of Problem 5, except for the following changes:

t_c (min)	PERCENT IMPERVIOUS (DIRECTLY CONNECTED)
150	20
120	40
90	60
70	80
60	90

6.6.3 Case Study

It is necessary to develop a runoff hydrograph using the excess hyetograph developed in the Chapter 5 case study. The hydrograph should be generated using a unit hydrograph developed with an SCS peak attenuation factor of 256 with a triangular shape. The watershed has a total area of 100 acres and a time of concentration of 1 hr. It is specified that 30-min time steps must be used.

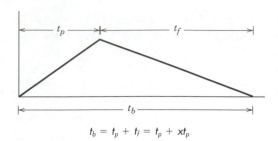

$$t_b = t_p + t_f = t_p + xt_p$$

FIGURE 6.43   Development of the SCS Unit Hydrograph.

The first step is to develop a unit hydrograph using the SCS shape factor. The unit hydrograph will be developed with discrete units on half-hour time steps.

where

t_b = hydrograph base time
t_p = hydrograph time to peak
t_f = hydrograph recession time

x is the ratio of t_f/t_p and is calculated by;

$$x = (1291/K) - 1 \qquad \text{for area in square miles}$$

In the example, t_p is equal to the time of concentration of 60 min and the ratio x is roughly equal to 4 (1291/256-1). Therefore the recession limb time t_f is roughly 4 hr.

	A	B
2	Time (hr)	Value
3	0.00	0.000
4	0.50	0.100
5	1.00	0.200
6	1.50	0.175
7	2.00	0.150
8	2.50	0.125
9	3.00	0.100
10	3.50	0.075
11	4.00	0.050
12	4.50	0.025
13	5.00	0.000

A normalized triangular unit hydrograph can be developed on $\frac{1}{2}$-hr time steps using the information (t_p = 1 hr, t_f = 4 hr). The values for this unit hydrograph are as shown. The sum of the values in this *discretized* unit hydrograph is 1.0. By using matrix algebra, this hydrograph can be multiplied by the excess hydrograph to yield a runoff hydrograph. Each step in the multiplication must be offset by the number of time steps.

The contents of each cell are found by multiplying the rainfall excess by respective value of the unit hydrograph as shown in Table 6.23.

```
Cell D3    D$2*$B3        Cell E3    E$2*$B3
Cell D4    D$2*$B4        Cell E4    E$2*$B4
```

TABLE 6.23 Multiplication of Rainfall Excess and Unit Hydrograph

	C	D	E	F	G	H	I	J	K	L	M	N	O
1	Time (hr)	.5	1.0	1.5	2.0	2.5	3.0	3.5	4.0	4.5	5.0	5.5	6.0
2	Excess (in.)	0	0.083	0.361	0.564	0.707	0.812	0.892	0.953	1.002	1.041	1.073	1.099
3		0	0.000	0.000	0.000	0.000	0.000	0.000	0.000	0.000	0.000	0.000	0.000
4		0	0.008	0.036	0.056	0.071	0.081	0.089	0.095	0.100	0.104	0.107	0.110
5		0	0.017	0.072	0.113	0.141	0.162	0.178	0.191	0.200	0.208	0.215	0.220
6		0	0.015	0.063	0.099	0.124	0.142	0.156	0.167	0.175	0.182	0.188	0.192
7		0	0.012	0.054	0.085	0.106	0.122	0.134	0.143	0.150	0.156	0.161	0.165
8		0	0.010	0.045	0.070	0.088	0.102	0.111	0.119	0.125	0.130	0.134	0.137
9		0	0.008	0.036	0.056	0.071	0.081	0.089	0.095	0.100	0.104	0.107	0.110
10		0	0.006	0.027	0.042	0.053	0.061	0.067	0.071	0.075	0.078	0.080	0.082
11		0	0.004	0.018	0.028	0.035	0.041	0.045	0.048	0.050	0.052	0.054	0.055
12		0	0.002	0.009	0.014	0.018	0.020	0.022	0.024	0.025	0.026	0.027	0.027
13		0	0.000	0.000	0.000	0.000	0.000	0.000	0.000	0.000	0.000	0.000	0.000

TABLE 6.24 Calculation of Runoff per Acre for Given Rainfall Excess

	A	B	C	D	E	F	G	H	I	J	K	L	M	N	O
1	Time (hr)	0.50	1.00	1.50	2.00	2.50	3.00	3.50	4.00	4.50	5.00	5.50	6.00	Flow	
2	Excess (in.)	0	0.083	0.361	0.564	0.707	0.812	0.892	0.953	1.002	1.041	1.073	1.099	in/hr	cfs/acre
3	0.00	0													0.00	0.00
4	0.50	0	0.000												0.00	0.00
5	1.00	0	0.008	0.000										0.01	0.01
6	1.50	0	0.017	0.036	0.000									.05	0.05
7	2.00	0	0.015	0.072	0.056	0.000								0.15	0.15
8	2.50	0	0.012	0.063	0.113	0.071	0.000							0.26	0.26
9	3.00	0	0.010	0.054	0.099	0.141	0.081	0.000						0.39	0.39
10	3.50	0	0.008	0.045	0.085	0.124	0.162	0.089	0.000					0.52	0.52
11	4.00	0	0.006	0.036	0.070	0.106	0.142	0.178	0.095	0.000				0.64	0.64
12	4.50	0	0.004	0.027	0.056	0.088	0.122	0.156	0.191	0.100	0.000			0.75	0.75
13	5.00	0	0.002	0.018	0.042	0.071	0.102	0.134	0.167	0.200	0.104	0.000		0.84	0.84
14	5.50		0.000	0.009	0.028	0.053	0.081	0.111	0.143	0.175	0.208	0.107	0.000	0.92	0.93
15	6.00			0.000	0.014	0.035	0.061	0.089	0.119	0.150	0.182	0.215	0.110	0.98	0.99
16	6.50				0.000	0.018	0.041	0.067	0.095	0.125	0.156	0.188	0.220	0.91	0.92
17	7.00					0.000	0.020	0.045	0.071	0.100	0.130	0.161	0.192	0.72	0.72
18	7.50						0.000	0.022	0.048	0.075	0.104	0.134	0.165	0.55	0.55
19	8.00							0.000	0.024	0.050	0.078	0.107	0.137	0.40	0.40
20	8.50								0.000	0.025	0.052	0.080	0.110	0.27	0.27
21	9.00									0.000	0.026	0.054	0.082	0.16	0.17
21	9.50										0.000	0.027	0.055	0.08	0.08
22	10.00											0.000	0.027	0.03	0.03
23	10.50												0.000	0.00	0.00

TABLE 6.25 Runoff Hydrograph Found Using SMADA with SCS Method and Peak
Attenuation of 256 for a 100-Acre Watershed

TIME (hr)	RAINFALL INCREMENT (in.)	INFILTRATION (in.)	EXCESS (cfs)	RUNOFF (cfs)
0.50	0.650	0.650	0.000	0.000
1.00	0.650	0.609	8.362	0.828
1.50	0.650	0.469	36.405	5.262
2.00	0.650	0.368	56.812	14.291
2.50	0.650	0.297	71.261	25.880
3.00	0.650	0.244	81.865	38.559
3.50	0.650	0.204	89.877	51.336
4.00	0.650	0.173	96.079	63.515
4.50	0.650	0.149	100.977	74.592
5.00	0.650	0.130	104.913	84.189
5.50	0.650	0.114	108.124	92.021
6.00	0.650	0.101	110.777	98.051
6.50	0.000	0.000	0.000	91.554
7.00	0.000	0.000	0.000	72.720
7.50	0.000	0.000	0.000	55.601
8.00	0.000	0.000	0.000	40.465
8.50	0.000	0.000	0.000	27.515
9.00	0.000	0.000	0.000	16.905
9.50	0.000	0.000	0.000	8.759
10.00	0.000	0.000	0.000	3.175
10.50	0.000	· 0.000	0.000	0.233

Offsetting each array multiplication by the time step of the original excess value
will give the following array, which can then be added across the rows to give the
runoff hydrograph as shown in Table 6.24.

The final column (runoff in cfs/acre) can be multiplied by the acreage of the
watershed to give the final runoff hydrograph in the desired units of cfs. This can be
compared with the results of the SMADA program, which uses the same calculations
in the SCS 256 method in Table 6.25.

6.7

REFERENCES

Capice, J. 1984. *Estimating Runoff Rates and Volumes from Flat High-Water-Table Watersheds,* Master's Thesis, University of Florida, Gainesville.

Chow, V. T. 1959. *Open-Channel Hydraulics,* McGraw-Hill, New York, pp. 108–114.

Clark, C. O. 1945. "Storage and the Unit Hydrograph," *ASCE Transactions,* **110,** pp. 1419–
1446.

Eagleson, Peter S. 1959. "Characteristics of Unit Hydrographs for Sewered Areas," American Society of Civil Engineers, Los Angeles, CA (unpublished).

Engman, E. T. 1983. "Roughness Coefficients for Routing Surface Runoff." *Proceedings of the Conference on Hydraulic Engineering,* American Society of Civil Engineers, pp. 560–565, New York.

Fleming, G. 1975. *Computer Simulation Techniques in Hydrology.* Elsevier, New York.

Florida Department of Transportation. 1986. *Drainage Manual.* Tallahassee, FL.

Golding, B. L. 1974. Area Water Control Plan, OCP Phase 6, Reynolds, Smith and Hills, Inc., Orlando, FL.

Golding, B. L. 1986, DABRO–Drainage Basin Runoff Model—A Computer Program, Hieldebrand Software, 8992 Islesworth Court, Orlando, FL 32819.

Hall, M. J. 1984. *Urban Hydrology,* Elsevier Science Publishers, London.

Horton, R. E. 1938, "The Interpretation and Application of Runoff Plot Experiments with Reference to Soil Erosion Problems," Proceedings of the Soil Science Society of America, **3,** pp. 340–349.

Kuichling, E. 1889. "The Relation Between the Rainfall and the Discharge of Sewers in Populous Areas," *Transactions of the American Society of Civil Engineers,* **20,** pp. 1–56.

Mitchi, C. 1974. "Determine Urban Runoff the Simple Way," *Water Wastes Engineers,* **10**(1) January.

Morgan, R. and Hullinhors, D. W. 1939. "Unit Hydrographs for Gauged and Ungauged Watersheds," U.S. Army Corps of Engineers, Engineers Office, Binghamton, NY.

Mulvaney, T. J. 1851. "On the Use of Self-Registering Rain and Flood Gauges," *Institute of Civil Engineering Transactions* (Ireland), **4**(2), pp. 1–8.

Pagan, A. R. 1972. "Rational Formula Needs Change and Uniformity in Practical Applications," *Water and Sewage Works.*

Pedersen, J. T., Peters, J. C., and Helwey, O. 1980. "Hydrographs by Single Linear Reservoir Model," *Journal of the Hydraulics Division,* **HY5,** ASCE, New York, May, pp. 837–851.

Ragan, R. M. 1971. "A Nomograph Based on Kinematic Wave Theory for Determining Time of Concentration for Overland Flow," Report #44, College Park, MD, University of Maryland, December.

Ramser, C. E. 1927. "Runoff from Small Agricultural Areas," *Journal of Agricultural Research,* **34**(9), pp. 797–823.

Schulz, E. F. 1980. *Problems in Applied Hydrology.* Water Resources Publications, Fort Collins, CO, 5th Printing, pp. 280–286.

Sherman, L. K. 1932. "Stream-Flow from Rainfall by the Unit-Graph Method," *Engineering News Record,* **108,** April, pp. 501–505.

Smisson, R. P. M. 1980. "The Single Pipe System for Stormwater Management," *Progressive Water Technology, Vol. 13.* International Air and Water Pollution Research/Pergamon Press, Oxford, England.

Snyder, W. M., Mills, W. C., and Stephens, J. C. 1970. "A Method of Derivation of Nonconstant Watershed Response Functions," *Water Resources Research,* **6**(1), February, pp. 261–274.

Stubchaer, J. M. 1975. "The Santa Barbara Urban Hydrograph Method," *Proceedings of the National Symposium of Hydrology and Sediment Control,* College of Engineering, University of Kentucky (Lexington, KY: ORES Publication).

U.S. Bureau of Reclamation. 1952. Unitgraph Procedures, Division of Planning, Denver, CO.

U.S. Department of Agriculture, Soil Conservation Service (USDA-SCS). 1975. *National Engineering Handbook,* Section 4, Washington, DC.

U.S. Department of Agriculture, Soil Conservation Service. 1986. *Urban Hydrology for Small Watersheds,* Technical Release No. 55, Washington, DC, June.

Walsh, T. B. and Wanielista, M. P. 1982. *Low Flow Analysis in Stormwater Management.* University of Central Florida, Orlando, FL.

Williams, G. R. 1950. "Hydrology," in *Engineering Hydraulics,* H. Rouse, Ed. Wiley, New York.

STREAMFLOW MEASUREMENTS AND FLOW CONTROL DEVICES

Flow measurements provide needed data for estimating runoff and streamflow volumes and flow rates from a particular precipitation event or over an extended period of time. Methods exist that directly measure volumes over time or indirectly measure surrogate variables. These surrogates are then used to estimate flow rates. Streamflow is a general term used to represent volumes or rates of flow in a river, creek, stream, or channel. Discharge is used primarily for rate measurements and applied to waters that flow from a specific watershed or pond area to a receiving water body. In this chapter, the more common methods for estimating flow rates are presented. Examples of their use in measuring streamflow in rivers and discharge from stormwater control facilities (primarily from ponds) are presented. The actual sizing of reservoirs and stormwater ponds are dependent on discharge measurements.

This chapter also presents information on water quality measurement (including sediment movement), sampling techniques and development of rating curves (i.e., streamflow or discharge versus elevation or stage; pollutant concentrations versus streamflow or discharge; and stage versus other parameters such as velocity, drainage area, and pollutant loads).

7.1
METHODS OF MEASUREMENT

The accuracy with which one can estimate flow rates over time (a hydrograph) for a stream location depends on the method used for estimation. This assumes an appropriate application of the procedures for measuring and interpreting the data. If the proper procedures are followed, the accuracy of the estimates of discharge will be limited only by the range of the instrument and human error.

Some methods for discharge estimation are classified as shown in Table 7.1.

7.1.1 Stage

Stage is the water surface elevation recorded relative to some horizontal datum elevation, frequently mean sea level. Stage is a reflection of all the hydrologic processes

TABLE 7.1   Classification of Flow Measurement Methods

A. Stage (water surface elevation)
   1. Visual observation
   2. Float
   3. Pressure sensor
   4. Electrical resistance
B. Discharge (nonstructural)
   1. Current meter
   2. Dilution
   3. Float
   4. Indirect via Manning's equation
C. Discharge (structural)
   1. Direct volume collection
   2. Weirs
   3. Flumes
   4. Orifices

---

and water transport characteristics of the watershed. Historical records of high water levels exist, and are either man-made or indirectly identified using the presence of water-related vegetation or stain marks on trees and structures.

Stage records are valuable for the definition of high and low water levels. Areal extent of flooding, history of the rate of fluctuation, and watershed hydrologic characteristics can be documented. The record of stage is called the stage hydrograph, primarily because it can be translated into flow rate (discharge) units.

A stage recorder can be as simple as a ruler along a bridge or other structure. It can be read periodically but is usually automatically recorded. The automation is achieved using a water float, pressure sensor, or change in electrical resistance caused by water contact. The float and sensors are attached by wire or chain to a chart that records all water level changes. Typical guide pulley assemblies for a recording device with a counterweight are illustrated in Figure 7.1. The float and the counterweight should not encounter interference. All systems must be maintained and calibrated to ensure accurate results. Since rapid flow rates may move or destroy the float and sensors, protection from potentially damaging forces must be provided. In addition, waves or short-period water surface disturbances caused by boating or man-made events are not generally desired and certainly not reflective of natural conditions. Thus, for stage-level recorders, some type of structure that damps out man-made surface disturbances is needed. The structure is placed adjacent to the watercourse and connected in such a way as to minimize friction losses. Since short-period water surface disturbances inside the structure are minimal, it is called a stilling well, shown in Figure 7.2. The well should be checked frequently for debris and other obstructions.

In rough terrain with limited access and where silt is a problem, a stilling well in a stream may not be feasible. Also, for short-period measurement, construction of a stilling well may not be economical. For these cases a bubble or manometer-servo water-level sensor is used. It uses dry nitrogen (116 ft^3 cyclinder) (Figure 7.3) and battery power. The pressure measured corresponds to the water head. A servo motor

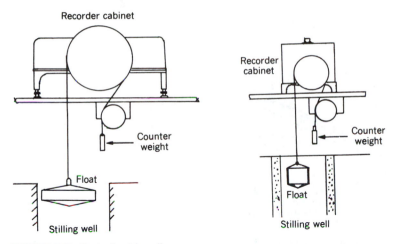

FIGURE 7.1 Typical guide pulley.

adjusts a recorder, such as the Stevens Type A also shown in Figure 7.3. Another picture of the recorder is shown in Figure 7.4. The bubbler and recorder can be located several hundred feet from the measuring point. Only tubing is required for connection to the stream.

There are other "stage detectors" available for use today. One device recognizes that water is a good conductor of electrical current. However, all systems must be maintained and calibrated to ensure accurate results.

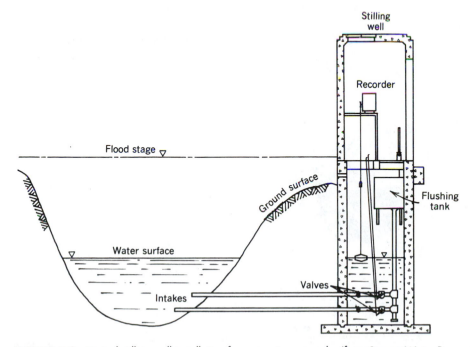

FIGURE 7.2 Typical stilling well installation for water stage recorder (from *Stevens Water Resource Data Book*, 4th Ed., Leapold and Stevens, January 1987).

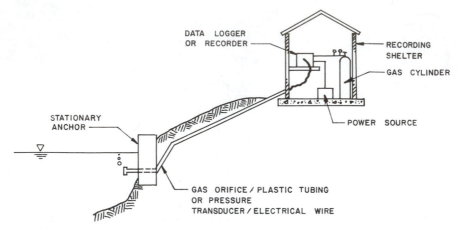

FIGURE 7.3   A gaging station with both a pressure transducer and nitrogen-gas-purge water-level sensing system.

7.1.2 Current Meter

A current meter is a device for sensing velocity. Older systems were revolution counting meters that converted angular velocity to linear velocity and consisted of a propeller or cup wheel, a revolution counter, shaft, earphones, weights, and rudder (see Figure 7.5). The meter with balancing weights has to be of sufficient weight so as not to be displaced during strong flow rates or must be anchored to a support. Normal propellers are designed to operate over a very wide range of velocities (0.03–10 m/s). The propeller reacts to flow velocity components in the axial direction only. Newer, electromagnetic meters remain hydrodynamically stable and function on an electromotive force (voltage) principle. The water, as a conductor, flows through a magnetic field.

Since velocity varies with location and time in a flowing stream or closed conduit, various areas and depths have to be used to estimate flow rates. Velocities near the

FIGURE 7.4   Stevens type A strip chart recorder (from Stevens, op. cit.).

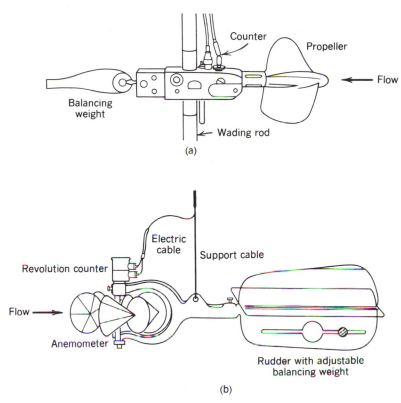

FIGURE 7.5 Current meters: (a) Propeller meter, and (b) Price current meter.

bottom of an open channel will be lower because of stream bed friction and higher velocities at other depths. When the depth of the watercourse is greater than 2 ft (0.6 m) the average velocity for the section can be estimated as the average of the velocities measured at depths equal to 0.2 and 0.8 times the total depth of the watercourse. If the depth of flow is less than 2 ft (0.6 m), the velocity meter is set at 0.6 times the depth of the watercourse and is measured from the water surface. The velocity at this depth is presumed to represent the average velocity for that section of stream. Guidelines for velocity measures with depth are shown in Table 7.2. A typical velocity profile with lines of equal velocity (isovels) is shown in Figure 7.6 and for pipe flow changing velocities perpendicular to the flow stream are shown. The number of sections across the channel must be sufficient enough to maintain desired accuracy. Usually, four measures are taken in a pipe.

Once the average velocity for a section is determined, the cross-sectional area is estimated and the flow rate can be determined using the deterministic form of the continuity equation, or

$$Q_i = A_i \overline{V}_i \qquad (7.1)$$

where

Q_i = volume flow rate for section i (cfs, cms)
A_i = cross-section area normal to the flow velocity in section i (ft^2, m^2)
$\overline{V}_i$ = average velocity for section i (fps, m/s)

TABLE 7.2 Depth Measure Points for Velocity and Averaging Equations

NUMBER OF MEASURES	DEPTH OF WATERCOURSE (ft)	OBSERVATION POINTS (MEASURED FROM WATER SURFACE)	AVERAGE VELOCITY ($\overline{V}$)
One	1–2	0.6D	$\overline{V} = V_{0.6}$
Two	2–10	0.2 and 0.8D	$\overline{V} = 0.5(V_{0.2} + V_{0.8})$
Three	10–20	0.2, 0.6, and 0.8D	$\overline{V} = 0.25(V_{0.2} + 2V_{0.6} + V_{0.8})$
Five	+20	1 foot, 0.2D, 0.6D, 0.8D, and 1 ft above bottom	$\overline{V} = 0.10(V_1 + 3V_{0.2} + 2V_{0.6} + 3V_{0.8} + V_B)$

where D = depth of water

The total flow rate is the sum of the flow rates of all sections across the stream. Sections are chosen to represent stream depths; thus, the number and the width of each section depend on the changes in stream depth and the degree of precision required. The depth of the stream is sounded and plotted as shown in Figure 7.6. The number of sections are determined (usually greater than 5 and less than 20). The velocity is measured at the center of each section and at depths recommended in

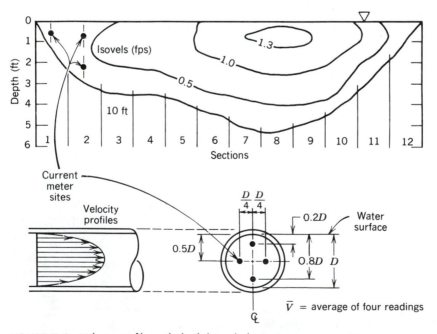

FIGURE 7.6 Velocity profiles with depth/area/velocity measurement points.

Table 7.2. This method is frequently called the velocity–area method

$$Q = \sum_i Q_i = \sum_i A_i \overline{V}_i \tag{7.2}$$

where Q = total streamflow (cfs).

Since the measurements for flow rates take time, changes in depth (stage) may occur. These changes must be documented. Steady-state (constant discharge and depth) conditions should exist for best results. Schutz (1980) presents other details and expands this discussion.

☐ **EXAMPLE PROBLEM 7.1**

For a particular stream, estimate the flow rate (runoff for this case) using the following data for velocities measured at two depths (0.2 and 0.8 of the total) and the cross-sectional area corresponding to the velocity measures.

SECTION	SAMPLE DEPTHS	1	2	3	4	5
Velocity (m/s)	0.2 D	0.4	0.8	1.2	1.0	0.6
	0.8 D	0.3	0.6	1.3	1.2	0.6
Area (m²)		3	6	10	8	4

Solution

$$Q = \sum_i Q_i = \sum_i A_i \overline{V}_i = \sum_i A_i(V_{0.2} + V_{0.8})_i/2$$

$$= 3(0.35) + 6(0.70) + 10(1.25) + 8(1.1) + 4(0.6)$$

$$= 1.05 + 4.20 + 12.50 + 8.8 + 2.4 = 29.00 \text{ m}^3/\text{s} \quad ☐$$

Since it takes considerable time to measure flow rates using a current-meter, it is best to estimate flow rates at various stages and then develop a graphical plot of flow rate versus stage, called a stage-discharge relationship. Discharges can then be estimated from stage measurements, which are easier to obtain than flow rates. The relationship also is referred to as the rating curve for the stream. Frequently, equations of best fit can be determined and computer aided solutions found. The value of the stage-discharge relationship is evident during high flow times where complete velocity measures are difficult. Extrapolation of the rating curve beyond the points of measurement is practiced but the adequacy of the estimate should be verified by other measures.

7.1.3 Dilution

In situations where conventional current meters are not useful (shallow flows, high velocities, inaccessible areas), the dilution method is used to estimate flow rates over a reach of stream. A known quantity of a substance can be injected at a point and water samples containing the substance withdrawn downstream. Ideally, the substance should be conservative, nonpolluting, and able to be detected in minute quantities. Materials that have been used in the past are common salt, sodium dichromate, oxygen 18 (^{18}O), deuterium (^{2}H), Rhodamine B, and Rhodamine W. The downstream sampling point must be chosen so that the substance has become nearly uniformly distributed across the stream cross section. An estimate for the downstream sampling point location is

$$L = 0.13C_Z[(0.7C_Z + b)/g][b^2/d] \tag{7.3}$$

where

L = length from point of injection (m)
C_Z = Chezy's roughness coefficient (See Equation 7.5)
b = average width of stream (m)
g = gravity constant (9.81 m/s^2)
d = average depth of flow (m)

Writing a mass-balance between the injection point and the downstream point as shown in Figure 7.7, one obtains the following (C'_1 and C'_2 are known and Q is unknown):

$$\text{Mass in} - \text{mass out} \pm \text{generation} = \text{accumulation}$$

For a conservative substance with no generation:

$$\text{Mass in} = \text{mass out}$$

$$C'_1 V = Q \sum_{0}^{t_2} C'_2 \, \Delta t \quad \text{and} \quad Q = (C'_1 V) \Big/ \left(\sum_{0}^{t_2} C'_2 \, \Delta t \right) \tag{7.4}$$

where

Q = unknown stream flow rate (L/s)
V = known injection volume (L)
C'_1 = injection concentration (mg/L)
C'_2 = downstream concentration (mg/L)
t_2 = time during which substance appears at location 2 (sec)

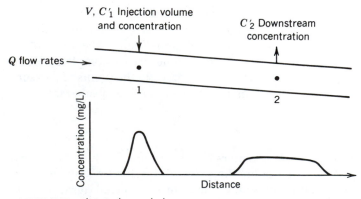

FIGURE 7.7 Chemical mass balance.

Similarly a continuous release of dye substance at a known concentration and discharge rate also permits determination of stream discharge:

$$Q = \frac{Q_1 C_1}{C_2} \tag{7.5}$$

Sufficient time must be allowed for the downstream concentration to reach equilibrium. In theory only one downstream sample is required as opposed to the extended sampling program as noted in the Example Problem 7.2.

□ EXAMPLE PROBLEM 7.2

At 07:00 hr, 400 kg of a tracer was injected immediately into a river. Two sampling points were used downstream at 14 km and 25 km. Tracer concentrations obtained from a sampling program are shown below. What is the flow rate downstream using the measurements at 14 km?

TIME (hr)	CONCENTRATION (mg/L) 14 km	25 km
0700	0	0
0800	0	0
0900	3	0
1000	10	0
1100	20	2
1200	15	9
1300	14	18
1400	7	15
1500	2	11
1600	2	7
1700	1	4
1800	1	3
1900		2
2000		1
2100		1
	$\Sigma = 75$	$\Sigma = 73$

Solution

Using the mass balance approach and Equation 7.4, at the 14 km mark, the cumulative concentration is 75 mg-hr/L and

$$Q_{14} = (400 \text{ kg} \times 10^6 \text{ mg/kg})/(75 \text{ mg-hr/L} \times 3600 \text{ sec/hr}) = 1480 \text{ L/s}$$

Note that the mass has decreased to 73 mg-hr/L at the 25-km marker. The flow rate estimate at 25 km would thus be greater than that at 14 km. The greater the loss of concentration, the greater the flow rate estimate. $Q_{25} = 1522$ L/s. □

7.1.4   Float

There are dangers to human life and inaccuracies in measuring velocities during floods. In these situations, specially prepared floats have been used to measure velocity. Two stations along the stream are marked and the distance between the points are recorded. The distance measured must be parallel to the centerline of the stream. The time it takes for the float to pass through the marked distance is recorded. For wide streams, additional floats are used. Generally, a boat or structure across the stream is used to place the floats in the stream upgradient from the first station.

The surface of the stream is at greater than average velocity. Thus, the float velocity is usually reduced by 20% to estimate the average section velocity. However, for shallow flows, the float may be submerged in the water deep enough to almost hit the stream bottom. In this situation, the average velocity is the velocity of the float.

7.1.5   Indirect Method Using Manning's Equation

This method uses equations to estimate velocities and flow rates given a measure of the physical characteristics of the stream.

One of the first uniform flow formulas for open channels was developed by Chezy (1769) (also see Chow, 1959) and is usually expressed as

$$\overline{V} = C_z R^{1/2} S^{1/2} \tag{7.6}$$

where
$\overline{V}$ = average velocity in cross-section (fps)
C_z = a roughness coefficient
R = hydraulic radius = flow area divided by the wetted perimeter (ft)
S = energy gradient, slope (ft/ft)

The most popular equation is Manning's:

$$\overline{V} = \frac{1.486}{n} R^{2/3} S^{1/2} \tag{7.7}$$

where
$\overline{V}$ = average section velocity (fps)
n = roughness coefficient for deep flow (see Table 7.3)
R = hydraulic radius = flow area divided by wetted perimeter (ft)
S = energy gradient, slope (ft/ft)

Comparing the Manning and Chezy equations, it is noted that

$$C_z = \frac{1.486 R^{1/6}}{n} \tag{7.8}$$

The metric equivalent of Equation 7.7 is

$$\overline{V} = \frac{1}{n} R^{2/3} S^{1/2} \tag{7.9}$$

where
$\overline{V}$ = m/s
R = m
S = m/m

Note that depth, width, stream profiles, and energy gradient estimates are needed. During an actual flow condition, this is time consuming. However, if the cross-sectional area is known, the use of Manning's equation is routine. For newly constructed open channels, this physical characteristic is usually known. The roughness coefficient used for newly constructed channels is the lower value of Table 7.3. For old operational channels, the higher value is generally more appropriate.

Several researchers have suggested values for Manning's n based on bed roughness (taken as the diameter of the typical bed material). Vischer (1987) quoted the earliest of these formulas due to Strickler (1923) as

$$n = \frac{d^{1/6}}{21.1} = 0.0474d^{1/6} \tag{7.10}$$

where d = particle diameter of median grain size of bed material (m)

Chow (1959) gives the typical result for most rivers in the United States as

$$n = 0.0342d^{1/6} \tag{7.11}$$

where d = particle diameter of median grain size of bed material (ft)

In many instances man-made channels and drainage canals are protected or stabilized by some form of grass or vegetative covering. The presence thereof results in increased turbulence, hence increased energy loss. Manning's n for grassed channels is typically referred to as the *retardance coefficient* (RV).

Ree and Palmer (1949) conducted a series of experiments on grass-lined channels and demonstrated that the retardance coefficient data were well correlated with the product of the mean velocity, and the hydraulic radius.

The degree of retardance is based primarily on the kind of vegetation, condition of growth, and "stand" (that is, density of grass measured as number of stems per unit area). Table 7.4 is provided as a guide. In addition, substituting for n in Equation 7.7, average velocity can be calculated by trial and error without calculating n provided the degree of retardance can be estimated.

7.1.6 Direct Measure

A direct measure of flow rates is done by the complete capture of a volume of water passing a point of reference. The time for the capture is known. Thus, the flow rate is the volume divided by time. Care must be exercised to capture the total volume of water. This limits the application of the direct measure to small flow rates found in some street inlets or laboratory models.

If the capture tank is used with some structural device for measuring depth or indirectly measuring flow rates, calibration of the structural device results. There

TABLE 7.3   Values of the Roughness Coefficient, n

TYPE OF CHANNEL AND DESCRIPTION	MINIMUM	NORMAL	MAXIMUM
A. Closed conduits flowing partly full			
a. Brass, smooth	0.009	0.010	0.013
b. Steel			
1. Lockbar and welded	0.010	0.012	0.014
2. Riveted and spiral	0.013	0.016	0.017
c. Cast iron			
1. Coated	0.010	0.013	0.014
2. Uncoated	0.011	0.014	0.016
d. Wrought iron			
1. Black	0.012	0.014	0.015
2. Galvanized	0.013	0.016	0.017
e. Corrugated metal			
1. 6 by 1 in. corrugations	0.020	0.022	0.025
2. 6 by 2 in. corrugations	0.030	0.032	0.035
3. Smooth wall spiral aluminum	0.010	0.012	0.014
f. Concrete			
1. Culvert, straight	0.010	0.012	0.013
2. Culvert with bends	0.011	0.013	0.014
3. Sewer with manholes, inlet, etc.,			
straight	0.013	0.015	0.017
g. Sanitary sewers	0.012	0.013	0.016

B. Channel conditions $n = (n_0 + n_1 + n_2 + n_3)m$ Values

a. Material involved	Earth		0.020
	Rock cut	n_0	0.025
	Fine gravel		0.024
	Coarse gravel		0.028
b. Degree of irregularity	Smooth		0.000
	Minor	n_1	0.005
	Moderate		0.010
	Severe		0.020
c. Relative effect of obstruction	Negligible		0.000
	Minor	n_2	0.010–0.015
	Appreciable		0.020–0.030
	Severe		0.040–0.060
d. Vegetation	Low		0.005–0.010
	Medium	n_3	0.010–0.025
	High		0.025–0.050
	Very high		0.050–0.100
e. Degree of meandering	Minor		1.000
	Appreciable	m	1.150
	Severe		1.300

Source: U.S. Department of Transportation, 1985, and W.L. Cowan, 1956.

TABLE 7.4 Guide in Selection of Retardance Coefficient*

STAND	AVERAGE LENGTH OF GRASS, IN.	DEGREE OF RETARDANCE		ESTIMATING EQUATION FOR COEFFICIENT
	>30	A	Very High	$n = .253(RV)^{-.53}$
	11–24	B	High	$n = .158(RV)^{-.51}$
Average conditions	6–10	C	Moderate	$n = .097(RV)^{-.46}$
	2–6	D	Low	$n = .073(RV)^{-.35}$
	<2	E	Very Low	$n = .048(RV)^{-.24}$

*Adapted from Ree et al.

are many structural devices used in practice. Some of the more common ones are presented here.

7.1.7 Weirs and Flumes

For measurements over a long period of time or for controlling discharges from a stormwater detention pond, a more permanent structural device is required. A detention pond attenuates (reduces) stormwater peak flows using an outflow control device. There are three main structures for gaging flow rates: thin-plate weirs, broad-crested weirs, and flumes. Generally, all barriers on the bottom of any channel that cause flow to accelerate while passing over are called weirs, whereas flumes are open conduits built such that the sides narrow the flow.

The thickness of the weir determines the type; narrow ones are called sharp crested, whereas thick weirs are called broad crested. The volume of the structure that is in the flow path determines the type of structure while the geometric profile determines the specific name of the structure (Figure 7.8). Table 7.5 lists weir-related terms. A schematic of a sharp-crested weir with staff gages is shown in Figure 7.9.

For all weirs, the bottom edge of the opening is called the crest or invert. For a sharp-crested weir, an energy balance can be developed to determine a general equation for flow rates. When water passes over the weir and is not obstructed, an energy balance can be written. The total available energy upstream is expressed as

$$E_{upstream} = y + V_1^2/2g \tag{7.12}$$

where y is the variable to represent changing "head" on the crest of the weir. It is measured upstream of the drawdown curve and V_1 is the approach velocity of the stream (again measured upstream of the drawdown). In some cases, the height P_1 is significant so that the approach velocity is near zero. This is also true for pond controlling weirs. The limits on y are zero to some depth H. The downstream energy is the velocity of flow. Equating energy up and downstream, one obtains by using the Bernoulli theorem with no losses:

$$0 + (V_1^2/2g) + y - \text{zero losses} = 0 + (V_{jet}^2/2g) + 0 \tag{7.13}$$

Thus, rearranging for the ideal no-loss case:

$$V_{\text{jet}} = \sqrt{2g(y + V_1^2/2g)} \tag{7.14}$$

an ideal

$$dQ = dA V_{\text{jet}} = (B \, dy) V_{\text{jet}} \tag{7.15}$$

TABLE 7.5 Comparison of Size and Flow Rate Range (Reproduced in part from Ackers et al., 1978)

TYPE	SIZE AND GEOMETRY			DISCHARGE RANGE
Thin-plate, full-width weir	B	P_1		
	0.15 m	0.2 m		0.8 L/s–100 L/s
	1.0 m	0.5 m		5.4 L/s–2.7 m³/s
	10 m[a]	1 m		50 L/s–77 m³/s
	B	P_1		
Thin-plate, side-contracted weir	0.15 m	0.2 m		1.4 L/s–67 L/s
	1.0 m	0.5 m		9.5 L/s–1.7 m³/s
	10 m[a]	1 m		90 L/s–49 m³/s
V-notch weir	$\theta = 20$[a]			0.2 L/s–330 L/s
	$\theta = 90$[a]			1.1 L/s–1.8 m³/s
	B	P_1		
Triangular-profile (Crump) weir, 1:2/1:5	0.3 m[b]	0.2 m		3 L/s–350 L/s
	1.0 m[b]	0.5 m		10 L/s–4.6 m³/s
	10 m[c]	1 m		0.3 m³/s–130 m³/s
	100 m[a,c]	1 m		3 m³/s–1300 m³/s
Triangular-profile weir, 1:2/1:2	B	P_1		
	0.3 m[b]	0.2 m		3 L/s–300 L/s
	1.0 m[b]	0.5 m		11 L/s–3.9 m³/s
	10 m[c]	1 m		0.3 m³/s–110 m³/s
	100 m[a,c]	1 m		3 m³/s–1100 m³/s
Flat-V weir, 1:2/1:5	P_1, P_2, P_v	Slope	B	
	0.2 m	1:10	4 m	14 L/s–5.0 m³/s
	0.5 m	1:20	20 m	27 L/2–180 m³/s
	1 m[a]	1:40	80 m	55 L/s–630 m³/s

continued

TABLE 7.5 *(Continued)*

TYPE	SIZE AND GEOMETRY			DISCHARGE RANGE
Flat-V weir,	P_1, P_2, P_r	Slope	B	
$1:2/1:2$	0.2 m	$1:10$	4 m	15 L/s–2.5 m³/s
	0.5 m	$1:20$	20 m	30 L/s–65 m³/s
	1 m[a]	$1:20$	40 m	30 L/s–330 m³/s
Rectangular-	B	P_1	L	
profile weir	0.3 m	0.2 m	0.8 m	8 L/s–180 L/s
	1 m	0.5 m	2 m	90 L/s–2.3 m³/s
	10 m[a]	1 m	2 m	1.5 m³/s–65 m³/s
	B	P_1	L	
Round-nosed	0.3 m	0.15 m	0.6 m	8 L/s–34 L/s
horizontal-	1 m	0.15 m	1 m	25 L/s–740 L/s
crested weir	10 m	1 m	5 m	1 m³/s–82 m³/s
	100 m[a]	1 m	5 m	10 m³/s–820 m³/s
Long-throated	B		L	
flumes	0.5 m		1 m	9 L/s–300 L/s
	0.1 m		1 m	3 L/s–290 L/s
	1 m[a]		4 m	270 L/s–41 m³/s
Parshall	B			
flumes	25.4 mm			0.1 L/s–5 L/s
	0.305 m			3 L/s–450 L/s
	2.438 m			0.1 m³/s–3.9 m³/s
	15.24 m			0.75 m³/s–93 m³/s

[a] There is no upper limit specified for the size of these structures.
[b] Lower limit of head assumed to be 0.03 m for smooth crest section.
[c] Lower limit of head assumed to be 0.06 m for concrete crest.

and integrating between $y = 0$ to H with C_f = friction loss coefficient:

$$Q = BC_f\sqrt{2g} \int_0^H (y + V_1^2/2g)^{1/2} \, dy$$

$$= 2/3 BC_f\sqrt{2g} \, [(H + V_1^2/2g)^{3/2} - (V_1^2/2g)^{3/2}] \tag{7.16}$$

and if V_1 approaches zero, with $C = \frac{2}{3}C_f$ (approximately)

$$Q = C\sqrt{2g} \, BH^{3/2} \tag{7.17}$$

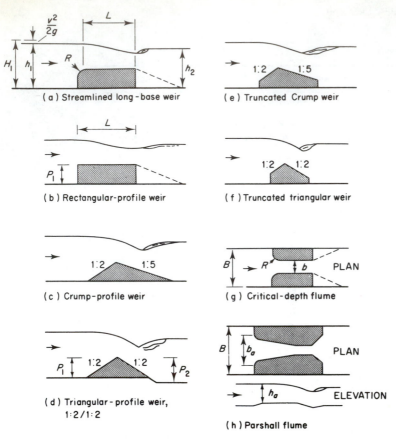

FIGURE 7.8 Comparison of volumes of gaging structures in lined rectangular channel: layout diagram (reproduced from Ackers et al., 1978, *Weirs and Flumes for Flow Measurement*).

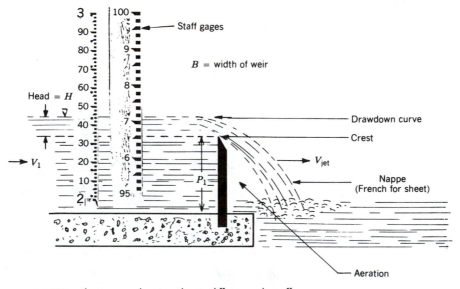

FIGURE 7.9 Sharp crested weir with two different style staff gages.

where

Q = flow rate (L^3/T)
B = width of weir (L)
H = head or depth of flow above weir crest (L)
g = gravitation acceleration constant (L/T^2)
C = loss coefficient for a weir

As noted by the air under the nappe of the discharge in Figure 7.9, there can be no downstream backwater effects on the weir. Also, note that the upstream water profile decreases as it approaches the crest, thus the head is measured at a point upstream equal to at least five times the weir width. This assumes an accuracy of measurement equal to ±1 percent.

For a rectangular weir extending the width of the channel, Equation 7.17 is rewritten as

$$Q = C_w BH^{3/2} \qquad (7.18)$$

where

Q = discharge (cfs, m^3/s)
C_w = 3.33 (U.S.), 1.84 (SI) and is dependent on P_1/H, L/H, N_R, shape, roughness (Note: this discharge coefficient includes $\sqrt{2g}$)
B = width of weir (ft, m)
H = head (ft, m)

The coefficient C_w has been related to other flow depths and can change by as much as 10% when the head varies (see Figure 7.16) For a rectangular weir not extended across the channel or one that has its ends contracted and negligible velocity head

$$Q = C_w H^{3/2}(B - 0.2H) \qquad (7.19)$$

Other types are also used and are listed as follows.

For low flows: 60° V-notch

$$Q = C_{60}H^{5/2} \qquad (7.20)$$

where C_{60} = 1.43 (U.S.), 0.80 (SI), (See Figure 7.17 for coefficient values as a function of head) and 90° V-notch

$$Q = C_{90}H^{5/2} \qquad (7.21)$$

where C_{90} = 2.50 (U.S.), 1.38 (SI) (Brater and King, 1980).

❑ **EXAMPLE PROBLEM 7.3**

A channel is known to produce a peak flow rate of 0.45 m^3/s. The upstream depth must not exceed 2.25 m. If you have a 1.5-m wide end contracted rectangular weir, how high should it be placed in the channel so as not to exceed the 2.25 m upstream depth?

Solution

Using the end-contracted weir formula and solving for the upstream depth (H):

$$Q = 1.84H^{3/2}(B - 0.2H) \qquad 0.45 = 1.84H^{3/2}(1.5 - 0.2H)$$

$$0.45 = (2.76H^{3/2}) - (0.368H^{5/2})$$

by trial and error, $H = 0.31$ m. Thus, the weir height above the bottom of the channel is

$$P_1 = 2.25 - 0.31 = 1.94 \text{ m or less} \quad \square$$

One of the oldest flumes and most widely used to estimate flow rate in irrigation and water treatment is the Parshall flume (Parshall, 1950). It is constructed to converge water through critical depth near the end of the convergence. The geometry of the flume is shown in Figure 7.10. The flume has the advantages of less head loss and solids deposition relative to a weir. There are 22 standard designs that estimate flow rates within ±3% accuracy. The dimensions of Figure 7.10 (A, B, C, and D) change with the overall flume size and the constants in Equation 7.17.

$$Q = 4Bh_1^{1.522B^{0.026}} \tag{7.22}$$

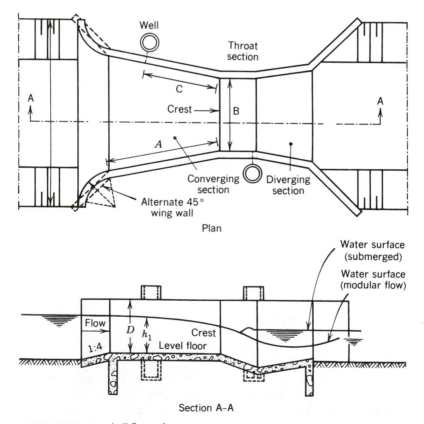

FIGURE 7.10 Parshall flume diagram.

where

Q = flow rate (m³/s, cfs)
h_1 = upstream depth (m, ft)
B = width of throat section (m, ft)

□ *EXAMPLE PROBLEM 7.4*

A Parshall flume of throat width equal to 2.0 ft will be used to measure peak discharge. The expected peak stage at the flume well is 2.5 ft. What is the peak discharge which this flume can estimate?

Solution

Calculate the flow rate using Equation 7.22.

$$Q = 4Bh_1^{1.522B^{0.026}}$$

or

$$Q = 4(2)(2.5)^{1.522(2)^{0.026}}$$

$$= 33.1 \text{ cfs} □$$

The type and size of weir or flume that is chosen for a particular application depends on the water quality (chemical composition and debris) and the range of discharge expected. Metals susceptible to corrosion should not be used for a weir or flume conveying water with a high corrosion potential. For very small discharges, V-notch weirs are used; for medium discharges, full-width thin plates and long-base weirs and flumes are used; and for large discharges, long-base structures are used. A comparison of discharge range, size, and geometry is shown for the different types of weirs and flumes in Table 7.5.

7.1.8 Other Gaging and Outlet Controls

With the growing use of stormwater storage facilities (detention ponds) to reduce peak flow rates, a wide range of flows must be considered in design and gaging. Rectangular weirs are commonly used to discharge and measure high flows and orifices are used to discharge and measure smaller flows. An orifice is an opening usually circular in shape. Typical outlet controls from a detention pond are shown in Figure 7.11. The side orifice in the riser pipe is used to slowly drain detained water, whereas the overflow weir and emergency spillway operate as weirs. Once the riser pipe is submerged, its discharge will be governed by the orifice equation rather than the weir equations. The equation for orifice flow is

$$Q = C_d A_0 \sqrt{2gH} \qquad (7.23)$$

where

Q = flow rate (cfs, m³/s)
C_d = discharge coefficient
A_0 = orifice area (ft², m²)

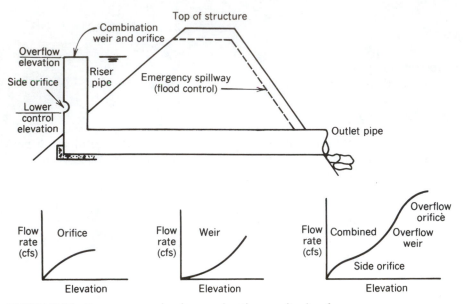

FIGURE 7.11   Stormwater pond outlet controls with generalized performance curves.

H = head (measured from center line of orifice) (ft, m)
g = gravitational constant (32.2 ft/sec², 9.81 m/s²)

The discharge coefficient will vary with the size, type of orifice (sharp edged, rounded, etc.), and the head. For a sharp-edged, 1-in. (25-mm) orifice, the coefficient varies in U.S. units between 0.609 and 0.594 for heads varying between 1 and 60 ft. For most applications, a discharge coefficient of 0.6 is used.

□   **EXAMPLE PROBLEM 7.5**

For a 4-in. orifice with a discharge coefficient of 0.60, what head results when measuring a discharge of 0.87 cfs?

Solution

Using Equation 7.23, the head can be calculated as follows.

$$Q = C_d A_0 \sqrt{2gH}$$

$$0.87 = 0.60(0.087)\sqrt{2(32.2)(H)}$$

$$H = 4.3 \text{ ft} \quad \square$$

7.2

NETWORK OF GAGES

Surface and groundwater gaging systems are costly and time consuming but affect many social activities. Gaging is therefore usually the responsibility of a government body such as the U.S. Geologic Survey in the United States or the regional water authority in Great Britain. The density of gaging stations depends on the nature of the terrain and the purpose of the station. In countries conducting streamflow measurements, there are:

1. Primary or principal stations defined as permanent stations to measure most ranges of discharges with records as accurate and complete as possible.
2. Secondary stations operating for short periods of time to obtain a satisfactory correlation with the record of a primary station; their function is usually to provide hydrological knowledge of streams likely to be used for future studies.
3. Special stations serving particular needs, such as reservoir sizing, determining runoff volumes from a small urban watershed, and dry weather flow stations for estimating abstractions.

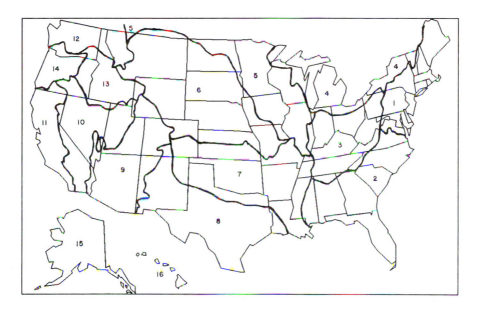

USGS Watersheds Corresponding to the Numbers
on the Map

WATERSHED	NAME	WATERSHED	NAME
1	North Atlantic	9	Colorado River
2	South Atlantic	10	The Great Basin
3	Ohio River	11	Pacific Slope California
4	St. Lawrence	12	Pacific Slope Washington
5	Hudson Bay and	13	Snake River
	Upper Mississippi	14	Pacific Slope Oregon
6	Missouri	15	Alaska
7	Lower Mississippi	16	Hawaii and Others
8	Western Gulf of Mexico		

FIGURE 7.12 Major river watersheds in the United States (from *U.S. Geological Survey*).

TABLE 7.6 Mean Discharge (cubic meters/second) at Selected Gaging Stations for River Basins in 1977 in the United Kingdom

STATION NUMBER	LOCATION OF FLOW MEASUREMENT STATION	WATER AUTHORITY/ RIVER PURIFICATION BOARD	GAGED FLOW												ANNUAL MEAN	LONG-TERM MEAN GAGED DISCHARGE	CATCHMENT AREA (SQUARE) (km²)
			JAN	FEB	MARCH	APRIL	MAY	JUNE	JULY	AUG	SEPT	OCT	NOV	DEC			
006007	Ness at Ness Side	Highlands	60.91	52.75	106.14	109.24	82.42	18.63	34.50	35.54	67.80	101.98	202.80	77.51	70.09	73.14	1,839.1
019001	Almond at Craigiehall	Fourth	10.13	13.74	5.42	4.11	3.72	3.58	1.10	1.68	4.34	11.64	14.32	5.85	6.58	5.06	369.0
025001	Tees at Broken Scar	Northumbrain	34.41	40.70	24.38	60.87	13.04	6.16	2.53	3.32	5.45	11.19	29.31	27.25	21.35	16.38	817.7
027001	Nidd at Hunsingore	Yorkshire	18.14	29.39	13.12	8.29	9.84	4.73	2.68	2.57	2.43	3.91	14.52	14.80	10.24	8.01	484.3
028009	Trent at Colwick	Severn-Trent	188.46	387.50	136.01	77.66	77.10	59.71	31.42	34.87	26.74	30.01	82.54	106.95	101.37	79.44	7,486.0
038005	Ash at Easneye	Thames	0.71	1.08	0.62	0.30	0.25	0.18	0.54	0.55	0.61	0.43	0.45	0.82	2.15	1.59	297.9
039001	Thames at Teddington	Thames	193.49	240.04	159.13	82.48	64.03	58.31	19.67	53.90	26.97	24.77	51.57	114.49	90.15	67.50	9,950.0
040011	Great Stour at Horton	Southern	8.06	7.29	6.29	4.32	3.29	2.31	1.78	2.01	1.65	1.53	2.61	4.44	3.28	3.30	345.0
052006	Yeo at Penn Mill	Wessex	6.56	7.55	5.40	2.05	1.84	1.095	0.68	0.85	0.61	0.61	1.62	6.10	2.88	2.46	213.1
045001	Seven at Bewdley	Severn-Trent	107.48	176.91	72.03	47.88	40.05	30.23	12.18	11.43	14.44	28.45	108.89	94.19	57.65	61.97	4,330.0
056001	Usk at Chain Bridge	Welsh	43.28	74.66	43.20	21.03	16.90	9.10	4.51	6.63	7.16	18.25	59.35	47.28	28.96	26.96	911.7
067015	Dee at Manley Hall	Welsh	43.37	83.99	34.81	26.54	19.34	13.83	10.66	9.50	11.39	23.23	76.62	43.70	32.67	27.66	1,019.0
079002	Nith a Friar's Carse	Solway	35.18	40.24	36.00	28.19	19.47	3.47	2.32	5.96	25.27	63.72	65.77	38.25	30.26	24.74	790.0
084013	Clyde at Daldowie	Clyde	56.18	74.81	57.79	49.77	39.12	15.62	13.36	17.46	50.01	101.51	108.99	50.72	52.73	36.90	1,903.0

Source: From United Kingdom Water Resources Data, 1979.

Some national, regional, and local agencies provide records of flow data from measurement programs. The U.S. Geological Survey (USGS) is one such agency. The USGS provides data for 16 major river watersheds in the United States, defined roughly as shown in Figure 7.12 and the accompanying legend. Each major watershed is divided into smaller ones and a vast quantity of data are available. In more recent history, smaller watersheds are being monitored and additional data are now available. Most of the streamflow records published by the USGS are found in two series of publications: the Water Resources Data and the Water Supply Papers.

Each structure for measuring flow is known as a gaging station. The printed records from the station are arranged in downstream order beginning with the main stream of the river. The record numbers are sequenced proceeding downstream until the main stream is joined by a tributary, which has a gaging station on it somewhere. The next record is then the most upstream gaging station on the tributary. The records of all the stations on the tributary are arranged in downstream order until the tributary joins the main stream. Other countries follow a similar ordering procedure. Gaging data for the United States are collected by the USGS and are accessible via the internet at http://h2o.usgs.gov/.

Mean discharge at some gaging stations in Great Britain are shown in Table 7.6. These gaging stations were selected to include major river basins with complete flow records for 1977 and to give a representative assessment of the variability of 1977 runoff for Great Britain. The annual means are the monthly totals divided by the total number of days in 1977. In most cases, long-term mean gaged discharges are based on daily records of five years or more. These type of records are common for most developed countries.

Groundwater gaging is done primarily to determine storage volumes and the direction of groundwater movement. Generally, water levels are recorded. Groundwater monitoring is becoming more and more important as surface waters are depleted and polluted, and wells are overpumped.

Major aquifers have been valuable sources of high quality water for many years with some areas relying totally on groundwater supplies. The lowering of regional water tables by the overdevelopment of the groundwater has led to a new appreciation of aquifers for water storage. When there are few measuring stations and data records are limited in length, then the analyses become more involved and statistical probabilities less assured.

7.3

WATER QUALITY MEASUREMENT

One of the major objectives of stormwater management is the reduction of pollutants within stormwater runoff that cause unwanted physical, chemical, and biological changes in the receiving waters (Wanielista and Yousef, 1993). As water moves across the land surface during or after a storm, it transports dissolved and suspended materials that have been picked up along the flow path. Sources of washoff materials include sediments, mineral salts, heavy metals, nutrients, pesticides, biodegradable organics, microbial contaminants, and a host of items due to man's activities.

These materials are in effect concentrated by precipitation and runoff and can cause significant changes in species and diversity of plant and animal populations. When these changes become socially unfavorable, the material causing the modification in water quality is called a pollutant. It is important for the hydrologist to recognize that water quality is inextricably linked to water quantity. Therefore, the integrity of receiving waters can be maintained and restored only as a result of thorough understanding of the water quality responses due to varied point and nonpoint sources of pollutants.

7.3.1 Sampling Techniques

Water quality sampling and analysis programs are necessary to determine the impact on receiving waters of the combination of various hydrologic events acting on various pollutant sources. Carefully designed sampling programs provide data essential to the understanding of the behavior of natural systems and the influence of human activity.

Sampling for nonpoint-source response curves should be carefully related to the hydrograph and concentration changes. Usually, the most reliable sampling is done by manual means, provided the labor force involved understands its tasks and is punctual. (Since storm events are difficult to predict it is easy to miss early parts of the hydrograph.) Thus instrumentation is often used to collect stormwater runoff and rainfall samples. Minimum instrumentation considered necessary for sampling should provide a record of rainfall, flow rates, concentrations, and mass loadings versus time. Continuous documentation of precise changes should not be expected due to the many variables related to storm runoff and the cost involved. The U.S. Geological Survey (1970, 1976, 1977) has published a series of manuals dealing with sampling and analysis procedures for stormwater runoff, receiving waters, and groundwater. Typical schematic diagrams are shown in Figure 7.13.

Grab sampling is often used, but it is not recommended. One instantaneous value is obtained, which most likely is not representative of the average conditions; and the variability of the concentrations and mass changes cannot be described. To estimate mass per event and average concentration per event, two methods are generally used: (1) volumes of stormwater are sampled and mixed together to obtain a flow-weighted composite sample; and (2) sequential discrete samples are taken at equal time intervals, and each sample concentration is determined. For the flow-weighted method, each sample is mixed with a volume of water directly proportional to the flow at the time of sampling. Samples taken at equal time intervals are weighted relative to the flow at the time of sampling. The event load is determined by multiplying each concentration and flow-weighted volume at the time of collection.

Similarly, equipment for sediment measurements varies widely. Instantaneous suspended-sediment samplers are often used but in effect are essentially grab sampling devices. Studies of their effectiveness indicate that they were not suitable for general field use (ASCE, 1975).

Alternatively, time-integrating, suspended-sediment samplers have come into widespread use. A time-integrated sampler is one that collects a sample during a finite time interval. Time-integrating samplers are of two types: depth-integrating and point-integrating. The depth-integrating sampler fills while it is moved at a uniform transit rate in a stream vertical. The point-integrating sampler fills while suspended at a selected point in a stream vertical.

Care must be exercised to coordinate transit rate with stream velocity, depth, air pressure in the sampler container, and nozzle size controlling flow into the sampler.

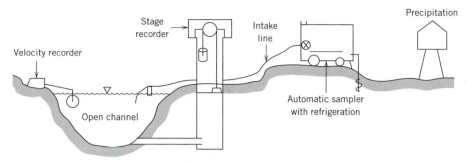

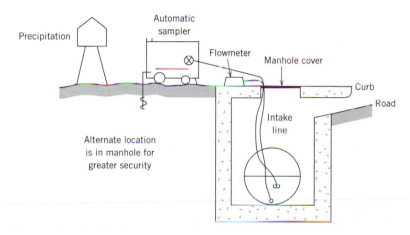

FIGURE 7.13 Schematic diagrams for sampling.

The DH-48 sampler with wading rod suspension is used in shallow streams when the product of depth, in feet, and velocity, in feet per second, does not exceed 10, is a typical example (see Figure 7.14).

7.3.2 Computation of Pollutant Load

The event mean concentration (EMC) is defined as the ratio of the event load for a specific contaminant divided by the event water volume:

$$\overline{C} = \frac{M}{R} \tag{7.24}$$

where
M = mass loading per event, mg
R = water volume per event, L
$\overline{C}$ = event mean concentration (EMC), mg/L

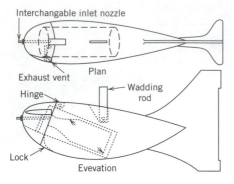

FIGURE 7.14 The DH-48 depth-integrating hand sampler for use on small streams.

The loading for an event is determined by summing the loadings during each sampling period, provided the flow rate (or volume) data and concentration data are available for the period. The following equation is used:

$$M = \sum_{i=1}^{n} R_i C_i \qquad (7.25)$$

where

R_i = volume proportional to flow rate at time i, L
C_i = concentration at time i, mg/L
n = total number of samples during a single storm event

A plot of concentration, flow rate, and loadings will produce response curves for the nonpoint sources. Typical continuous plots are shown in Figure 7.15.

The event mean concentration is a statistical parameter used to represent a flow-weighted concentration of a desired water quality parameter during a single storm event. If sequential water samples collected, the *EMC* can be measured as the concentration in a flow-weighted composite sample with the volume of each fraction directly proportional to the flow at time of collection. On the other hand, if sequential discrete samples are collected over the hydrograph, the *EMC* value can be determined by calculating the cumulative mass of pollutant (loadograph) and dividing it by the volume of runoff (area under the hydrograph).

The long-term average loading rate was given by DiToro et al. (1979) as

$$\overline{M} = \frac{MD}{\Delta} \qquad (7.26)$$

where

D = average duration of storms
Δ = average time between storms (interevent time)

7.3.3 Sediment Load Computations

The presence of sediment in streams, rivers, ponds, and reservoirs has its origin in soil erosion. Sediment production, known as gross erosion, is the result of disintegration

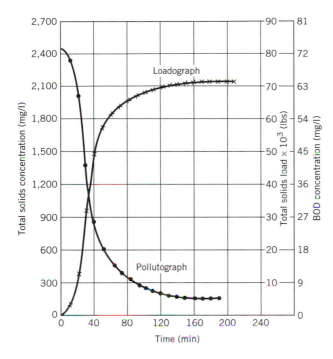

FIGURE 7.15 Typical plot of pollutant loading for a storm event.

and decomposition of earth materials. Disintegration includes processes by which materials are broken into smaller bits and pieces without substantial chemical change. All the geomorphological processes (weathering, large temperature changes, flow of water, wind movement, erosional forces, mechanical abrasion, etc.) are involved. Decomposition refers to the breaking down of mineral components of the earth materials by chemical reaction. The geochemical processes of solution, hydration, oxidation, and carbonation are included.

These processes have the effect of land degradation and loss of productivity. It is estimated that there is an irreversible loss in productivity on about 6 milion hectares of fertile land each year worldwide, Lal (1994). Oldeman (1991) estimated that human-induced soil degradation has affected about 24% of the inhabited land area. The present rate of cropland erosion in the United States is of the order of 13 metric tons per hectare per year (Crosson, 1995).

All eroded material does not enter the stream system. Some is deposited at natural or man-made barriers, or may be deposited within stream channels or their flood plain. Thus sediments that are produced are in transit througout the watershed. Sediment eroded at one location may be stored or deposited temporarily at various locations downstream and subsequently remobilized several times before reaching the watershed outlet. Frequently, large amounts of sediment are trapped in ponds or reservoirs and therefore immobilized and tending to reduce the water storage capacity of the pond or reservoir.

The portion of eroded material which does travel through the drainage network to a downstream measuring point is referred to as the sediment yield. The concept of a sediment delivery ratio, SDR, has been used to relate the gross erosion rates to the delivery of sediments, or sediment yield, to the watershed outlet:

$$SDR = \frac{SY}{A} \qquad (7.27)$$

where

SY = sediment yield at downstream point (t/km²/yr)

A = gross erosion or sediment production (t/km²/yr)

The revised universal soil loss equation (RUSLE) developed by the Natural Resource Conservation Service (NRCS, 1995) based on the earlier work of Wischmeier and Smith (1978) is widely used to predict gross erosion rates:

$$A = R \, K \, L \, S \, C \, P \qquad (7.28)$$

where

A = the computed soil loss due to sheet, rill and gully erosion (tons/acre/year, or metric tons/km²/year)

R = is the rainfall-runoff erosivity factor (taken as the average value of the annual sum of the products of kinetic energy and maximum 30-minute rainfall intensities.

K = is a soil erodibility factor based on USDA soil types defined as the soil loss per unit of rainfall from a standard unit plot.

L = is the slope length factor.

S = slope steepness factor.

C = is the crop-management factor, i.e., vegetative cover and management practice.

P = an erosion-control-practice factor, e.g. ratio of soil loss under a given practice (such as contouring or strip-cropping to that under straight-row farming.

T = LS are frequently combined into one topographic factor.

The multiplication of Equation 7.28 by the sediment delivery ratio of Equation 7.27 adjusts the estimated soil loss to an estimate of sediment yield. The general trend of the sediment delivery ratio with drainage area is given as

$$SDR = 27.04 \ A^{-.23} \qquad (7.29)$$

where

SDR = is in percent

A = drainage area in square miles

It is noted that the SDR varies from essentially 100% for watersheds of less than 1 sq mi to less than 5% for watersheds larger than 1000 sq mi. Sampling remains as the surest method of determining the sediment load being transported in streams and deposited in reservoirs due to the many hydrologic variables involved.

7.4
SUMMARY

Flow rate measurements provide data for an accounting of water in the hydrologic cycle. Accurate estimates are needed and there are at least three general methods: stage, nonstructural discharge, and structural discharge.

- The most widely used method for continuous data collection is stage recording. Stage is depth of flow and is converted to flow rate.
- Weirs, orifices, and flumes have been used to measure flow rate and equations developed which relate head (stage) to discharge. Weir-related terms are shown in Table 7.7.

TABLE 7.7 Weir-Related Terms

Broad crested	A weir having a substantial length of crest parallel to the direction of flow of water over it, on which the nappe is supported for an appreciable length, and which produces no bottom contraction of the nappe.
Cipolletti	A contracted weir of trapezoidal shape, in which the sides of the notch are given a slope of 1 horizontal to 4 vertical, in order to compensate as far as possible for the effect of end contractions.
Contracted	A weir whose crest extends only part way across the channel in which it is installed, and is terminated by extensions in the same plane as the crest, these extensions rising above the water level on the upstream side of the weir, their effect being to produce a contraction in the width of the stream of water as it leaves the notch.
Crib	A low diversion weir built of log cribs filled with rock.
Diverting	A weir placed in a sewer for the purpose of diverting storm flow. Also called an overflow, overfall, or side-flow weir in combined sewers.
Draw-door	A diversion weir fitted with doors or gates capable of being raised vertically so as to retain water when desired.
Effluent	A weir at the outflow end of a hydraulic structure.
Flat crested	A weir whose crest is horizontal in the direction of flow, and of appreciable length when compared with the depth of water passing over it.
Free	A weir that is not submerged; a weir in which the tail water is below the crest or where the flow is not in any way affected by tail water.
Influent	A weir at the inflow end of a hydraulic structure.
Irregular	A weir whose crest is not of standard or regular shape.
Leaping	An opening or gap in the invert of a combined sewer through which the dry-weather flow will fall to a sanitary sewer and over which a portion or all of the storm flow will leap.

TABLE 7.7 *(Continued)*

Log	A low weir, of triangular cross section, built of layers of logs placed side by side with butt ends downstream.
Parabolic	A weir with a notch parabolic in shape with the axis of the parabola vertical.
Rectangular	A weir whose notch is rectangular in shape.
Rounded crest	A weir whose crest is convex upward in the direction of flow over the weir. The term is also applied to weirs where the center of the crest may be flat, and the corners may be rounded.
Sharp crested	A weir whose crest, usually consisting of a thin plate (generally of metal), is so sharp that the water in passing over it touches only a line.
Submerged	A weir which, when in use, has the water level on the downstream side at an elevation equal to, or higher than, their weir crest; the rate of discharge is affected by the tailwater.
Suppressed	A weir with one or both sides, flush with the channel of approach. This prevents contraction of the nappe adjacent to the flush side. The suppression may occur on one end, or both ends.
Sutro	A weir with at least one curved side and horizontal crest, so formed that the head above the crest is directly proportional to the discharge.
Trapezoidal	A weir whose notch is trapezoidal in shape.
Triangular	A weir whose notch is triangular in shape, usually used to measure very small flows.
V-notch	Another term for Triangular.
Wide crested	Another term for Broad Crested.

Source: Glossary—Water and Sewage Control Engineering, Joint Task Force. Published by American Society of Civil Engineers, New York, updated 1955.

- The Manning formula (Equation 7.7) is in general use for pipe and regular shaped open channels. Again, stage or depth of flow is needed along with other physical estimates of channel characteristics.
- Stormwater detention ponds have weir and orifice combination outlet control structures. Flow rate versus storage volume or depth curves can be developed that reflect an estimate for discharge rate as a function of pond depth and storage volume.
- All flow rate measurement methods should be calibrated on a periodic basis. The bed depth of a river changes and devices malfunction. The accuracy of the flow rate data is only as good as the calibration.
- There exists a network of streamflow gages, the data from which may be of some value for other sites.
- Water quality measurements (including estimates of sediment yields) are becoming increasingly important. Sampling techniques are discussed, and pollutant load computations are outlined.

7.5

PROBLEMS

7.5.1 Hand Problems

1. Recalculate the discharge for Example Problem 7.1 if the area of Section 4 were doubled.

2. A 3-m-wide rectangular weir is placed across an open channel. There are end contractions. Graph a flow rate diagram for the catchment which drains into the channel, given the following data on time and the corresponding head on the weir (meters).

Time (min)	0	20	40	60	80	100	120	150	200
Head (meters)	0	0.4	0.9	1.2	1.0	0.8	0.6	0.3	0

3. For the catchment that produced the flow rate diagram for Problem 2, the land use was changed and a 1520-mm concrete pipe on a longitudinal slope of 0.05 replaced the open channel. Graph the flow rate for the storm event that produced the following pipe flow depth, area, and hydraulic radius. Assume that flow rate starts at time zero.

Time (min)	0	20	40	60	80	100
Flow depth (m)	0	0.76	1.52	0.91	0.46	0
Flow area (m^2)	0	0.90	1.81	1.13	0.46	0
Hydraulic radius (m)	0	0.38	0.38	0.42	0.26	0

4. Assume the storms that produced the flow rate graphs of Problems 2 and 3 were similar and in fact the rainfall volume was 100,000 m^3 over the catchment. Estimate the runoff coefficient for each problem. Discuss the differences in hydrograph shapes and runoff coefficients.

5. a. A conservative chemical was added to an open channel to determine runoff in the channel. The chemical dosage rate was 6000 mg/L and 400 L were added as a slug. At a downstream station, the center of chemical mass was measured 4 min after release upstream and the average concentration was 10 mg/L. Estimate the runoff rate. State your assumptions.

   b. If the cross-sectional area at the measuring point downstream is 2 m^2 and the distance from the upstream injection point to the downstream point is 120 m. Estimate the runoff flow rate. The time of travel remains at 4 min.

6. Assume a 100-mm-diameter orifice is used to drain a 2-m^2 tank holding stormwater from the roof of a small building. The coefficient of discharge is 0.65. How long does it take to lower the tank from 2.5 m to 1.0 m above the center line of the orifice if the tank area varies as $A = 2 - (y/3)$ (m^2) where $y =$ depth? *Hint*:

$$q_t = \frac{1}{C_d A_0 \sqrt{2g}} \int_{y_1}^{y_2} A y^{-1/2} \, dy$$

7. For the stream cross section shown in Figure 7.6, provide some reasonable velocity measures for each of the 12 sections and estimate the flow rate for the complete cross section.

8. For Example Problem 7.2, what is the discharge estimate at the 25-km marker? What percentage of the substance was lost between the two stations? Comment on how this loss affects your discharge estimate.

9. Specify the size of orifice in a riser pipe to measure a flow rate of 0.5 cfs from a storage reservoir when the head on the orifice is 3 ft and the coefficient of discharge is 0.60.

10. What depth of flow above a weir crest can be expected upstream of a sharp-crested rectangular weir extending the width of a rectangular channel if the channel width is 22 ft and the expected discharge is from an 8-acre impervious watershed located in Orange County, Florida with time of concentration equal to 20 min. The storm used for the depth analysis is the 1-in-5-year event. Also, what is the depth of the channel if the design velocity is one foot per second?

11. A rectangular sharp crested weir with end contractions is 1.5 m wide. How high should it be placed in a channel to maintain an upstream depth of 2.25 m for 0.45 m^3/s flow?

12. Using Figures 7.16 and 7.17, what size and type of weir would you use to measure flows in a drainage ditch 3 ft wide and 2 ft deep if the expected flows were about

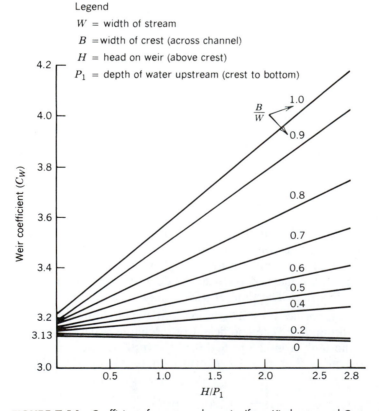

FIGURE 7.16 Coefficients for rectangular weirs (from Kindsvater and Carter, 1959).

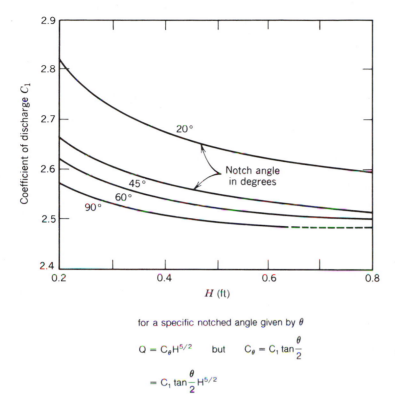

for a specific notched angle given by θ

$$Q = C_\theta H^{5/2} \qquad \text{but} \qquad C_\theta = C_1 \tan\frac{\theta}{2}$$

$$= C_1 \tan\frac{\theta}{2} H^{5/2}$$

FIGURE 7.17 Discharge coefficients for sharp-crested V-notch weirs (from Brater and King, 1980).

0.5 to 1.5 cfs? The flow depth in the channel cannot exceed 2 ft. Base your decision on the accuracy of measuring depth.

13. A 20-in. diameter pipe is placed in a vertical position at the end of a stormwater pond. The pipe acts as a weir for heads up to 1.5 ft. At heads of 0.1 ft and 1 ft, the flow rates are 0.52 cfs and 10.5 cfs, respectively. Using $Q = C_W B H^n$ as the general equation form for a weir, estimate the weir coefficients C_W and n.

14. a. What depth of flow can one expect using a 4-ft-wide contracted rectangular weir whose crest is 2 ft off the channel bottom and the discharge was 25 cfs?

b. The maximum depth above the crest of a 60° V-notch weir is 2 ft. What is the maximum discharge that can be measured using this weir?

c. How large of an impervious area (acres) can be measured for the 10-year storm if the design intensity is 8 in./hr using the 60° V-notch weir and the results of part (b)?

15. a. For an 8-in. concrete pipe ($n = 0.012$) carrying 1 cfs, what is the slope to maintain a velocity of 2 ft/sec? The pipe is flowing full.

b. What is the invert downstream if the pipe length is 1000′ and the invert upstream is 80.0′?

16. Calculate the upstream invert of a 200-ft pipe flowing full at a design velocity of 2 fps if the downstream invert elevation is 10.5 ft. The pipe is discharging 10 cfs and is made of concrete. Also, what is the pipe diameter (best commercial size)?

TABLE 7.8 Watershed Data 10B Lake Okeechobee and the Everglades Area 02264495 Shingle Creek at Campbell, FL

DAY	OCT	NOV	DEC	JAN	FEB	MAR	APR	MAY	JUN	JUL	AUG	SEP
					DISCHARGE (ft^3/sec), WATER YEAR OCTOBER 1972 TO SEPTEMBER 1973							
1	53	39	71	76	298	115	88	41	36	97	293	319
2	63	35	65	72	281	105	101	41	36	128	366	384
3	78	32	65	73	359	94	101	30	36	75	332	389
4	84	42	62	69	297	94	121	41	33	134	301	369
5	71	73	62	69	256	89	124	41	30	237	279	386
6	70	66	62	65	233	78	108	40	30	217	278	401
7	60	49	65	65	214	84	85	37	36	281	305	403
8	60	49	59	54	196	73	136	33	33	220	332	396
9	60	42	59	75	189	79	120	49	30	168	330	406
10	66	39	55	67	445	30	111	53	33	151	303	417
11	73	39	52	80	352	80	90	41	30	127	281	414
12	73	39	55	189	291	73	81	37	30	119	274	428
13	63	39	52	161	270	72	75	33	27	139	260	440
14	56	43	41	130	274	60	83	36	27	128	251	404
15	56	49	41	112	286	60	82	32	30	125	241	391
16	53	42	63	108	274	54	81	38	36	103	208	387
17	53	45	49	112	252	58	72	31	30	91	184	354
18	53	42	50	121	239	62	62	27	24	82	164	306
19	53	39	44	121	246	68	65	33	33	115	149	284
20	65	42	44	118	214	62	64	32	36	118	150	264

Day												
21	63	49	37	115	193	50	55	32	47	88	157	240
22	53	45	106	154	178	53	51	31	97	82	183	240
23	49	45	97	397	163	54	50	28	131	100	381	231
24	49	42	86	476	157	59	38	28	144	91	375	219
25	46	43	86	412	147	49	34	37	97	88	313	220
26	46	64	82	337	142	80	38	31	81	85	285	415
27	35	61	82	360	136	79	27	25	70	97	280	890
28	25	57	83	436	126	73	41	22	59	112	276	791
29	28	68	79	501	—	78	41	24	51	112	271	681
30	32	69	83	426	—	73	37	33	70	117	263	624
31	35	—	80	347	—	78	—	36	—	149	277	—
Total	1,724	1,428	2,017	5,898	6,713	2,266	2,262	1,073	1,483	3,977	8,342	12,093
Mean	55.6	47.6	65.1	190	240	73.1	75.4	34.6	49.4	128	269	403
Max	84	73	106	501	445	115	136	53	144	281	381	890
Min	25	32	37	54	126	49	27	22	24	75	149	219
ac-ft	3,420	2,830	4,000	11,700	13,320	4,490	4,490	2,130	2,940	7,890	16,550	23,990

Cal yr	1972 Total 35,641	Mean 97.4	Max 472 Min 18	ac-ft 70,690	
Wtr yr	1973 Total 49,276	Mean 135	Max 490 Min 22	ac-ft 97,740	

Source: U.S. Geological Survey, 1973.
Location—Lat 28°16′01″, long 81°26′53″, in SE quarter sec. 31, T.25 S., R.29 F., Oscola County, near left bank at downstream side of bridge on county road, 100 ft (30 m) downstream from Atlantic Coast Line Railroad bridge, 0.8 mi (1.3 km) northeast of Campbell, and 2.5 (4.0 km) upstream from Lake Tohopekaliga.
Drainage Area—180 mi^2 (466 km^2) approximately; includes part of watershed in Reedy Creek Swamp.
Period of Record—October 1968 to current year.

17. If a smooth, straight, cut rock rectangular channel drops 3 ft for every mile in length, what is the flow (in cfs) in the channel if the normal depth is 2 ft? The width of the channel is 10 ft.

18. Develop a discharge–storage curve similar to the generalized performance curve of Figure 7.11 for a 3-in. diameter orifice 1 ft below the overflow elevation in a 2 ft diameter riser pipe. The discharge–storage curve starts at the control elevation and ends 2 ft above the overflow elevation.

19. It should be noted that for steady uniform flow, the slope of the energy grade line, the slope of the water surface, and the slope of the bed are all equal. Select a control volume for a prismatic channel, and recalling that the component of the gravity force causing the flow must equal the resisting force (assumed to be proportional to the average cross-section velocity squared), show that Chezy's equation results.

20. Develop a sediment rating curve that best fits the data given below. Discuss which form is the best and why.

Dependent Variable Sediment Discharge (mt/day)	Independent Variable Discharge Rate (m³/sec)
9.5	0.15
22.0	0.21
85.0	0.29
180.0	0.35
650.0	0.72

21. Given monthly average stream flows (USGS data for Blues Creek) determine an estimate for the average sediment load, mt/day, passing the gaging station during the year. Assume the rating curve of Problem 20 is applicable.

Month	m³/sec	Month	m³/sec
J	0.052	J	0.201
F	0.069	A	0.037
M	0.035	S	0.021
A	0.047	O	0.017
M	0.010	N	0.015
J	0.075	D	0.015

7.5.2 Computer-Assisted Problems

1. Using the REGRESS program, estimate an equation for the coefficient of discharge for a weir with no end contractions as a function of H/P_1. Use Figure 7.16 and "pick off" at least 10 points to fit the equation. For the 60° V notch in Figure 7.17, estimate an equation for the coefficient of discharge as a function of head (H). Comment on how you determine the "best" equation.

2. Develop a computer program to solve for the volume of flow for any month given daily streamflow data. Data are available from Table 7.8 if not available elsewhere.

3. Write a computer program to develop a listing of cumulative mass flow or the mass curve (plot) for any rainfall data.

7.6
REFERENCES

Ackers, P. et al. 1978. *Weirs and Flumes for Flow Measurement,* Wiley, New York.

American Society of Civil Engineers. 1975. Sedimentation Engineering, ASCE, New York.

Brater, E. F. and King, H. W. 1980. *Handbook of Hydraulics,* McGraw-Hill, New York.

Chow, V. T. 1959. *Open Channel Hydraulics,* McGraw-Hill, New York.

Cowan, W. L. 1956. "Estimating Hydraulic Roughness Coefficients," *Agricultural Engineering,* **37**(7), 473–475, July.

Crosson, Pierre. 1995. "Soil Erosion Estimates and Costs," *Science,* **269,** 28 July.

DiToro, D. M. 1979. "Statistics of Receiving Water Response to Runoff." Paper presented at the Urban Stormwater and Combined Sewer Overflow Impact on Receiving Water Bodies, National Conference, University of Central Florida, Orlando, November.

Dooge, James C. 1989. "Robert Manning (1816–1897)," at Centennial of Manning's 1889 paper, December 6, Institution of Engineers of Ireland, Dublin.

Kindsvater, C. E. and Carter, R. W. 1959. "Discharge Characteristics of Rectangular Thin-Plate Weirs," ASCE Transactions, 772–822.

Lal, R. 1994. *Soil Erosion Research Methods,* St. Lucie Press, Del Rey Beach, FL.

Manning, Robert. 1891. "On the flow of water in open channels and pipes," Transactions Institution of Civil Engineers of Ireland, **20,** 161–207, Dublin.

Oldeman, L. R. 1991–92. "Global Extent of Soil Degradation." Biannual report, International Soil reference and Information Center, Wageningen, Netherlands.

Parshall, R. L. 1950. Measuring Water in Irrigation Channels with Parshall Flumes and Small Weirs, Soil Conservation Circular No. 843, USDA, Washington, DC, May.

Ree, W. O. 1949. "Hydraulic Characteristics of Vegetation and Vegetated Waterways," *Agricultural Engineering,* **30**(4), April.

Ree, W. O. and V. J. Palmer. 1949. "Flow of water in channels protected by vegetated linings," U.S. Soil Conservation Service, Technical Bulletin No. 767, February.

Schultz, E. F. 1980. "Problems in Hydrology," Water Resources Publications, Fort Collins, CO, pp. 280–286.

Stillwater Outdoor Hydraulic Laboratory, *Handbook of Channel Design for Soil and Water Conservation,* U.S. Soil Conservation Service, SCS-TP-61, March 1947; revised, June 1954.

Strickler, A. 1923. "Beitrage zur Frage Geschwindigkeitsformel und der Rauhiggkeitszahlen fur Strome, Kanal und geschlossene Leitungen," Mitteilung Nr.16 des Amtes fur Wasserwirtschaft, Bern, Switzerland.

United Kingdom Water Resources Data. 1979. Her Magestic Surface Water Office, Water Data Unit, Wallingford, U.K.

U.S. Department of Agriculture. 1995. *USDA Agriculture Handbook 703.* Washington, D.C.

U.S. Department of Transportation. 1985. Hydraulic Design of Highway Culverts, Report No. FHWA-IP-85-150 Federal Highway Administration, McLean, VA, September, p. 34.

U.S. Geological Survey. 1970. *Methods for Collection and Analysis of Water Samples for Dissolved Minerals and Gases,* Book 5, Chap. A1.

U.S. Geological Survey. 1973. *Water Resources Data for Florida Water Year-1973.* Water Resources Division, Tallahassee, Florida.

U.S. Geological Survey. 1976. *Guide for Collection, Analysis, and Use of Urban Stormwater Data.* Conference Report by the Engineering Foundation, published by ASCE.

U.S. Geological Survey. 1976. *Guidelines for Collection and Field Analysis of Ground-Water Samples for Selected Unstable Constituents,* Book 1, Chap. D2.

U.S. Geological Survey. 1977. *Methods for Collection and Analysis of Aquatic Biological and Microbiological Samples,* Book 5, Chap. A4.

U.S. Geological Survey. 1978. "Sediment," Chapter 3 in *National Handbook of Recommended Methods for Water Data Acquisition.* USGS, Reston, Virginia.

Vischer, D. 1987. "Strickler Formula, a Swiss Contribution to Hydraulics," Separat druck aus wasser, energie, luft. On occasion of centennial of Strickler's birth, Baden, Switzerland.

Wanielista, M. P. and Yousef, Y. A. 1993. Stormwater Management. John Wiley & Sons, New York.

Wischmeier, W.H. and D. D. Smith. 1978. "Predicting rainfall erosion losses," *USDA Agriculture Handbook 537,* U.S. Dept. of Agriculture.

FLOW ROUTING

When we examine streamflow data from two points on the same stream, we see that hydrograph peaks are usually different. Without any additional tributary inputs between an upstream and downstream point, the upstream hydrograph peak is frequently larger than the downstream peak. The hydrograph peak reduction is called attenuation. The possible causes of attenuation relate to the transport characteristics including the storage volume between the upstream and downstream locations. Flow routing can be described as a procedure for predicting the temporal and spatial hydrograph.

Flow routing procedures route hydrographs over land, through conduits and through reservoirs. Flow routing methods are used to analyze the effects of conduit modifications, detention or reservoir storage, spillway sizing, pumping stations, changes in land use, and overtopping of highway embankments. Flood warning systems use some form of flood routing to predict flood stages before the actual stages are realized. Also, it should be noted that the models are useful for applications in stormwater detention pond design.

The theoretical basis for both hydrologic and hydraulic routing are presented. Hydrologic models have a closed-form solution equation, while hydraulic models usually require some form of numerical integration with a finite difference approach being illustrated. Hydrologic models are more commonly used and the most popular ones are presented.

8.1

THEORETICAL BASIS

Most of the hydrograph models of the previous chapters use a unit-graph approach, but frequently it is difficult to develop adequate relationships between physical watershed parameters and the unit-graph shape. Routing procedures may produce more accurate results.

Routing procedures are generally classified as hydrologic and hydraulic. Hydrologic procedures use the continuity equation and mathematical relationships between discharge and storage. Many investigators and most data suggest the discharge/storage relationship can be assumed to be linear, at least over a small storage differential volume. However, nonlinear relationships are also used. Hydraulic procedures use both the continuity and momentum equations.

The fundamental laws that govern and describe fluid flow are described by the momentum and continuity equations (Chow, 1959).

$$\frac{\partial y}{\partial x} + \frac{v}{g}\frac{\partial v}{\partial x} + \frac{1}{g}\frac{\partial v}{\partial t} = S_0 - S_f \tag{8.1}$$

$$\frac{\partial A}{\partial t} + \frac{\partial Q}{\partial x} = q \tag{8.2}$$

where

y = depth (ft)
v = velocity (ft/s)
x = longitudinal distance (ft)
t = time (sec)
g = gravitational acceleration (ft/sec^2)
S_0 = ground slope (ft/ft)
S_f = friction slope (ft/ft)
Q = flowrate (cfs)
A = flow area (ft^2)
q = discharge per unit length (cfs/ft)

Methods for solving the dynamic (Equation 8.1) and the continuity (Equation 8.2) relationships are approximate because generally more exact or closed-form solutions do not exist (Yevjevich, 1968 and Chow, 1959). Equations 8.1 and 8.2 have been approximated by the following equations (Manning's Equation, 8.3; Continuity Equation, 8.4):

$$V = (1.486/n)R^{2/3}S_0^{1/2} \tag{8.3}$$

$$Q = VA \tag{8.4}$$

where

A = average flow area (ft^2)
R = hydraulic radius (ft)
V = velocity (ft/sec)
n = Manning's coefficient (Table 5.3)
S_0 = ground slope (ft/ft)
Q = flowrate (ft^3/sec)

One way to solve the continuity equation is to use finite difference equations to determine the depth of flow at each time period. The resulting depth of flow as a function of rainfall excess depth, $R(t)$, is

$$R(t + \Delta t) = R(t) + \frac{[Q(t + \Delta t) - Q(t)]}{A}(\Delta t) \tag{8.5}$$

Knowing the input flow rate to a drainage component, overland flow, gutter, sewer, etc., the output depth of flow or hydrograph can be calculated. Steady-state approximations can be made and hydrographs approximated, but the procedure is time consuming. Thus, computer programs were developed (Hydrologic Engineering Center, 1979). These programs will aid in flow routing using either hydrologic or hydraulic routing procedures.

8.2

KINEMATIC WAVE

Hydrographs and flood waves can be described as either dynamic or kinematic. Dynamic waves are determined by mass, inertial forces, and pressure force while kinematic waves do not include mass and forces. Kinematic waves are determined by the weight of the fluid flowing downhill in response to gravity. Flows will remain approximately uniform along the channel, thus kinematic flows are classified as unsteady uniform flows. The flow is unsteady because velocity at a point can change with time. A uniform flow is one for which the velocity at a specific point in the channel does not change for a given time period.

The equations used to express unsteady state flows were developed in 1870 and are typically referred to as the St. Venant equations. Derivations of the equations from the basic principles of mass momentum can be found in Chow (1959) or Henderson (1966). When the dynamic terms in the momentum equation (Equation 8.1) are minimal, this implies that the slope of the bed is about equivalent to the friction slope $(S_0 = S_f)$ and with no backwater effects, the discharge can be described as a function of flow depth if one assumes the bed slope and bed material to remain constant.

$$Q = \alpha A^m \tag{8.6}$$

where

Q = flow rate (cfs)
α, m = kinematic wave constants
A = cross-sectional area (ft^2)

The momentum equation has been reduced to a functional relationship and what remains is the movement of flow using the continuity equation. Substituting Equation 8.6 into Equation 8.2, one obtains

$$\frac{\partial A}{\partial t} + \alpha m A^{(m-1)} \frac{\partial A}{\partial x} = q$$

and

$$\alpha m A^{(m-1)} \partial A / \partial x \equiv \alpha m (\overline{A}^{m-1}) \frac{\Delta A}{\Delta X} \tag{8.7}$$

This equation can now be solved for the only dependent variable (A) by using a numerical technique. There are several methods in use to solve the equation (Mahmood and Yevjevich, 1975). The finite difference method will be presented because of its general acceptability (Hydrologic Engineering Center, 1985).

The finite difference method is a "point approximation" to the partial differential equation (Kersten, 1969). It uses simple difference equations to replace the partial differential equations for an array of space and time points (Figure 8.1). It is known as an explicit solution method. The discharge and water surface elevations are computed for each intersection point. Computations advance along the downstream direction for each time step (Δt) until all the flows and stages are calculated over the entire time. Substituting the area differences into Equation 8.7, one obtains

$$\frac{A_{i,j} - A_{i,j-1}}{\Delta t} + \alpha m [(A_{i,j-1} + A_{i-1,j-1})/2]^{m-1}$$
$$* \left(\frac{A_{i,j-1} - A_{i-1,j-1}}{\Delta x} \right) = \bar{q} \tag{8.8}$$

where

$$\bar{q} = (q_{i,j} + q_{i,j-1})/2 \tag{8.9}$$

j = time step, and i = the space position. The only unknown is $A_{i,j}$, thus rearranging Equation 8.8 one obtains

$$A_{i,j} = A_{i,j-1} + \bar{q}(\Delta t) - \alpha m (\Delta t / \Delta x)[(A_{i,j-1} + A_{i-1,j-1})/2]^{m-1}$$
$$* [A_{i,j-1} - A_{i-1,j-1}] \tag{8.10}$$

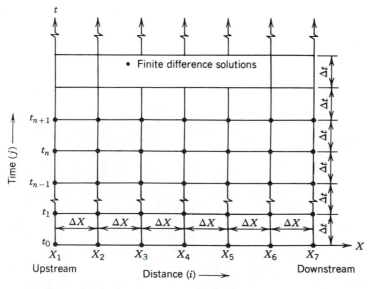

FIGURE 8.1   The characteristic curves of a fixed $X - t$ grid.

Once $A_{i,j}$ is known, $Q_{i,j}$ is calculated from

$$Q_{i,j} = \alpha(A_{i,j})^m \tag{8.11}$$

This standard form is used when the average wave celerity, $\bar{c}$, is less than the space-to-time ratio.

$$\bar{c} < \frac{\Delta x}{\Delta t} \tag{8.12}$$

The average celerity for a reach is based on the average flow area for the time step and is calculated using

$$\bar{c} = \alpha m(\overline{A})^{m-1} \tag{8.13}$$

The choice of Δx and Δt should be done to identify the peak discharge and develop a convergence for the solution procedure. In general, the following guidelines should be used (Hydrologic Engineering Center, 1985).

$$\frac{L}{50} \leq \Delta x \leq \frac{L}{2} \tag{8.14}$$

where

$$L = \text{total length of flow}$$

The time step is the time interval at which the ordinates of the inflow hydrograph are represented. The time step should be sufficiently short to not only define the hydrograph and peak flow within the model accuracy but also provide for numerical stability of the mathematics involved in the routing procedure. A guideline for the selection of the time step is

$$\Delta t \leq t_p/5 \tag{8.15}$$

where

$$t_p = \text{time to peak}$$

However, sometimes Δt is controlled by the time interval during which precipitation data were collected.

If wave celerity $\geq \Delta x/\Delta t$, it is possible for a flood wave to propagate more rapidly through space and time than the numerical technique can estimate by the standard equations. Numerical stability can be gained by rewriting the continuity equation and solving for flow rate, or

$$\frac{(A_{i-1,j} - A_{i-1,j-1})}{\Delta t} + (Q_{i,j} - Q_{i-1,j})/\Delta x = \bar{q}$$

Solving for $Q_{i,j}$, the only unknown

$$Q_{i,j} = Q_{i-1,j} + \bar{q}\,\Delta x - \left(\frac{\Delta x}{\Delta t}\right)(A_{i-1,j} - A_{i-1,j-1}) \tag{8.16}$$

and the area is calculated from

$$A_{i,j} = (Q_{i,j}/\alpha)^{1/m} \tag{8.17}$$

Equations 8.16 and 8.17 are known as the conservative form of the kinematic equations.

8.2.1   Overland Flow

For an overland flow situation of unit width and using Manning's equation for shallow depth (y_0), the hydraulic radius is simply equal to depth and the area of flow is depth, thus, Equations 8.3 and 8.4 can be combined with $R = y_0$, and $A = y_0$ and q_0 as discharge per unit width:

$$q_0 = \frac{1.486}{N} S_0^{1/2} y_0^{5/3} \tag{8.18}$$

or

$$q_0 = \alpha_0 y_0^{m_0} \tag{8.19}$$

where the subscripts (0) to overland flow and the Manning's roughness parameter is defined as N because it usually has a larger value than the channel or pipe roughness factor n. Suggested values were shown in Table 5.1. If a section of overland flow is constant and the bed consists of the same surface material, α_0 and m_0 are defined by equating q_0 in Equations 8.18 and 8.19 as (Hydrologic Engineering Center, 1979):

$$\alpha_0 = \frac{(1.486 S_0^{1/2})}{N}$$

and

$$m_0 = \tfrac{5}{3} \tag{8.20}$$

The continuity Equation 8.2 can be rewritten for shallow flows of unit width as

$$\frac{\partial y_0}{\partial t} + \frac{\partial q_0}{\partial x} = r \tag{8.21}$$

where

$$r = \text{rainfall excess rate (cfs/ft}^2)$$

Substituting Equation 8.19 into Equation 8.21 and considering overland flow:

$$\frac{\partial y_0}{\partial t} + \alpha_0 m_0 y_0^{(m_0-1)} \frac{\partial y_0}{\partial x} = r \tag{8.22}$$

which is an equation parallel in structure to Equation 8.7 and also can be solved by finite difference methods. For overland flow calculations, rainfall excess rate (ft/s or cfs/ft²) is used and replaces $\bar{q}$.

8.2.2 Kinematic Parameters for Channel Shapes

The concept of hydrograph generation using the kinematic wave technique can be applied to overland and conduit flows as represented in Figure 8.2. Depicted are four basic elements of a flow system and their kinematic parameters. For a trapezoidal section, it is not possible to derive a simple formula for α and m, however, given a depth (y_c), the Manning equation can be solved as

$$Q_c = \frac{(1.486 S_c^{1/2})}{n} (\bar{A}_c)^{5/3} \left\{ \frac{1}{[w + zy_c(1 + z^2)^{1/2}]} \right\}^{2/3} \tag{8.23}$$

and m is approximately 5/3. An example problem is beneficial to display the calculations, however, computer solutions are more reasonable because of the time involved and the choice of Δx and Δt are not always known.

Overland Flow

$$\alpha_0 = \frac{(1.486 S_0^{1/2})}{N}$$

$$m_0 = 5/3$$

Triangular Channel

$$\alpha_c = (0.94/n)S^{1/2}[Z/(1 + Z^2)]^{1/3}$$

$$m_c = 4/3$$

Pipe (Circular)

$$\alpha_c = (0.804/n)S^{1/2}D^{1/6}$$

$$m_c = 5/4$$

Rectangular

$$\alpha_c = (1.486/n)S^{1/2}w^{2/3}$$

$$m_c = 5/3$$

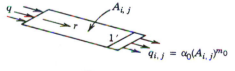

$$q_{i,j} = \alpha_0(A_{i,j})^{m_0}$$

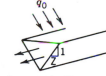

$$Q_c = \alpha_c(A_{i,j})^{m_c}$$

$$Q_c = \alpha_c(A_{i,j})^{m_c}$$

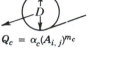

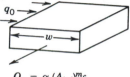

$$Q_c = \alpha_c(A_{i,j})^{m_c}$$

FIGURE 8.2 Elements and parameters.

□   *EXAMPLE PROBLEM 8.1*

A 1/5-acre parking lot is about 60 ft wide and 150 ft long with the lot sloping toward one side. It is completely impervious with no initial abstraction or infiltration. A rainfall of 1 in./hr is expected during the first 5 min, then 4 in./hr for the next 5 min and 2 in./hr for the last 5 min. There is a triangular ditch on one side of the lot that collects the overland flow of water (see Figure 8.3). A time step of 5 min (300 sec) is chosen because of the precipitation data interval. The space interval is set at 30 ft for overland flow and 50 ft for channel flow. Estimate the peak discharge and a hydrograph shape using the kinematic equations. Additional data are shown in Figure 8.3.

Solution

First calculate the parameters (α, m) for both the overland and channel flow equations. Next, calculate the flow to the side channel from overland flow, being careful to substitute rainfall excess (cfs/ft²) for average unit width flow rate. Some of the calculations are detailed in Table 8.1. The standard formula is assumed to be accurate, however, celerity is checked and the formula adjusted. The celerity by time step is

$$\bar{c} = \frac{\Delta x}{\Delta t} = \frac{30}{300} = 0.10 \text{ ft/s}$$

Size: $\frac{1}{5}$-Acre Parking Lot

Overland Flow

    Flow = 60 ft

   Slope = 0.05 ft/ft

     N = 0.15

    Δx = 30 ft

Δt = 5 min = 300 sec

    m_0 = 1.67

Triangular Channel

   Flow = 150 ft

 Slope = 0.005 ft/ft

    n = 0.025

Side slope = 1 on 2

   Δx = 50 ft

   m_c = 1.33

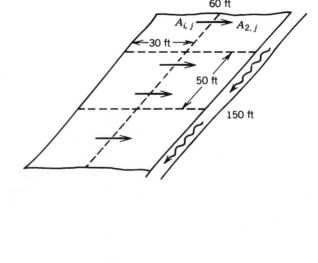

FIGURE 8.3   Example Problem 8.1 schematic.

TABLE 8.1 Overland Flow Calculations—Example Problem 8.1 (let $A_{i,j} = y_{i,j}$)

| TIME | | RAINFALL EXCESS RATE $\bar{r}$ | $A_{1,i}$ | $A_{2,i}$ | CHANNEL FLOW |
STEP (j)	(min)	(cfs/ft^2)	(ft^2)	(ft^2)	$q_{2,i}$(cfs/ft)
0	0	0	0	0	0
1	5	[a]2.31×10^{-5}	[b]0.006930	0.006930	0.000550
2	10	9.26×10^{-5}	[c]0.018130	[d]0.024001	0.004436
3	15	4.63×10^{-5}	0.011962	0.020450	0.003352
4	20	0	0	0.010935	0.001178
5	25	0	0	0	0

[a]1 in./hr $\times$ 1 ft/12 in. $\times$ 1 hr/3600 sec = 2.31(10^{-5}) ft/sec or cfs/ft^2
[b]Using Equation 8.10: for $\bar{c} < \Delta x/\Delta t$, and $\bar{r} = \bar{q}$

$$A_{1,1} = A_{1,0} + \bar{r}(\Delta t) - \alpha_0 m_0 \left(\frac{\Delta t}{\Delta x}\right)\left[\frac{(A_{1,0} + A_{0,0})}{2}\right][A_{1,0} - A_{0,0}]$$

$$= 0. + 2.31(10^{-5})(300) - 2.22(1.67)(300/30)[(0 + 0)/2](0 - 0)$$

$$= 0.00693 \text{ ft}^2$$

Check celerity:

$$\bar{c} = \alpha_0 m_0 (\bar{A})^{m_0 - 1} = 2.22(1.67)[(0 + 0.00693)/2]^{0.67}$$

$$= 0.08 \text{ ft/sec}$$

thus $\bar{c} < \Delta x/\Delta t$ or $0.08 < 0.10$ ft/s and using Equation 8.11,

$$q_{1,1} = \alpha_0 [A_{1,1}]^{m_0} = 2.22[0.00693]^{1.67}$$

$$= 0.000550 \text{ cfs/ft}$$

$$[c]A_{1,2} = A_{1,1} + \bar{r}(\Delta t) - \alpha_0 m_0 (\Delta t/\Delta x)[(A_{1,1} + A_{0,1})/2]^{m_0-1}[A_{1,1} - A_{0,1}]$$

$$= 0.00693 + 9.26 \times 10^{-5}(300) - 2.22(1.67)(300/30)[(0.00693 + 0)/2]^{0.67}$$

$$\times (0.00693 - 0)$$

$$= 0.00693 + .02778 - .005773 = 0.028937$$

Check celerity:

$$\bar{c} = 2.22(1.67)[(0.00693 + 0.028937)/2]^{0.67} = 0.25 \text{ ft/s}$$

continued

since $\bar{c} > \Delta x/\Delta t$, use conservative Equation 8.16.

$$q_{1,2} = Q_{0,2} + \bar{r}(\Delta x) - (\Delta x/\Delta t)[A_{0,2} - A_{0,1}]$$

$$= 0 + 9.26 \times 10^{-5}(30) - (30/300)(0 - 0)$$

$$= 0.002778 \text{ cfs/ft}$$

$$A_{1,2} = (0.0022778/2.22)^{0.6} = 0.01813 \text{ ft}^2$$

$$^d A_{2,2} = A_{2,1} + \bar{r}(\Delta t) - \alpha_0 m_0(\Delta t/\Delta x)[(A_{2,1} + A_{1,1})/2]^{m_0-1}[A_{2,1} - A_{1,1}]$$

$$= 0.00693 + 9.26 \times 10^{-5}(300) - 2.22(1.67)(300/30)$$

$$\times [(0.00693 + 0.00693)/2]^{0.67}(0)$$

$$= 0.00693 + 0.02778 - 0 = 0.03471 \quad \text{and} \quad \bar{c} > \Delta x/\Delta t$$

$$q_{2,2} = Q_{1,2} + \bar{r}(\Delta x) - (\Delta x/\Delta t)[A_{1,2} - A_{1,1}]$$

$$= 0.002778 + 9.26 \times 10^{-5}(30) - (30/300)(0.018131 - 0.00693)$$

$$= 0.002778 + 0.002778 - 0.0011220 = 0.004436 \text{ cfs}$$

$$A_{2,2} = (0.004436/2.22)^{0.6} = 0.024001 \text{ ft}^2$$

Next, calculations for flow in the triangular ditch are completed and are shown in Table 8.2.

α Calculations

For Overland Flow
$$\alpha_0 = (1.486/N)S_0^{1/2}$$
$$= (1.486/0.15)(.05)^{1/2}$$
$$= 2.22$$

For Triangular Channel
$$\alpha_c = [(0.94 S^{1/2})/n][Z/(1 + Z^2)]^{1/3}$$
$$= [(0.94(0.005)^{1/2})/0.025][2/(1 + 2^2)]^{1/3}$$
$$= 1.96$$

TABLE 8.2   Kinematic Calculation Triangular Channel—Example Problem 8.1

i	t (min)	q (cfs/ft)	$A_{1,i}$ (ft^2)	$A_{2,i}$ (ft^2)	$Q_{3,i}$ (cfs)
0	0	0	0	0	0
1	5	0.000550	0.04077	0.06211	0.06535
2	10	0.004436	0.19511	0.31376	0.59863
3	15	0.003352	0.15813	0.28767	0.54415
4	20	0.001178	0.07218	0.13229	0.21696

The final shape of the hydrograph can be more accurately defined if the time step were made smaller, say $\Delta t = 2$ min. The time to peak is about 10 min and a guideline for the selection of Δt requires $\Delta t \leq t_P/5$.

Rainfall

1 in./hr for first 5 min or 0.08 in. in 5 min
4 in./hr for next 5 min or 0.33 in. in 5 min
2 in./hr for last 5 min or 0.17 in. in 5 min
no rainfall past 15 min

❑

8.3

ROUTING BY THE INVENTORY EQUATION

There exist in common use, at least two hydrologic flood-routing methods for routing an inflow hydrograph-through storage in a reservoir, river, or stormwater detention basin. Routing flows through channels or streams is considered to have storage between the inlet and outlet structures or mile markers and, thus, is similar to reservoir/detention basin storage. The inventory method and the Muskingum formula are two common ways of flood routing using the continuity equation. Both methods assume a relationship between the inflow and outflow hydrographs with the outflow being dependent on previous outflow and inflows. Consider the hydrographs of Figure 8.4 that show the primary characteristics of routing. The outflow hydrograph is determined from a given inflow hydrograph. The hydrograph shape is modified as water is stored. The maximum storage occurs when the inflow and outflow rates are equal. Up to this point in time, storage was increasing because the rate of inflow exceeded outflow rate. Beyond the time of maximum storage, the outflow rate exceeds inflow rate, thus storage decreases.

By using the continuity Equation 8.2 and considering $\partial Q/\partial x$ is the change in flow per unit channel length (ft^2/sec) or inflow minus outflow, $\partial Q/\partial x$ is

$$\frac{\partial O}{\partial x} = \frac{I - O}{\Delta x} \qquad (8.24)$$

where

I = inflow, L^3/T
O = outflow, L^3/T

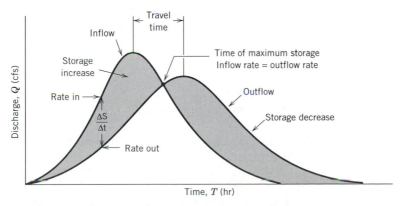

FIGURE 8.4 Inflow and outflow hydrographs.

The change of area per unit time, $\Delta A / \Delta t$ multiplied by channel length can be considered as a change in channel storage volume, or

$$\frac{\Delta A\, \Delta x}{\Delta t\, \Delta x} = \frac{\Delta S}{(\Delta x)(\Delta t)} \tag{8.25}$$

where

ΔS = storage volume, L^3
Δt = time interval, t

Substituting Equations 8.24 and 8.25 into Equations 8.2 and multiplying by Δx and Δt yields a form of the inventory equation:

$$I(\Delta t) - O(\Delta t) = \Delta S \tag{8.26}$$

recognizing that ΔS includes $q\, \Delta x\, \Delta t$ term. Dividing by Δt and using average values,

$$\bar{I} - \bar{O} = \frac{\Delta S}{\Delta t} \tag{8.27}$$

and by taking the limit as $\Delta t \to O$

$$\bar{I} - \bar{O} = \frac{dS}{dt} \tag{8.28}$$

Most of the more commonly used methods of flood routing are based solely on the solution of the inventory equation, which provides for conservation of mass. Typically, these methods employ a storage discharge relationship in a repetitive manner to solve the inventory equation to determine the ordinates of the outflow hydrograph for a given inflow hydrograph. The inventory equation as applied to a reservoir routing states that the "volume of inflow minus the volume of outflow over a given time interval is equal to the change in volume stored over that time interval." Therefore, the inventory equation can be written in the following form:

$$S_2 - S_1 = \bar{I}(\Delta t) - \overline{O}(\Delta t) = \Delta S = \left[\frac{(I_1 + I_2)}{2}\right](\Delta t) - \left[\frac{(O_1 + O_2)}{2}\right](\Delta t) \tag{8.29}$$

where

$\bar{I}$ = the average inflow during the time step, Δt, (L^3/T)
$\overline{O}$ = the average outflow during the time step, Δt, (L^3/T)
ΔS = the change in storage volume during the time step, Δt, (L^3)

Collecting the unknowns of Equation 8.29 on one side of the equation, and letting $(I_1 + I_2)/2 = \bar{I}$ (average inflow), N_2 is a variable defined as

$$N_2 = S_2 + \left(\frac{O_2}{2}\right)\Delta t = S_1 - \left(\frac{O_1}{2}\right)\Delta t + \bar{I}\, \Delta t \tag{8.30}$$

To solve the problem one must find a relationship between outflow and storage, called an O–S relationship.

Thus, make

$$N = S + \left(\frac{O}{2}\right)\Delta t \qquad \text{(volume units)}$$

or in another form

$$N = \left(\frac{S}{\Delta t}\right) + \left(\frac{O}{2}\right) \qquad \text{(flow units)}$$

An N–O relationship can be developed from the O–S relationship, since N is a function of S and O.

Equation 8.30 can be written in another form if we add $(O_1 \Delta t)$ to both sides of the equation (letting $i - 1 = 1; i = 2$):

$$O_{i-1}\Delta t + S_i + \left(\frac{O_i}{2}\right)\Delta t = S_{i-1} + \left(\frac{O_{i-1}}{2}\right)\Delta t + \bar{I}\,\Delta t$$

$$N_i = N_{i-1} + \bar{I}\,\Delta t - O_{i-1}\,\Delta t \qquad \text{(volume units)} \qquad (8.31)$$

or

$$N_i = N_{i-1} + \bar{I} - O_{i-1} \qquad \text{(flow units)} \qquad (8.32)$$

Since Equation 8.29 is the basis of many hydrologic routing methods, its application in a particular routing method is demonstrated in Example Problem 8.2.

☐ EXAMPLE PROBLEM 8.2

Given an inflow hydrograph and a detention volume discharge relationship, determine the outflow hydrograph using an inventory equation.

INPUT HYDROGRAPH		BASIN DISCHARGE	
TIME (hr)	FLOW (cfs)	VOLUME (ft^3)	DISCHARGE
0	0	$\leq 10{,}000$	0
1	25	$> 10{,}000$	1 ft^3/sec/10,000 ft^3
2	20		
3	16		
4	12		
5	9		
6	6		
7	3		
8	0		

Solution

S–O Relationship with $S = 10,000$ ft³ at $Q = 0$

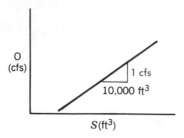

$$O = (S - 10,000)/10,000 \qquad \text{for } S > 10,000 \text{ cf and}$$

$$S = 10,000(O) + 10,000 \qquad \text{if } O \text{ is } > \text{zero}$$

N–O Relationship:

$$N = S + \frac{O}{2}(\Delta t) = 10,000(O) + 10,000 + \left(\frac{O}{2}\right)(3600)$$

for $\Delta t = 1$ hr:

$$N = 11,800(O) + 10,000$$

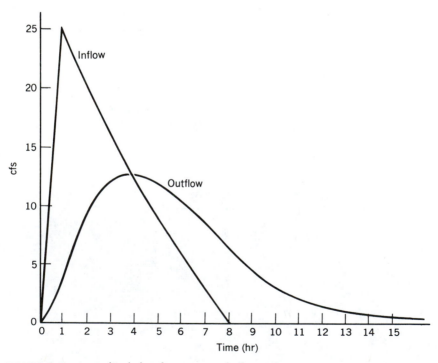

FIGURE 8.5   A graphical plot of input–output hydrograph.

and

$$N - 10,000 = 11,800(O)$$

$$O = \frac{(N - 10,000)}{11,800} \qquad \text{if } O \text{ is } > \text{zero}$$

The solution is tabulated in Table 8.3 and the graphs are shown in Figure 8.5. ❑

TABLE 8.3 Calculations for Example Problem 8.2
ASSUME AT TIME ZERO, 10,000 ft³ OF WATER IS IN STORAGE.
SOLUTION: $t = 1$ hr $= 3600$ sec   FROM $N - O$ and $S - O$ RELATIONSHIPS

								$N_i = i\Delta t + S_{i-1}$		
TIME	I_{i-1}	I_i	i	$i\Delta t$	S_{i-1}	O_{i-1}	$O_{i-1}\dfrac{\Delta t}{2}$	$-O_{i-1}\dfrac{\Delta t}{2}$	O_i	S_i
(hr)	(cfs)	(cfs)	(cfs)	(ft³)	(ft³)	(cfs)	(ft³)	(ft³)	(cfs)	(ft³)
0	—	0	0	0	10,000	0	0	10,000	0	10,000
1	0	25	12.5	45,000	10,000	0	0	55,000	3.814	48,140
2	25	20	22.5	81,000	48,140	3.814	6,865	122,275	9.515	105,150
3	20	16	18.0	64,800	105,150	9.515	17,127	152,823	12.104	131,036
4	16	12	14.0	50,400	131,036	12.104	21,787	159,649	12.682	136,821
5	12	9	10.5	37,800	136,821	12.682	22,828	151,800	12.016	130,164
6	9	6	7.5	27,000	130,164	12.016	21,629	135,535	10.639	116,386
7	6	3	4.5	16,200	116,386	10.639	19,132	113,454	8.767	97,673
8	3	0	1.5	5,400	97,673	8.767	15,780	87,292	6.550	75,502
9	0	0	0	0	75,502	6.550	11,790	63,712	4.552	55,519
10	0	0	0	0	55,519	4.552	8,194	47,325	3.163	41,632
11	0	0	0	0	41,632	3.163	5,693	35,939	2.198	31,980
12	0	0	0	0	31,980	2.198	3,956	28,024	1.527	25,274
13	0	0	0	0	25,274	1.527	2,749	22,525	1.062	20,615
14	0	0	0	0	20,615	1.062	1,912	18,703	0.738	17,375
15	0	0	0	0	17,375	0.738	1,328	16,047	0.512	15,125
16	0	0	0	0	15,125	0.512	922	14,203	0.356	13,562
17	0	0	0	0	13,562	0.356	641	12,921	0.248	12,476
18	0	0	0	0	12,476	0.248	446	12,029	0.172	11,720
19	0	0	0	0	11,720	0.172	310	11,410	0.120	11,200
20	0	0	0	0	11,200	0.120	216	10,984	0.0834	10,834
21	0	0	0	0	10,834	0.0834	150	10,683	0.058	10,580
22	0	0	0	0	10,580	0.058	104	10,475	0.040	10,402
23	0	0	0	0	10,402	0.040	72	10,330	0.028	10,280
24	0	0	0	0	10,280	0.028	50	10,230	0.020	10,195
25	0	0	0	0	10,195	0.020	36	10,159	0.013	10,134
26	0	0	0	0	10,134	0.013	23	10,111	0.009	10,094
27	0	0	0	0	10,094	0.009	16	10,078	0.007	10,066
28	0	0	0	0	10,066	0.007	13	10,053	0.004	10,045

Checking: $\Sigma I \Delta t = 327,600$ ft³ $\Sigma O_{i-1}(\Delta t/2) = 327,538$ ft³.
$< 5\%$ error in acceptable.
The input and output hydrographs are shown in Figure 8.5.

In many situations the relationship between storage, inflow, and outflow must be developed from the geometric characteristics of a pond. The developed relationship is called a *stage–storage–discharge* (SSD) relationship. *Stage* refers to the water elevation within the pond with respect to some datum. *Storage* is the volume of water stored in the pond at any given time. Using the SSD relationship, flows can be routed through the pond. The SSD relationship can be developed knowing an area–stage relationship for the pond and information about the pond outlet, such as weir information. Example Problem 8.3 illustrates the use of pond area–stage information to develop an SSD relationship and route flows through a pond.

❑ EXAMPLE PROBLEM 8.3

A circular pond has a bottom area of 1.0 acre at a stage of 100 ft and a side slope of 6:1. A 90° V-Notch weir is located in the pond with an invert at 102 ft. Develop an outflow hydrograph from the pond if the inflow hydrograph of Example Problem 8.2 flows into the pond and the initial depth in the pond is 101.5 ft.

Solution

A stage–area relationship must be developed from the geometry of the pond. From the side slope of 6:1 it is known that the pond radius increases by a factor of 6 for each increase in stage. The original pond radius can be determined assuming a circular pond:

$$(1 \text{ acres})(43{,}560 \text{ ft}^2/\text{acre}) = \pi R^2$$

Therefore the pond radius at the base of the pond is 117.75 ft. Pond area can then be found at any stage using the relationship

$$A = \pi(117.75 + 6(Y - 100))^2$$

where Y is the stage at the area being calculated.

Outflow at any stage can be developed from the 90° V-Notch weir equation

$$Q = 2.5 \; H^{2.5}$$

where Q is flowrate in cfs and H is water height above the invert in feet. Storage is calculated as the average of the area between the two steps multiplied by the difference in stage.

$$S_i = \frac{A_i + A_{i-1}}{2}(Y_i - Y_{i-1})$$

where
S = storage
A = area
Y = stage

A spreadsheet for the stage–storage–discharge relationship is shown in Table 8.4.   ❑

TABLE 8.4 Development of the Stage Storage Discharge Relationship for Example Problem 8.3

	A	B	C	D	E	F
1	Stage	Radius	Area	Area	Storage	Discharge
2	(ft)	(ft)	(sq. ft)	(acres)	(ac-ft)	(cfs)
3	100.0	117.75	43560	1.00	0.00	0.00
4	100.5	120.75	45808	1.05	0.51	0.00
5	101.0	123.75	48112	1.10	1.05	0.00
6	101.5	126.75	50473	1.16	1.62	0.00
7	102.0	129.75	52891	1.21	2.21	0.00
8	102.5	132.75	55365	1.27	2.83	0.44
9	103.0	135.75	57895	1.33	3.48	2.50
10	103.5	138.75	60482	1.39	4.16	6.89
11	104.0	141.75	63126	1.45	4.87	14.14
12	104.5	144.75	65826	1.51	5.61	24.71
13	105.0	147.75	68583	1.57	6.38	38.97

The stage–storage–discharge relationship can then be used to route the inflow to the outflow. To preserve numerical accuracy the time step of the inflow was reduced to 15 minutes and a linear relationship between time steps is assumed. Using the inventory equation and directly calculating discharge using stage, the following result shown in Table 8.5 are determined. Pond stage is found by linear interpolation between pond storage values in the stage–storage–discharge relationship. If the discharge equation were not known, the discharge could also be determined using a linear interpolation in the SSD relationship.

EXAMPLE PROBLEM 8.4

The procedure outlined in the previous examples based on Equation 8.30 (the inventory equation) is also referred to as the *storage indication* method, and the term "N" is called the storage indicator quantity. This result is often extended for use in routing flood waves through major multipurpose reservoirs. There is both a lag and an attenuation as the flood wave passes through the reservoir.

Let us rearrange Equation 8.29 as follows:

$$I_1 + I_2 + \left(\frac{2S_1}{\Delta t} - O_1\right) = \frac{2S_2}{\Delta t} + O_2 \tag{8.33}$$

All terms on the left-hand side are known quantities, and a value of the right-hand side can be computed. The corresponding value of O_2 can be determined from a previously prepared "routing" curve. That is, for a given reservoir, data are known for elevation-storage and elevation-discharge curves, see Figure 8.6.

The routing computation then proceeds step by step for succeeding time periods. A typical solution is presented in Table 8.6 based on the routing curves of Figure 8.6. Note that $\frac{2S}{\Delta t} - O$ is easily computed from known values of $\frac{2S}{\Delta t} + O$ and O.

TABLE 8.5   Results of Routing Hydrograph through the Pond of Example Problem 8.3

	A	B	C	D	E	F	G	H
1			Inflow	Pond	Pond	Pond	Pond	Outflow
2	Time	Inflow	Volume	Storage	Storage	Stage	Outflow	Volume
3	(hrs)	(cfs)	(cf)	(cf)	(ac-ft)	(ft)	(cfs)	(cf)
4	0.00	0.00	0.00	70468	1.62	101.50	0.00	0.0
5	0.25	6.25	5625.00	76093	1.75	101.61	0.00	0.0
6	0.50	12.50	11250.00	87343	2.01	101.83	0.00	0.0
7	0.75	18.75	16875.00	104218	2.39	102.15	0.02	18.7
8	1.00	25.00	19125.00	123324	2.83	102.50	0.44	418.3
9	1.25	23.75	21375.00	144281	3.31	102.87	1.77	1992.4
10	1.50	22.50	20250.00	162539	3.73	103.18	3.82	5031.0
11	1.75	21.25	19125.00	176633	4.05	103.42	6.04	8870.6
12	2.00	20.00	18112.50	185875	4.27	103.58	7.83	12478.8
13	2.25	19.00	17100.00	190496	4.37	103.69	9.24	15364.1
14	2.50	18.00	16200.00	191332	4.39	103.70	9.45	16818.8
15	2.75	17.00	15300.00	189813	4.36	103.68	9.08	16670.9
16	3.00	16.00	14400.00	187542	4.31	103.63	8.54	15859.7
17	3.25	15.00	13500.00	185182	4.25	103.59	8.01	14898.8
18	3.50	14.00	12600.00	182883	4.20	103.55	7.51	13967.6
19	3.75	13.00	11700.00	180616	4.15	103.51	7.03	13089.8
20	4.00	12.00	10912.50	178438	4.10	103.47	6.60	12267.7
21	4.25	11.25	10125.00	176296	4.05	103.44	6.18	11499.4
22	4.50	10.50	9450.00	174246	4.00	103.40	5.80	10781.9
23	4.75	9.75	8775.00	172240	3.95	103.36	5.44	10114.1
24	5.00	9.00	8100.00	170225	3.91	103.33	5.09	9477.3
25	5.25	8.25	7425.00	168173	3.86	103.29	4.75	8858.6
26	5.50	7.50	6750.00	166065	3.81	103.26	4.42	8251.3
27	5.75	6.75	6075.00	163888	3.76	103.22	4.09	7652.8
28	6.00	6.00	5400.00	161636	3.71	103.18	3.76	7062.4
29	6.25	5.25	4725.00	159298	3.66	103.14	3.44	6480.3
30	6.50	4.50	4050.00	156868	3.60	103.09	3.12	5907.3
31	6.75	3.75	3375.00	154335	3.54	103.05	2.81	5344.4
32	7.00	3.00	2700.00	151691	3.48	103.00	2.51	4793.0
33	7.25	2.25	2025.00	148923	3.42	102.97	2.35	4371.3
34	7.50	1.50	1350.00	145902	3.35	102.92	2.02	3932.3
35	7.75	0.75	675.00	142644	3.27	102.86	1.71	3358.2
36	8.00	0.00	337.50	139624	3.21	102.80	1.44	2835.7
37	8.25		0.00	136788	3.14	102.75	1.22	2395.5
38	8.50			134393	3.09	102.71	1.05	2038.5
39	8.75			132354	3.04	102.67	0.91	1762.6
40	9.00			130591	3.00	102.64	0.80	1545.1
41	9.25			129046	2.96	102.61	0.72	1369.8
42	9.50			127677	2.93	102.58	0.64	1225.6
43	9.75			126451	2.90	102.56	0.58	1105.4
44	10.00			125346	2.88	102.54	0.53	1003.6
45	10.25			124342	2.85	102.52	0.49	916.6
46	10.50			123425	2.83	102.50	0.45	841.5
47	10.75			122584	2.81	102.49	0.41	776.1
48	11.00			121808	2.80	102.47	0.38	718.7
49	11.25			121089	2.78	102.46	0.36	668.0
50	11.50			120421	2.76	102.45	0.33	622.9
51	11.75			119798	2.75	102.44	0.31	582.7
52	12.00			119215	2.74	102.42	0.29	546.5

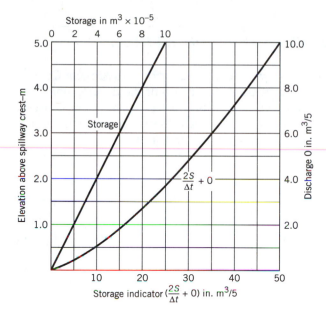

FIGURE 8.6 Routing curve for uncontrolled reservoir.

Further, if a means exists to regulate reservoir outflow, (e.g., values, gates, power discharges) Equation 8.33 may be modified to include the given controlled outflow:

$$I_1 + I_2 - 2O_R + \left(\frac{2S_1}{\Delta t} - O_1\right) = \left(\frac{2S_2}{\Delta t} + O_2\right) \tag{8.34}$$

where O_R is the regulated outflow. ❑

TABLE 8.6 Routing Table Based on Curve of Figure 8.6

DAY	HOUR	I	$\frac{2S}{\Delta t} - O$	$\frac{2S}{\Delta t} + O$	O	S
0	noon	1.0	8.6	10.0	0.7	200,800
	mn	1.0	9.0	10.6	0.8	211.680
1	noon	2.5	10.1	12.5	1.2	244,080
	mn	6.0	13.4	18.6	2.6	345,600
2	noon	10.3	20.9	29.7	4.4	546,480
	mn	12.6	27.2	43.8	8.3	766,800
3	noon	9.0	29.0	48.8	9.8	842,400
	mn	5.1	26.9	43.1	8.1	756,000
4	noon	3.0	23.4	35.0	5.8	630,720
	mn	1.8	19.8	28.2	4.2	518,400
5	noon	1.1	16.7	22.7	3.0	455,520
	mn	1.1	14.8	18.8	2.3	356,400

All flows in m^3/sec and S in m^3.

8.4

ROUTING BY THE MUSKINGUM METHOD

The Muskingum method for flood routing was developed for the Muskingum Conservancy District flood control study in the 1930s. A discussion of the development with assumptions will illustrate its use. A stream length is chosen that has near constant geometric properties, and at the beginning of the reach, the inflow and storage are assumed to be related to depth by the following equation forms.

$$I = ay_u^n \tag{8.35}$$

and

$$S_I = by_u^m \tag{8.36}$$

where

$$I = \text{inflow rate, L}^3/\text{T}$$
$$S_I = \text{storage inflow, L}^3$$
$$y_u = \text{depth of flow upstream, L}$$
$$a, b, m, n = \text{constants for the stream reach}$$

Similar equations are developed for the downstream point, or

$$O = ay_d^n \tag{8.37}$$

and

$$S_0 = by_d^m \tag{8.38}$$

where

$$O = \text{outflow rate, L}^3/\text{T}$$
$$S_0 = \text{outflow storage, L}^3$$
$$y_d = \text{depth of flow downstream, L}$$

Equating depths upstream and downstream, the following equations result.

$$S_I = (b/a^{m/n})(I^{m/n}) \tag{8.39}$$
$$S_0 = (b/a^{m/n})(O^{m/n}) \tag{8.40}$$

Now, postulate that the storage (S) within the reach is a weighting of both the input and output, or

$$S = cS_I + (1 - c)S_0 \tag{8.41}$$

where

$$c = \text{weighting factor between 0 and 0.5}$$

When $c = 0.0$, maximum attenuation is achieved, and when $c = 0.5$, the input hydrograph is not attenuated.

Let $K = b/a^{m/n}$ and $x = m/n$, such that Equation 8.41 is rewritten as

$$S = K[cI^x + (1 - c)O^x] \tag{8.42}$$

where

K = storage time constant for the reach, units of time. (Also has physical interpretation as the travel time in the reach.)

For rectangular channels, $n = 5/3$, $m = 1$ and $x = 0.6$ have been used. For natural channels, x has been calculated to be larger, and in most cases x is assumed equal to 1.0, as was done in the Muskingum method. Thus, the Muskingum formula is

$$S = K[cI + (1 - c)O] \tag{8.43}$$

The constants of Equation 8.43 must be determined for a reach of stream. Field streamflow data are generally used for gaged sites and the experiences of the investigators are used for ungaged sites. Applications using streamflow data generally show that K is reasonably close to the travel time in the reach. When field data are available, Equation 8.43 can be written in difference form over successive time periods of analysis, Δt, where Δt is chosen to be less than K and greater than $2Kc$ (Chow, 1964).

$$S_2 - S_1 = K[c(I_2 - I_1) + (1 - c)(O_2 - O_1)] \tag{8.44}$$

Combining the inventory Equation (8.29) with Equation 8.44 to eliminate the storage variables and solving for the unknown O_2, given I_2, I_1, and O_1, the following Equation (8.45) results

$$O_2 = c_0 I_2 + c_1 I_1 + c_2 O_1 \tag{8.45}$$

in which

$$c_0 = \frac{-Kc + 0.5(\Delta t)}{K - Kc + 0.5(\Delta t)}$$

$$c_1 = \frac{Kc + 0.5(\Delta t)}{K - Kc + 0.5(\Delta t)}$$

$$c_2 = \frac{K - Kc - 0.5(\Delta t)}{K - Kc + 0.5(\Delta t)}$$

and

$$c_0 + c_1 + c_2 = 1.0$$

Note that the time units for K and Δt must be the same. If K and c are known with an upstream hydrograph, routing can be accomplished.

8.4.1 Muskingum Routing Constants *K* and *c*

When no flow or streamflow data are available, it is common to use the travel time in the stream reach to approximate K and $c = 0.2$. If inflow and outflow hydrograph records are available, better estimates of K and c are possible. Since storage and outflow are assumed to be related by Equation 8.43, an acceptable value of c would be one that gives a linear relationship. After finding the linear relationship, K is the reciprocal of the slope or

$$K = \frac{\Delta S}{\Delta O} \tag{8.46}$$

Equating K values of Equation 8.46 and the Muskingum equation (Equation 8.43), one obtains a relationship

$$\frac{S}{cI + (1-c)O} = \frac{\Delta S}{\Delta O} \tag{8.47}$$

□ **EXAMPLE PROBLEM 8.5**

Determine an estimate for K and c for a river segment (reach) that is believed to have relatively constant cross-sectional area and slope, if inflow and outflow hydrographs are available as shown in Table 8.7.

Solution

It is assumed that the water surface profile is uniform or unbroken. The inventory equation (Equation 8.29) is used to solve for the cumulative storage, and the denominator of Equation 8.47 is used to calculate the weighted discharge for various assumed values of c. Graphs are drawn for the rising and falling limbs of

TABLE 8.7 Calculations for Example Problem 8.5

Day	$\bar{I}$ (cfs)	$\bar{O}$ (c = 0) (cfs)	S^a (cfs-days)	WEIGHTED DISCHARGE (cfs)[b] c = 0.1	c = 0.2	c = 0.3
1	0.0	0.0	0.0	0.0	0.0	0.0
2	35.0	10.0	25.0	12.5	15.0	17.5
3	95.0	50.0	70.0	54.5	59.0	63.5
4	60.0	60.0	70.0	60.0	60.0	60.0
5	35.0	50.0	55.0	48.5	47.0	45.5
6	20.0	35.0	40.0	33.5	32.0	30.5
7	10.0	25.0	25.0	23.5	22.0	20.5
8	5.0	15.0	15.0	14.5	13.0	12.0
9	0.0	10.0	10.0	9.0	8.0	7.0
10	0.0	5.0	5.0	4.5	4.0	3.5
11	0.0	0.0	0.0	0.0	0.0	0.0

[a]$S_i = S_{i-1} + (\bar{I}_i - \bar{O}_i)$.
[b]$O_i = c\bar{I}_i + (1-c)\bar{O}_i$.

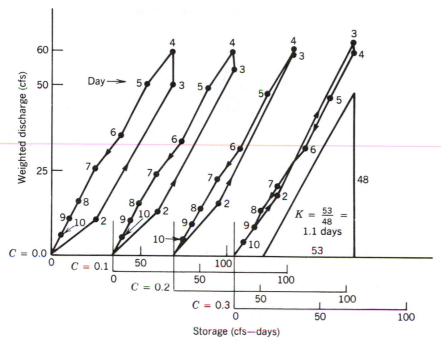

FIGURE 8.7 A discharge–storage curve—Example Problem 8.5.

the outflow hydrograph, and the loop that best approximates a straight line is chosen (see Figure 8.7). A straight line results because of Equations 8.40 and 8.43. The estimates for K and c are 1.1 days and 0.3, respectively. ❑

8.4.2 Additional Methods for Estimating K and c

When streamflow data are available, a graphic display of the upstream and downstream hydrograph produces insight on the translation and by proportional analysis an estimate of K.

From Figure 8.8, a relationship between the inflow and outflow hydrographs can be developed. This empirical relationship is generally chosen in time units. Similar

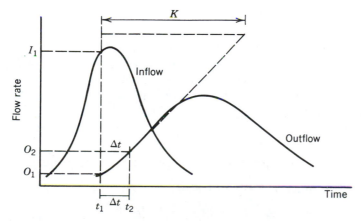

FIGURE 8.8 The graphical convex method.

triangles result in a relationship to estimate K using inflow and outflows at two consecutive periods of time.

$$\frac{O_2 - O_1}{\Delta t} = \frac{I_1 - O_1}{K} \tag{8.48}$$

and

$$\frac{\Delta t}{K} = \frac{O_2 - O_1}{I_1 - O_1} \tag{8.49}$$

The choice of Δt is made such that it is less than 20% of the time to peak. Thus, from stream flow data, the outflow and inflow hydrograph values are known, and K can be estimated.

Other equations for estimating the constants c and K have been developed from empirical equations, such as

$$K = \frac{LA}{3600Q} \tag{8.50}$$

where
L = length between points (ft)
A = average cross-sectional area (ft²)
Q = steady-state average discharge (ft³/sec)

and

$$c = \frac{V}{V + 1.7} \tag{8.51}$$

where
V = steady-state velocity (ft/sec)

□ **EXAMPLE PROBLEM 8.6**

A flowrate of 5 ft³/sec passes an upstream point in a stream and travels 5280 ft downstream in a rectangular channel and at the same time; the flowrate downstream is 3 ft³/sec. Neglecting evaporation and infiltration, what is the next flowrate downstream if the channel is 8 ft wide with a channel slope of 0.001 and the roughness coefficient is 0.029?

Solution

Calculate the upstream conditions first using continuity and Manning's equation.

$$Q = AV = (1.486/0.029)(8d)[8d/(8 + 2d)]^{2/3}(0.001)^{1/2} = 5$$
$$d = 0.6 \text{ ft}; \quad \text{and} \quad A = bd = 8(0.6) = 4.8 \text{ ft}^2$$

$$V = \frac{Q}{A}$$

$$= \frac{5}{4.8} = 1.04 \text{ fps}$$

now

$$c = \frac{V}{(V + 1.7)} = \frac{1.04}{(1.04 + 1.70)} = 0.38$$

For the downstream conditions and the next time period:

$$O_2 = cI_1 + (1 - c)O_1$$

$$= (0.38)(5) + (0.62)(3)$$

$$= 3.76 \text{ cfs} \quad \square$$

☐ **EXAMPLE PROBLEM 8.7**

This problem was adopted from the U.S. Department of Transportation (FHWA, 1984) and is typical of highway design problems. A 6-mi reach of a channelized river is flooding and destroying homes. A channel improvement is proposed that will cut off the channelized river and reducing the length of channel to 5 mi (Figure 8.9). What effect will this channel improvement have on the peak discharge experienced at the roadway at point B? A synthetic hydrograph at point A is estimated for a 25-yr design discharge and is shown in Figure 8.10.

Solution

The average discharge for this hydrograph is 2200 cfs (71 cms). Using the cross-sectional data, the average travel time is computed by using a value of .025 for Manning's n and the following formulas.

$$V = \left(\frac{1.486}{.025}\right) R^{2/3} S_0^{1/2}$$

In the original 6-mi reach, the average travel time is computed to be 0.70 hr. For the modified 5-mi reach, the average travel time is computed to be 0.55 hr. For

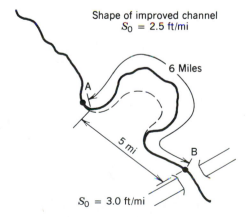

Shape of improved channel
$S_0 = 2.5$ ft/mi

6 Miles

A

5 mi

B

$S_0 = 3.0$ ft/mi

FIGURE 8.9 A proposed channel improvement.

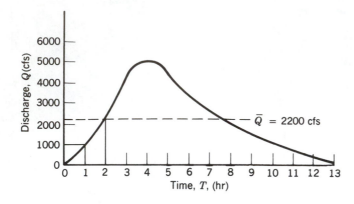

FIGURE 8.10 Hydrograph at point A.

the original reach, the coefficients c_0, c_1, and c_2 are first computed using $\Delta t = 1$ hr, an assumed value of $c = 0.2$, and $K = 0.70$ hr as follows:

$$c_0 = [-0.70(0.2) + 0.5(1)]/[0.70 - 0.70(0.2) + 0.5(1)] = 0.3396$$

$$c_1 = [0.70(0.2) + 0.5(1)]/[0.70 - 0.70(0.2) + 0.5(1)] = 0.6038$$

$$c_2 = [0.70 - 0.70(0.2) - 0.5(1)]/[0.70 - 0.70(0.2) + 0.5(1)] = 0.0566$$

These values can be checked as follows:

$$c_0 + c_1 + c_2 = 0.3396 + 0.6038 + 0.0566 = 1.0000$$

The outflow hydrograph ordinates can now be computed with Equation 8.45.

TABLE 8.8 Existing Reach—
Example Problem 8.7
Outflow Calculations

t (hr)	I (cfs)	O (cfs)
0	0	0
1	1000	340
2	2000	1303
3	4200	2707
4	5200	4455
5	4400	4886
6	3200	4020
7	2500	3009
8	2000	2359
9	1500	1851
10	1000	1350
11	700	918
12	400	610
13	0	276
14	0	16
15	0	1

TABLE 8.9 Modified Reach—
Example Problem 8.7
Outflow Calculations

t (hr)	I (cfs)	O (cfs)
0	0	0
1	1000	415
2	2000	1452
3	4200	2948
4	5200	4695
5	4400	4900
6	3200	3870
7	2500	2867
8	2000	2269
9	1500	1775
10	1000	1275
11	700	858
12	400	565
13	0	223
14	0	0
15	0	0

Beginning at $t = 1$ hr:

$$O_2 = C_0I_2 + C_1I_1 + C_2O_1 = 0.3396(1000) + 0.6038(0) + 0.0566(0)$$
$$= 340 \text{ cfs } (9.6 \text{ cms})$$

and

$$O_2 = 0.3396(2000) + 0.6038(1000) + 0.0566(340) = 1303 \text{ cfs } (36 \text{ cms})$$

These values along with the remaining calculations are tabulated in Table 8.8. The same procedure is used to route the hydrograph through the modified reach. The routing coefficients are recomputed using $K = .55$, the travel time through the modified reach. The new coefficients are

$$c_0 = 0.4149$$
$$c_1 = 0.6489$$
$$c_2 = -0.0638$$
$$c_0 + c_1 + c_2 = 1.0000$$

The results of the hydrograph routing through the modified reach are summarized in Table 8.9. The peak discharge at the bridge for the original channel is 4886 cfs (138 cms) and for the shorter channel is 4900 cfs (139 cms). The difference is not significant and the channel modification will have minimal effect on the peak discharge experienced at the bridge. ❑

8.5

SUMMARY

Hydraulic routing procedures use both the momentum and continuity equations, while hydrologic routing procedures use the continuity equation. Included in this chapter are three commonly used routing procedures: the kinematic, inventory, and Muskingum equations.

- The kinematic procedure uses both momentum and continuity principles with solution by finite differences.
- The kinematic equations require a detailed description of the watershed physical parameters, such as slope and cross sections, but the hydrograph shape does not have to be specified as in the unit–graph methods.
- The time and distance steps selected for modeling should define hydrograph and peak discharge within the accuracy expected.
- The inventory method is based on the continuity equation, and must include an outflow–storage (O–S) relationship to route hydrographs.
- The Muskingum method uses the continuity principle and assumes a weighting of inflow and outflow for a stream reach. The parameters of the method should be estimated from existing data or past experiences.
- There are numerous calculations required for routing methods (see example problems). It is beneficial to do some calculations to learn the sensitivity of the methods and the details of the calculations.

8.6

PROBLEMS

8.6.1 Hand Problem

1. Explain in your own words the theoretical basis for the development of hydrologic models and compare this to the development of hydraulic models. Give an example of a method which fits each classification.

2. For an overland flow situation, develop the finite difference equations to solve for unit width flow rates (similar to Equation 8.10).

3. In Table 8.1, show the calculations for $A_{1,3}$, $A_{2,3}$, and $q_{2,3}$.

4. What is the relative change (sensitivity) in the values of $Q_{3,1}$ (Table 8.2) if the channel roughness coefficient changed is to 0.015?

5. If rainfall excess of Example Problem 8.1 were constant at 4 in./hr, what is an estimate of peak discharge using kinematic equations? Compare this discharge with an estimate using the rational formula.

6. Using the input hydrograph of Example Problem 8.2, determine the size (volume) of a detention basin if the outflow hydrograph is

STORAGE (ft³)	BASIN DISCHARGE (ft³/sec)
≤ 10,000	0
> 10,000	10 ft³/sec (constant value)

Assume the detention basin is initially empty or the 10,000 ft³ of initial storage will eventually percolate into the ground.

7. What are the rainfall excess increments in 30-min intervals and the inflow rates (cfs) for a 100-ac impervious area for the first 2 hr from a 4-in. storm over 6 hr by using the NRCS dimensionless rainfall Mass Curve Type II if the maximum soil storage (S') is 0.5 in.? Now route this rainfall excess through the new channel of Example Problem 8.7.

TIME	ΔP (in.)	INFLOW (cfs)	OUTFLOW (cfs)
0	0.00		
30	0.17		
60	0.20		
90	0.26		
120	1.89		
150	1.34		
180	0.34		
210	0.29		
240	0.30		
270	0.22		
300	0.15		
330	0.18		
360	0.10		

What is the "routed" flow (output hydrograph) at 120 min if $t_c = \Delta t = 30$ min? No other information is available.

8. Using Manning's equation, calculate the discharge for a natural channel cut from rock that is severely irregular and not meandering or vegetation. The depth of flow is 4 ft, and the slope of the channel is 0.005.

9. Derive Equation 8.42 from Equations 8.39 and 8.40 being careful to state your assumptions and define all variables.

10. Using the coefficients in the Muskingum routing equation as developed in Example Problem 8.5 solve for the discharge hydrograph if the input hydrograph were

Time (days)	0	1	2	3	4	5	6	7	8	9	10	11	12
Average input (cfs)	0	100	250	380	300	200	130	80	50	30	15	5	0

11. If the outflow hydrograph of Example Problem 8.5 were attenuated further, or for the following average flow values (cfs) starting at Day 1, what are estimates of c and K?

Day	1	2	3	4	5	6	7	8	9	10	11	12
Outflow	0	10	40	50	45	35	30	20	15	10	5	0

12. Given the following inflow and outflow hydrographs for a stream reach with a relatively constant cross-sectional area and slope, determine the Muskingum routing constants K and c.

DAY	I (cfs)	O (cfs)
1	0	0
2	33	6
3	99	50
4	61	61
5	33	50
6	17	40
7	6	25
8	0	11
9	0	3
10	0	0

13. What is the discharge hydrograph for a 50-ac watershed with a rational coefficient of .4 and a storm intensity of 2 in./hr over 2 hr if the time of concentration is 60 min and the rational hydrograph is routed through a reservoir with the following characteristics? (Assume that the pond level is at control elevation.)

   Permanent pool (dead storage) = 43,560 ft³ (at control elevation)

   Discharge/storage = 4 ft³/sec/20,000 ft³

   Show all work for a 3-hr time period using Δt = 10 min and Δt = 15 min.

14. Specify the volume of a detention pond to attenuate the runoff from a constant intensity 3-in. rainfall event in 20 min on a 12-acre, completely impervious watershed. The precondition peak discharge is 24 cfs and occurs at 40 min, and the shape of the precondition hydrograph is the typical SCS 484 triangular. The time step for analysis is 10 min. Use the inventory equation. The attenuation must meet the precondition peak discharge.

15. Size a detention pond (maximum size in acre-feet) if the design pond input hydrograph is for a 10-min rainfall of constant intensity equal to 5 in./hr on a 20-acre, impervious area. The time of concentration is 40 min. The pond discharges 1 cfs for every 5000 cf of storage above the control elevation. The pond is at the control elevation at the start of the rainfall. Use the rational method to generate the input hydrograph with a computation interval of 5 min.

16. Determine the Muskingum storage time constant for a river segment that has a relatively constant cross-sectional area and slope if the weighting factor is equal to (a) 0.3, and (b) 0.2 using the following inflow and outflow data.

	A	B	C
1	Day	I	O
2		(cfs)	(cfs)
3	0	0	0
4	1	18	5
5	2	47	25
6	3	30	30
7	4	16	25
8	5	10	15
9	6	5	12
10	7	2	7
11	8	0	5
12	9	0	2
13	10	0	0

8.6.2 Computer Problems

1. Write a computer program to solve Example Problem 8.2 and generate a table similar to Table 8.3. Make the program as general as possible by having the capability of reading any input hydrograph and storage–discharge relationship as one linear piecewise approximation.

2. Using the SMADA computer program obtain a print out for an outflow hydrograph. The watershed and routing conditions are

   $A = 100$ ac
   % impervious $= 60$
   $DCIA = 50\%$
   $CN = 70$ (pervious area)
   Santa Barbara method
   $t_c = 200$ min

   The printout must illustrate about 55% reduction in the peak discharge. No pollution control should be done. Use a rainfall volume and time distribution of your choice. This problem illustrates the sensitivity of discharge as related to a storage discharge relationship.

3. For the previous problem, what duration storm would produce the largest detention volume? Use at least five different durations.

4. For Problem 2, use a 24-hr NRCS Type II storm to generate the hydrograph for the given watershed. If the percent impervious increases to 100% with all impervious area directly connected and if the time of concentration is 150 min, what happens to the peak discharge? Using SMADA, design a circular detention pond that reduces the peak discharge to the precondition value. Specify the stage–

storage–discharge relationship for the pond, weir type and placement, and initial stage.

8.7
_____

REFERENCES

Chow, V. T. 1959. *Open-Channel Hydraulics.* McGraw-Hill, New York.

Chow, V. T. (Editor). 1964. *Handbook of Applied Hydrology.* McGraw-Hill, New York.

Federal Highway Administration (FHWA). 1984. *Hydrology,* Hydrologic Engineering Center Circular No. 19, McLean, VA.

Henderson, F. M. 1966. *Open Channel Flow.* Macmillan, New York.

Hydrologic Engineering Center. 1979. *Introduction and Application of Kinematic Wave Routing Techniques Using HEC-1,* McLean, VA.

Hydrologic Engineering Center. 1985. *Flood Hydrograph Package, Users Manual,* Training Document No. 10, Devries and MacArthur, U.S. Army Corps of Engineers, Davis, CA (May).

Kersten, R. D. 1969. Engineering Differential Systems, McGraw-Hill Book Company, New York.

Mahmood, K. and Yevjevich, V. 1975. *Unsteady Flow in Open Channels.* Water Resources Publications, Fort Collins, CO.

Wanielista, M. P., 1977. *Manual of Stormwater Management Practices.* Report submitted to the East Central Florida Regional Planning Council, Winter Park, November.

Yevjevich, V. 1968. "Computer and Observed Unsteady Water–Surface Profiles in a Circular Cross-Section," *American Society of Civil Engineers Hydraulics Division 16th Specialty Conference,* August.

CHAPTER

9

GROUNDWATER HYDROLOGY

Groundwater hydrology is a study area of hydrology concentrating on the storage and movement of water beneath the ground surface. It is an important study area for many reasons. One reason is that some streams and lakes are fed primarily by groundwaters. Another reason is the management of stormwater can be accomplished by on-site infiltration and regional ponds for infiltration. To encourage groundwater recharge in some areas, on-site and areawide infiltration is mandated by municipal regulations. Still another reason is that vast storage volumes of water are available under the ground and pipes (wells) can penetrate the storage areas to extract water for beneficial uses. Over 25% of nonsaline water used in the United States is extracted from groundwater. This does not include the hydroelectric power generation use of water. Additional information relative to groundwater quality considerations is also given. The study of the hydrologic cycle would not be complete without an understanding of the exchanges of water between ground and surface supplies. This chapter presents information to aid in quantifying groundwater storage and movement with definition of terms relating to groundwater.

9.1

THE OCCURRENCE OF GROUNDWATER

Groundwater is water resident beneath the ground surface. It results from waters that infiltrate from the ground surface and percolates to the underlying strata. Infiltrated groundwaters pass through an unsaturated zone in route to a saturated zone. The water table separates the two zones. The pressure at the water table is atmospheric.

There can be a capillary fringe to which water will rise above the water table. In fine silty sand, the capillary rise can be as high as 50 cm, while the rise in gravel is only about 2 or 3 cm. It is called the soil–moisture region. The water in this region fluctuates in quantity as the vegetation uses moisture and percolation from infiltated waters occurs. Percolated water (gravity water), capillary water, and air exist among the soil particles. The percolated water moves downward in primarily the larger soil pores. The smaller the pore spaces, the less gravity movement of waters resulting from infiltration.

Soil moisture is measured as the loss in weight of soil after being oven dried at 103 to 105°F. Higher temperatures may burn off organics. After percolated water has passed through the soil, that soil moisture remaining is defined as the field capacity and is essentially the water retained in the soil at a tension pressure of about 0.33 atm (Colman, 1947). When plants can no longer use water from the soil, the remaining soil moisture is said to be at the wilting point. The soil moisture difference between field capacity and wilting point is called available water moisture. This is useful water storage. Commonly reported moisture values for different soils are shown in Table 9.1.

The water table level will fluctuate and may be directly related to precipitation. When reporting levels, the time and date of measurement should be reported. The quantity of water that can be stored under the surface depends on the porosity of the subsurface strata, the type of liquid and adjacent underground soil and water table conditions. The media in which groundwater moves are characterized by many factors, two of which are void spaces (porosity) and resistance (permeability).

9.1.1 Porosity

Porosity (n_p) is defined as the ratio of void volume to total volume and may range from a small fraction to about 0.90.

$$n_p = \frac{V_v}{V} = \frac{(V - V_m)}{V} = 1 - \frac{V_m}{V} \tag{9.1}$$

where

n_p = porosity (subscript p because n is used in Manning's equation)
V_v = volume of the voids
V = total volume of sample
V_m = volume of the soil (material)

The volume of the soil can be determined by dividing the weight of the "dry" soil by the specific weight (soil solid specific gravity times unit weight of water). In a granular mass composed of uniform spheres with the loosest possible packing, $n_p = 47.6\%$, and with the densest possible, $n_p = 26\%$.

Regional water-bearing strata, called aquifers, often consist of unconsolidated material like sandstones and limestones. These strata usually have porosity values ranging from 5 to 15%. Gravel and sand aquifers have higher porosities. Limestone itself is relatively impervious but is soluble in water and so frequently has wide joints and solution passages that make the rock similar to a very porous rock in its capacity

TABLE 9.1 Common Moisture Values as a Percent of Dry Soil Weight

SOIL TYPE	FIELD CAPACITY (AFTER PERCOLATION)	WILTING (AFTER PLANT USE)	AVAILABLE (FIELD WILTING)
Sand	5	2	3
Sandy loam	12	5	7
Loam	19	10	9
Silty loam	22	13	9
Clay loam	24	15	9

TABLE 9.2 Representative Porosity Values for Common Soils

SOIL TYPE	POROSITY %	TYPICAL GRAIN SIZE (mm)
Tilled Soil	50–60	(1)
Clay	45–55	<0.005
Silt	40–50	0.05 − 0.005
Fine to Med. Sand	30–35	0.25 − 0.05
Medium Sand	30–40	1.0 − 0.25
Med. to Coarse Sand	35–40	2.0 − 1.0
Gravel	30–40	4.0 − 2.0
Coarse Gravel	15–25	>4.0
Mixed Gravel & Sand	20–35	(1)
Sandstone	10–20	(2)
Shale	1–10	(2)
Limestone	1–10	(2)

Adapted from Todd (1980) and ASTM specifications. (1) Depends on degree of mixing and tillage practice; (2) depends on weathering and solution cavities.

to hold water over a wide area. For a material to be permeable, the pores must be connected.

Representative values of porosity are presented in Table 9.2. These values may be used with Equation 5.18 to estimate infiltration rates, as well as in Equation 9.7 to estimate seepage rates.

9.1.2 Permeability

The water in the pores of an aquifer is subject to gravitational forces and so tends to flow downward through the connected pores of the material. The resistance to this underground flow varies widely. The rate at which water moves through the pores of the material (permeability or hydraulic conductivity) is a measure of this resistance. Aquifers with large pores such as coarse gravels are said to have a high permeability, and those with very small microscopic pores such as clay, have a low permeability.

If two piezometers are placed in the groundwater, the velocity of the groundwater can be calculated with the idealized conditions of Figure 9.1 using the relationship of change in head per length of flow or

$$v = K \left[\frac{(h_1 + z_1) - (h_2 + z_2)}{L} \right] \tag{9.2}$$

where
K = permeability (length/time)
L = distance between piezometers (length)
h = pressure heads (length)
z = elevation heads (length)
v = Darcy velocity (length/time)

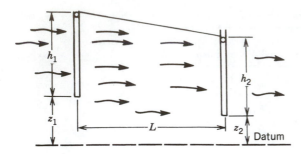

FIGURE 9.1   One-dimensional groundwater flow.

Equation 9.2 and modifications are called the Darcy equations after the French hydrologist Henri Darcy, who discovered that velocity was proportional to hydraulic gradient. The factor of proportionality is the permeability.

Permeability can be determined in the laboratory using permeameters. Two types are used: constant head where constant pressure heads are maintained at the inlet and outlet, and falling head where constant head is maintained only at the outlet end. Schematics of the equipment are shown in Figure 9.2. The constant head is suitable for estimating the permeability of soild like sands and gravels, whereas the falling head is more suitable for soils of relatively low permeability, like fine silt.

For the constant head permeameter the volume flow rate through a given cross-section area perpendicular to flow is given by the continuity equation ($Q = Av$). Rearranging Equation 9.2 for $K = vL/\Delta H$ yields Equation 9.3 with sample area equal to πR^2 and H equal to ΔH.

$$K = \frac{LQ}{H\pi R^2} \tag{9.3}$$

where
K = permeability (m/day, ft/day)
H = head loss (m or ft)

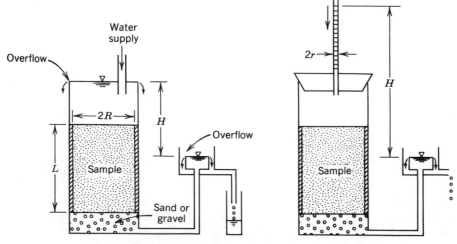

FIGURE 9.2   Permeameters—constant head on left, falling head on right.

L = sample height (m or ft)
Q = outlet flow rate (m³/day, ft³/day)
R = radius of sample (m or ft)

The permeability from the falling-head permeameter also can be expressed by an equation. In the small, narrow standpipe, flow rate can be expressed as

$$Q = \pi r^2 \frac{dh}{dt} \tag{9.4}$$

where

$$r = \text{radius of standpipe (m, ft)}$$

and on the basis of flow through the sample as

$$Q = K\pi R^2 \frac{H}{L} \tag{9.5}$$

Equating the two expressions for flow rate and integrating, one obtains

$$K = \left(\frac{Lr^2}{tR^2}\right)\ln\left(\frac{H_1}{H_2}\right) \tag{9.6}$$

where
H_1 = starting height of water column (m, ft)
H_2 = ending height of water column (m, ft)
t = time required for water to drop (day)

The rate of water movement through a soil matrix itself is called seepage or pore velocity and is greater than the Darcy velocity because the flow area is reduced. The unrestricted flow rate measured using Darcy velocity must be equal to the flow rate through the soil media, thus the two flow rates can be equated as

$$Q_{\text{Darcy}} = Q_{\text{Seepage}}$$
$$vA = v_s A_s$$

where
v_s = seepage velocity (m/day)
v = Darcy velocity (m/day)
A = total area (m²)
A_s = seepage area (m²)

and the seepage area is expressed as

$$A_s = n_p A$$

Thus the seepage velocity can be calculated from the Darcy velocity by substituting for A_s:

$$vA = v_s n_p A$$

or

$$v_s = v/n_p \qquad\qquad (9.7)$$

The residence time in the soil is less than the residence time considering the total flow area because the seepage velocity of flow in the soil is greater.

9.1.3   Factors Affecting Permeability

Permeability estimates may vary locally due to variability in soil materials. If soil changes are expected, a suitable number of samples should be taken. Solution cavities and other groundwater channels may also affect estimates.

If the soil materials do not change, then other factors also may affect permeability, such as temperature, ionic composition, and presence of entrapped air. Temperature has an effect on water viscosity, and thus on permeability. Higher temperatures mean lower viscosity. Thus, the water will have less resistance for movement through the soil, and the permeability will increase. The change in permeability is linear. Values of permeability are normally expressed at 20°C and values of permeabilities at other temperatures are calculated using

$$K_t = \left(\frac{\mu_{20}}{\mu_t}\right) K_{20} \qquad\qquad (9.8)$$

where

K_t = permeability at temperature t°C
μ_{20} = absolute viscosity at 20°C [N·sec/m²]
μ_t = absolute viscosity at t°C [N·sec/m²]
K_{20} = permeability at 20°C

As water passes through clay materials, the pores may be very small causing larger ions in the water to be removed. Thus, the soil behaves like a selective membrane. At high water gradients, monovalent cations are retarded more so than the divalent ones. McNeal (1968) developed a sodium absorption ratio (SAR) with calcium and magnesium:

$$SAR = \frac{Na^+}{\sqrt{\dfrac{Ca^{++} + Mg^{++}}{2}}} \qquad\qquad (9.9)$$

where
SAR = sodium absorption ratio (SAR)
Na^+ = sodium concentration (meq/L)
Ca^{++} = calcium concentration (meq/L)
Mg^{++} = magnesium concentration (meq/L)

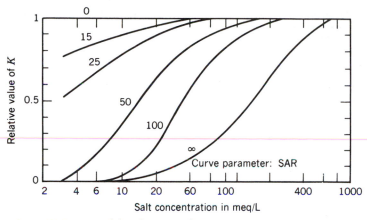

FIGURE 9.3 Permeability changes with SAR.

The concentration of an ion can be expressed as the number of equivalents per weight, such as milligram equivalents per liter. Thus, meq/L = concentration (mg/L) per equivalent weight. Equivalent weight is the atomic weight divided by the ionic charge. (See Appendix B.) At a given salt concentration, permeability decreases with increasing SAR, as shown in Figure 9.3.

Completely saturated soils are used to determine permeabilities in the laboratory. However, entrapped air in the soils physically reduces the permeability. In sandy soils, the permeability in the unsaturated zone may be only about one-half the permeability at saturation (Bouwer, 1966). Entrapped air may also occur after a rapid rise or fall of the groundwater table.

9.1.4 Common Permeability Values

The common units of permeability are meters/day or feet/day. However, units of centimeter/hour or inches/hour also are useful for faster moving waters and other units such as gallons per minute/square feet and feet/second are used where rate values are in similar units.

Investigators have tried to relate the permeability of a soil to its physical properties (grain size, density, porosity) and found some empirical formulas which were primarily useful for site specific results. One such formula often identified as specific or intrinsic

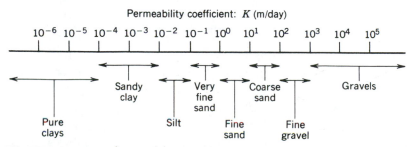

FIGURE 9.4 Range of permeability in soils.

TABLE 9.3 Range of Permeability for Common Soils

SOIL MATERIAL	PERMEABILITY m/day
Gravel	2592 to 25.92
Coarse Sand	518 to 7.78×10^{-2}
Medium Sand	51.8 to 7.78×10^{-2}
Fine Sand	17.3 to 1.73×10^{-2}
Silt	1.73 to 8.64×10^{-5}
Till	0.173 to 8.64×10^{-8}
Clay	4.32×10^{-4} to 8.64×10^{-7}
Karst Limestone	1728 to 8.64×10^{-2}
Sandstone	0.518 to 2.59×10^{-5}
Shale	1.73×10^{-4} to 8.64×10^{-9}
Basalt	3.4×10^{-2} to 1.73×10^{-6}
Weathered Granite	2.59×10^{-1} to 4.32

permeability was developed for clean water supply filter sands and is

$$K = C(d_{10})^2 \tag{9.10}$$

where

K = permeability coefficient (m/day)
d_{10} = grain size (mm) of sand for which 10% is finer
C = regression constant (average value = 1000)

Well-mixed (heterogeneous) soil often displays the simple relationship given by Equation 9.10 and is solely a function of the porous medium. Possible range of K's for various natural soils are shown in Figure 9.4 and Table 9.3. These are to be used only as representative values.

Table 9.4 also illustrates some values for permeabilities, porosity, and specific

TABLE 9.4 Approximate Average Porosity, Specific Yield, and Permeability of Various Soil Materials (partly adopted from Bouwer, 1966 and Todd, 1980)

MATERIAL	POROSITY (%)	SPECIFIC YIELD (%)	AVERAGE PERMEABILITY K (gpd/ft^2)	(m/day)
Clay	45	3.0	0.01	0.0004
Sand	35	25.0	1,000.00	41.0000
Gravel	25	22.0	100,000.00	4,100.0000
Gravel & sand	20	16.0	10,000.00	410.0000
Sandstone	15	8.0	100.00	4.1000
Limestone, shale	5	2.0	1.00	0.0410
Quartzite, granite	1	0.5	0.01	0.0004

yield for various soil materials. Specific yield is the ratio of the volume of water released to a unit volume of saturated, unconfined aquifer material. Expressed as a percentage, it is calculated from the ratio of volume of water removed to total volume of the soil multiplied by 100. If expressed as a fraction, it is called storage coefficient. The permeability numbers should be considered as only general estimates.

9.2

AQUIFERS AND SPRINGS

An aquifer is a water-bearing soil or rock that can release its water in sufficient quantities to make it economically feasible to develop for water supply. The aquifer can be either confined or unconfined, depending on whether or not a water table or free water surface exists under atmospheric pressure. Springs result when water either drains by gravity or is forced from the ground. Water enters the ground and is forced by gravity to percolate into lower strata.

As groundwater percolates into the aquifer, it will reach a level at which the aquifer is saturated. This saturation surface may approach the ground resulting in apparent horizontal flow. Waters closest to the ground surface or surficial waters can fluctuate greatly during the course of a year. The surficial waters fall during dry time periods and rise in rainy weather. The upper water surface in the deep aquifers and the surficial waters are usually moving slowly toward the nearest free water surface such as a lake or river, or the sea. However, if there is an impermeable layer underlying an aquifer and this layer outcrops on the ground surface, then the groundwater will appear on the surface and is called a spring. There are at least four different classes of springs as shown in Figure 9.5. The U.S. Geological Survey classes major springs, as shown in Table 9.5.

It is equally possible for a groundwater aquifer to become overlain by impermeable material and thus be under pressure. These aquifers, receiving water from distant sources, are called confined aquifers and the surface to which the water would rise in

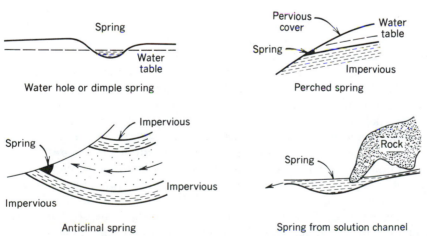

FIGURE 9.5 Typical spring classifications.

TABLE 9.5 USGS Classification of Springs

MAGNITUDE	DISCHARGE
First	>100 cfs
Second	10–100 cfs
Third	1–10 cfs
Fourth	100–448.8 gpm
Fifth	10–100 gpm
Sixth	1–10 gpm
Seventh	1 pt/min–1 gpm
Eighth	<1 pt/min

a standpipe if it could, is called the piezometric surface. Wells drilled into such confined aquifers are called artesian wells, and the aquifers are called artesian aquifers. If the piezometric surface is above ground level and an artesian well penetrates the aquifer, a flowing well results. A fracture or flow in the impermeable overlay of an artesian aquifer will result in an artesian spring. Sometimes areas of impermeable material may exist in an aquifer. This may happen through geological faulting or through a lens of clay occurring in an otherwise sandy area. A small local water table, called a perched water table, may result, and this can often be a long way above the true piezometric surface.

9.2.1 Aquifers—An Example

The aquifers described above can be found in many regions of the world. They can provide in these regions much of the fresh water used for public, industrial, and agricultural use. As an example, the state of Florida contains abundant groundwater resources. The average freshwater withdrawals from the four major aquifers of the state amounted to 3800 million gallons per day (MGD) in 1980. Irrigation withdrawals alone accounted for 42% of this averagae daily quantity. The four principal aquifers are shown in Figure 9.6. The Biscayne aquifer of South Florida consists of limestone, sandstone, and sand. It is an unconfined aquifer. Total water withdrawn is in the order of 500 MGD. The intermediate aquifer is limestone and discontinuous with shell beds and clay layers. The largest of the four is the Floridan. This consists of limestone and dolomite. It is confined in deep areas and unconfined in outcrop areas and currently provides over 500 million gallons per day. The sand and gravel aquifer of Northwest Florida is unconfined near upper-surface areas and confined in deeper areas. Springs are abundant with 27 of the United States' 78 first-magnitude springs resident in Florida. The groundwater reservoirs are primarily recharged from infiltrated and percolated rain waters. In the Biscayne aquifers, a recharge potential is maintained by the storage of runoff waters (fresh water) in canals. Maintenance of a high water table in this area is important to minimize saltwater intrusion. Annual recharge rates for the four major Florida aquifers have been related to rainfall and found to vary yearly from less than one inch to a maximum of about 5 in.

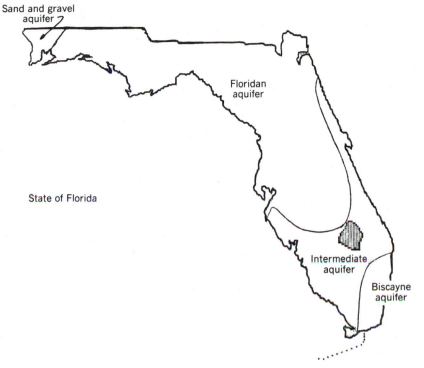

FIGURE 9.6   An example of principal aquifers. *Source:* National Water Summary—Florida, 1985.

9.3

MOVEMENT OF GROUNDWATERS

Except in large caverns and fissures, groundwater flow is almost exclusively laminar. It has been shown that the velocity of flow in capillary tubes is proportional to the slope S of the energy gradient. Darcy confirmed the applicability of this principle to flow in uniform sands, and Equation 9.2 can be rewritten as

$$V = KS \qquad (9.11)$$

Assuming the continuity equation is valid for a particular part of an aquifer, of area A, flow rate is estimated from

$$Q = VA = KAS \qquad (9.12)$$

where Q is the flow rate (volume per unit time) through a cross-sectional area A of aquifer. Note that the velocities are relatively slow compared to surface flows. Thus, velocity heads are negligible in groundwater flow. The slope, S, is either the slope of the water table or the piezometric surface.

The flow through an aquifer with width Y and depth D in gallons per day can be written if area is equal to YD:

$$Q = YDKS \qquad (9.13)$$

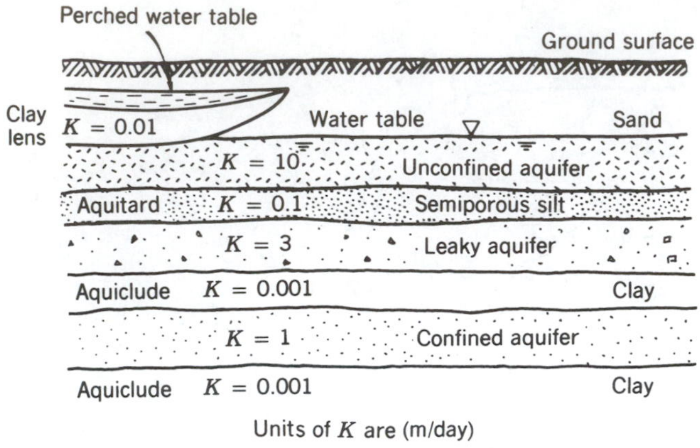

FIGURE 9.7 Aquifer definitions (with sample K values—metric units).

Equation 9.13 is similar to the equation for the flow of electricity (Ohm's law) where S is analogous to the gradient of the voltage drop, KYD to the conductance of the circuit, and Q to the amperage.

The transmissivity of an aquifer is the flow in gallons per day (cubic meters per day) through a section of aquifer 1 ft (1 m) wide under a hydraulic gradient of unity. The unit of measure commonly used in the United States is gallons per day per foot width. The product KD in Equation 9.13 is often combined into a parameter called transmissivity with Equation 9.13 rewritten as

$$Q = YTS \qquad\qquad (9.14)$$

where T is the aquifer transmissivity.

Once recharge water has infiltrated, its percolation downwards to the groundwater storage depends on the geological structure as well as on the rock composition. Figure 9.7 shows a section through a series of sedimentary rocks in which it is most usual to find productive aquifers—beds of rock with high porosity that are capable of holding large quantities of water. In general, the older the rock formation, the more consolidated is the rock material and the less likely it is to contain water. Igneous and metamorphic rocks are not good sources of groundwater unless weathered and/or fractured. The sedimentary rock strata have different compositions and porosities. In the much simplified diagram, layers of porous sands of limestones are subdivided by less porous material such as silt or clay that inhibits water movement. Semiporous beds that allow some seepage of water through them are known as aquitards; they slow up percolation to the porous layers below, which are called leaky aquifers since they can lose as well as gain water through an aquitard. The clay beds that are mainly impermeable are called aquicludes and the porous layers between them are confined aquifers in which the water is under pressure. Water pressure between aquifers will determine water movement. In the top sandy layer, the water table at atmospheric pressure marks the variable upper limit of the unconfined aquifer, although locally a lens of clay can hold up the groundwater to form a perched water table.

◻ **EXAMPLE PROBLEM 9.1**

An area of land is underlain by glacial soils, which include a thick sand horizon. Under the soils is a rock layer. Both the sand and rock layers contain groundwater. The water from the rock horizon is abstracted for a public water supply. It is

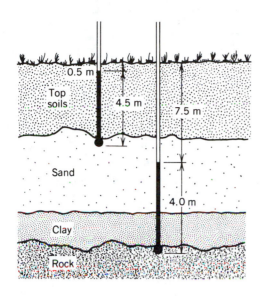

FIGURE 9.8 Piezometer installation for liquid waste disposal proposal.

proposed that contaminated effluent be disposed of by pumping it into the sand horizon.

Would this lead to contamination of the public water supply source? Assume that the density of groundwater in the rock horizon is 1000 kg/m³, that of the fluid in the sand layer (after mixing with the contaminated fluid) is likely to be 1040 kg/m³ and that site exploration drilling has shown the pressure head in the two layers (see Figure 9.8).

Solution

Flow occurs from high pressure to lower pressure. If the pressure in the sand horizon is greater than that in the rock layer, then the polluted water may be carried into the public water supply.

Pressure intensity (sand horizon) $= (\rho)(g)(h)$

$$= 1040 \times 9.81 \times 4.5$$

$$= 45.91 \text{ kN/m}^2$$

Pressure intensity (rock layer) $= 1000 \times 9.81 \times 4.0$

$$= 39.24 \text{ kN/m}^2$$

Pressure $= \gamma h = \rho g\, h$

Thus, a slightly higher pressure intensity exists in the sand and if there are breaks in the clay, the potential for the flow of contaminated fluid to the rock aquifer exists. ❏

9.4

FLOW IN A CONFINED AQUIFER

An aquifer can be confined if its ceiling and floor are impermeable over an area of study. The pressure in this artesian aquifer forces water to be recorded at piezometric

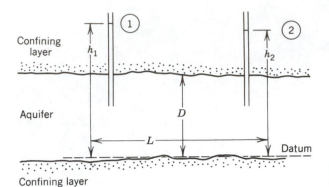

FIGURE 9.9   Schematic of confined aquifer.

levels. If piezometric levels are measured at the same time for different locations, the gradient of flow can be established. Using this gradient, an estimate of flow can be made assuming no other inputs or outputs across the boundary of the aquifer. Darcy's law for one-dimensional flow can be used to solve simple systems of lateral and vertical groundwater flow. Vertical-flow components are often neglected where groundwater moves primarily in the lateral direction (DeWiest, 1965). Consider the schematic of the confined aquifer as shown in Figure 9.9. Construction of an energy balance, i.e., Bernoulli equation, yields Equation 9.15.

$$\frac{V_1^2}{2g} + \frac{P_1}{\gamma} + h_1 = \frac{V_2^2}{2g} + \frac{P_2}{\gamma} + h_2 + \Delta h \text{ (head loss)} \tag{9.15}$$

where

V_1, V_2 = velocities
g = gravitational constant
γ = specific weight of fluid
h = piezometric heads
L = length of artesian aquifer between piezometric measurements (Figure 9.9)
D = thickness of aquifer (Figure 9.9)
Δh = head loss
P_1, P_2 = pressure

Normally, V_1 and V_2 are similar and the difference in pressure and elevation is measured by the piezometer, or:

$$\Delta h = \frac{P_1}{\gamma} + z_1 - \frac{P_2}{\gamma} - z_2$$

or for length ΔL,

$$S = \Delta h / \Delta L \text{ (energy gradient slope)} \tag{9.16}$$

and using the definition of permeability

$$V = KS \tag{9.17}$$

substituting Equation 9.16 into Equation 9.17 for a differential length (ΔL) gives

$$V = K(\Delta h/\Delta L) \qquad (9.18)$$

which is an expression of Darcy's law with $\Delta h/\Delta L$ a positive value for decreasing head. A negative sign may be present on the right-hand side of the equation if Δh is negative.

Now, let $Q = AV$, thus Equation 9.18 becomes $Q = KA(\Delta h/\Delta L)$ and per unit width $(A = D)$, thus $Q' = DV$ and

$$Q' = KD(\Delta h/\Delta L) \qquad (9.19)$$

and assuming head varies linearly:

$$\Delta h/\Delta L = \frac{(h_1 - h_2)}{L} \qquad (9.20)$$

9.5

FLOW IN AN UNCONFINED AQUIFER

For an unconfined aquifer, a similar equation can be developed. Figure 9.10 illustrates the schematic of an example unconfined aquifer.

$$Q = KA \, dh/dL \qquad (9.21)$$

Area varies along the path of flow and for unit width $A = y(1)$ thus $Q' = Ky \, dy/dL$ with y the variable on depth, thus integration of

$$Q' \int_0^L dL = K \int_{h_2}^{h_1} y \, dy$$

results in

$$Q' = \frac{K(h_1^2 - h_2^2)}{2L} \qquad (9.22)$$

which is known as the Dupuit equation.

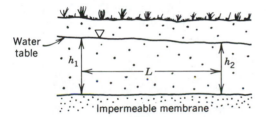

FIGURE 9.10 Unconfined aquifer schematic.

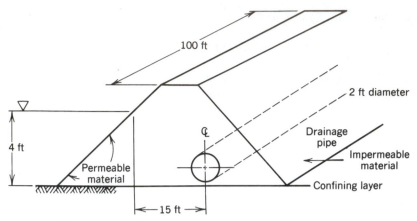

FIGURE 9.11 Reservoir seepage.

☐ **EXAMPLE PROBLEM 9.2**

Estimate the discharge from a large reservoir through a permeable sand/clay mixture with $K = 40$ gal/ft²-day. The face of the permeable dam is 4 ft deep and the discharge pipe is 2 ft in diameter. A schematic of the physical situation is shown in Figure 9.11.

Solution

$$Q' = (K/2L)(h_2^2 - h_1^2) = \frac{40 \text{ gal/ft}^2\text{-day}}{2(15)\text{ft}} (4^2 - 2^2)$$

$$= (40/30)(16 - 4) = 16 \text{ gal/ft-day and for 100 feet } Q = 1600 \text{ gal/day}$$

$$Q = 1600 \text{ gal/day} \times 1 \text{ } CF/7.48 \text{ gal} = 214 \text{ ft}^3\text{-day}$$

Decrease in pond elevation if pond has a surface area of 1 acre:

$$[214 \text{ ft}^3/43,560 \text{ ft}^2] \times 12 \text{ in./ft} = 0.06 \text{ in./day} \quad ☐$$

9.6

UNIFORM INFILTRATION AND DRAINAGE

An unconfined aquifer with the saturated zone above and sloping toward a surface water body can contribute groundwater flow to the surface source. Consider an unconfined aquifer with a uniform rate of infiltration being applied over an area. Once the percolated waters reach the superficial water table, drainage moves towards a free standing surface water body, as shown in Figure 9.12. Again, using Darcy's law with lateral flow and velocity at distance x equal to v_x, one can write

$$v_x = K\left(\frac{\Delta h}{\Delta x}\right) \tag{9.23}$$

where $\Delta h/\Delta x$ = slope of the water table, taken as a positive value.

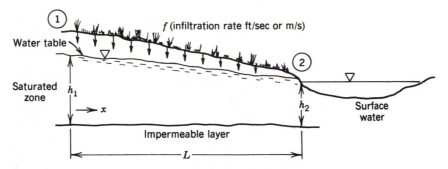

FIGURE 9.12 Uniform infiltration with drainage to surface waters.

The flow per unit width of aquifer is

$$Q = KA(\Delta h/\Delta x) \quad \text{and} \quad A = h \text{ for unit width}$$

thus

$$Q' = Kh(\Delta h/\Delta x)(\text{units are: ft}^3/\text{ft-sec, m}^3/\text{m-s}) \tag{9.24}$$

also, note $Q' = fx$, thus one can write from a mass balance:

$$fx = Kh\left(\frac{\Delta h}{\Delta x}\right)$$

and

$$Kh \, \Delta h = fx \, \Delta x \tag{9.25}$$

Integration between the limits of infiltration produces

$$K(h_1^2 - h_2^2) = fL^2$$

and

$$h_1 = (h_2^2 + fL^2/K)^{1/2} \tag{9.26}$$

where
h_1 = height of water table at boundary upgradient
h_2 = height of surface water table at pond
L = distance of point 1 from the surface water
f = infiltration rate
K = permeability

9.7

WELL SYSTEMS

A well is a system of pipes for removing or injecting fluids into a subsurface area. The subsurface hydraulics are altered or controlled. The construction of a well can use drilling equipment, casings, grouting, well screens, pipe fittings, and the pump facility. Some injection wells may operate by gravity and thus do not require a pump. Also, not all wells use all the other equipment during construction. Examples of two completed wells are shown in Figure 9.13. For the consolidated soils, casing is not needed; however in unconsolidated soils (typically sand and gravel) casing is necessary. The well screen is used when water is being removed from the aquifer. The drilling of the well is usually done by one of the following methods (Campbell and Lehr, 1977).

1. *Cable Tool.* A cable with bottom bit that crushes the formation material. The cuttings are usually removed by bailing or recirculating a slurry.
2. *Rotary Hydraulic.* A rotating bit usually using a water-based slurry with dry material. Cuttings are removed from the outside of the casing by continuously circulating slurry and removed fluid.
3. *Reverse Circulation.* Same as (2) but cuttings are removed from the inside of the casing. Slurry is injected on the outside of the casing.

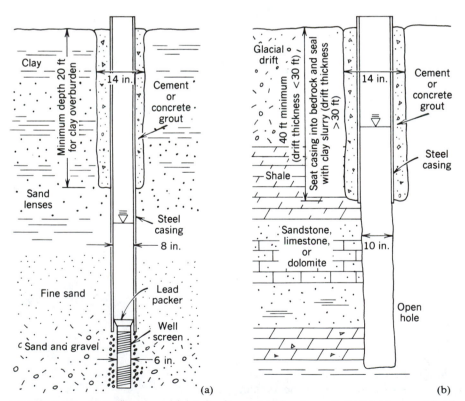

FIGURE 9.13 (*a*) Well in unconsolidated sediments (sand and gravel). (*b*) Well in consolidated sediments (Campbell and Lehr, 1977).

4. *Air.* Same as (2) except forced air is used.

5. *Air Percussion.* A percussion mechanism is attached to the bit and rotary motion persists.

6. *Hollow Rod.* Same as (1) except cuttings travel up the casing and shorter more rapid strokes are noted.

7. *Jet.* Same as (6) except evacuated cuttings return up the outside of the casing.

8. *Driven.* The well is driven using repeated blows on a plugged pipe.

Well-point systems are composed of many wells placed close enough to each other to be able to intercept flow from the same area. Well points are used to remove contaminated ground waters or lower the water table in shallow aquifers.

Aquifer permeability and the storage coefficient are used to describe the hydraulic characteristics. The storage coefficient is the volume of water yielded per unit horizontal area and unit drop in water table (unconfined aquifer) or piezometric surface (confined aquifer). Thus, if a water table aquifer yields 2 m³ of water during a drop of 4 m in a horizontal area of 10 m², the storage coefficient is 0.05 or 5%. The yield or performance of a well can be determined under equilibrium or nonequilibrium conditions. For equilibrium conditions to be met, the cone of depression caused by pumping must be stabilized.

9.7.1 Yield of a Confined Aquifer

When the well fully penetrates a horizontal confined aquifer (Figure 9.14), flow to the well comes from all directions (i.e., radial two-dimensional flow). By pumping the

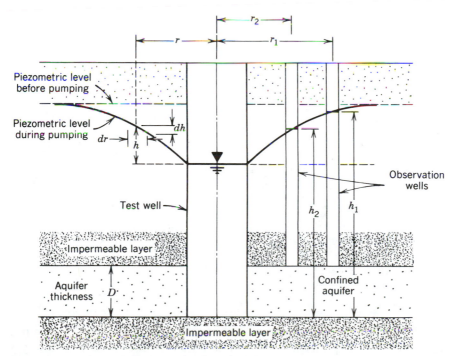

FIGURE 9.14 Pumping from a confined aquifer.

well at a steady rate and waiting until the well level is constant, observation of the drawdown levels h_1 and h_2 at observation wells at a known distance r_1 and r_2, from the pumped well allows estimation of the permeability of the aquifer. If the permeability of an aquifer is multiplied by the thickness (or height) of an aquifer, then the term known as transmissivity can be calculated. As presented earlier, the dimension of transmissivity is length^2/time, for example m^2/day or gal/day-ft (1 m^2/day = 80.5 gpd/ft). In practice, the groundwater levels from two or more observation wells at different radii are more useful than water levels in the well itself. To ensure a steady flow, there must be continuous recharge to the aquifer from sources distant to the well. Assuming also that the aquifer is homogeneous and isotropic and is not affected by compression in dewatering, the equilibrium flow to the well at any radius r can be expressed by

$$Q = (2\pi r)\frac{KD\,dh}{dr} \qquad (9.27)$$

where

$\quad Q$ = pumping rate (ft^3/s)
$2\pi r$ = perimeter of a cylindrical shell (ft)
$\quad K$ = permeability (ft/s)
$\quad D$ = thickness of aquifer (ft)

Integrating, one obtains

$$Q = 2\pi DK \frac{(h_1 - h_2)}{\ln(r_1/r_2)} \qquad (9.28)$$

and $T = KD$

$$Q = 2\pi T \frac{(h_1 - h_2)}{\ln(r_1/r_2)} \qquad (9.29)$$

Rearranging and changing units in Equation 9.28, the coefficient of permeability may be determined from

$$K = \frac{[528Q\log_{10}(r_1/r_2)]}{[D(h_1 - h_2)]} \qquad (9.30)$$

where

$\qquad K$ = permeability (gpd/ft^2)
$\qquad Q$ = flow (gpm)
D, r, h = units of feet

□   **EXAMPLE PROBLEM 9.3**
_____

For an artesian aquifer 150 ft thick and composed of fine sand, a well is pumped to equilibrium at 1000 gpm. The drawdown at an observation well 400 ft away is 2 ft and at 40 ft it is 12 ft. What is the permeability?

Solution

Using Equation 10.30 with $Q = 1000$ gpm, $D = 150$ ft, and the differential drawdown is 10 ft.

$$K = \frac{[528(1000)\log(400/40)]}{150(10)} = 352 \text{ gpd/ft}^2 \quad \square$$

9.7.2 Yield of an Unconfined Aquifer at Equilibrium

The yield and permeability of an unconfined aquifer can be determined by performing a pumping test from a well with observation wells similar to that done for the confined aquifers. The physical situation is shown in Figure 9.15. Here, flow is assumed to be radial with a horizontal water table. The rate of flow is written as

$$Q = (2\pi r)(K)(h)(dh/dr) \tag{9.31}$$

Integration yields

$$Q \int_{r_2}^{r_1} dr/r = (2\pi K) \int_{h_2}^{h_1} h(dh)$$

or

$$Q = \frac{K\pi(h_1^2 - h_2^2)}{\ln(r_1/r_2)} \tag{9.32}$$

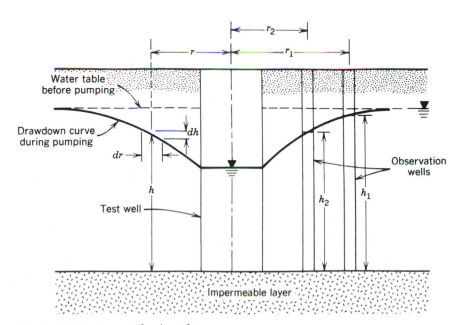

FIGURE 9.15 An unconfined aquifer.

If Q, r_1, r_2, h_1, and h_2 are determined from field measurements, then K can be calculated, or if K is known, the discharge can be calculated for specified drawdown conditions. For a change in flow rate units to gpm, Equation 9.32 is rewritten as

$$K = \frac{1055 Q \log_{10}(r_1/r_2)}{[h_1^2 - h_2^2]} \tag{9.33}$$

where

Q = gpm
K = gpd/ft^2
r, h = ft

In practice, Equations 9.32 and 9.33 are considered valid as long as the drawdown equilibrium conditions does not exceed one-half the original aquifer thickness.

❏ **EXAMPLE PROBLEM 9.4**

Calculate the steady-state discharge if the drawdown at observation wells remains constant at 20 ft and 15 ft corresponding to observation wells 100 ft and 200 ft from the proposed well location. The unconfined aquifer permeability is 70.0 gpd/ft^2 and the aquifer thickness is 80 ft.

Solution

Rearrange Equation 9.33 with

$$h_1 = 80 - 15 = 65 \text{ ft}$$

$$h_2 = 80 - 20 = 60 \text{ ft}$$

$$Q = \frac{[70.0(65^2 - 60^2)]}{[1055 \log(200/100)]} = 137.8 \text{ gpm} \quad ❏$$

9.7.3 Unsteady Flow

Equilibrium equations usually overestimate permeability and thus the yield of a well when equilibrium is not obtained. This is the usual situation in practice because when a well is first pumped, a large quantity of water comes from the initial cone of depression and equilibrium usually takes a significant time to achieve. There are two methods for solution namely one by Theis (Todd, 1980) and the other by Cooper and Jacob (1946). The Theis method is usually solved by evaluating an infinite series equation form. This procedure involves the graphical superimposition of two curves. Often, this procedure can be shortened, simplified, and more accurate when the observation distance from the well is small and the time for analysis is large. For these situations, Cooper and Jacob (1946) found an expression for T as follows:

$$T = (264Q/\Delta h) \log_{10}(t_2/t_1) \tag{9.34}$$

where

  T = transmissivity (gpd/ft)
  Q = well discharge (gpd)
  Δh = drawdown per time period (ft)
  t_2, t_1 = time periods for analysis

The field data for drawdown are plotted on semilog paper as shown in Figure 9.16. As time increases, a straight line relation results. If t_2 and t_1 correspond to one log cycle, then Equation 9.34 reduces to

$$T = \frac{264Q}{\Delta h} \tag{9.35}$$

and Δh equals the drawdown distance over one log cycle (feet).

  The slope and the intercept (t_0) permits computation of the variables used in the storage Equation 9.36.

$$S_c = 0.3T\frac{(t_0)}{r^2} \tag{9.36}$$

where

  S_c = storage coefficient (volume yield per unit horizontal area and per unit drop in entire area water table (unconfined) or piezometric surface (confined aquifer))
  t_0 = time corresponding to zero drawdown (days)
  r = distance from test to observation well (feet)

The storage coefficient for an unconfined aquifer is the specific yield.

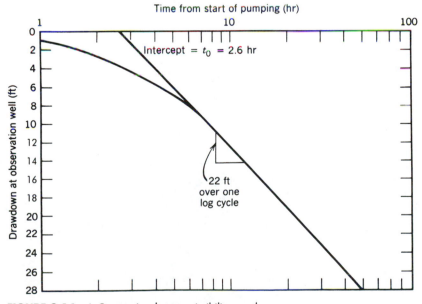

FIGURE 9.16 A Cooper-Jacob transmissibility graph.

❏   ***EXAMPLE PROBLEM 9.5***

Using the data of Figure 9.16, calculate the transmissivity and storage coefficient if the well discharge were 900 gpm and the distance of the observation well from the well is 100 ft.

Solution

$$T = \frac{264Q}{\Delta h} = \frac{264(900)}{22.0} = 10,800 \text{ gpd/ft}$$

and

$$S_c = \frac{0.3(10,800)(2.6/24)}{100^2} = 0.035 \quad ❏$$

9.7.4 Salt Water Intrusion

Under natural conditions, hydrostatic equilibrium develops in proportion to the density differences between the salt- and freshwater zones along coastal areas. Since the specific gravity of seawater is about 2.5% greater than fresh water, a "bubble" of fresh water "floats" on the seawater. (See Figure 9.17*a*.)

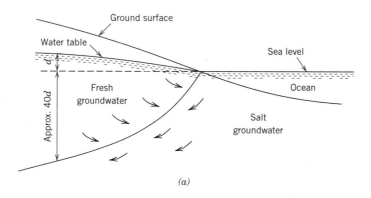

(a)

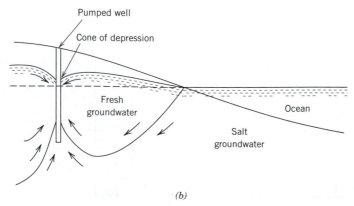

(b)

FIGURE 9.17   Illustration of saltwater intrusion.

TABLE 9.6 Viscosity versus Salt Content @ 20°C

NaCl% BY WT.	C_s (g/L) CONCENTRATION	ABSOLUTE VISCOSITY	
		N sec/m^2	lb sec/ft^2
—	—	1.002×10^{-3}	2.089×10^{-5}
0.50	5.0	1.011×10^{-3}	2.108×10^{-5}
1.00	10.1	1.020×10^{-3}	2.127×10^{-5}
1.50	15.1	1.028×10^{-3}	2.143×10^{-5}
2.00	20.2	1.036×10^{-3}	2.160×10^{-5}
2.50	25.4	1.044×10^{-3}	2.177×10^{-5}
3.00	30.6	1.052×10^{-3}	2.194×10^{-5}
3.50	35.8	$*1.060 \times 10^{-3}$	$*2.210 \times 10^{-5}$
4.00	41.1	1.068×10^{-3}	2.227×10^{-5}
4.50	46.4	1.076×10^{-3}	2.244×10^{-5}
5.00	51.7	1.085×10^{-3}	2.262×10^{-5}
5.50	57.1	1.094×10^{-3}	2.281×10^{-5}
6.00	62.5	1.104×10^{-3}	2.302×10^{-5}

*Approximate value for normal seawater. It is noted that the viscosity of saltwater deviates from the viscosity of seawater by less than 1% for concentrations up to about 40 g/L.

This natural equilibrium between fresh- and saltwater will be disrupted by the existence of any groundwater table gradient and/or the alteration of the natural interfaces between salt- and freshwater due to excessive pumping and groundwater removal (See Figure 9.17b). The change in hydrostatic balance will then allow salt water intrusion. The intrusion of salt water into the freshwater aquifer is a good example of the change of viscosity due to chemical makeup of the water, hence a change in permeability occurs leading to altered flow rates in the movement of groundwater. Table 9.6 illustrates the change in viscosity with change in salt content.

9.8

GROUNDWATER QUALITY CONSIDERATIONS

Water is a solvent for many salts and some types of organic matter. As groundwater moves along flowlines from recharge to discharge areas, its chemistry is altered by the effects of a variety of geochemical processes (Freeze and Cherry, 1979). These processes may include many chemical reactions, dissolution of limestone, oxidation-reduction reactions, ion-exchange processes, decomposition of aquifer rocks, transport of various leachates, industrial and municipal waste products, mining wastes, and saltwater intrusion.

Natural water is not entirely pure. Many of the dissolved natural substances contribute to human health by providing essential nutrients while many contaminants introduced in the natural hydrologic system by humans or otherwise are associated with a wide range of potential environmental impact problems. The following terms help focus the discussion:

- *Water Quality.*   Water quality is the relative term used to convey the idea of the potential usability of groundwater or surfacewater for a particular purpose.
- *Contaminant.*   All suspended particulate matter, solutes (dissolved substances), and immiscible substances that are introduced into the hydrologic system (regardless of the level of concentration) are considered contaminants.
- *Pollutant.*   A pollutant is a contaminant present in sufficiently high concentrations so as to constitute a health hazard.

Because water is such a powerful solvent it contains some measure of dissolved constituents even in a pristine environment. The type and amount of these constituents determine the suitability of natural sources of water for various uses. Groundwater, due to its long residence time in contact with geologic materials, generally contains higher concentrations of dissolved solids than does surfacewater.

Some important factors contributing to quality considerations of groundwater are: (1) the mineral composition of aquifer rocks; (2) the general geohydrologic framework of the area; (3) the potential for groundwater mixing and the occurrence of interactive geochemical processes; (4) and the activities of man relative to the recycling capability of the aquifers. Some recommended maximum concentration levels are indicated in Table 9.7.

TABLE 9.7   US Environmental Protection Agency Drinking-Water Standards

CONTAMINANT	RECOMMENDED MAXIMUM CONCENTRATION LEVEL
Chloride	250 mg/L
Color	15 color units
Copper	1 mg/L
Corrosivity	Noncorrosive
Fluoride	2.0 mg/L
Foaming Agents	0.5 mg/L
Iron	0.3 mg/L
Manganese	0.05 mg/L
Odor	3 threshold odor number
pH	6.5–8.5
Sulfate	250 mg/L
Total dissolved solids	500 mg/L
Zinc	5 mg/L

9.9
CONJUNCTIVE USE

Conjunctive use occurs when both surface and groundwater sources are used to supply water. An engineer may have a choice of sources. It may be both technically and economically wise to develop both sources and use them jointly to minimize cost. The advantage of using two sources of supply is that the variations in the quantity and quality of surface water do not usually coincide with those of groundwater. Thus, a more economical and reliable supply can be maintained by switching from one source to the other when needed. Groundwater is usually a more reliable source when available surface water tends to be at a minimum, yet the surface water can be drawn on first during rainy seasons when rainfall excess is at a maximum. This allows the groundwater storage to be replenished naturally by infiltration and percolation.

There are several conjunctive use schemes being planned and practiced and their success has been documented. Engineers and planners should study all sources to determine the relative economic worth. Computer modeling is usually practiced to determine operating rules for both sources.

Conjunctive use schemes can also refer to the optimal use of two or more surface reservoirs and the reuse of wastewater. For example, multiple reservoirs with inflows from different watersheds and with different yields may be operated together to maximize the available water. A single reservoir may be used in connection with a diversion from another river, or a surface water source may be supplemented by the output from a desalination or wastewater plant.

When we apply the techniques of mathematical programming with the masses of data such as inflows, storage capacities, pumping rates, and the like, we can evaluate the minimum cost yields for any combination of sources.

9.10
SUMMARY

Groundwater hydrology is an important study because of its potential application areas. Streamflow, stormwater infiltration, potable water from groundwaters, saltwater intrusion, groundwater recharge, and waste injection are a few areas where hydrology concepts and equations are used.

- The storage and movement of groundwater can be quantified using the equations of this chapter.
- In the unsaturated zone of an unconfined aquifer, soil moisture exists as both percolated (gravity) water and capillary water. Water available to plants is determined by defining the difference between field and wilting moisture content.
- Permeability is a measure of the rate at which water moves in the soil whereas porosity is a measure of void space. Specific yield is the volume of water released per unit volume of soil. These properties vary with the type of soil.

- Flow rate computation equations are available for both the confined aquifer (Equation 9.19) and the unconfined aquifer (Equation 9.22).

- Seepage of groundwater from the surficial aquifer to a surface water body can be estimated using the unconfined aquifer equation. Also, if a constant infiltration to the water table can be defined, Equation 9.26 can be used to determine the rise in the groundwater table and then the unconfined aquifer equation can be used.

- Leaky aquifers exist where the confining strata are not completely impervious. The equations for this are beyond the scope of this work but are found in other publications (Hantush, 1960; Walton, 1970).

- Equations to estimate the yield from confined (Equation 9.29) and unconfined (Equation 9.32) aquifers have been developed assuming a steady-state condition has been achieved.

- When equilibrium is not obtained, methods to estimate yield were developed. A method by Cooper and Jacob (1946) was presented.

- Groundwater quality considerations were presented, including the problem of saltwater intrusion.

9.11

PROBLEMS

1. Calculate the coefficient of permeability for a soil with a 10% finer grain size of 0.14 mm. Then classify the soil by its texture. What is an approximate value for the soil porosity and specific yield?

2. A laboratory premeability test is used to estimate the coefficient of permeability for an aquifer. It is a standard test at a temperature of 60°F. If the aquifer temperature varies from 40°F to 50°F during the year, what is the lowest estimate of actual permeability in the aquifer? If the laboratory estimate is 200 gpd/ft^2?

3. A well is pumped at an equilibrium flow of 454 gpm to determine a confined aquifer permeability. The thickness of the aquifer is 100 ft and the drawdown in observation wells is noted as 3 ft and 9 ft at 100 ft and 10 ft, respectively, from the pumped well. What is the permeability?

4. Using the same data from Problem 3 except consider an unconfined aquifer that was totally saturated before pumping, what is the estimate of permeability?

5. An estimate of groundwater flow into a lake is required to complete a water volume balance on the lake. Runoff, outflow, precipitation, evaporation, and transpiration are known. An unconfined aquifer exists with an impermeable membrane at approximate elevation of 30 ft. The soil permeability has been estimated at 6 m/day. An observation well is established 20.2 ft from the lake and the groundwater in the surficial aquifer is measured at elevation 52.13 ft. The lake level at this time was 50 ft. What is the inflow estimate in units of m^3/s and cfs if the circumference of the lake is 400 m?

6. Calculate the transmissivity and the storage coefficient using the unsteady flow method given the following data. Water was pumped from a well at a rate of 950 gpm. The observation well was 80 ft from the pumped well.

ELAPSED TIME (hr)	DRAWDOWN (ft)
1	1.0
2	2.6
3	4.7
4	5.7
5	6.5
6	8.0
8	10.3
10	12.4
12	15.0
18	18.0
25	21.2
35	24.0

7. What is the Darcy velocity of groundwater flowing through a sand with a permeability of 10 m/day given that two piezometers, placed 10 m apart, give the following data.

	PIEZOMETER 1 (m)	PIEZOMETER 2 (m)
Elevation head	0	−3
Pressure head	0.2	0.1

8. The water table 3000 ft away from a lake has an elevation of 64 ft. Use Darcy's law to find the velocity of the groundwater 3000 ft away from the lake. The elevation of the lake is 60 ft, and the soil surrounding the lake is a fine sand.

9. A soil sample weighing 1.05 lb fills a 500-cm³ container. The sample is placed in a 103°F oven overnight and is weighted again the next day. If the weight after drying is 0.85 lb and the specific gravity of the soil granuals is 2.65, what is the porosity of the soil sample?

10. What is the porosity ratio of a soil if a 1-m³ sample contains 10% voids by volume?

11. If the permeability of a sand is found to be 10 m/day at 30°C, what is its permeability at 20°C?

12. Determine the travel distance for groundwater from a stormwater detention pond to a lake. Filtration material has been placed in the ground between the pond and the lake and has a permeability of 10^{-2} ft/sec with a porosity of 0.2. The difference in elevation between the water surface in the pond and the water surface of the lake is 10 ft. If the residence time is 5.8 days, what is the travel path length? Also, determine the concentration of a chemical discharged to the lower lake if the initial concentration is 5000 mg/L and the rate of removal through the filtration column is one-half for every day of residence time (half-life is one day).

13. A sample of groundwater has the following concentrations of ions:

Cations	mg/L	Anions	mg/L
Ca^{++}	56.0	HCO_3^-	256.0
Mg^{++}	16.0	SO_4^{--}	43.0
Na^+	85.0	Cl^-	82.0

(a) What is the Sodium Absorption Ratio (SAR)? (b) If the SAR limit for agricultural use is a value of 10 or less, would you recommend this water for irrigation?

14. The laboratory coefficient of permeability, K, of a soil sample is 400 gpd/ft^2. (a) What would be the permeability of the same material at 45°F? (b) What would be the permeability if the groundwater salt content was 2.0% by weight?

15. Use the mass balance (water budget) techniques of Chapter 1 to develop a version of the hydrologic/storage equation for analysis of leachate seepage from a landfill into local groundwater. Explain why each term is included.

9.12

REFERENCES

Bedient, P. B., Rifai, H. S., and Newell, C. J. 1994. Groundwater Contamination. PTR Prentice Hall, Englewood Cliffs, NJ.

Bouwer, H. 1966, "Rapid Field Measurement of Air Entry Value and Hydraulic Conductivity of Soil," *Water Resources Research,* **2,** pp. 729–738.

Campbell, M. D. and Lehr, J. H. 1977. "Well Cost Analysis," *Water Well Technology,* 4th ed. McGraw-Hill, New York.

Chow, V. T. 1964. *Handbook of Applied Hydrology.* McGraw-Hill, New York.

Colman, E. A. 1947. "A Laboratory Procedure for Determining Field Capacity of Soils," *Soil Science,* **63,** p. 277.

Cooper, H. H., Jr. and Jacob, C. E. 1946. "A Generalized Graphical Method for Evaluating Formation Constants and Summarizing Well-Field History," *Transactions of the American Geophysico Union,* **27,** pp. 526–534.

DeWiest, R. J. M. 1965. "History of the Dupuit–Forchheimer Assumptions in Groundwater Hydraulics," *Transactions American Society of Agricultural Engineers,* **8,** pp. 508–509.

Freeze, R. A. and Cherry, J. A. 1979. *Groundwater,* Prentice Hall. Englewood Cliffs, NJ.

Hantush, M. S. 1960. "Modification of the Theory of Leaky Aquifers," *Journal of Geophysical Research,* **65,** pp. 3713–3725.

McNeal, B. L. 1968. "Prediction of the Effects of Mixed-Salt Solutions on the Soil Hydraulic Conductivity," *Proceedings of the Soil Science Society of America,* **32,** pp. 190–193.

National Water Summary—Florida. 1985. "Florida Groundwater Resources," U.S. Geological Survey Water-Supply Paper 2275.

Todd, D. K. 1980. *Ground Water Hydrology.* Wiley, New York.

Walton, William C. 1970. *Groundwater Resource Evaluation.* McGraw-Hill, New York.

CHAPTER

10

STORMWATER MANAGEMENT

This chapter presents applications of hydrologic concepts for the management of runoff rates, volume of runoff, and the quality of the runoff. Both surface and groundwater hydrology principles are used. Some of the contents of this chapter could be used after reading any of the first nine chapters to demonstrate applications.

Most of the chapter materials are useful in determining the combination of best management practices (BMPs). A best management practice is a stormwater control system that is economically and technically feasible as a means of reducing the quantity of pollutants to meet water quality goals.

The applications and additional concepts presented in this chapter are:

1. Surface water availability from streamflow records.
2. Design storm concepts.
3. Stormwater detention pond designs.
4. Culverts as outflow control devices.
5. Stormwater off-line retention pond design.
6. Swale design.
7. French drains to lower water tables and as outflow control devices.
8. Stormwater re-use.
9. Storm sewer design.
10. Water quality improvement.

10.1

SURFACE WATER AVAILABILITY FROM STREAMFLOW

Streamflow records and measurement were presented in Chapter 7. One of the applications for streamflow data is the determination of volume and water discharged at the point of measurement over a period of time and the expected yield (rate of removal) from an impoundment reservoir at the site.

The networks of ground and surface water gages provide quantity data that help determine water volume availability. The future water needs of such users as municipalities, agricultures, and industries can be established. Then the investigator must estimate the availability of water.

For a surface reservoir site to be developed, the sources of water are identified. The average annual rainfall is a first indicator of possible water availability. The existence of perennial (constantly flowing) rivers is a significant indication of the magnitude of surface sources. Also, a search for suitable groundwater aquifers can be made at the same time.

The yield (volume of water per time period) of the source needs to be investigated before a supply can be relied on for the design life of a project. Once a suitable source of water has been found, some form of storage is essential to guarantee continuous supplies. If the supplies are to come from surface waters, a storage reservoir must be constructed. Frequently, natural lakes can be used or modified for use. Water from groundwater sources is usually already in some type of reservoir storage. A feasibility study for evaluating a possible surface reservoir site would include:

1. A watershed survey to determine watershed and reservoir areas and storage volumes.
2. A geological survey of the sites for the storage structure (dam or well).
3. Hydrological assessments of the flows at the sites for yield (volume/time) and storage (volume).
4. Appraisal of any land use changes and the future amenity value of existing or future structures.
5. Preliminary design and cost estimation.

Feasibility studies are done for several possible sources in order to determine the source with sufficient yield and minimum cost.

10.1.1   Volume and Yield of a Surface Water Reservoir

How much water can be withdrawn per time period (yield or draft) from a given size of reservoir? How much volume must be stored in a reservoir to achieve a desired yield over a period of time? These are among the basic questions that a designer and operator of a reservoir must ask.

For a fixed storage capacity in a reservoir the yield during a drought period of great severity must be estimated. If a higher yield is required than that supplied by the reservoir, the capacity would need to be supplemented by other sources. To ensure adequate storage for a specified yield or to evaluate the yield of an assumed storage, a study of all net inputs at the reservoir site is needed. Expected losses by evaporation, transpiration, and infiltration from the reservoir must be used to reduce available reservoir storage. The sum of inflow minus reservoir losses is net inflow.

Design and evaluation is done using deterministic (fixed) yields and net inputs or stochastic net inputs and yields. Discharges with time that are known with certainty and representative of all flows are used for deterministic models. For stochastic models, data are synthesized (generated from probability distributions) with the length of generated data usually exceeding the design life of the structure.

10.1.2   Mass Curve (Ripple's Method)

This is a deterministic model using a series of streamflow data to help calculate the volume and yield of a proposed reservoir site. For each time sequence, the cumulative

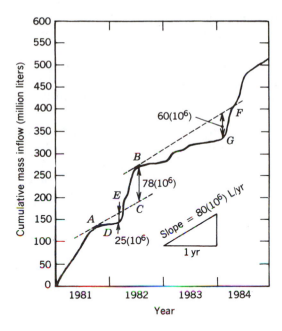

FIGURE 10.1 The use of a mass curve to determine the reservoir capacity required to produce a specific yield.

input volumes are plotted against time. The volumes are cumulated from streamflow. The cumulative volume curve is expressed as a function of streamflow as

$$\text{Volume} = \int Q \, dt = \sum Q(\Delta t) \tag{10.1}$$

The technique is demonstrated using Figure 10.1. The difference between any two points on the cumulative volume curve (solid line) is a storage volume for that period of time, assuming no losses from the reservoir. The yield is the rate of demand for water (dotted line). It is represented in Figure 10.1 as a constant value but in practice it changes with time. An example are yields for agriculture uses which change with crop production. The yield changes are reflected by changes in the slope of the yield line.

The method assumes the reservoir is full at the start of a drought period. A drought period is defined as one in which the inflow rate is less than the yield. The maximum difference between the yield and the cumulative volume curve is the reservoir size needed before the drought period. Several drought periods are evaluated and a decision on reservoir size and yield is made from the several periods. The results are only as accurate as the data input. This usually requires a long-term streamflow record.

❑ **EXAMPLE PROBLEM 10.1**

To obtain a water yield of $80(10^6)$ L/yr for the inflows shown in Figure 10.1, what reservoir capacity is required?

Solution

The solution requires tangents drawn to the mass curve. At points A and B the tangents have slopes equal to the yield of $80(10^6)$ L/yr. The maximum departure between yield and mass occurs at G and is $60(10^6)$ L, which is the required reservoir capacity. The reservoir would be empty at point G, and at F it would be full again. The reservoir is assumed to be full at A, depleted by $25(10^6)$ L of storage at D, and full again at E. When the reservoir is full, all inflow in excess of the demand would be discharged downstream. If the reservoir is full at point A, it will fill again at point E and between points C and B there will be an excess of $78(10^6)$ L.   ❑

The yield that can be expected given a reservoir capacity may also be determined using mass curves (Figure 10.1). To accomplish this, lines are drawn tangent to the high points of the mass curve (A and B) so that their maximum departure from the mass curve does not exceed the specified reservoir capacity. The resulting slopes estimate the yields that can be expected each year with the specified storage capacity. Also note that the demand line must intersect the mass curve at some point, otherwise the reservoir will not refill.

❑   **EXAMPLE PROBLEM 10.2**
_____

If a reservoir of $30(10^9)$ L capacity is built for which the mass curve of Figure 10.2 applies, what yield will be available?

Solution

In this problem, the storage level is specified first, then the yield (slope) is calculated. In Figure 10.2, the tangents to the mass curve are drawn to specify a maximum departure from the mass curve equal to $30(10^9)$ L. The tangent from B has the least slope, $60(10^9)$ L/yr, so this is the minimum yield. The tangent at A indicates a possible yield of $100(10^9)$ L/yr, but this demand could not be satisfied between points B and C without storage in excess of $30(10^9)$ L.   ❑

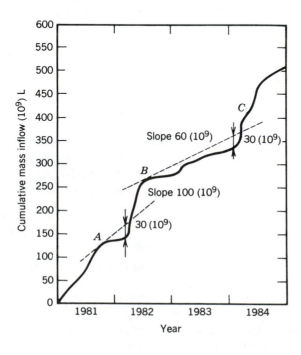

FIGURE 10.2 The use of a mass curve to determine the possible yield from a reservoir of specified capacity.

10.1.3 Modification of Storage Volumes

Cumulative mass inflow can be modified to incorporate supplemental sources of input such as additions from groundwater. Also, losses over extended periods of time, such as those from evapotranspiration and seepage can be used to reduce cumulative mass. Evapotranspiration losses may be significant and thus affect the storage capacity. The lower portion of Figure 10.3 illustrates decreasing cumulative mass in a reservoir during low or zero surface water input times. This decrease is contrasted to a situation with no loss in a reservoir as shown in the upper portion of Figure 10.3. When losses and supplemental sources are incorporated into the analysis, storage capacity may and usually will change or yield will have to be adjusted.

10.1.4 Stochastic Models

In mass curve analysis, it was assumed that flow rates would repeat themselves every time period (year) without deviation. However, the patterns are not exactly repeated every year. Thus, more years of data or other methods which preserve the history of the streamflow are useful, such as (1) random generation and (2) serial correlation matrices (Markov processes). To initiate an understanding of the methods, terms are defined in Table 10.1.

Since streamflow, evapotranspiration, and seepage are stochastic processes, it is possible to assign a probability to a reservoir being full or empty. All hydrologic data can be used and probability distributions estimated for net inflow. In addition, a probability distribution for demand rates (yields) can be developed.

When probability distributions are used to extend streamflow or rainfall records, the hydrologic records will be preserved. These distributions are useful to preserve the history of the hydrologic events and to extend the record length. This extension and others are sometimes known as techniques of operational hydrology.

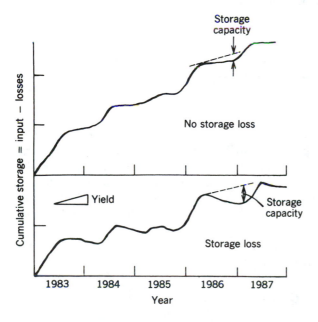

FIGURE 10.3 Cumulative storage with and without storage losses.

TABLE 10.1 Stochastic Models

1. *Serial Correlation.* A variable at time (t) is dependent on past data $(t - n)$: $n = 1 \ldots k$ (i.e., correlated in time). Example: Streamflow data in the Oklawaha River at time $(t - n)$. Usually, n is of the orders of hours or days.

2. *Cross Correlation.* One variable correlated with another variable at the same time (t) but at different places. Example: Streamflow in Boggy Creek correlated with streamflow in the Little Econ River at time (t) (i.e., correlated in space).

3. *Conditional Probability.* $\Pr(A|B)$, used to describe both serial and cross correlations, read as probability of event A given event B.

4. *Discrete Probability Distributions.* Approximations (in most cases) for the continuous probability distributions of streamflow, reservoir release volumes, and other hydrologic events. Discrete distributions are usually made to simplify mathematical manipulations.

Let

$$\Pr_i = \text{probability of flows}$$

$$\Pr_i = \frac{\text{flow within interval } i \text{ (number of times)}}{\text{Total number of data points}}$$

$$\sum_{i=1}^{n} \Pr_i = 1.0$$

5. *Markov Process.* Stochastic model utilizing conditional probabilities. As an example, consider the serial correlation of daily streamflow.

Let

$$\Pr_{ij} = \text{conditional probability of the current (any time) daily}$$
$$\text{streamflow being within interval } j \text{ given last time}$$
$$\text{interval's streamflow } i$$

$$= \frac{\text{number of flows in } j \text{ following } i}{\text{total number of flows in } i}$$

then

$$\sum_{j=1}^{m} \Pr_{ij} = 1 \qquad \text{(for all } i\text{)}$$

A probability distribution is a realistic expression of the "chances" or probability of satisfying demands. Using probability distributions, decisions can be made that better reflect the actual historical nature of the hydrologic process. It is more acceptable to the decision-making community (voters, elected officials, appointed government people) to make a statement, such as "the reservoir as designed will satisfy the water needs of our projected populations 99% of the time." This statement can be made using stochastic models.

The operation of a reservoir can be considered on a discrete time basis. The contents of the reservoir at any time (t), is given by

$$V_t + Q_t(\Delta t) - Y_t = V_{t+1} \tag{10.2}$$

where V_t is the reservoir volume at time t, $Q_t(\Delta t)$ is the volume of inflow during the time period, and Y_t is the volume demand or releases (yields over time) during the same time interval.

Equation 10.2 also is used to estimate the size of a reservoir as a function of inflows and yields. Constraints on the maximum size of the reservoir are used and realistic results can be determined.

As an example of stochastic methods, Hardison (1966) developed estimates for reservoir size and yield based on probability of annual streamflow data. He used probability frequency distributions to fit the empirical data. Thus, he was able to determine a quantitative index on the variability of flow and storage requirements. His work was applied to 22 major watersheds in the contiguous United States (Lof and Hardison, 1966). The variability of annual surface water yields were measured by the coefficient of variability or the index of variability. The coefficient of variability is the standard deviation of annual yield divided by the average annual yield, while the index is the standard deviation of the common logarithms. The coefficient of variability ranged from a low of .2 in the New England area to a high of 1.0 in the South Pacific. The maximum yield (net flow adjusted for losses) for the 22 watersheds for 98 of 100 years was about 3660 billion L/day (about 1000 billion gal/day). This requires a storage capacity of 3.6 billion ac-ft but only 2 billion ac-ft of storage can be made available. These studies indicate a yield and required reservoir storage with a probability of .98 (98 out of 100).

10.1.5 Random Generation (Independent Events)

The use of random generation procedures assumes that successive hydrologic events are independent and a known probability distribution can be found that represents the true forces generating the hydrologic events. Consider as an example, streamflow data. Certainly, on a large river system, the flows in a short time period, days or hours, appear to be serial correlated with previous time period flows or the probability of one flow rate given another is

$$\text{Probability } \{Q_{t+1}|Q_t\} \neq 0 \qquad (10.3)$$

where

Q_{t+1} = average flow in time period $(t + 1)$, L³/t
Q_t = average flow in time period (t), L³/t

If the probability expressed by Equation 10.3 were zero or near zero, then one could state the flows to be independent. Independent events are defined as

$$\text{Probability } \{A|B\} = 0 \qquad (10.4)$$

where A and B = hydrologic events.

❏ EXAMPLE PROBLEM 10.3

Given the following streamflow data during low flow and the beginning of a rising limb of a hydrograph, determine the probability of a 30-m³/s flow following a 20-m³/s flow.

TIME	AVERAGE FLOW	TIME	AVERAGE FLOW
(DAYS)	(m^3/s)	(DAYS)	(m^3/s)
1	20	5	20
2	15	6	30
3	20	7	35
4	20	8	40

Solution

The event specified is the transition of flow from 20 to 30 m³/s. Thus, the beginning flow value must be 20 m³/s. This value occurs four times in the very limited data sets. The number of times the flow of 30 m³/s follows 20 m³/s is once, therefore, the probability is calculated as

$$\text{Probability } \{Q_{t+1}|Q_t\} = \text{probability } \{30/20\} = \tfrac{1}{4}$$

Also, the probability of 20 m³/s given 20 m³/s is .5 and the probability of 15 m³/s given 20 m³/s is .25. Thus, the probability of flow values of any magnitude given 20 m³/s must add up to 1.0 which is the case. Note that this is a very small amount of data and are used only to illustrate the calculation of probability that demonstrates dependence of flow in one period of time on another. ❏

The period of time necessary to use a random generation procedure for streamflows is usually seasonal average data or yearly averages. Rainfall data separated by 4 to 6 hr of non-rainfall usually constitute sufficient time to justify independence in rainfall.

Consider again the problem of specifying the size of a reservoir, but this time use a random generation of streamflow into the reservoir, call it $Q_{i,t}$ where i = average flow interval and t = time period. A discrete probability distribution is developed for streamflow and is graphically shown in Figure 10.4. The general inventory equation (mass balance) is rewritten to specify the random streamflow and releases (see Figure 10.5)

$$V_t + Q_{i,t}(\Delta t) - Y_t = V_{t+1} \qquad (10.5)$$

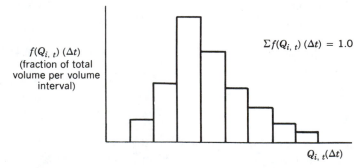

FIGURE 10.4   A discrete probability distribution on streamflow.

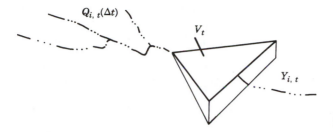

FIGURE 10.5   A reservoir schematic.

and

$$\text{Max}\{V_i\} = V_{\text{max}} \qquad \forall i \text{ (for all values of } i\text{)} \tag{10.6}$$

subject to:

1. Starting reservoir volume, V_0 is assumed equal to the minimum.

$$V_0 = V_m$$

2. A minimum reservoir volume, V_m, must be maintained, so that Y_i may equal zero during some time period i.

$$V_i \geq V_m \qquad \forall i$$

3. No negative releases or reservoir sizes.

$$V_i \geq 0 \qquad Y_i \geq 0 \qquad \forall i$$

Also the probability of a specified volume of reservoir at any time can be determined as

$$\Pr\{V_{i,t}\} = \frac{x_{V_{i,t}}}{n_t} \tag{10.7}$$

where

$x_{V_{i,t}}$ = number of times reservoir volume is equal to the interval V_i
n_t = total number of time periods.

In schematic form, Equation 10.5 can be shown as Figure 10.5.

The streamflow data must be estimated relative to the history of its probability distribution. The more streamflow data, the more accurate the empirical probability distribution, and a continuous theoretical distribution with unbounded limits can be estimated. This would allow the estimation of very large and very small streamflow values. These distributions were presented in Chapter 2. Our intent here is to develop the procedure for generating flow rates from discrete probability distributions of independent flow rates. Since the generation of flow rates do not depend on previous flow rates, and the distribution (Figure 10.4) of flows must be reproduced over a long period of time, a technique that is not biased toward the choice of flow or is strictly random in its choice must be used. Random numbers are numbers that have an equal probability of choice. Furthermore, to reproduce a probability distribution, a proportional number of random numbers can be assigned to the fraction of the total per interval. For example,

If
$$Q_{it} = 20 \text{ m}^3/\text{s with 4\% frequency}$$
$$= 30 \text{ m}^3/\text{s with 8\% frequency}$$
$$= 40 \text{ m}^3/\text{s with 15\% frequency}$$

then given 100 random numbers, 4 numbers would be assigned to 20 m³/s, 8 numbers to 30 m³/s, and 15 numbers to 40 m³/s. This technique of streamflow generation is called Monte Carlo.

10.1.6 Serial Correlation

For shorter time periods, the serial correlation of going from one hydrologic state to another can be determined. When using streamflow data, it is advantageous to divide the streamflow into periods of time with similar hydrologic characteristics such as low flow, increasing streamflow values, decreasing streamflow values, wet season, and dry season. The quantity of data needed are usually in the order of 10 yr or more of daily values. The number of years depends on how accurately one can define the conditional discrete probability distributions or estimate from the discrete distribution the form of a theoretical one. One must define the probability of going from one streamflow to another. Since there are more than one streamflow, a matrix of probabilities is defined: it is called a conditional probability matrix.

Given two streamflow states ($i = 1, 2$) and four transitions states ($j = 1, 2, 3, 4$), the two discrete distributions result as shown in Figure 10.6. The information given on these distributions can be expressed in a matrix of conditional probabilities, called a transition probability matrix (Figure 10.7).

Random generation can then be used to generate the flow data given the transition matrix. In practice, transition matrices are developed for a rising limb, a falling limb, and base flow. In addition, the timing for the beginning of the rising limb is determined from rainfall data or an interevent probability distribution on streamflow. Peak flows can be related to rainfall excess or rainfall. The usual way of generating the flow-rate data is by means of a heuristic computer program and most recently by expert systems.

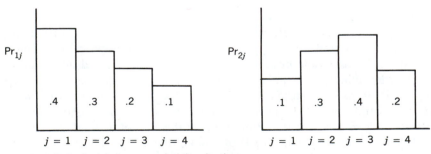

FIGURE 10.6 Serial discrete probability distributions.

Streamflow—state j (time $y + 1$)

		1	2	3	.	.	.	m
Streamflow state i (time y)	1	.4	.3	.2				
	2	.1	.3	.4	.	.	.	
	.		.					
	.		.					
	.		.					
	n							

FIGURE 10.7 A transition probability matrix.

10.2

DESIGN STORMS

For the calculation of peak discharge, questions of major significance revolve around the choice of the design storms. It is significant because of the time involved in calculating the peak discharge. Not all storm events can be tested to determine the worst discharge. An economic analysis with the calculation of cost/benefit ratios, indicates a best choice for the frequency (return period) for a design storm. The design storm is chosen, and the runoff volume (rainfall excess) and peak discharge are calculated usually based on intensity/duration/frequency curves given for the particular area. The peak of the rainfall excess is important for design of conduits (open or closed) and rainfall excess is used for design of storage systems. The input and output hydrographs provide the needed information to design and operate a water detention facility. If an outlet structure exists, then we can determine an outflow hydrograph. Then, the input hydrograph for a number of storm durations can be used to determine the size of a storage basin. One of the objectives in this chapter is to develop rainfall excess (volume) and runoff hydrographs (flow rates), which relate to both conduit and detention basin design.

TABLE 10.2 Design Frequencies for Highway Structures

	USUAL DESIGN FREQUENCIES IN YEARS	
TYPE OF STRUCTURE	MOTORWAYS, INTERSTATE AND CONTROLLED ACCESS HIGHWAYS	OTHER HIGHWAYS AND FRONTAGE ROADS[a]
Inlets and sewers	10	2–5
Inlet for depressed roadways	50	2–10
Culverts	50	2–10
Small bridges	50	10–50
River crossings	50	10–50

Source: Florida Department of Transportation, Drainage Manual, 1986.
[a] Varies with state and local regulations.

10.2.1 Design Storm Frequency

The selection of a design storm rainfall frequency is best made only after economic analyses of benefits and costs are performed. However, these studies are expensive and not a routine undertaking. In addition, the results frequently depend on many variables such as land use, soil types, topography, economic activity, and meteorological factors. Therefore, most municipalities, state and federal agencies will specify the design frequency for most drainage design, such as the 10- or 25-yr frequency rainfall based on studies conducted elsewhere, preferably on similar watersheds. Table 10.2 shows some examples of usual design frequencies.

Rainfall frequency–intensity information for use in drainage design is readily available in such publications as the U.S. Weather Bureau Technical Paper No. T.P. 40 (U.S. Weather Bureau, 1961), textbooks, and state (e.g., Florida DOT 1986) drainage manuals. The rainfall frequency–intensity data in these publications were intended for design by the Rational Method, in which the rainfall duration is assumed equal to or greater than the time of concentration in the watershed. Also, a uniform rainfall distribution is assumed to occur throughout the rainfall duration. Of course, these assumptions are rarely good for large watersheds, so other procedures must be examined.

10.2.2 Design Duration and Distribution

When a situation requires the development of a hydrograph, the duration of the particular frequency rainfall and the distribution of the rainfall must be known. Stan-

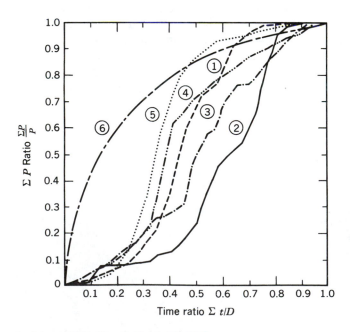

1. Orlando W.B. Airport, March 15, 1960
2. Orange City, September 10–11, 1960
3. Orlando W.B. Airport, October 15–16, 1956
4. SCS Type II
5. Orange County
6. Weather Bureau

FIGURE 10.8 Dimensionless mass curves of rainfall. *Source:* Golding, 1973.

dard rainfall duration and/or distribution curves have been adopted by many government and consulting organizations for specific areas. For instances, Orange County, Florida, has an unofficial 25-yr frequency, 24-hr design rainfall developed after the floods of 1960. Golding (1973) plotted dimensionless mass curves of rainfall to compare major rainfalls in central Florida with Orange County's standard distribution and the NRCS's Type II distribution. In addition, the U.S. Weather Bureau published suggested storm distributions for the southeastern United States (Figure 10.8). Golding concluded that Orange County's standard distribution was compatible with major storms in the central Florida area and with the NRCS Type II distribution.

Using a rainfall of specific duration and volume, we can find the amount of rain falling during specified intervals using some standard or selected distribution. Applying the NRCS's Type II standard distribution to a 25-yr, 6-hr rainfall of 6 in. results in the rainfall at 15-min intervals shown in Table 10.3.

TABLE 10.3 Design Rainfall—25-yr Frequency, 6-hr Duration, 15-min Increments

TIME (min)	TIME (hr)	ΣP (in.)	ΔP (in.)	ΣR^a (in.)	ΔR (in.)
0	0	0	0	0	0
15	0.25	0.10	0.10	0.00	0.00
30	0.50	0.21	0.11	0.02	0.02
45	0.75	0.33	0.12	0.07	0.05
60	1.00	0.48	0.15	0.15	0.08
75	1.25	0.64	0.16	0.27	0.12
90	1.50	0.81	0.17	etc.	
105	1.75	1.08	0.27		
120	2.00	1.38	0.30		
135	2.25	2.46	1.08		
150	2.50	3.60	1.14		
165	2.75	3.90	0.30		
180	3.00	4.20	0.30		
195	3.25	4.44	0.24		
210	3.50	4.68	0.24		
225	3.75	4.86	0.18		
240	4.00	5.01	0.15		
255	4.25	5.16	0.15		
270	2.50	5.28	0.12		
285	4.75	5.40	0.12		
300	5.00	5.52	0.12		
315	5.25	5.64	0.12		
330	5.50	5.76	0.12		
345	5.75	5.88	0.12		
360	6.00	6.00	0.12		

[a]For $CN = 95$.

TABLE 10.4 Design Rainfall—Orange County 25-yr Frequency, 24-hr Duration, 30-min Increments

TIME (hr)	ΔP (in.)	ΣP (in.)	TIME (hr)	ΔP (in.)	ΣP (in.)
0	0	0	12.5	0.18	7.16
0.5	0.01	0.01	13.0	0.18	7.34
1.0	0.01	0.02	13.5	0.12	7.46
1.5	0.02	0.04	14.0	0.11	7.57
2.0	0.02	0.06	14.5	0.05	7.62
2.5	0.05	0.11	15.0	0.05	7.67
3.0	0.06	0.17	15.5	0.04	7.71
3.5	0.11	0.28	16.0	0.04	7.75
4.0	0.12	0.40	16.5	0.04	7.79
4.5	0.17	0.57	17.0	0.04	7.83
5.0	0.17	0.74	17.5	0.04	7.87
5.5	0.18	0.92	18.0	0.04	7.91
6.0	0.19	1.11	18.5	0.03	7.94
6.5	0.50	1.61	19.0	0.03	7.97
7.0	0.50	2.11	19.5	0.03	8.00
7.5	0.52	2.63	20.0	0.03	8.03
8.0	0.52	3.15	20.5	0.02	8.05
8.5	0.85	4.00	21.0	0.02	8.07
9.0	0.86	4.36	21.5	0.02	8.09
9.5	0.58	5.44	22.0	0.02	8.11
10.0	0.57	6.01	22.5	0.02	8.13
10.5	0.30	6.31	23.0	0.02	8.15
11.0	0.30	6.61	23.5	0.01	8.16
11.5	0.19	6.80	24.0	0.01	8.17
12.0	0.18	6.98			

Applying Orange County's Standard Distribution to a 25-yr frequency, 24-hr rainfall of 8.17-in. results in the rainfall increments (30 min) shown in Table 10.4. Both the 6-in. and 8.17-in. storms were reduced because of the large size of the watershed. For larger watersheds, the point measured maximum rainfall is rarely found over the entire watershed. The average volumes for the entire watershed is less. Thus, an aerial reduction factor can be developed. A graph for allowable reduction in point measure rainfall is given in T.P. No. 40 (U.S. Weather Bureau, 1961). For example, consider a 3-hr storm, the point measured rainfall is reduced by 10% for a 13,000-ha (50 mi^2) watershed and by 20% for a 52,000-ha (200 mi^2) watershed.

☐ **EXAMPLE PROBLEM 10.4**

To illustrate the calculation of rainfall excess using the curve number procedure, consider the precipitation record of the 25-yr, 6-hr design rainfall shown in Table

10.3. Rainfall excess is determined on a cumulative basis using cumulative rainfall, then time incremental values of rainfall are determined by subtracting the cumulative values of rainfall excess before and after the time increment. First calculate or determine the curve number, say $CN = 95$, a highly impervious area. Therefore, the maximum storage is

$$S' = \frac{100}{95} - 10 = 0.53 \text{ in.}$$

and the maximum excess is $6 - 0.53 = 5.47$ in.

Solution

The following example calculations assume that rainfall excess is zero if cumulative rainfall is less than 20% of the maximum soil saturation. However, this is not common with a high curve number, like $CN = 95$ but is used to illustrate calculations.

At 15 min

$$P = .10 \qquad R = \frac{[0.10 - 0.2(0.53)]^2}{[0.10 + 0.8(0.53)]} = 0.00 \text{ in.}$$

At 30 min

$$P = .21 \qquad R = \frac{[0.21 - 0.2(0.53)]^2}{[0.21 + 0.8(0.53)]} = 0.02 \text{ in.}$$

At 45 min

$$P = .33 \qquad R = \frac{[0.33 - 0.2(0.53)]^2}{[0.33 + 0.8(0.53)]} = 0.07 \text{ in.}$$

$$\text{and} \qquad 0.07 - 0.02 = 0.05 \text{ in. rainfall excess}$$

At 60 min

$$P = .48 \qquad R = \frac{[0.48 - 0.2(0.53)]^2}{[0.48 + 0.8(0.53)]} = 0.15 \text{ in.}$$

$$\text{and} \qquad .15 - .07 = 0.08 \text{ in. rainfall excess}$$

At 75 min

$$P = .64 \qquad R = \frac{[0.64 - 0.2(0.53)]^2}{[0.64 + 0.8(0.53)]} = 0.27 \text{ in.}$$

$$\text{and} \qquad 0.27 - 0.15 = 0.12 \text{ in. rainfall excess}$$

and so on.   ❏

To calculate rainfall excess from precipitation records, curve numbers may be used, as mentioned previously. For various CN values, a variety of rainfall excess histories are tabulated or graphed as shown in Figures 10.9 through 10.12. The problem can be completed by additional (somewhat extensive) hand calculations, or by use of a computer program. The reader is encouraged to do some additional calculations by hand and to execute a few computer runs using SMADA.

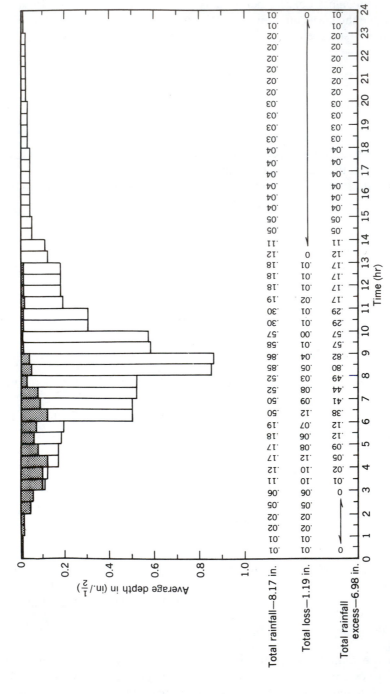

FIGURE 10.9 A hyetograph of 25-yr design rainfall, curve number 90, 24-hr duration. *Source:* Golding, 1974.

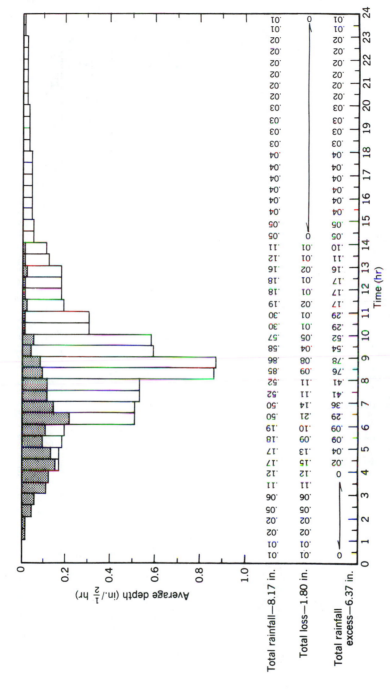

FIGURE 10.10 A hyetograph of 25-yr design rainfall, curve number 85, 24-hr duration. *Source:* Golding, 1974.

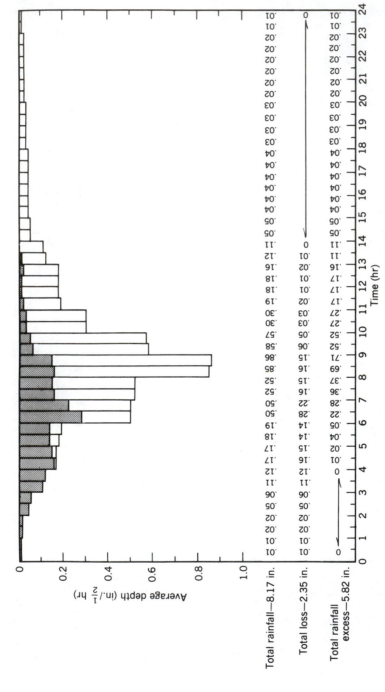

FIGURE 10.11 A hyetograph of 25-yr design rainfall, curve number 80, 24-hr duration. *Source:* Golding, 1974.

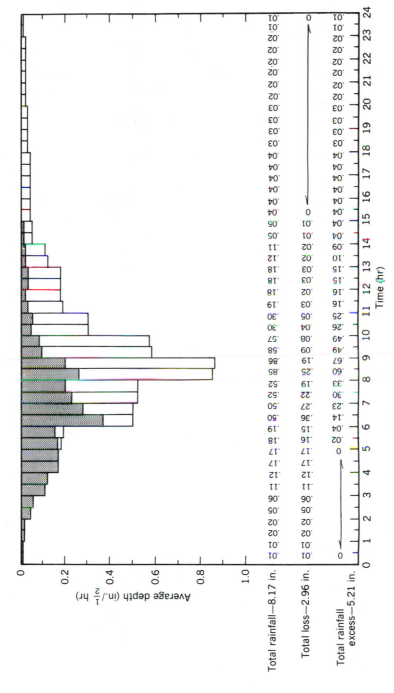

FIGURE 10.12 A hyetograph of 25-yr design rainfall, curve number 75, 24-hr duration. *Source*: Golding, 1974.

Recall that the type of ground cover and soil types will affect rainfall excess. Figure 10.13 illustrates the reduction in rainfall excess with decreasing curve numbers. These curve numbers must be estimated for each runoff area. Therefore, adequate information on soil type and land use must be available.

Rainfall excess is important for the design of urban water transmission and storage systems. The designer should compare various duration storms to determine which storm is most critical and therefore should be used. In the computation of design hydrographs, generally the shorter duration design rainfalls will result in a higher peak flow because of the short time period and higher-intensity rainfall. However, the longer duration storms must be studied to validate this general statement.

For the design of a detention basin, generally a longer-duration rainfall event should be used. The total volume of rainfall is usually greater and will require a larger-sized basin than from a short duration event of the same frequency.

Some water management regulations for developing lands specify that the peak discharge before development must equal the peak discharge after development. The choice of rainfall duration will affect the volume of storage needed to meet this requirement. To illustrate this variable sizing of detention systems, consider a simplified explanation using the rational formula to estimate hydrographs using the IDF (Intensity–Frequency–Duration) curves of Figure 10.14. Note from Figure 10.15 that the peak discharge decreases as the duration increases. By fixing the outflow hydrograph rate, the detention storage area can be estimated as the difference between the input and output hydrograph.

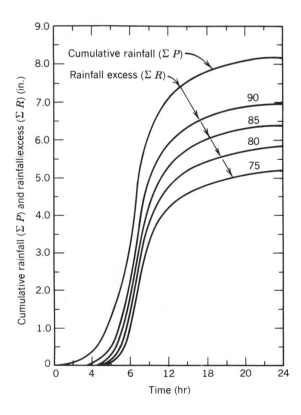

FIGURE 10.13   Mass curves of rainfall and direct runoff, 25-yr frequency, 24-hr rainfall for different curve numbers.

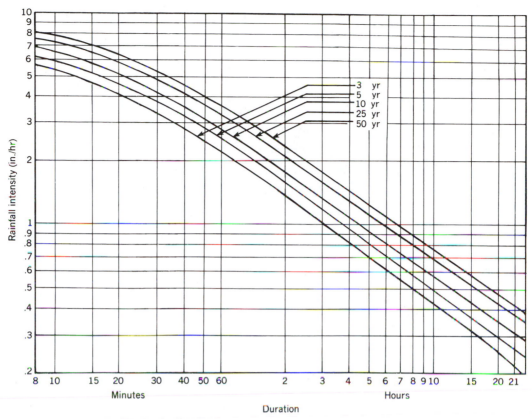

FIGURE 10.14 Rainfall intensity-duration Zone 3, Jacksonville, FL. *Source:* Florida State Department of Transportation.

Peak flow as computed by the Rational Method is closely related to a hydrograph with a base length equal to two times the time of concentration. This situation holds true only if the duration of the rainfall is assumed equal to the time of concentration. If the rainfall duration is longer than the time of concentration, a trapezoidal shaped hydrograph results with a peak flow again equal to CiA as was previously derived. However, the peak flow of the trapezoidal hydrograph is smaller than that of the triangular hydrograph as the intensity of rainfall i decreases with the duration of the rainfall. A rainfall–frequency–intensity curve is shown on Figure 10.14. Thus, for any rainfall intensity–duration–frequency curve, one can draw a family of hydrographs as shown in Figure 10.15; each one with a larger rainfall duration. The triangular hydrograph shown in Figure 10.15 (in which it is assumed that the duration of the rainfall is equal to the time of concentration) gives, as previously stated, the maximum value of Q ($= Q_p$) and is normally used by the engineer to size pipes and other systems. However, the other hydrograph (trapezoidal) result in greater volumes of runoff, the areas under each trapezoidal hydrograph being the volumes of runoff in each case.

Once such a family of hydrographs is developed, a detention basin size can easily be computed if the allowable outflow rate from the basin is established. In some regulations, the outflow rate is set as the natural flow in the receiving stream prior to any development.

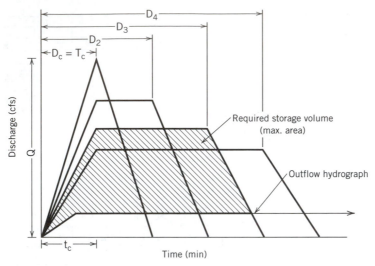

FIGURE 10.15   A family of hydrographs, rational method procedure.

If the allowable outflow rate is plotted on the family of hydrographs, the required storage volume can be computed. The maximum area (storage volume) below one of the trapezoidal hydrographs and above the plotted allowable outflow rate (cross-hatched area in Figure 10.15) can be estimated.

The obvious advantages of computing the storage area using a hydrograph by the Rational method procedure are ease and simplicity. The obvious disadvantage is possible inaccuracy. However, where small homogeneous watersheds are involved (i.e., ≤ 10 ac or less), this method will generally produce satisfactory results.

Considering the rainfall–intensity–duration curves, the short-duration (approximately 2 hr or less) storms should be used for design of transport and possibly detention facilities in smaller areas. The long-duration storms are more useful in design of detention facilities for larger watersheds. However, many short- and long-duration storms must be used to calculate detention basin size to determine the worst case (maximum storage volume).

10.3

HYDROGRAPH ATTENUATION IN A STORMWATER DETENTION BASIN

One of the major problems associated with new urban development is the increased volumes and rates of stormwater runoff generated within previously natural watersheds. Runoff volume is increased when natural pervious land surfaces are covered by such impervious structures as buildings, roadways, and parking lots and when natural depressions are removed, which serve as storage areas for ponded surface runoff in their natural state. The rate of runoff, including peak flow rates, is significantly increased when structural drainage systems such as storm sewers, swales, and ditches which greatly reduced the time of concentration of runoff, are constructed. Figure

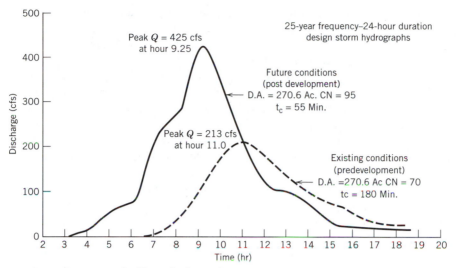

FIGURE 10.16   Typical subbasin hydrographs.

10.16 illustrates existing and future condition runoff hydrographs for a drainage basin resulting using a 24-hr duration storm. The existing condition hydrograph represents the runoff generated within the basin in its natural developed state, whereas the future condition hydrograph represents the runoff generated after development has taken place. As indicated by these hydrographs, both the peak rate and the volumes of runoff are significantly increased as a result of development within the watershed.

Increased volumes and rates of runoff generated by urban developments most often intensify downstream flooding problems, and, in some cases, especially in areas of flat terrain where backwater effects are significant, upstream flooding problems are also intensified, due to the limited conveyance and storage capacities of natural drainageways to handle the increased problem and rate of runoff from contributory watersheds.

Many urban governments and other public agencies responsible for stormwater drainage and local flood control have been giving increasing attention to practical and economical means of reducing the losses and inconveniences caused by flooding created by increased surface runoff. In many areas, local governments are realizing the importance of responsible stormwater management and some have adopted policies, especially in areas with identified flooding problems, that require that land developers take appropriate steps to ensure that peak runoff flow rates after development are not in excess of those that prevailed prior to development and that best management practices are used.

To meet such requirements, various on-site detention facilities are being incorporated into many planned developments, the most common of which include roof top storage, parking-log storage, and multipurpose detention and retention basins and ponds (Poertner, 1974). On-site detention of any type generally refers to storage of excess runoff on the site prior to its discharge into downstream drainage systems and gradual release of the temporarily stored runoff after the peak of the runoff inflow has passed.

The effect of a detention facility on runoff flow rates can be shown most descriptively by observing the inflow and outflow hydrographs for a detention pond, as shown in Figure 10.17. Note that although the total volume of runoff (area under the curves) is not reduced, the flow rate leaving the detention facility is significantly lower than the inflow rate. The objective, then, of any on-site detention facility is simply to regulate the runoff from a given rainfall event and to control discharge rates to reduce the impact on downstream drainage systems either natural or man made. Generally, detention facilities will not reduce the total volume of runoff but will simply redistribute the rate of runoff over a certain period of time by providing temporary "live" storage of a certain amount of the runoff. The volume of temporary live storage provided is the volume indicated by the area between the inflow and outflow hydrographs as shown in Figure 10.17.

The major benefit derived from properly designed and operated detention facilities is the reduction in downstream flooding problems. Other benefits include reduced costs of downstream drainage facilities, reduction in pollution of receiving streams, and even enhancement of aesthetics within a development area by providing the core of blue-green areas for parks and recreation.

Although the beneficial effects of stormwater detention are generally recognized, standard engineering design methods have not been developed that are universally accepted by designers of facilities or by the agencies that review and approve the designs. However, there do exist practical conventional techniques and procedures that provide useful guidelines to engineers who have not had extensive experience in the design of detention facilities. General guidelines are presented to demonstrate technically accepted methods and techniques that can be used in the design of such facilities.

The basic purpose of a stormwater detention basin is to reduce the rate of runoff from a contributory subbasin by providing temporary storage of excess runoff. Therefore, the major design considerations are, of course, the volume of storage required and the maximum permitted release rate. Although the storage volume required can

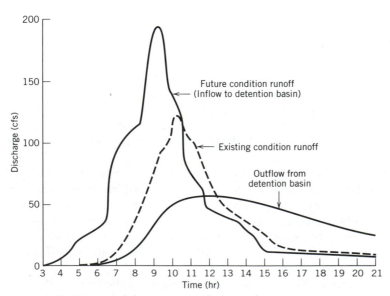

FIGURE 10.17 Typical detention system hydrographs.

be determined in a number of ways, the most commonly used criterion is to provide sufficient storage to limit the maximum rate of outflow from the detention basin to a maximum permitted release rate (Poertner, 1974). The volume of storage required and the maximum permitted release rate are, therefore, ultimately related. A more detailed discussion of this interrelationship of storage and outflow will be presented later in Section 10.3.1. Before discussing the more technical aspects and design procedures involved in the engineering design of detention basin facilities, some of the more general design considerations that should be considered throughout the design phases are mentioned briefly.

Throughout the design process, the designer should be committed to considering the potential impacts of the completed facility. Such impacts can be positive or negative and can be as broadly classified as social, economic, political, and environmental. Designers can often influence the positive or negative aspects of these impacts by their careful evaluation and decisions made in the design process. Generally speaking, the completed facility should provide for safety to people and wildlife and protection of real property and wildlife habitats. From an economical standpoint, the facility should provide for the lowest capital and annual costs attainable within project constraints. Multipurpose use of the facility and esthetic enhancement of the general area should also be major considerations. Above all, the facility should function in such a manner as to be compatible with overall drainage systems both upstream and downstream and to promote a best management practice approach to providing stormwater drainage and flood control.

This section does not emphasize general design considerations such as these just mentioned since they can often be quite intangible as well as site specific; rather, we present some of the generally accepted hydrologic design procedures and concepts as a basis for the application of general hydrologic and hydraulic principles involved in the engineering design of detention basin facilities. However, the general design considerations should be considered seriously through the design process of any detention basin facility.

Many management practices exist to attenuate (reduce peak flow) the hydrograph. Source controls, such as reducing impervious areas, catch basin design modifications, and on-site ponding all serve to reduce the hydrograph peak flow. The quantity of pollutants removed are, however, different and must also be examined before decisions are made on the procedures to reduce hydrograph peak flow.

10.3.1 Storage and Outflow Hydrographs

The common way to reduce the peak runoff from a particular watershed is to construct a holding pond (detention) to store the runoff (i.e., provide a volume (reservoir or pipe storage) to store a portion of the volume of runoff from the watershed and then to slowly remove the detained water from the pond). An outlet control structure is provided to allow the temporarily stored water to flow out within a specified period of time so that part of the detention storage is again available when the next storm rainfall occurs. Thus, both the storage volume and outlet device control the rate of discharge. Unfortunately, in flat terrain there are few, if any, natural reservoir sites. However, there are usually lakes and borrow pits available that can be used as holding ponds and can "control" the contributary drainage areas thereto.

In sewer systems, larger underground pipes with outlet control devices (inlet control to sewer) can be used for peak flow reduction. A self-regulating device with

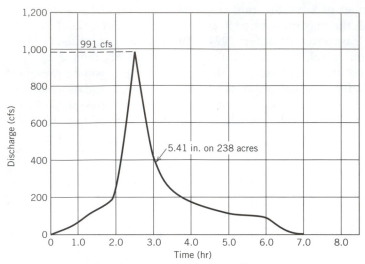

FIGURE 10.18 The 6-hr design storm hydrograph, curve No. 95.

no moving parts and no energy consumption would be ideal. Such devices are marketed under the name "Hydro-Brake" (shown in Figure 10.19). The principle of operation is to essentially use the static head in storage to create back pressures that reduce flow rates. As the pressure at discharge increases, the flow patterns within the device start a violent swirling motion, forming a vortex that creates the back pressure. Relative to a straight orifice, discharge rates have been reduced by 60 to 90% (H.I.L. Technology Inc., Scarlborough, Maine).

Consider as an example the basin that produced the runoff hydrograph shown in Figure 10.18 (Golding, 1974). Located in this particular drainage basin is an existing large borrow pit lake, which is ideally situated for reducing peak flow. Using the available storage therein and using a 36-in. reinforced concrete pipe culvert as the outflow spillway, the runoff hydrograph was attenuated to a peak of only 49 cfs. The

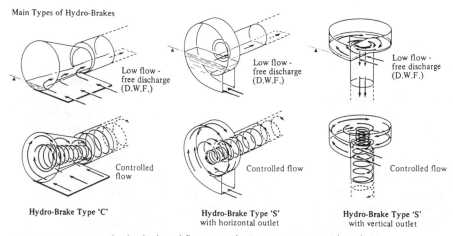

FIGURE 10.19 Hydro-brake liquid flow control (D.W.F. = Dry Weather Flow).

routing procedures of Chapter 8 were used to route through the reservoir storage available in the borrow pit lake.

In this particular case, the peak of the hydrograph was reduced from 991 to 49 ft^3/sec, admittedly an unusual situation. However, peak flows can be drastically reduced (up to 90%) by the use of available reservoir storage in existing lakes, borrow pits, etc. Generally, approximately 10 to 20% of small drainage basins up to 1.0 mi^2 area with corresponding water depths from 5 to 20 ft are required to significantly reduce peak discharge. For the situation reported by Golding (1974) the area of the borrow pit detention reservoir, 25 ac, amounted to approximately 10% of the contributory 238-ac watershed. The depth at maximum storage was approximately 5 ft.

Of interest in this particular situation was the fact that tailwater caused by a downstream culvert actually delayed flow out of this lake until hour 5 and reduced the outflow to approximately 22 ft^3/sec until hour 7, at which time critical depth at the culvert outlet took over and increased flow to 49 ft^3/sec. Normally, substantial outflow would probably start at hour 2.

For the hydrograph attenuation reported by Golding (1974), much time has elapsed for the runoff to flow out of the reservoir storage basin. Assuming that outflow through the pipe culvert spillway remained constant at 49 ft^3/sec (actually it decreased with time because the head on the culvert decreased as water flowed out of the holding pond), it would take a little over one day for the culvert to empty the pond. A convenient conversion factor to remember is that 1 ft^3/sec for one day is equal to about 2 ac-ft (ft^3/sec days $\times$ 1.98 = ac-ft). Therefore, 49 ft^3/sec for one day is equal to approximately 98 ac-ft, which compares with 107 ac-ft in the basin.

10.3.2 Case Studies

To demonstrate the preliminary design procedures of the previous sections, a simplified example design problem will be solved to illustrate the application of technically acceptable hydraulic and hydrologic methods and techniques. Assume that a detention pond is to be designed for a proposed development for which runoff hydrographs were computed. Assume that the proposed development is in an area in which the local governing agency requires that future peak runoff rates resulting from a 25-yr frequency, 24-hr duration design storm do not exceed those that exist under predevelopment conditions for the same design storm.

The 25-yr, 24-hr runoff hydrograph for the watershed under natural, predevelopment conditions has been computed using the SCS unit hydrograph method (shown in Figure 10.20 along with the future conditions hydrograph as computed by the same procedure). As indicated by the existing-conditions hydrograph, the peak runoff flow rate is 213 ft^3/sec, which will then be the maximum permissible release rate from the detention basin as required by the governing agency. The inflow hydrograph to the proposed detention basin is, then, the future conditions hydrograph shown in Figure 10.20. Using the inventory equation, the cumulative volume of inflow minus the cumulative volume of outflow at any time is equal to the volume stored at the time. Therefore, an estimate of the total volume required can be determined by measuring the area between the inflow hydrograph and the assumed outflow hydrograph. As indicated in Figure 10.20, an estimated volume of 944 ft^3/sec-hr or 78 ac-ft is determined.

Based on an evaluation of the downstream drainage facilities, the normal control water surface elevation should not be lower than elevation 92 mean sea level (MSL); based on the existing topography around the site of the proposed pond, the maximum

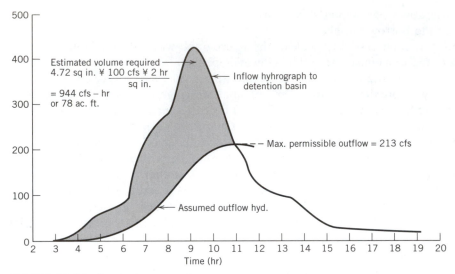

FIGURE 10.20   Required storage volume.

design stage should not exceed elevation 96, not including an allowance for a 1-ft freeboard. Therefore, the maximum depth of live storage for the detention basin is the difference in these two elevations, or 4 ft.

The average area required for the detention basin can then be approximated by dividing the estimated storage required by the depth of live storage. The estimated average area required is, therefore, 78 ac-ft divided by 4 ft or 19.5 ac. Note that this area only approximates the area required at approximately the average depth of live storage, which, in this case, is approximately 2 ft above the control elevation or about elevation 94 MSL. Since the area is only an approximation, it should be used only as a starting point for development of the storage–elevation relationship.

To simplify the computation of the storage–elevation relationship in the preliminary design of relatively large irregularly shaped detention basins with relatively constant and flat side slopes of about 10 horizontal to 1 vertical flatter, it is usually sufficient to approximate the change in area for every foot of change in depth by the following relationship:

$$\Delta A_d = \frac{(SS \times SL_d)}{43,560}$$

where
$\quad \Delta A_d$ = the change in area in acres for every foot of change in water depth from the average depth, d
$\quad\quad SS$ = the ratio of horizontal to vertical length of the side slopes
$\quad\quad SL_d$ = the shore length in feet at the average depth, d
$\quad 43,560$ = square feet per acre

A preliminary layout of the detention basin can then be made such that the area at elevation 94 MSL is about 19.5 ac as shown in Figure 10.21. Then, by scaling the shoreline length and applying the above relationship, a storage–elevation curve can

Preliminary layout—Detention basis:

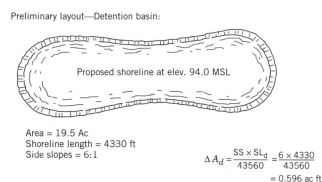

Proposed shoreline at elev. 94.0 MSL

Area = 19.5 Ac
Shoreline length = 4330 ft
Side slopes = 6:1

$$\Delta A_d = \frac{SS \times SL_d}{43560} = \frac{6 \times 4330}{43560}$$

$$= 0.596 \text{ ac ft}$$

1	2	3	4	5	6	7	8
Elev. (ft)	Depth (d) (± ft)	$\Delta A_d \times d$ (acres)	Area (acres)	Avg. area (acres)	Δd (ft)	Δ Storage (Ac-Ft)	Σ Storage (Ac-Ft)
92.0	−2	−1.193	18.31	18.60	1.0	18.6	0.0
93.0	−1	−0.596	18.90	19.20	1.0	19.2	18.6
94.0	0	0.0	19.50	19.80	1.0	19.8	37.8
95.0	+1	+0.596	20.10	20.40	1.0	20.4	57.6
96.0	+2	+1.193	20.69	20.84	0.5	10.4	78.0
96.5	+2.5	+1.491	20.99				88.4

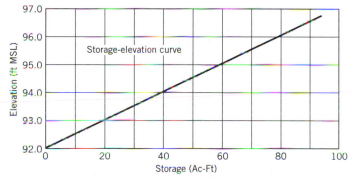

FIGURE 10.21   The computation of a storage-elevation curve.

be computed and plotted as shown in Figure 10.21. Note that the areas at various elevations in column 4 of the computation table are computed by adding to the average area the product of ΔA_d (the change in area per unit change in depth) and d (the depth from the average elevation either positive or negative). Average areas between each elevation are then computed in column 5; incremental volumes of storage between the changes in volume in column 7; and the storage–elevation curve is then plotted from the points in columns 1 and 8.

The assumptions and approximations used in developing the storage–elevation relationship in this manner will, of course, limit the accuracy of the results to some extent. However, in the preliminary design of large detention systems of this type, sufficient accuracy is usually obtained by making such approximations. At any rate, on completion of the final design of the detention system, an actual storage–elevation curve should be computed using a final grading plan for the basin that can be compared to the curve used in the preliminary design computations.

The next step is to determine a trial size for the outfall structure. In this problem, a weir spillway with a crest elevation at 92 MSL will be used as the control structure, and it will be assumed that a free outfall condition will exist at all times. Knowing the maximum permissible discharge rate and the maximum allowable stage or headwater elevation, the structure can be sized by applying the rectangular weir formula to determine the required weir length. Assuming a weir discharge coefficient of 3.2, the required length is computed as follows:

$$B = Q/CH^{3/2} = 213/[3.2(96 - 92)^{3/2}] = 8.32 \text{ ft}$$

and rounding off: use $B = 8.5$ ft. Other types of flow control devices can be used. The theoretical flow characteristics are represented as follows.
    For a circular orifice (Equation 7.22)

$$Q = CA_o\sqrt{2gh}$$

where
    Q = orifice discharge (ft³/sec)
    C = coefficient of discharge
    A_o = orifice cross-sectional area (ft²)
    g = gravitational acceleration constant = 32.2 ft/sec²

Note: When water surface behind orifice falls below the top of the orifice opening, the equation for a simple weir opening should be used (Equation 7.17).

$$Q = CBH^{3/2}$$

where
    C = equivalent weir coefficient of discharge
    B = equivalent weir length (ft)
    H = hydraulic head above the weir invert (ft)

A rating curve of discharge vs. stage for the proposed structure can then be computed by again applying the rectangular weir formula solving for Q at various stages in the detention basin. The required computations and the plotted rating curve for this structure are shown in Figure 10.22.
    As we stated previously, the inventory equation as applied in reservoir routing states that the volume of inflow minus the volume of outflow over a given time interval is equal to the change in volume stored over the time interval. In most applications of the inventory equation, the flow and storage variables are expanded and were developed in Chapter 8 (Equation 8.29).

$$\bar{I} = \frac{(I_1 + I_2)}{2} \qquad \bar{O} = \frac{(O_1 + O_2)}{2} \qquad \text{and} \qquad \Delta S = S_2 - S_1$$

where
    I_1 and I_2 = inflow rates at time 1 and subsequent time 2
    O_1 and O_2 = outflow rates at time 1 and subsequent time 2
    S_1 and S_2 = volumes stored at time 1 and time 2, respectively

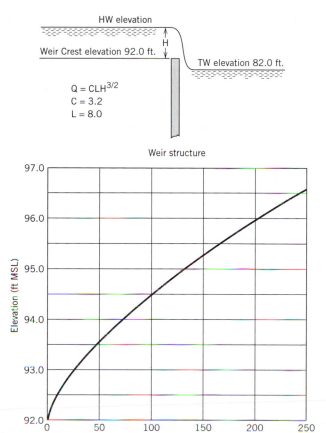

FIGURE 10.22 Computation of hydraulic rating curve.

Dividing both sides of the inventory equation by Δt and substituting terms:

$$\bar{I} - \bar{O} = \frac{\Delta S}{\Delta t}$$

$$\bar{I} - \frac{(O_1 + O_2)}{2} = \frac{(S_2 - S_1)}{\Delta t}$$

$$\bar{I} - \frac{O_1}{2} - \frac{O_2}{2} = \frac{S_2}{\Delta t} - \frac{S_1}{\Delta t}$$

Rearranging terms, the following equation results:

$$\frac{S_2}{\Delta t} + \frac{O_2}{2} = \bar{I} + \frac{S_1}{\Delta t} - \frac{O_1}{2}$$

Since both storage (S) and outflow (O) are functions of the water surface elevation in the detention basin and if Δt is held constant, then there exists a relation for outflow, O, as a function of the compound variable ($S/\Delta t + O/2$). Furthermore, if the average inflow, $\bar{I}$, outflow, O, and storage, S, are known for a given time, then the equation can be rewritten with only one unknown variable, that is, $(S_2/\Delta t + O_2/2)$ or rewriting Equation 8.30:

$$N_2 = \left(\frac{S_2}{\Delta t} + \frac{O_2}{2}\right) = \bar{I} + \left(\frac{S_1}{\Delta t} + \frac{O_1}{2}\right) - O_1$$

This then is the working form of the inventory equation. Other hydrologic methods typically rearrange the terms of the inventory equation in much the same manner as done here to obtain the equation in a working form for the actual routing procedure.

Having the inventory equation in the form of Equation 8.30 allows for the development of a relationship between O_2 and $(S_2/\Delta t + O_2/2)$, which if plotted in graphic form is generally referred to as the storage-indicating working curve for solution of the inventory equation in the actual routing process.

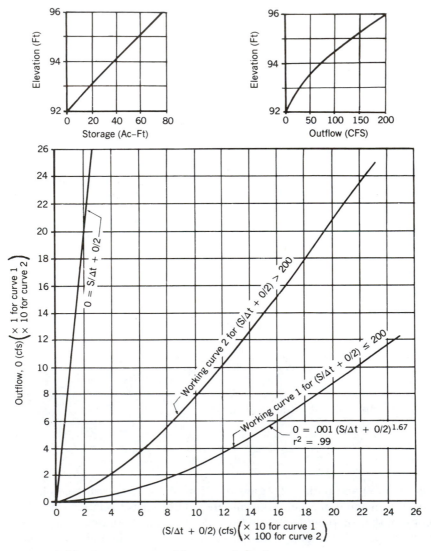

FIGURE 10.23   The computation of the storage-indicating curve.

Using the storage–elevation curve for the detention basin and the hydraulic rating curve for the outlet structure, the points of the storage-indicating working curve for the example problem can be computed and plotted as shown in Figure 10.23 for a routing interval, Δt, of 30 min. Due to mathematical instability, usually caused by large changes in inflow rate, large changes in stage result over one routing time step.

TABLE 10.5 Reservoir Routing Operations Table (Storage-Indicating Method)
$(S_2/\Delta t + O_2/2) = \bar{I} + (S_1/\Delta t + O_1/2) - O_1$

1 TIME (hr)	2 INFLOW, I (ft^3/sec)	3 AVG. INFLOW, $\bar{i}$ (ft^3/sec)	4 $S_2/\Delta t + O/2$ $= \bar{i} + S_1/\Delta t$ $+ O_1/2 - O_1$ (ft^3/sec)	OUTFLOW, O^a (ft^3/sec)
2.5	0	0	0	0
3.0	0.7	0.4	0.4	0.0
3.5	6.6	3.6	4.0^b	0.1
4.0	18.6	12.6	16.5^c	0.2
4.5	35.8	27.2	43.5	0.5
5.0	56.0	45.9	88.8	2.2
5.5	69.9	63.0	149.6	5.6
6.0	79.5	74.7	218.7	10
6.5	130.4	105.0	313.7	15
7.0	206.8	169.8	468.3	25
7.5	247.6	228.2	671.5	43
8.0	268.7	258.2	886.7	66
8.5	328.6	298.6	1119.3	92
9.0	413.1	370.8	1398.1	125
9.5	403.7	408.4	1681.5	163
10.0	348.0	375.8	1894.3	191
10.5	283.9	316.0	2019.3	206
11.0	209.8	246.8	2060.1	210
11.5	162.0	185.8	2036.0	208
12.0	122.3	142.2	1970.2	202
12.5	107.5	114.9	1883.1	190
13.0	102.3	104.9	1798.0	179
13.5	90.0	96.2	1715.2	168
14.0	72.6	81.3	1628.5	156
14.5	54.5	63.6	1536.1	143
15.0	34.2	44.4	1437.5	130
15.5	26.4	30.3	1337.8	118
16.0	23.5	25.0	1244.8	106
16.5	22.6	23.1	1161.9	96
17.0	22.3	22.4	1088.3	88
17.5	22.2	22.2	1022.5	81
18.0	22.2	22.2	963.7	74
18.5	20.4	21.3	911.9	68
19.0	17.4	18.9	865.2	64

[a]Estimate from plot.
[b]$3.6 + 0.4 - 0 = 4.0$.
[c]$12.6 + 4.0 - 0.1 = 16.5$.

Thus, portions of the outflow hydrograph can be distorted, and negative rates may occur during recession periods if O_2 is greater than $(S_2/\Delta t + O_2/2)$. It is, important, therefore, when using this method to be certain that a small enough time interval is selected so that numerical instability problems do not interfere with the accuracy of the routing computations.

An easy way to determine whether a small enough time step has been selected is to plot a line of equal values on the working curve that represents $O_2 = (S_2/\Delta t + O_2/2)$. If the working curve falls below this line, then a small enough time interval is being used, since O_2 will always be less than $(S_2/\Delta t + O_2/2)$, as shown in Figure 10.23.

Once the working curve for the detention system has been developed and the ordinates of the inflow hydrograph are available at time intervals of Δt, a routing operations table can be developed as shown in Table 10.5. As shown in this table, the ordinates of the inflow hydrograph are placed in column 2 and the average inflow rates, $\bar{I}$, are computed and placed in column 3 by arithmetically averaging the inflow rates between each time step. Also, the values of $(S/\Delta t + O/2)$ and outflow O at the initial time are taken as zero.

With the operations table in this form, the mathematical routing procedure can begin. The value of $(S/\Delta t + O/2)$ can be computed for the second time step. Therefore, $(S_2/\Delta t + O_2/2) = \bar{I} + (S_1/\Delta t + O_1/2) - O_1 = 0.4 + 0.0 - 0.0 = 0.4$. Then, from the storage-indicating working curve, the value of O_2 can be determined from the value of $(S_2/\Delta t + O_2/2) = 0.4$. From the working curve, a value of O_2 at time = 0.5 h is determined to be zero. The equation gives a more exact value of 0.0002.

The outflow rate for the next time step is then determined in the same manner. Again, applying the continuity equation, we get $(S_2/\Delta t + O_2/2) = 3.6 + 0.4 - 0.0$ or 4.0. From the working curve with $(S/\Delta t + O/2) = 4.0$ one gets an outflow rate O_2 at time = 1 hr of 0.1 ft³/sec.

This procedure is then repeated for each successive time step until all the ordinates of the outflow hydrograph have been computed. Table 10.5 shows the completed operations table for this routing procedure.

From the ordinates of the outflow hydrograph from the detention pond for the

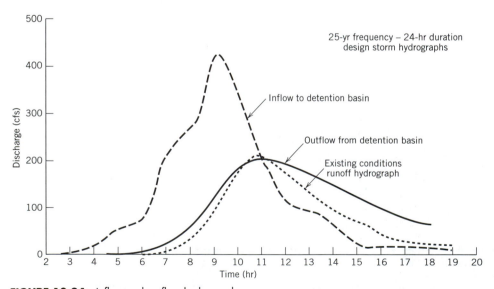

FIGURE 10.24   Inflow and outflow hydrographs.

design storm being used, we can plot the hydrograph and evaluate the performance of the first trial design of the detention system. As shown in Figure 10.24, the inflow and outflow hydrographs are plotted with the existing conditions runoff hydrograph for the drainage pond. Note here that the peak outflow rate from the detention pond is 210 ft³/sec as compared to a peak existing conditions runoff rate of 213 ft³/sec. Based on the criteria of the governing agency, our system is slightly overdesigned; however, the effects of timing of the flood flow rates should also be considered.

This example problem has been simplified. Complications resulting from neglect of infiltration hydrographs, quality transformations, quality removal efficiencies, costs, and variable tailwater conditions can require additional work.

10.4

CULVERTS

Stormwater detention ponds are frequently connected together with culverts. A typical culvert is a hydraulically short (generally, less than a few hundred feet) conduit that conveys stormwaters from one detention pond to another, or through an embankment, or past some other type of flow obstruction. When water passes through the culvert, friction forces, inlet losses, and exit losses force water on the upstream side to pond at a deeper elevation. Thus, various inlet configurations and a variety of materials are used in culvert design to decrease head losses. In addition, there are a variety of culvert shapes as shown in Figure 10.25. The selected shape is based on construction cost, limitation on upstream water surface elevation, embankment heights, and flow-rate limitations.

Flow rates through a culvert may be improved by reducing friction losses at the inlet side. Since upstream ponded areas and natural channel characteristics are usually wider than the culvert width (barrel width), there is a flow contraction at the culvert. A more gradual flow transition will lessen the energy loss; leveled edges are more efficient than square ones, and side or slope tapered inlets further reduce the flow contraction. The different types of inlet contractions are shown in Figures 10.26, 10.27, 10.28 (Federal Highway Administration, 1985). Other factors affect flow rates and headwater conditions, many of which must be examined when determining culvert sizes, inlet construction details, and headwater elevations. These factors are listed in

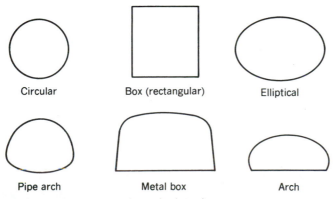

Circular   Box (rectangular)   Elliptical

Pipe arch   Metal box   Arch

FIGURE 10.25   Commonly used culvert shapes.

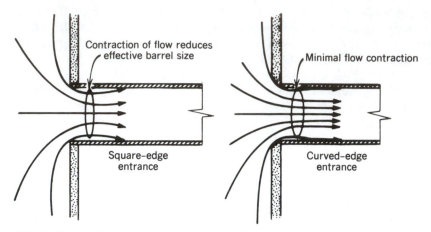

FIGURE 10.26   Entrance contractions (schematic).

Table 10.6. Inlet control occurs when the culvert barrel is capable of passing more flow than the inlet will accept. Outlet control occurs when the culvert barrel is not capable of passing as much flow as the inlet will accept.

When free surface flow (otherwise known as open channel flow) exists, the flow control is determined by knowing critical depth and a dimensionless ratio using average velocity and depth of flow, known as Froude number, F_r.

$$F_r = \frac{\overline{V}}{\sqrt{gy}}$$ (10.8)

TABLE 10.6   Factors Influencing Culvert Performance

FACTOR	INLET CONTROL	OUTLET CONTROL
Headwater elevation	X	X
Inlet area	X	X
Inlet edge configuration	X	X
Inlet shape	X	X
Barrel roughness		X
Barrel area		X
Barrel shape		X
Barrel length		X
Barrel slope[a]		X
Tailwater elevation		X

Source: Federal Highway Administration (1985).
[a]Barrel slope affects inlet control performance to a small degree but may be neglected.

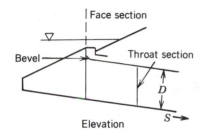

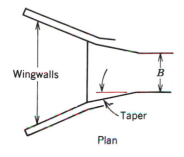

FIGURE 10.27 A side-tapered inlet. *Source:* The Federal Highway Administration, 1985.

where

$\overline{V}$ = average velocity = Q/A (ft/s) (m/s)

Q = flow rate (ft³/s) (m³/s)

A = cross-sectional area of flow (ft²) (m²)

y = hydraulic depth of flow (ft) (m)

g = gravitational acceleration, (32.2 ft/sec²) (9.8 m/sec²)

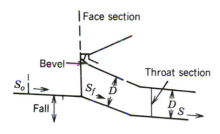

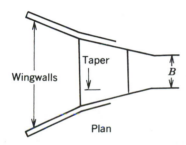

FIGURE 10.28 A slope-tapered inlet. *Source:* The Federal Highway Administration, 1985.

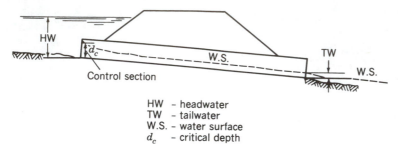

HW – headwater
TW – tailwater
W.S. – water surface
d_c – critical depth

FIGURE 10.29   A typical inlet control, flow conditions.

and

$$y = \frac{A}{w} \qquad\qquad (10.9)$$

where

$$w = \text{width of the free surface (ft) (m)}$$

When $F_r > 1.0$, flow is supercritical; if $F_r < 1.0$, flow is subcritical; and if $F_r = 1.0$, flow is critical. The characterization of subcritical and supercritical determine the location of a control section and the type of control. The control section of a culvert (control by the inlet) is located just inside the entrance, because critical depth (minimum specific energy) is located at or near this location and the flow downstream is supercritical. A typical inlet control is shown in Figure 10.29. Important variables are defined in Figure 10.29 such as headwater, tailwater, and critical depth. If subcritical flow exists in the culvert and a control section exists downstream, or critical depth is found at the end of the culvert, outlet control conditions exist. Typical outlet controls are shown in Figure 10.30.

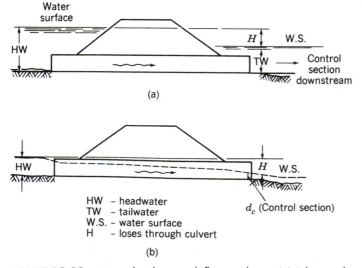

HW – headwater
TW – tailwater
W.S. – water surface
H   – loses through culvert

d_c (Control section)

(b)

FIGURE 10.30   A typical outlet control, flow conditions. (a) Submerged. (b) Unsubmerged.

The total head loss through the culvert can be estimated using the following equation which is the sum of head losses at the entrance, barrel friction, and exit losses assuming near zero velocity downstream. If bends or other minor losses were involved, they would also be included.

$$H = (k_e + (29n^2 L/R^{1.33}) + 1)(V^2/2g) \qquad (10.10)$$

where
k_e = entrance loss coefficient (see Table 10.7)
n = Manning's roughness coefficient (see Table 7.3)
L = conduit length (ft)
R = hydraulic radius (ft)

If tailwater conditions downstream result in a significant downstream velocity V_d, then the exit loss is modified as

$$H_{exit} = 1.0 \left(\frac{V^2}{2g} - \frac{V_d^2}{2g} \right) \qquad (10.11)$$

Thus, headwater elevation can be calculated given the tailwater condition, conduit shape, conduit material, flow rate, length of pipe (L) and slope of pipe (S_0).

$$HW = TW + H + LS_0 \qquad (10.12)$$

where
HW = headwater depth (ft)
TW = tailwater depth (ft)
H = headloss through culvert (ft)

A storage discharge relationship must be developed, but a culvert can function under inlet or outlet control depending on flow conditions, tailwater, and headwater (stage)

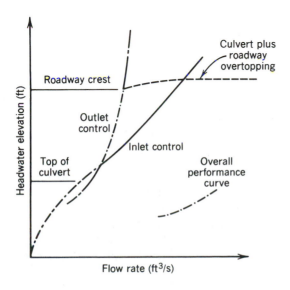

FIGURE 10.31 A culvert performance curve with roadway overtopping.

TABLE 10.7   Entrance Loss Coefficients: Outlet Control, Full or Partly Full Entrance Head Loss

TYPE OF STRUCTURE AND DESIGN OF ENTRANCE	COEFFICIENT k_e
Pipe, Concrete	
Projecting from fill, socket end (groove end)	.2
Projecting from fill, sq. cut end	.5
Headwall or headwall and wingwalls	
Socket end of pipe (groove end)	.2
Square edge	.5
Rounded (radius = $1/12 D$)	.2
Mitered to conform to fill slope	.7
End section conforming to fill slope[a]	.5
Beveled edges, 33.7° or 45° bevels	.2
Side- or slope-tapered inlet	.2
Pipe, or Pipe Arched, Corrugated Metal	
Projecting from fill (no headwall)	.9
Headwall or headwall and wingwalls square edge	.5
Mitered to conform to fill slope, paved or	
unpaved slope	.7
End section conforming to fill slope[a]	.5
Beveled edges, 33.7° or 45° bevels	.2
Side- or slope-tapered inlet	.2
Box, Reinforced Concrete	
Headwall parallel to embankment (no wingwalls)	
Square edged on 3 edges	.5
Rounded on 3 edges to radius of $1/12$ barrel	
dimension, or beveled edges on three sides	.2
Wingwalls at 30° to 75° to barrel	
Square edged at crown	.4
Crown edge rounded to radius of $1/12$ barrel	
dimension, or beveled top edge	.2
Wingwall at 10° to 25° to barrel	
Square edged at crown	.5
Wingwalls parallel (extension of sides)	
Square edged at crown	.7
Side- or slope-tapered inlet	.2

Source: Federal Highway Administration (1985).
[a]"End section conforming to fill slope," made of either metal or concrete, are the sections commonly available from manufacturers. From limited hydraulic tests they are equivalent in operation to a headwall in both *inlet* and *outlet* control. Some end sections, incorporating a *closed* taper in their design have a superior hydraulic performance.

conditions. Thus, both inlet and outlet control relations must be developed as shown in Figure 10.31. Culvert discharge curves are similar to the orifice curves that act both as a weir and orifice. However, a culvert can have one more control section (inlet, outlet, and throat of a contraction) plus embankment overtopping. An overall discharge curve would be made up of the controlling curves producing the least discharge. Using this combined discharge, it is now possible to determine headwaters for any flow rate since headwaters determine flow rates.

10.5

OFF-LINE RETENTION (INFILTRATION PONDS)

Stormwater volume and pollutants in the stormwater can be controlled by diverting the stormwater to infiltration ponds adjacent to the sewer line. The quantity diverted can be expressed in terms of the depth of runoff over the entire watershed or the diversion of the runoff from the first quantity (depth) of precipitation. Examples of both criteria are to (1) divert the first half-inch of runoff (over the total watershed) and (2) divert the runoff from the first inch of rainfall. If the runoff coefficient for a watershed is .5, then the runoff from the first inch of rainfall is the same as the first half-inch of runoff. Thus, the criteria may be consistent with one another.

The percent volume control can be determined from a cumulative distribution on storm event rainfall volume (Figures 2.1b and 2.15). From Figure 2.1b, the frequency of storms diverted as a function of storm volume can be determined (i.e., 90% of all events have less than 1 in.). From Figure 2.15, which is derived from Figure 2.1b, the percent of yearly volume diverted as a function of storm volume can be determined. As an example, 80% of yearly volume is diverted if the diversion is set to capture the first inch of runoff (runoff = rainfall) for each and every storm.

An easy to use design criteria formula (State of Florida, 1988) to control 80% of the yearly volume of runoff is

$$DV = \frac{CPA}{12} \tag{10.13}$$

where

DV = diversion volume (ac-ft)
$\quad C$ = runoff coefficient (≥ 0.5)
$\quad P$ = 1 inch rainfall volume (if Figure 2.15 is applicable)
$\quad A$ = watershed area (ac)
$\quad 12$ = conversion factor (in./ft)

The calculations using Equation 10.13 require that the runoff coefficient always be greater than .5 or the largest runoff coefficient be used consistent with published values (Table 6.5).

A cumulative frequency curve on rainfall volume and a diversion frequency curve for yearly volume should be developed for specific areas and percentage control should be noted to determine precipitation volume (inches). The runoff coefficient used is usually the greatest for a particular land-use designation because of the fact that the complete infiltration volume may not be recovered before the next rainfall runoff event. This stochastic event of rainfall may be compensated for by building a larger pond. By simulating the rainfall runoff events over a long period of time, the largest pond volume to accumulate the diversion of the runoff from each and every event can be calculated.

Wanielista (1977) simulated the rainfall–runoff events and diversion into percolation ponds for a 20-yr historical record. The rainfall record was for the central Florida area. The percolation pond had a minimum infiltration rate of 1 in./hr. For a completely impervious area and a percolation pond at a maximum depth of 5 ft, the efficiency of diversion and design volume of the percolation pond is calculated using the simulation (stochastic) equations in Table 10.8.

If the watershed area is not completely impervious and a composite curve number can be estimated for the watershed, the following equation is used to calculate the volume of a percolation pond that has a maximum depth of 5 ft.

$$V_5 = V_I[0.59 + 0.37(CN_c/100)] \qquad\qquad (10.14)$$

where

V_5 = volume of 5-ft-deep percolation pond (ac-ft)
CN_c = composite curve number
V_I = volume for 100% impervious area, Table 10.8 (ac-ft)

The percolation pond can be built at less depth but occupy greater surface area for a given volume of runoff. However, the less deep pond will infiltrate a depth (amount) of stormwater in less time than a deeper pond because of the lesser depth. An infiltration rate of 1 in./hr can drain a 2-ft-deep pond in 24 hr, whereas a 4-ft-deep pond will take more than 24 hr. Therefore, less deep ponds are more likely to recover storage capacity before the next runoff event and the size of the pond to store the first specified amount of runoff from each and every storm even will be less. A formula developed for the meteorological conditions of Florida that relates depth of percolation pond to storage volume is

$$V_D = V_M + \left[\frac{V_5 - V_M}{4}\right](D - 1) \qquad \text{for } D < 5 \qquad\qquad (10.15)$$

where

V_D = pond volume for depth D (ac-ft)
D = water depth (ft)
$V_M = (A \times DI)/12$ and DI = diversion depth (in.)
$\qquad\qquad\qquad\qquad\qquad A$ = watershed area (ac)
$\qquad\qquad\qquad\qquad\qquad 12$ = conversion factor (in./ft)

TABLE 10.8 Pond Volume for Diverted Runoff as a Function of Diversion Volume

DIVERSION VOLUME (in.) OF EACH AND EVERY STORM[a]	RUNOFF DIVERTED[b] Δ_m = 4-24-72	V_I VOLUME OF 5-ft-DEEP POND FOR 100% IMPERVIOUS AREA (ac-ft)
0.25	48-25-16	$0.016A^{1.28}$
0.38	55-36-25	$0.032A^{1.21}$
0.50	55-43-30	$0.046A^{1.18}$
0.75	68-56-42	$0.09A^{1.11}$
1.00	80-65-54	$0.14A^{1.07}$
1.50	90-77-66	

where A = total watershed area in acres

[a]Usually equal to the fraction of runoff from 1 in. of rainfall.
[b]Based on Figure 2.15.

☐ *EXAMPLE PROBLEM 10.5*
_____

A 24-ac multiunit family residential area with an average lot size of 1/4 ac is required by state law to treat the runoff from the first inch of rainfall using off-line retention (diversion to percolating ponds). What size of pond do you specify if the previous area is classified as Hydrologic Soil Group B and the maximum pond depth is 4 ft? Use both the state of Florida design criteria and the equations derived from simulating rainfall-runoff.

Solution

Using state of Florida criteria and a runoff coefficient for multifamily residential of .6 (Table 6.5), Equation 10.13 is used. Figure 2.15 is applicable for the design.

$$DV = \frac{CPA}{12}$$
$$= (.6)(1)(24)/12$$
$$= 1.2 \text{ ac-ft}$$

Using the simulation equations, the percent directly connected area is 38% (Table 5.8) and $CN_c = 75$. The diversion volume must be equal to or greater than 0.38 in. (use equation from Table 10.8) with $A = 24$ ac.

$$V_I = .032A^{1.21}$$
$$= .032(24)^{1.21}$$
$$= 1.50 \text{ ac-ft}$$

and

$$V_M = (A \times DI)/12$$
$$= [(24)(.38)]/12 = 0.76 \text{ ac-ft}$$
$$V_5 = V_I(.59 + .37CN_c/100)$$
$$= 1.50[(0.59 + 0.37(75)/100] = 1.30 \text{ ac-ft}$$

for 4-ft-deep pond:

$$V_4 = V_M + \left[\frac{V_5 - V_M}{4}\right](D - 1)$$

$$= 0.76 + \left[\frac{1.30 - 0.76}{4}\right](4 - 1) = 1.17 \text{ ac-ft}$$

Thus, by using a simplified design formula or the simulation formulas, the results are comparable for this specific watershed. ☐

10.6

SWALE DESIGN

Swales are vegetated open channels that infiltrate and transport runoff waters. By incorporating the hydrologic processes of runoff and infiltration, a swale design based on quantity is possible. Low velocities are important to prevent particle transport and loss of soil. The vegetation within the swales are very effective for the removal of solids and retention of soil (Lord, 1986). Erosion can be lessened by a vegetative area immediately after construction.

Wanielista et al. (1988) reported on the design of swales that had a seasonal high water table at least 1 ft below the bottom of a swale. They conducted 20 field tests doing a mass balance on runoff input to a swale and its output. The difference in input and output was the infiltration volume, from which an infiltration rate was calculated. Limiting infiltration rates during actual swale operation were estimated at 5 to 7.5 cm/hr (2–3 in./hr). The soils at the sites were classified according to the American Association of State Highway and Transportation Officials (AASHTO) soil classification systems as A-3 and A-2-4, which are fine sand and silty sand soils, respectively. For the same swales, the double-ring infiltrometer was used to estimate the limiting infiltration rates. They were recorded at 12.5 to 50 cm/hr (5–20 in./hr).

Using a mass balance of input and output waters in a swale system, Equation 10.16 was developed to estimate the length of a swale necessary to infiltrate all the input rainfall excess from a specific storm event using a triangular shaped cross-sectional area. Another equation was developed for a trapezoidal cross-sectional shape (see Appendix E for a complete derivation).

$$L = \frac{KQ^{5/8}S^{3/16}}{N^{3/8}f} \tag{10.16}$$

where

L = length of swale (m or ft)
Q = average runoff flow rate (m³/s or cfs)
S = longitudinal slope (m/m or ft/ft)
N = Manning's roughness coefficient (for overland flow) (Table 5.1)
f = infiltration rate (cm/hr or in/hr)
K = constant that is a function of side slope parameter Z(1 vertical/Z horizontal) and is defined as

Z (SIDE SLOPE) (1 VERTICAL/Z HORIZONTAL)	K (SI UNITS) (f = cm/hr, Q = m³/s)	K (ENGLISH UNITS) (f = in./hr, Q = cfs)
1	98,100	13,650
2	85,400	11,900
3	71,200	9,900
4	61,200	8,500
5	54,000	7,500
6	48,500	6,750
7	44,300	6,150
8	40,850	5,680
9	38,000	5,285
10	35,670	4,955

For most watersheds, the length of a swale necessary to infiltrate 3 in. of runoff waters was found to be excessive or at least twice the distance available. Thus, some type of swale block (berm) or on-line detention/retention may be more helpful. Pitt (1986) indicated the most cost effective solution for the reduction of runoff volume, residual solids, and bacteria was infiltration. This result is similar to that of Yousef et al. (1985) who recommended that swale blocks should be considered to reduce further the chemical constituents and runoff volumes.

Using as a design criteria, the runoff volume for 7.5 cm (3 in.) of rainfall and storage of noninfiltrated runoff, Wanielista et al. (1988) have developed swale block designs for highway applications. Basically, the swale block volume can be calculated for a fixed length of swale and a triangular cross section using

$$\text{Volume of runoff} - \text{volume of infiltration} = \text{swale block volume}$$

$$Q(\Delta t) - Q_f(\Delta t) = \text{volume of swale}$$

$$Q(\Delta t) - \left[\frac{LN^{3/8}f}{KS^{3/16}}\right]^{8/5}(\Delta t) = \text{volume of swale} \qquad (10.17)$$

where

Q_f = average infiltration rate (m^3/sec)
Δt = runoff hydrograph time (sec)

Swale volume must be available to contain the runoff waters. In highway designs for high speed situations, safety must be considered, thus a maximum depth of water equal to 0.5 m (about 1.5 ft) and flow line slopes on the berms of 1 vertical/20 horizontal are recommended. Along lower-speed highways or in some residential/commercial urban settings, steeper flow line berm slopes (1 on 6) are acceptable.

❑ **EXAMPLE PROBLEM 10.6**
_____

Design of a swale is shown in this example. Consider as an example, a swale section along Interstate 4 near Orlando, Florida. The parameters of Equation 10.17 are

$$N = 0.05$$

$$S = 0.0279$$

$$Q = 0.0023 \text{ m}^3/\text{sec (0.08 cfs)} \qquad \text{for} \qquad \Delta t = 100 \text{ min}$$

$$f = 7.5 \text{ cm/hr (3.0 in./hr)}$$

$$Z = 7$$

Solution

a. What swale length would be necessary to infiltrate all the waters (using Equation 10.16)?

$$L = [44,300(0.0023)^{5/8}(0.0279)^{3/16}]/[(0.05)^{3/8}(7.5)] = 208 \text{ m}$$

b. Only 76 m (250 ft) were available, thus how much storage volume is necessary (using Equation 10.17)?

$$(0.0023)(60)(100) - \left[\frac{(76)(0.05)^{3/8}(7.5)}{44,300(0.0279)^{3/16}} \right]^{8/5} 60(100) = \text{volume}$$

and volume of storage $= 11.1 \text{ m}^3$. ❏

10.7

FRENCH DRAINS

A french drain is a perforated or slotted pipe placed in select backfill aggregate material with a fabric filter surrounding the aggregate as shown in Figure 10.32. Also, a french drain may consist solely of aggregate materials. The purpose of a french drain is to lower the water table or remove stormwater through the bank of a stormwater pond (peak rate and volume control).

Sometimes, open channels can be used to lower a water table, but open ditches are not always acceptable in developed areas. The banks of an open ditch also may be eroded from a high groundwater table. Thus, an underground piping system (french drain) that removes the water without removing the soil particles may be necessary. Excessive soil loss reduces the carrying capacity of the soil and can cause subsidence. Some types of protective filter material and select aggregate placed adjacent to the erodable soils can provide a solution to soil loss. A fabric filter must be specified to prevent piping, reduce fabric filter clogging, and pass the water. These criteria must all be present.

The retention ability of a filter material to retain soil is a function of its openings and the size of the soil particles. Cedergren et al. (1972) reports on a criteria that states that the equivalent opening size of the fabric (standard test method for fabrics) divided by the nominal diameter of soil particles for which 85% of the soil degradation is finer must be less than 2 to retain the soil and prevent excess piping. Particles larger than the fabric pore size will be retained, and some of the smaller ones will interact with the fabric filter and will be impacted. Other smaller ones will be "bridged" between the fabric and the soil particles and further retention is possible. Clogging of the filter can be minimized by specifying the expected hydraulic gradient of the groundwater during the selection process. The permeability of the fabric filter should be at least 10 times greater than the protected soil, which should allow the soil water to drain.

Proper specification of the size of pipe and trench depth below the expected water

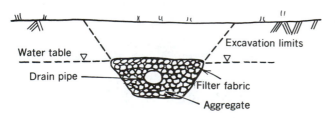

FIGURE 10.32 A French drain.

table is necessary if the water table is to be lowered. Darcy's law (see Chapter 9) can be used to approximate the seepage from saturated soil into the french drain. Rewriting Equation 9.13 as

$$\frac{Q}{L} = KAS \qquad (10.18)$$

where

Q/L = seepage rate, CFD/foot of trench
  A = area normal to seepage flow for a 1-ft-long trench (ft²/ft)
  K = permeability (ft/day)
  S = slope of flow line (ft/ft)

Note that the flow rate units are cubic feet per day (CFD) rather than the customary cubic feet per second, which implies relatively low flow rates.

Since the water table will fluctuate, the area normal to flow will fluctuate. The maximum water table elevation is chosen, and the flow rate (size of trench) determined from this condition. Permeability of the soil can be estimated using laboratory or field permeability tests. Generally, lower values are chosen for determining drawdown time and higher values for determining pipe sizes.

For the aggregate that is used around the pipe, its permeability must be specified to ensure that it is greater than the predicted soil. Typical aggregate permeabilities as they relate to grain-size distribution (sieve) analysis are shown in Figure 10.33. Furthermore, the aggregate must not deteriorate over time. Some limestones do break down and form an impermeable material. Cross-sectional area and hydraulic gradients are usually determined by site conditions, thus permeabilities of rock should be specified for desired flow conditions. During seepage, some fine soil particles do enter the aggregate, thus reducing the permeability.

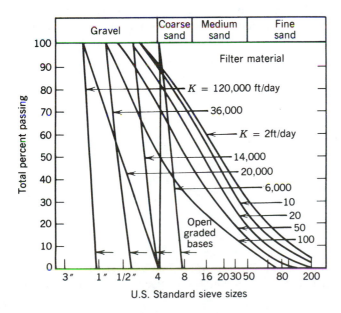

FIGURE 10.33 Aggregate permeabilities. *Source:* Cedergren et al., 1972.

◻ *EXAMPLE PROBLEM 10.7*

Consider a french drain that has a 4-ft wide × 3-ft-deep cross section. It must discharge seepage water at a rate of 25000 ft³/day for the 100 ft of drain. The drain slope is 1 ft/100 ft, and the length of drain is 100 ft. What must the permeability of the backfill aggregate be to ensure proper drainage?

Solution

Use Darcy's equation $Q/L = KAS$ where $Q/L = 25000/100 = 250$ ft³/day-foot of drain. Thus, $250 = K(3 \times 4)(.01)$ and $K = 2083$ ft/day. Since permeability will probably decrease over time, an aggregate with greater than 2083 ft/day should be selected. An aggregate with a permeability of about 3000 ft/day or greater should be specified.   ◻

10.8

STORMWATER RE-USE

An alternative to discharge of stormwater from wet detention pond systems is the reuse of the stormwater over the watershed for irrigation (Livingston et al., 1994). The reuse of stormwater reduces the volume of discharged stormwater, therefore decreasing the loss of a potential freshwater resource and decreasing the pollutant discharge from the system. The stormwater can be used for a number of purposes including (1) irrigating open lands, (2) recharging groundwater, (3) supplementing water used for cooling purposes, (4) supplementing car wash water, (5) enhancing and creating wetlands, and (6) supplying water for agricultural use.

The re-use pond differs from the typical detention pond in that instead of the temporary storage being depleted using a discharge device, such as a weir or orifice, it is drawn down using a re-use system. Even though the re-use system can be used to draw down the pond, a discharge structure is still necessary for flood control.

Stormwater re-use systems can be designed by the procedures recommended by Wanielista et al. (1991). Continuous simulation of the re-use pond using rainfall records was used in a mass balance (Equation 10.19) with the results being represented in a rate-efficiency-volume (*REV*) curve.

$$R + G - RU - D = \Delta S \qquad (10.19)$$

where

R = rainfall excess or runoff volume
G = supplemental water volume
RU = re-use volume on the effective impervious area
D = discharge
S = storage in pond

Re-use rate is the rate in volume per time at which the system re-uses the stormwater. Typically the volume is expressed in inches over the watershed's effective impervious area. Re-use efficiency is the percentage of water which is re-used and not discharged.

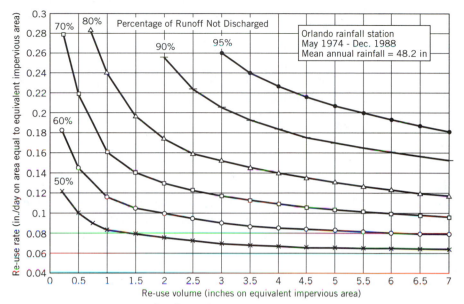

FIGURE 10.34   Some examples of REV charts.

Re-use volume is the volume in the pond that is available for re-use. The *REV* curve for Orlando, Florida, is shown in Figure 10.34.

Effective impervious area as used in the curve is defined as the total watershed area multiplied by the rational coefficient. The rational coefficient is calculated based on the runoff resulting from the average rainfall event volume.

$$EIA = CA \qquad (10.20)$$

where

EIA = effective impervious area
C = rational coefficient of watershed
A = total watershed area

EXAMPLE PROBLEM 10.8

It is assumed that a detention pond which will be used for re-use water removes 50% of the pollutant mass effluent from a 4-ha impervious area. Regulations require that the pond remove 90% of the pollutant mass. If the maximum re-use storage available in the pond is equal to the volume from a 7.6-cm rainfall event, what is the re-use rate required to meet the pollutant removal.

Solution

The discharged percentage of pollutants from the pond is 100 − 50 = 50%, which must be decreased to 100 − 90 = 10% by decreasing the volume discharged from the pond. If it is assumed that the pond will still remove 50% of the pollutants from the pond, then the percentage of volume allowed to discharge can be calculated using a mass balance.

$$0.5V_{\text{discharged}} = 0.1V_{\text{total}} \tag{1}$$

$$V_{\text{total}} = V_{\text{discharged}} + V_{\text{re-used}} \tag{2}$$

$$1 = V_{\text{discharged}}/V_{\text{total}} + V_{\text{re-used}}/V_{\text{total}}$$

$$1 - E/100 = V_{\text{discharged}}/V_{\text{total}}$$

combining equations gives

$$0.1/0.5 = 1 - E/100$$

$$100(1 - 0.2) = E$$

or re-use efficiency = 80%. From the Orlando *REV* curve re-use rate is a function of re-use efficiency (80%) and re-use volume (7.6 cm). Re-use rate is roughly 0.152 in./day on *EIA* per day or 0.386 cm/day on *EIA*. This can be converted to a volume 154 m³/day, as shown.

$$RU = (0.386 \text{ cm/day})(EIA)(4 \text{ ha}/EIA) = 154 \text{ m}^3/\text{day} \quad \square$$

10.9

DESIGN OF STORM SEWER SYSTEMS

The design of storm sewer systems is a direct application of the principles from both hydraulics and hydrology. *IDF* curves are used to specify rainfall intensities. Watershed characteristics are used to estimate the volume and flow rate of runoff from the rainfall. Flow equations are used to calculate pipe or channel sizes necessary to convey the calculated rates of flow. States and municipalities may have specific regulations pertaining to the use of hydrologic data in the storm sewer design. These are usually available in the form of a drainage manual. This section covers the fundamentals of the storm sewer design process using the rational method.

10.9.1   Determination of Storm Sewer Flow Rates

A storm sewer is typically designed for a specific return period storm, usually 10 or 25 yr. This return period is used in conjunction with a local *IDF* curve to determine a rainfall intensity. A more detailed discussion of this process is included in this text in Chapter 3. The duration used in the determination of the rainfall intensity is equal to the time of concentration of the contributing watershed. Storm sewer design regulations usually specify a minimum time of concentration, and if the watershed time of concentration is less than the specified minimum, the specified minimum is used rather than the watershed time of concentration. In cases where a storm sewer inlet has upstream piping, the maximum of the watershed time of concentration or the accumulated upstream travel time is used.

$$D = \max \begin{cases} \text{Watershed } TC \\ \text{Minimum Required } TC \\ \text{Accumulated Upstream } TC \end{cases}$$

where

D = duration used for calculation of rainfall intensity
TC = time of concentration

The intensity (i) of the rainfall is used in the rational equation ($Q = CiA$) to determine the flow rate at the storm sewer inlet. The rational equation is believed applicable because in most storm sewer design situations, the entire watershed is broken up into a number of smaller subwatersheds each influent to a manhole or inlet. Flows into each inlet are then summed in a downstream direction. These flow rates are used to determine the size of the storm sewer piping. Pipe sizes should be selected to meet or exceed the required capacity for the design storm. The use of the rational formula for each inlet and the sum of these flows are considered a conservative estimate for pipe sizing.

10.9.2 Hydraulic Grade Calculations

Once flow rates have been calculated throughout the system, the hydraulic grades in the systems also should be calculated. Using the principles of hydraulics and a known tailwater elevation at the outlet along with other physical parameters of the systems the hydraulic grades can be calculated. For an in-depth discussion the reader is referred to a number of hydraulics texts (Chow, 1959; French, 1985). Several texts devoted to storm sewer design principles also exist and cover these topics in detail (American Iron and Steel Institute, 1980; ASCE and WEF, 1970 and 1992). These principles are also covered in state and federal drainage manuals (e.g., FDOT, 1986; AASHTO, 1991).

Within storm sewer pipe systems the slope of the hydraulic grade line (HGL) can be calculated using Manning's equation.

$$V = \frac{1.49}{n} R^{2/3} S^{1/2} \tag{10.21}$$

where

V = velocity in feet per second
R = hydraulic radius in feet
S = slope of the hydraulic grade line as a fraction (ft/ft)

In the case of open channel flow this slope assumes that flow is at the normal depth and the slope of the HGL is parallel to the slope of the pipe. A discussion of gradually varied flow, where the HGL is not parallel to channel slope is beyond the scope of this text.

With a known slope of the HGL, it is possible to calculate upstream hydraulic grade with a known downstream hydraulic grade.

$$HG_{upstream} = HG_{downstream} + SL \tag{10.22}$$

where

HG = hydraulic grade
S = slope of the HGL
L = length of piping

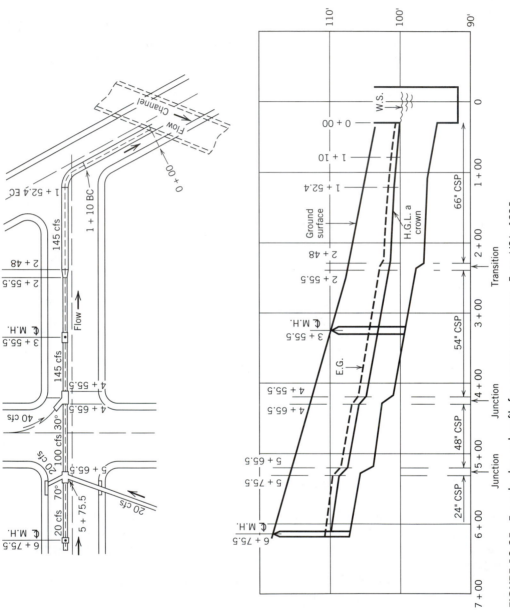

FIGURE 10.35 Example plan and profile for a storm sewer system. *From AISA, 1985.*

The term SL is also the headloss in the pipe section. Headloss at junctions, inlets, or manholes can be calculated using the equation;

$$h_L = K\frac{V^2}{2g} \tag{10.23}$$

where

h_L = headloss in feet
K = headloss coefficient dependent upon geometry
V = maximum velocity influent to junction (fps)
g = gravitational constant

By summing headlosses in an upstream direction the hydraulic grade at any point within the system can be calculated. A plot of these hydraulic grades for a sample system is shown in Figure 10.35.

10.9.3   Constraints for Storm Sewer Design

Storm sewer design follows a number of design constraints which must be met. Minimum velocities inside the storm sewer piping must usually be greater than 2 to 2.5 fps. This ensures that materials do not deposit in the piping. The velocity must also usually be less than 10 to 15 fps to prevent scouring. The pipe crown (or soffit), or the top of the pipe, must be (typically) 3 ft below the level of the ground to prevent crushing or collapsing of the pipe under loads. The distance between the pipe soffit and the ground elevation is referred to as *pipe cover*. Pipes must be sized large enough to prevent flooding. Surcharge conditions are often discouraged in design.

❏   **EXAMPLE PROBLEM 10.9**

A storm sewer system must be designed to convey storm water from three watersheds to a pond with a free discharge tailwater condition. The following watershed information applies:

	Watershed 1	Watershed 2	Watershed 3
Area	1.25 acres	0.67 acres	0.94 acres
C	0.78	0.75	0.85
TC	10 min	10 min	10 min
Ground Elev	98.7 ft	97.4 ft	94.5 ft

The inlet for watershed 1 is 300 ft from the inlet for watershed 3. The inlet for watershed 2 is 250 ft from the inlet for watershed 3. The inlet for watershed 3 is 450 ft upstream of the outlet pond. The IDF curve for the region for a 25-yr return period is (in equation form)

$$i = \frac{64}{(12 + D)^{0.8}} \tag{10.24}$$

where

i = intensity (in/hr)
D = duration (min)

The following constraints must be met in the design of the system

	Minimum	Maximum
Velocity	2.5 fps	15 fps
Cover	3 ft	10 ft

Assume that the pipe roughness (Manning's n) is 0.02 for all pipes. The ground elevation at the outlet is 93.1 ft.

Solution

It is advisable to draw a diagram of the system before starting the design process. A simple diagram of the system is shown in Figure 10.36. The flowrate of watershed 1 and watershed 2 can be calculated using $Q = CiA$ and the equation for rainfall intensity.

$$i = \frac{64}{(12 + 10)^{0.8}} = 5.4 \text{ in./hr}$$

$$Q = CiA, Q_{\text{watershed 1}} = 5.3 \text{ cfs}, Q_{\text{watershed 2}} = 2.7 \text{ cfs}$$

Using the calculated flow rates and knowing the ground elevations at inlet 1, inlet 2, and inlet 3 we can try pipe sizes and invert elevations for the piping from inlet 1 to inlet 3 and inlet 2 to inlet 3. Use the pipe calculator available in the Routines option of the OPSEW program to find the best pipe size and slope. This pipe size should minimize cover while still meeting the cover constraints. The slope of the pipe can match the slope of the ground; therefore the slopes can be calculated

$$S_{1 \rightarrow 3} = \frac{98.7 - 94.5}{300} = 0.014$$

$$S_{2 \rightarrow 3} = \frac{97.4 - 94.5}{250} = 0.0116$$

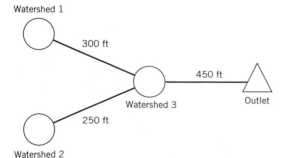

FIGURE 10.36 Example nodal diagram for a storm sewer system.

An 18-in. pipe from inlet 1 to inlet 3 has a normal depth of 0.88 ft (10.6 in.) and flow velocity of 4.89 fps at this depth. A 15-in. pipe from inlet 2 to inlet 3 will have a normal depth of 0.69 ft (8.3 in.) and a flow velocity of 3.86 fps. The pipe invert should be placed so that the cover constraint is met.

$$\text{Upstream invert pipe 1} = 98.7 - 3 - 1.5 \,(18 \,\text{in.}) = 94.2 \,\text{ft}$$

$$\text{Downstream invert pipe 1} = 94.5 - 3 - 1.5 \,(18 \,\text{in.}) = 90.0 \,\text{ft}$$

$$\text{Upstream invert pipe 2} = 97.4 - 3 - 1.25 \,(15 \,\text{in.}) = 93.15 \,\text{ft}$$

$$\text{Downstream invert pipe 2} = 94.5 - 3 - 1.25 \,(15 \,\text{in.}) = 90.25 \,\text{ft}$$

Time of flow in each pipe must be calculated—knowing the pipe length and the velocity we can determine this flowrate.

$$\text{Time}_{\text{Pipe 1}} = 300 \,\text{ft}/4.91 \,\text{fps} = 61.1 \,\text{sec} = 1.02 \,\text{min}$$

$$\text{Time}_{\text{Pipe 2}} = 250 \,\text{ft}/3.86 \,\text{fps} = 64.8 \,\text{sec} = 1.08 \,\text{min}$$

The time of concentration to use for inlet 3 is therefore 10 min. + 1.08 min. or 11.08 min. This duration gives an intensity of 5.2 in./hr and a flow rate of 4.2 cfs. The total flow rate through pipe 3 will be the sum of the inlet flow rates and the influent flow rate.

$$Q_{\text{pipe 3}} = 4.2 \,\text{cfs} + 5.3 \,\text{cfs} + 2.7 \,\text{cfs} = 12.2 \,\text{cfs}$$

If the slope for pipe 3 follows the ground slope then $S_{\text{pipe 3}} = 0.0031$. A 30-in. pipe at this slope and flow rate will have a normal depth of 1.72 ft and a velocity of 3.38 fps. The pipe inverts should be placed so that cover constraints are met.

$$\text{Upstream invert pipe 3} = 94.5 - 3 - 2.5 \,(30 \,\text{in.}) = 89.0 \,\text{ft}$$

$$\text{Downstream invert pipe 3} = 93.1 - 3 - 2.5 \,(30 \,\text{in.}) = 87.6 \,\text{ft}$$

Assuming there is not a backwater condition at the outlet and that flow will exist at the normal depth with no backwater or drawdown, the hydraulic grade in the system can be calculated. Junction losses at inlet 3 can be estimated using the junction headloss equation. The maximum influent velocity is 4.89 fps, using an assumed $K = 0.5$, the headloss is $h_L = 0.5 \,(4.89)^2/64.4 = 0.19$ ft.

$$HG_{\text{outlet}} = 87.6 + 1.72 = 89.32 \,\text{ft} \qquad HG_{\text{outflow inlet 3}} = 89.0 + 1.72 = 90.72 \,\text{ft}$$

$$HG_{\text{inflow inlet 3}} = 90.72 + 0.19 = 90.91 \,\text{ft}$$

$$HG_{\text{outflow pipe 2}} = \text{maximum of 90.91} \qquad \text{and} \qquad 90.25 + 0.69 = 90.94 \,\text{ft}, \,90.94 \,\text{ft}$$

$$HG_{\text{outflow pipe 1}} = \text{maximum of 90.91} \qquad \text{and} \qquad 90.0 + 0.88 \,\text{ft} = 90.88 \,\text{ft}, \,90.91 \,\text{ft}$$

The outflow of pipe 1 has a mild backwater condition. Since the backwater is not severe we will assume that the flow reaches normal depth before reaching the upstream invert.

$$HG \text{ inflow pipe 1} = 94.2 + 0.88 = 95.08 \,\text{ft}$$

$$HG \text{ inflow pipe 2} = 93.15 + 0.69 = 93.84 \,\text{ft}$$

This pipe system will meet all the required constraints.

10.10

WATER QUALITY

A major concern with stormwater runoff is the quality of the runoff water. This is a concern in areas where land use changes bring about a greater pollutant load on natural systems than the pre-existing condition; for example, a wooded area being developed into an industrial site. The water quantity changes due to the area changing from pervious to highly impervious can be a major concern. BMPs must be considered to reduce the pollution load to the receiving water body.   ❑

10.10.1   Pollutographs

Just as a hydrograph describes the flow rate of water over time for a storm event, a pollutograph describes the rate at which pollutants dissolved or suspended in the runoff are transported. The basis of the pollutograph model is a first-order reaction, i.e., the amount of pollutants transported is proportional to the amount of pollutants remaining on the watershed:

$$\frac{-dP}{dt} = kP \tag{10.25}$$

which can be integrated from an initial pollutant mass (Po) to give the equation

$$P = P_o e^{-kt} \tag{10.26}$$

and adjusted to give

$$P_o - P = P_o(1 - e^{-kt}) \tag{10.27}$$

where
P_o = initial loading (lb or kg)
P = mass remaining after time t (lb or kg)
k = transport rate constant ($1/t$)
t = time
$P_o - P$ = mass transported in time t

The pollutant mass can also be related to the rainfall excess R to give equation

$$\frac{P_o - P}{P_o} = 1 - e^{-cR} \tag{10.28}$$

where
c = transport rate coefficient (dependent upon land use and pollutant, $1/in.$)
R = cumulative rainfall excess at time t (in.)
$(P_o - P)/P_o$ = Fraction of pollutant transported at time t

TABLE 10.9 Transport Rate Coefficient c (per inch) for Six Different Pollutants and Four Different Locations with Land Use and Characteristics

POLLUTANT	LAND USE			
	RESIDENTIAL	HIGHWAY	COMMERCIAL	APARTMENT
Total nitrogen	2.83	2.24	2.63	2.07
Total phosphorus	2.64	2.30	2.74	1.66
Total carbon	2.40	2.62	2.44	1.92
Chemical O_2 demand	2.56	2.72	2.82	2.32
Suspended solids	2.61	2.04	2.74	2.97
Total lead	—	2.17	2.95	2.11
Average	2.61	2.35	2.72	2.18
Total area (acres)	40.8	58.3	20.4	14.7
EIA (acres)	2.4	10.5	20.0	6.5
t_c (min)	110.	13.0	7.0	4.0

Values for this coefficient can be developed based on land use and pollutant (Wanielista and Yousef, 1993) as shown in Table 10.9.

10.10.2 Loadographs

The cumulative fraction or mass of pollutant transported is referred to as a *loadograph* (see Figure 10.37). The volume of rainfall excess in units of volume (ft³, m³) can be

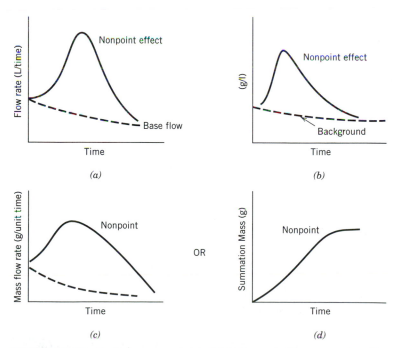

FIGURE 10.37 Typical response curves: (a) hydrograph; (b) concentration changes with time or pollutograph; (c) loading rate with time; (d) loadograph (cumulative mass).

calculated from the rainfall excess in inches by multiplying by the watershed area. The excess rate (in./hr, cfs) can be calculated by dividing the volume of excess by the time step (Δt). The pollutant mass load at any time step can be found as the difference between the mass load at that time step and the next time step. Using these calculations, the pollutant concentration at any time step (Δt) can be found.

$$C_t = \frac{P_{t+\Delta t} - P_t}{Q_{avg}\,\Delta t}$$ (10.29)

where

C_t = concentration of pollutant at time t
P_t = mass of pollutant transported at time t
$P_{t+\Delta t}$ = mass of pollutant transported at time $t + \Delta t$
Q_{avg} = average hydrograph runoff rate between time t and $t + \Delta t$
Δt = time step for hydrograph

The plot of pollutant concentration (C_t) versus time is known as a *pollutograph* (see Figure 10.36).

☐ EXAMPLE PROBLEM 10.10

A transport rate coefficient of 2.83 in.$^{-1}$ has been recorded for a residential area for total nitrogen (TN). If the total mass of total nitrogen on a 100-acre watershed is 45,000 lb, what is the rainfall excess at which 80% of this will be removed from the watershed? Graph the pollutograph for a 6-hr constant rainfall of 1 in./hr if the curve number (CN) for the site is 85 and the time of concentration is 60 min.

Solution

Using Equation 10.28, the rainfall excess for 80% removal can be estimated

$$\frac{P_o - P}{P} = 1 - e^{-cR}$$

$$0.8 = 1 - e^{-2.83R}$$

therefore $R = 0.57$ in.

To develop the pollutograph for this example, the rainfall excess must first be developed. For a homogeneous watershed with a curve number of 85 and a constant rainfall intensity, a spreadsheet can be developed to determine rainfall excess. For curve number 85, $S' = 1000/85 - 10 = 1.765$. A time step of 15 min is chosen, and the equation $R = (P - 0.2S')2/(P + 0.8S')$ for $P > 0.2S'$ is used to calculate the cumulative rainfall excess.   ☐

10.11
SUMMARY

This chapter presented material useful for both (1) reservoir design and operation and (2) stormwater pond volume and peak flow management. Stormwater management systems can be used to attenuate hydrographs. The degree of attenuation can be accomplished by both volume and discharge control. Additional information was presented for stormwater re-use, storm sewer design, and water quality considerations.

- Yield is the rate of water released from a reservoir.
- Mass diagrams for streamflow or runoff events are useful to estimate retention volumes (sizing of reservoirs) and yields.
- Stochastic yields, evapotranspiration, seepage, and streamflow can be used to evaluate reservoir sizes and yield.
- Detention pond outflow rates can be adjusted using weirs, orifices, and culverts.
- Design storm frequency is usually specified based on an economic analysis. The largest size of detention pond is determined using this frequency and a number of storms of different duration.
- French drains remove water from the ground and are used in high water table areas and beneath stormwater ponds to draw down the ponded water.
- Off-line retention ponds can be used to divert a fraction of yearly runoff water for treatment.
- Swales can be designed to infiltrate a volume of runoff water.
- Stormwater re-use systems design alternatives are outlined.
- Applications of principles of hydraulics and hydrology to the design of storm sewers are presented.
- Additional water quality considerations are presented leading to pollutograph and loadograph studies.
- Best management practices (BMPs) are those stormwater control systems used to reduce pollution.

10.12

PROBLEMS

10.12.1 Hand Problems

1. Explain how a mass diagram is developed from streamflow data assuming that only one year of streamflow data are available. Discuss and illustrate your answer on a graph indicating how the size of a reservoir and yield can be determined.
2. Explain why you would use a stochastic model for determining yield for a specific size reservoir.
3. What are the precipitation increments in 30-min intervals and the cumulative precipitation resulting from a 4.2-in. storm using the NRCS Rainfall Distribution Type II? Assume a 6-hr-duration storm.
4. What shape does the rainfall excess hydrograph have (draw cumulative rainfall excess graphs) for the rainfall distribution of Problem 3 if the CN were 60 and 80.
5. The following data were gathered for a 100-min storm on a 2-ac watershed. Assume that infiltration capacity is reached at 100 min and that unused infiltration capacity is available for future rainfall storage at a specific time. Estimate the rainfall excess (in inches and in ac-ft) at the end of 100 min.

TIME (min)	HYETOGRAPH PRECIPITATION INTENSITY (in./hr)	INFILTRATION CAPACITY (in./hr)
0	0.0	0.0
5	0.1	1.2
10	0.3	1.0
15	0.5	1.0
20	0.6	0.9
25	0.9	0.8
30	1.2	0.8
35	1.4	0.6
40	1.6	0.5
45	1.3	0.5
50	1.2	0.4
60	0.8	0.4
70	0.7	0.3
80	0.5	0.3
90	0.2	0.3
100	0.1	0.3

6. Solve Problem 5, assuming that unused infiltration capacity at one time is not available for rainfall excess storage in the future.

7. For the flow rate data of Table 7.6, select a station and plot the monthly flow rates. Estimate the total flow volume in cubic meters and cubic feet. Also draw a mass curve for the year. Compare your answer to the annual volume using the annual mean of Table 7.6.

8. For the following empirical probability distribution on streamflow, generate two years of data assuming that the flow rates are independent.

FLOW RATES/MONTH (AVERAGE cfs)	FREQUENCY (FRACTION)
20	.05
30	.15
40	.25
50	.20
60	.15
70	.10
80	.05
90	.03
100	.02

9. Use the generated streamflow data of Problem 8 to calculate the maximum size of a reservoir if the release rate is at 40 cfs or the previous period storage volume, whichever is the least amount. The net loss from evaporation and gain from

infiltration and direct precipitation is zero. The starting volume is 50 cfs. Also, what is the maximum release rate if the maximum size of the reservoir is 150 cfs?

10. A vertical wall reservoir tank outside of Miami, Florida, has a very dense clay lining, no vegetation, and three intermittent flowing canals that lead into it. During one winter, there is no rainfall recorded in the month of January. The depth of the reservoir at the beginning of January is 22 ft. If the surface area of the reservoir is 5 ac and the daily demand is 500,000 gal/day, what is the depth of the reservoir (using data from this text) on February 1? Be sure to consider all inputs and outputs. Also, comment on how the management of this reservoir might be improved.

11. Using the annual mean volume discharge from any station in Table 7.6, estimate the rainfall excess per year for the catchment. Present your answer in units of centimeters and inches.

12. Estimate the surface water storage at the end of one year for a watershed area, using the U.S. Geological Survey Water-Data Report and U.S. Department of Commerce Climatological Data. Use monthly increments. What assumptions must be made to estimate the end-of-year storage? Example water quantity and precipitation data are attached at the end of these questions (Tables 10.10 and 10.11 can be used).

13. Develop a mass curve for the following stream flow volume data. Next, determine the size of reservoir necessary to ensure a water yield of 4000 m^3/month over the 2-yr period. Discuss your assumptions used to solve the problem.

MONTH/YEAR	1000 m^3	MONTH/YEAR	1000 m^3
Jan. 1985	5.0	Jan. 1986	3.8
Feb. 1985	7.4	Feb. 1986	9.4
March 1985	8.3	March 1986	6.8
April 1985	3.6	April 1986	0.2
May 1985	2.1	May 1986	3.1
June 1985	1.6	June 1986	0.4
July 1985	0.4	July 1986	1.1
Aug. 1985	2.8	Aug. 1986	8.4
Sept. 1985	6.9	Sept. 1986	8.9
Oct. 1985	12.4	Oct. 1986	6.4
Nov. 1985	3.4	Nov. 1986	1.7
Dec. 1985	0.3	Dec. 1986	3.6

14. For the months of May through September of Problem 13, assume evaporation losses from the reservoir of 400 m^3/mon. How does this affect your yield?

15. Plot the flow rates for 2 mon from any streamflow records or using Table 10.10, plot the time period from January 11 through March 10, 1973. Estimate the volume flow rate under this streamflow and then report the answer in units of cubic feet, cubic meters, and inches over the entire watershed area of 180 mi^2.

TABLE 10.10 Watershed Data for Shingle Creek at Campbell, FL (U.S. Geological Survey, 1973)

DISCHARGE (ft³/sec), WATER YEAR OCTOBER 1972 TO SEPTEMBER 1973

DAY	OCT	NOV	DEC	JAN	FEB	MAR	APR	MAY	JUNE	JULY	AUG	SEP
1	53	39	71	76	298	115	88	41	36	97	293	319
2	63	35	65	72	281	105	101	41	36	128	366	384
3	78	32	65	73	359	94	101	30	36	75	332	389
4	84	42	62	69	297	94	121	41	33	134	301	369
5	71	73	62	69	256	89	124	41	30	237	279	386
6	70	66	62	65	233	78	108	40	30	217	278	401
7	60	49	65	65	214	84	85	37	36	281	305	403
8	60	49	59	54	196	73	136	33	33	220	332	396
9	60	42	59	75	189	79	120	49	30	168	330	406
10	66	39	55	67	445	80	111	53	33	151	303	417
11	73	39	52	80	352	80	90	41	30	127	281	414
12	73	39	55	189	291	73	81	37	30	119	274	428
13	63	39	52	161	270	72	75	33	27	139	260	440
14	56	43	41	130	274	60	83	36	27	128	251	404
15	56	49	41	112	286	60	82	32	30	125	241	391
16	53	42	63	108	274	54	81	38	36	103	208	387
17	53	45	49	112	252	58	72	31	30	91	184	354
18	53	42	50	121	239	62	62	27	24	82	164	306
19	53	39	44	121	246	68	65	33	33	115	149	284
20	65	42	44	118	214	62	64	32	36	118	150	264
21	63	49	37	115	193	50	55	32	47	88	157	240
22	53	45	106	154	178	53	51	31	97	82	183	240
23	49	45	97	397	163	54	50	28	131	100	381	231
24	49	42	86	476	157	59	38	28	144	91	375	219

25	46	43	86	412	147	49	34	37	97	88	313	220
26	46	64	82	337	142	90	38	31	81	85	285	415
27	35	61	82	360	136	79	27	25	70	97	280	890
28	25	57	83	436	126	73	41	22	59	112	276	291
29	28	68	79	501	—	78	41	24	51	112	271	681
30	32	69	83	426	—	73	37	33	70	117	263	624
31	35	—	80	347	—	78	—	36	—	149	277	—
Total	1,724	1,428	2,017	5,898	6,713	2,266	2,262	1,073	1,483	3,977	8,342	12,093
Mean	55.6	47.6	65.1	190	240	73.1	75.4	34.6	49.4	128	269	403
Max	84	73	106	501	445	115	136	53	144	281	381	890
Min	25	32	37	54	126	49	27	22	24	75	149	219
ac-ft	3,420	2,830	4,000	11,700	13,320	4,490	4,490	2,130	2,940	7,890	16,550	23,990

Cal yr	1972	Total 35,641	Mean 97.4	Max 472	Min 18	ac-ft 70,690
Wtr yr	1973	Total 49,276	Mean 135	Max 490	Min 22	ac-ft 97,740

Location—Lat 28°16'01'', long 81° 26'53'', in SE quarter sec. 31, T.25 S., R.29 E., Osceola County, near left bank at downstream side of bridge on country road, 100 ft (30 m) downstream from Atlantic Coast Line Railroad Bridge, 0.8 mi (1.3 km) northeast of Campbell, and 2.5 (4.0 km) upstream from Lake Tohopekaliga.

Drainage area—180 mi^2 (466 km^2) approximately; includes part of watershed in Reedy Creek Swamp.

Period of record—October 1968 to 1973.

Gage—Water-stage recorder. Datum of gage is at mean sea level. Water-stage recorder for Lake Tohopekaliga used as auxiliary gauge for this station.

Average discharge—5 yr, 109 cfs (3.087 m^3/s), 78,970 ac-ft/yr (97.4 hm^3/yr).

Extremes—Current year: Maximum discharge, 906 cfs (25.7 m^3/s), September 27 (gage height, 58.10 ft or 17.709 m); minimum daily discharge. 22 cfs (0.62 m^3/s) May 28; minimum gage height, 51.85 ft (15.804 m) June 15.

Period of record: Maximum discharge, 1,510 cfs (42.8 m^3/s) October 5, 1969 (gage height, 59.72 ft or 18.203 m); minimum daily discharge, 31 cfs (0.088 m^3/s) May 9, 10, 1971; minimum gage height, 49.78 ft (15.173 m) May 10, 1971.

Remarks—Records fair. Considerable flow diverted from Reedy Creek Swamp into Shingle Creek above station.

TABLE 10.11 Monthly Precipitation Records (U.S. Department of Commerce) Total Precipitation (in.)

YEAR	JAN	FEB	MAR	APR	MAY	JUNE	JULY	AUG	SEPT	OCT	NOV	DEC	ANNUAL
1934[a]	1.04	3.37	4.33	4.58	8.08	13.35	9.00	1.27	3.14	1.50	0.09	0.55	50.30
1935	1.37	2.79	0.70	2.26	2.42	2.47	10.13	7.61	9.79	4.07	0.85	4.81	49.27
1936[a]	4.11	6.29	2.90	1.58	3.58	11.28	2.63	4.95	5.81	3.07	2.21	1.77	52.18
1937	0.97	5.00	2.97	3.78	4.47	5.22	5.14	13.14	9.37	4.55	3.67	0.82	59.10
1938	0.73	0.81	1.74	0.34	6.30	4.49	4.81	4.36	5.30	3.88	1.49	0.30	34.55
1939	1.21	0.35	1.75	4.97	4.87	15.64	6.34	8.90	5.24	1.67	0.39	1.07	52.42
1940	2.14	2.89	4.23	4.44	1.72	6.67	10.14	8.04	7.35	0.37	0.22	5.81	54.02
1941	4.69	4.16	2.47	5.53	2.73	8.18	9.44	6.45	4.76	5.33	3.61	2.29	52.65
1942	2.32	3.03	5.83	2.32	1.17	10.57	2.01	6.71	4.17	0.24	0.12	2.80	41.29
1943	1.19	0.50	3.92	1.53	5.42	3.66	5.17	5.85	7.18	3.04	0.87	1.28	39.61
1944	2.14	0.10	3.69	4.07	2.83	6.43	11.04	5.39	4.52	8.53	0.11	T	48.84
1945	3.86	0.11	0.54	1.47	2.93	13.70	7.06	5.28	15.87	1.61	1.00	2.52	55.95
1946	2.24	2.96	1.15	0.81	4.24	7.78	8.56	10.06	7.75	3.32	0.97	0.28	50.13
1947	0.87	4.78	5.55	4.98	2.81	11.61	13.90	6.71	8.87	4.83	1.90	0.66	67.47
1948	6.44	1.84	4.05	1.08	0.97	1.97	8.76	12.30	10.81	2.55	0.45	1.31	52.53
1949	0.31	0.47	0.29	3.02	2.54	7.97	6.05	8.83	8.25	1.51	1.22	3.82	44.28
1950	0.15	0.48	3.44	4.82	2.93	5.55	8.27	3.48	7.93	14.51	0.09	4.30	55.95
1951	0.52	2.28	0.96	5.99	1.40	5.08	14.51	7.84	9.34	3.08	4.86	2.06	57.92
1952	0.70	5.47	6.67	2.88	2.45	2.32	4.43	6.51	4.94	3.69	0.74	0.65	41.45
1953	2.86	2.89	3.03	6.18	1.87	6.28	6.85	15.19	8.84	3.50	4.78	3.58	65.86
1954	0.45	1.16	0.99	4.44	3.55	5.81	13.64	4.39	3.99	5.07	2.68	1.80	47.97
1955	2.00	1.12	1.59	1.36	3.13	4.73	6.88	6.65	6.97	4.10	2.17	1.56	42.26

Year													
1956	1.66	0.90	0.16	4.03	3.70	5.41	3.88	6.10	6.27	8.24	1.26	0.30	43.91
1957	0.91	1.93	3.76	4.74	8.58	4.39	4.35	9.45	7.47	1.68	0.82	2.85	50.93
1958	4.49	2.83	6.16	3.79	2.68	3.83	9.93	3.40	1.65	7.27	2.48	2.69	51.20
1959	2.78	4.55	7.69	4.91	4.44	7.95	8.02	6.77	8.33	5.97	0.99	1.37	63.77
1960[a]	1.49	5.64	10.54	2.55	0.50	9.50	19.57	3.20	11.21	3.17	0.30	1.07	68.74
1961	1.75	2.82	2.21	0.28	0.43	8.08	9.93	6.99	4.84	2.87	0.92	0.66	41.78
1962	1.11	2.08	3.55	1.58	2.74	3.11	12.77	5.11	12.24	1.90	2.46	1.70	50.35
1963	3.17	4.76	2.69	1.23	3.56	6.67	3.83	3.54	6.72	0.46	6.39	2.26	45.28
1964	6.18	3.4	4.65	2.14	2.74	6.11	6.68	9.00	9.47	1.64	0.45	1.91	54.39
1965	1.79	3.67	3.02	0.66	0.52	7.36	11.55	5.49	5.99	4.06	1.06	2.23	47.40
1966	4.45	6.31	2.57	1.92	6.57	9.77	6.73	7.76	6.25	1.98	0.09	0.99	55.39
1967	0.84	5.49	1.31	0.28	1.69	11.16	4.63	6.83	5.84	0.35	0.03	2.42	40.91
1968	0.65	2.76	2.27	0.30	3.72	18.26	5.60	3.44	5.91	5.47	2.82	0.88	52.10
1969	2.22	3.30	5.52	2.38	1.40	5.04	6.73	7.17	6.44	9.45	0.87	4.66	55.18
1970	4.05	6.77	3.66	0.45	4.08	4.92	5.97	5.91	3.25	2.60	0.24	2.05	43.96
1971	0.45	2.98	1.46	1.52	4.31	4.39	8.29	7.51	2.98	3.06	1.21	1.93	40.09
1972	0.99	4.96	5.06	1.39	3.76	6.33	3.98	16.11	0.43	2.34	4.11	1.89	51.35
1973	4.82	2.73	4.13	2.82	4.74	6.63	6.24	7.33	11.53	1.10	0.74	2.56	55.37
Record Mean[b]	2.18	2.97	3.43	2.57	3.14	6.83	8.25	7.08	7.17	3.97	1.58	1.88	51.05

[a]Indicates a break in the data sequence during the year or season due to a station move or relocation of instruments. See station location table.
[b]Record mean values above (not adjusted for instrument location changes listed in the station location table) are means for the period beginning in 1943.

16. What are the rainfall excess increments in 30-min intervals and the inflow rates (cfs) for a 100-ac watershed area for 6 hr from a 5.44 in storm over 6-hr using the NRCS/CN procedure if the maximum soil storage (S') is 0.5 in.? Now route this rainfall excess through the new channel of Example Problem 8.6.

TIME	P (in.)	INFLOW (in.)	INFLOW (cfs)	OUTFLOW (cfs)
0	0.00			
30	0.17			
60	0.20			
90	0.26			
120	1.89			
150	1.34			
180	0.34			
210	0.29			
240	0.30			
270	0.22			
300	0.15			
330	0.18			
360	0.10			

What is the "routed" flow (output hydrograph) at 120 min if $t_c = \Delta t = 30$ min? No other information is available.

17. Using the Weather Bureau hyetograph dimensionless mass curve and the NRCS Type II, which curve would produce maximum peak discharge for a highly impervious urban area with zero initial abstraction? Describe and show some calculations.

18. Determine the depth of a french drain (saturated depth) if the width is 3 ft, the effective permeability of the aggregate is 2000 ft/day, and the rate of discharge for 100 ft of trench is 1000 ft³/day. The trench slope along the flow line is 0.001. Discuss your answer.

19. A detention pond design for your local area is needed to store the runoff from a 1-hr design storm with a 3-yr return period. The watershed area is 100 ac with a composite $CN = 80$. If you do not have IDF curves for your area, use Orange County curve in Appendix C.

    a. Size the pond (acres and acre-feet) if the pond is 3 ft deep and assume a rectangular pond.

    b. How long (to the nearest hour) will it take to drain 90% of the pond water if an infiltration test yielded the following?

$$\text{Initial infiltration} = 5 \text{ in./hr} = f_0$$

$$\text{Final infiltration} = 1 \text{ in./hr} = f_c$$

$$\text{Recession constant} = 2/\text{hr} = K$$

20. a. A 20-ac residential area has a runoff coefficient of 0.50. Size an off-line retention pond with length = width with water depth of 4 ft and side slopes of 1 on 1 to accommodate the runoff from 1 in. of rainfall. Specify the surface area of the pond and the bottom area, along with the total volume.

b. Add another 1-ft depth of pond as freeboard. Also, if a 10-ft-wide buffer and maintenance area is needed around the pond, what is the total acreage of the off-line retention area?

21. A pre- versus postdevelopment hydrograph analysis for a 100-ac watershed must be completed. The precondition watershed has a runoff coefficient of 0.2 and a hydrograph shape with a peak attenuation factor of 0.31 with t_c = 120 min. The postcondition is most likely to have a hydrograph shape similar to the standard NPCS–peak attenuation factor, a runoff coefficient of 0.4, a time of concentration of 1 hr, and a rainfall intensity of 5 in./hr. Draw both hydrographs and estimate the peak discharge. Next, estimate the storage volume that approximates the condition of pre- versus postpeak discharge. Express storage in terms of cubic feet.

22. a. Determine a length of swale with Bermuda grass and an infiltration rate of 4 in./hr. It is scheduled to be a triangular section with one on six side slopes and a longitudinal slope of 0.005. The swale must effectively percolate all the runoff that comes from a half-acre impervious area and a constant rainfall intensity of 0.5 in./hr.

b. If only 400 ft of swale are available and the runoff hydrograph lasts for 200 min, how much additional storage is necessary?

23. A 20-ac subdivision has as a design rainfall intensity specification of 3.7 in./hr. The area is 38% impervious with all the impervious areas being directly connected. The soil type defined by the NRCS is A. The pervious area does not contribute runoff. The time of concentration for the watershed is 60 min. One may wish to substitute an intensity from an IDF curve (25-yr return frequency) from your own area.

a. What is the peak discharge using the rational formula?

b. What is the rainfall excess in liters?

c. What is the approximate lot size?

24. A corrugated metal pipe under a roadway passes 30 cfs at design conditions. The downstream pond at design invert elevation is 250.5 ft. What is the elevation of the upstream pond if the connecting 3-ft-diameter pipe is 100 ft long with a Manning's coefficient of 0.015 and a tail water depth of 6 ft? There is no headwall on the corrugated metal pipe.

25. Obtain a year of flow rate data or use the data of Table 10.10 to generate a serial transition matrix for daily data. Be careful to identify at least three different flow periods and at least five flow states for each flow period. It is frequently helpful to plot the data first. Next, generate a hydrograph with one peak discharge.

10.12.2 Computer Problems

1. An area of the county requires use of the 1- in 1-yr storm frequency for detention basin design (use appropriate IDF curves). What duration (1 or 6 hr) produces the largest detention basin if the watershed area is 50 ac, the rational coefficient is 0.8, the output hydrograph starts at 1 hr at a constant 10 ft^3/sec, and the time of concentration is 1 hr? Compare only the two duration storms and use the rational formula.

2. For a 25-yr, 2-hr storm from any IDF curve, develop the incremental storm volumes (10-min increments) according to a dimensionless cumulative mass rainfall curve. For a 270.6-ac watershed that is 75% impervious, 100% directly connected impervious area, has a curve number of 80 for pervious area and time of concentration of 55 min, develop two hydrographs using the NRCS hydrograph procedure with $K = 484$ and $K = 300$. Comment on the differences.

3. Change the soil type in Problem 2 to a curve number of 53.5 and comment on the hydrograph relative to Problem 2.

4. Assume another completely different dimensionless cumulative rainfall distribution and compare all the hydrographs for the assumptions of Problems 2 and 3.

5. Using the SMADA program and the NRCS unit–time unit–hydrograph procedure, develop the resulting hydrograph using the watershed data of Problem 2, a hydrograph attenuation factor of 250, and a constant rainfall of 2 in. over 4 hr. Discuss input assumptions, use $t = 10$ min.

6. Increase the attenuation factor of Problem 5 to 350 with the same input assumptions. Now change the hyetograph to 0.25 in. for the first hour, 0.50 in. for the second hour, 1 in. for the third hour, and 0.25 in. for the fourth hour. How does the resulting hydrograph plot compare to the others? Discuss peak and shape.

7. Develop two instantaneous hydrographs for the watershed of Problem 2. Use the hyetograph of your choice. The percentage of the impervious area that is directly connected is 60 and should be used in the analysis. One hydrograph is estimated without any stormwater management. The second hydrograph is estimated considering 1 in. of initial abstraction over the total area (stormwater management). Compare resulting hydrograph shapes and comment on your results.

8. Increase the total rainfall volume of Problem 2 by 1.5 in. and compare answers.

9. Using a 24-hr storm, 100-yr return period and the watershed description of Problem 2 except a directly connected area of 80% and a time of concentration of 5 hr, develop a hydrograph at 1-hr intervals using the NRCS Type II rainfall distribution. State all assumptions.

10. Using the working curve 2 of Figure 10.23, develop a best fit equation using 10 data points from the graph. Also develop two linear equations to "best" fit the line.

10.12.3 Case Studies

1. Complete the detailed calculations of Example 10.10 and plot the hydrograph, loadograph, and pollutograph.

2. Determine the quantities and peak flow rates that must be accommodated in a small sub-division stormwater design. The sub-division is composed of three irregular "blocks" with the following characteristics:

Block	Acres	t_c min	W. S. coeff.	Travel Time (min.)	From to
1	8.0	10	.35	10	(1 to 2)
2	5.6	8	.40	6	(2 to 3)
3	6.5	9	.30	11	(3 to out.)

Use the rational method and assume the local design code requires a minimum time of concentration of 15 min. Compare the results using individual subareas with one that allows using a weighted runoff coefficient and rainfall intensity based on the longest time of concentration.

3. What reservoir storage capacity would be required to produce a yield 600 acre-ft/month for a site where the monthly inflows during a critical low-flow period and the monthly rainfall and pan evaporation rates are as shown.

Month	Flows cfs	Evaporation (in.)	Rain (in.)
J	29	2.2	3.5
F	36	2.9	5.3
M	20	4.3	4.8
A	6	5.2	6.0
M	3	5.7	3.0
J	4	5.3	2.8
J	6	5.3	1.7
A	3	4.9	0.3
S	7	4.4	0.2
O	10	3.8	0.5
N	17	2.7	2.0
D	12	2.1	3.5

Assume that the average reservoir area is 640 acres, the runoff coefficient for the land that will become part of the reservoir is 0.30, and the owner will be required to release a minimum of 3.0 cfs at all times.

4. A nearby chicken farm proposes to spread chicken manure on adjacent land. A typical 2.0-in. rainfall washes 40% of the weekly accumulation of wastes into a nearby stream. The farm produces 1 million chickens per week, 52 weeks per year. It takes 8 weeks to raise chickens to an average weight (mass) of 1 kg/bird. Manure is produced at the rate of 53 g/kg/day/bird and phosphorus is produced at a rate of 0.40 g/kg/day/bird.
   a. What is the transport rate coefficient, c?
   b. What is the average daily production of manure in kg?
   c. What is the average daily production of P in kg?
   d. If this production is spread over nearby pasture land and the average annual rainfall is 48 in. with a watershed coefficient of 0.5, what land area in acres is required to assure that the resulting concentration of P in the runoff is 0.13 mg/L or less?

10.13

REFERENCES

American Society of Civil Engineers and Water Environment Federation. 1970. "Design and Construction of Sanitary and Storm Sewers," New York. *Manual of Engineering Practice No. 37,*

American Society of Civil Engineers and Water Environment Federation. 1992. "Design and Construction of Urban Stormwater Management Systems," New York. *Manual of Engineering Practice No. 77,*

American Iron and Steel Institute. 1980. "Modern Sewer Design," Washington, DC.

Carroll, Robert G., Jr. 1983. "Geotextile Filter Criteria," prepared for the *1983 Symposium on Geotextiles,* Transportation Research Board, U.S. Federal Highway Administration, Washington, DC, January.

Cedergren, H. R., O'Brien, K. H., and Arman, J. A. 1972. "Guidelines for the Design of Subsurface Drainage Systems for Highway Structural Sections," Report No. FHWA-RD-72-30, prepared for the Federal Highway Administration, June, p. 25.

Chow, V. T. 1959. *Open Channel Hydraulics,* McGraw-Hill Publishing Company, New York.

Federal Highway Administration (FHWA). 1985. *Hydraulic Design of Highway Culverts, Hydraulic Design Series No. 5,* McLean, VA.

French, R. H. 1985. *Open Channel Hydraulics,* McGraw-Hill, New York.

Golding, B. L. 1974. "Master Drainage Plan Florida Center," Reynolds, Smith and Hills, Inc., Jacksonville, FL, Vol. 1.

Hardison, C. H. 1966. "Storage to Augment Low Flow," *Proceedings of Reservoir Yield Symposium,* Water Research Association, Wallingford, England.

Lof, G. O. and Hardison, C. H. 1966. *Storage Requirements for Water in the United States,* Resources for the Future. Washington, DC.

Lord, B. N. 1986. "Effectiveness of Erosion Control," *Urban Runoff Technology,* Engineering Foundation Conference, New England College, Henniker, NH, June, p. 281–290.

Pitt, R. 1986. "The Incorporation of Urban Source Area Controls in Wisconsin's Priority Watershed Projects," *Urban Runoff Technology,* Engineering Foundation Conference, New England College, Henniker, NH, June, p. 290–313.

Poertner, H. H. 1974. "Practices in Detention of Urban Stormwater Runoff," *Special Report No. 43,* American Public Works Association, Chicago.

State of Florida. 1988. *Administrative Code, Chapter 17.25* Tallahassee, FL.

U.S. Geological Survey. 1973. *Water Resources Data for Florida,* Water Resources Division. Tallahassee, FL, p. 63.

U.S. Weather Bureau. 1961. "Rainfall Frequency Atlas of the United States," T.P. No. 40, U.S. Department of Commerce, Washington, DC (May).

Wanielista, M. P. 1983. *Stormwater Management: Quantity and Quality.* Ann Arbor Sciences Publishers, Ann Arbor, MI, pp. 245–252.

Wanielista, M. P., Yousef, Y. A., and Avellaneda, E. 1988. *Alternatives For The Treatment of Groundwater Contaminants: Infiltration Capacity of Roadside Swales,* Florida Department of Transportation, FL-ER-38-88, Tallahassee, Florida, April.

Yousef, Y. A., Wanielista, M. P., Harper, H. H., Pearce, D. B., and Tolbert, R. D. 1985. *Removal of Highway Contaminants by Roadside Swales,* Florida Department of Transportation, FL-ER-30-85, Tallahassee, Florida, July.

State of Florida Department of Transportation. 1986. *Drainage Manual,* Tallahassee, FL.

Wanielista, M. P. 1977. Off-line Retention Pond Design In: *Proceedings of Stormwater Retention/Detention Basin Seminar,* Y. A. Yousef, Ed., University of Central Florida, Orlando, FL (Italy) pg 48–71.

NOTATION

A	Watershed area	BOD	Biochemical oxygen demand
A	Cross-sectional area	b	Limiting value
A	Soil hydrologic group	b	Constant
A	Gross erosion or sediment production	b	Gumbel parameter
		b	Regression coefficient
A	Computed soil loss due to erosion	b	Stage-storage coefficient
		b	Bottom width of channel
Ac	Acres	C	Crop management factor
A_i	Area between isochrones	C	Empirical coefficient
A_x	Cross-section area	C	Constant shape factor
AC	Average cost	C	Pollutant concentration
ADP	Antecedent dry period	C	Soil hydrologic group
ADT	Average daily traffic	C'	Constituent concentration
ALOSS	Potential rain loss rate	C	Concentration
AMC	Antecedent moisture condition	C	Runoff coefficient
		C	Cropping management factor in universal soil loss equation
A_o	Area of orifice		
A_s	Water surface area	C_1	Constituent input to a system
$\overline{A}_s$	Average surface area	C_2	Constituent output from a system
A_u	Drainage area of u stream order		
		C_d	Discharge coefficient
a	Estimate of skewness	C_i	Runoff coefficient— impervious surface
a	Regression constant		
a	Stage-discharge coefficient	CA	Contributing area
BI	Boundary input	CCN	Composite curve number
BO	Boundary output	C_N	Rectangular weir coefficient
B	Soil hydrologic group	CN	Rainfall excess curve number
B	Width of Parshall flume throat	C_o	Initial organic concentration
		C_e	Equilibrium concentration
B	Width of weir	C_P	Runoff coefficient—pervious surface
B	Water surface width		
B	Subsurface flow	C_v	Coefficient of variation

C_e	Storage coefficient	e_a	Vapor pressure of air
C_Z	Chezy roughness coefficient	e_s	Vapor pressure of snowpack
C_{60}	60° triangular weir coefficient	e_0	Saturation vapor pressure
C_{90}	90° triangular weir coefficient	e	Symbol for exponential of
$\overline{C}$	Composite concentration	F	Accumulated mass infiltration
CRF	Capital recovery factor	F	Force
CUML	Cumulative rain loss	F	Frequency factor
c	Wave celerity	F	Flood event
c_s	Snowpack specific heat	F_r	Froude number
c_1	Storage coefficient	FW	Future worth
c_2	Storage coefficient	$F(t)$	Infiltration volume as a function of time
D	Discharge		
D	Duration of rainfall	$F(x)$	Cumulation distribution function
D	Soil hydrologic group		
D	Dustfall	$F(x, y)$	Cumulative joint distribution function
D	Depth of pond or water		
D'	Incremental rainfall excess duration	f	Infiltration rate
		f	Relative humidity
DA	Drainage area	f	Function symbol
DCIA	Directly connected impervious area	f	Pipe friction factor
		f_c	Ultimate infiltration rate
DD	Dust and dirt accumulation	f_0	Initial infiltration rate
DI	Diversion volume (inches)	f(t)	Infiltration rate as a function of time
DO	Dissolved oxygen		
DV	Diversion volume (Ac-Ft)	$f(x)$	Probability density function
D_W	Dustfall on watershed	f'	First derivative of function
d	Depth of flow	G	Population skew coefficient
d	Depth of gutter flow	$G(x)$	Exceedance probability
d'	Fraction directly connected impervious area	G_s	Specific gravity
		g	Acceleration constant of gravity
d_s	Depth of snowpack		
d_s	Particle diameter	g	Skew coefficient
E	Evaporation	g	Grams
E_S	Depth of evaporation from snow	ΔH	Energy change
		H	Reservoir depth above spillway
E	Washoff decay coefficient		
E_P	Pan evaporation	H	Head
E_L	Lake evaporation	HG	Hydraulic grade
E	Saturated vapor pressure	HGL	Hydraulic grade line
EFF	Effectiveness	H_{exit}	Exit loss coefficient
EIA	Effective impervious area	h	Head
ET	Evapotranspiration	h	Curb inlet height
EMC	Event mean concentration	h	Depth of unconfined aquifer
EMV	Expected monetary value	I	Identity matrix
ERAIN	Watershed coefficient	I	Inflow
EVAP	Long-term evaporation	I	Inflow rate
E_0	Relative flow fraction	I_A	Initial rainfall abstraction
e	Vapor pressure	I_0	Initial infiltration rate

I_c	Ultimate or constant infiltration rate	L	Channel length
IDF	Intensity-Duration-Frequency Curve	L	Length of gutter
		L	Slope length factor
IR	Instantaneous runoff hydrograph	L_f	Latent heat of fusion
		L_p	Depth of percolating water
IUH	Instantaneous unit hydrograph	LR	Pollutant loading rate
		L_T	Length of slotted pipe
i	Rainfall intensity	L_u	Depth of unsaturated soil column
i	Interest rate	M	Mass
$i(t-\tau)$	Rainfall intensity at time $(t-\tau)$	M	Carbon weight
		MC	Marginal cost
K	A recession constant	MAX	Maximum
K	Frequency factors	MIN	Minimum
K	Headloss coefficient	M_p	Pollutant mass
K	Hydraulic conductivity	m	Kinematic parameter
K_T	Thermal conductivity	m	Mass
K	Muskingum storage coefficient	m	Aquifer thickness
		m	Stage-storage coefficient
K	Soil erodibility factor	m	Regression coefficient
K	Swale constant	m	Plot position
K	Storage coefficient	N	Outflow–storage relationship
K	Baseflow recession constant	N	Manning's overland flow coefficient
K	Permeability		
K	Horton depletion coefficient	N	Number of future trials
K_r	Routing constant for Santa Barbara	N	Annual precipitation
		N_D	Number of days without runoff
K_s	Hydraulic conductivity at saturation		
		N_R	Reynolds number
K_i	Recession limb coefficient	N_u	Number of stream segments of order u
K'	Composite storage coefficient		
k'	Constant	n	Number of storm hydrograph ordinates
k	Saturated permeability		
k	Snowmelt watershed constant	n	Number of years or observations
k	Consumptive use coefficient		
k	Proportionality constant	n	Retardance (Kerby formula)
k	Convolution recession constant	n	Number of events
		n	Regression parameter
k	Pearson Type II deviate	n	Stage-discharge coefficient
k_e	Culvert entrance loss and snow evaporation constant	n	Manning's roughness coefficient
		n	Shape parameter
k_s	Seasonal consumptive use coefficient	n	An empirical constant
		n_p	Porosity
L	Length of curb opening	O	Outflow volume
L	Stream length	O_i	ith outflow rate
L	Length	O_p	Peak outflow rate
L	Lag time	OR	Overflow rate
L	Length of overland flow		

P	Precipitation volume	q	Runoff rate
P	Erosion control practice factor	q	Flow rate
P	Erosion control practice factor in USLE	q	Overland flow discharge rate
		q	Specific adsorption capacity
P	Precipitation depth	q_0	Initial discharge
P	Accumulated rainfall depth	R	Risk
P	Loading	R	Rainfall excess
P_a	The antecedent precipitation index	R	Return flow
		R	Hydraulic radius
P_G	Perimeter of grate opening	R	Rainfall-runoff erosivity factor
P_0	Initial amount of pollutant		
P_{OW}	Annual wet weather loading	$\underline{R}$	Runoff depth
P_p	Piezometric potential	$\overline{R}$	Weighted hydraulic radius of the main sewer flowing full
Pr	Probability		
P_t	Amount of pollutant remaining on ground at time t	RA	Ratio of impervious area to total area
P_i	Isohyetal cell average precipitation	R_I	Runoff rate from impervious surface
		R_M	Efficiency of sediment removal
PD	Population density		
PW	Present worth	RU	Reuse volume of stormwater
$\overline{P}$	Mean of 24-hr annual maximum rain depths	RAIN	Long-term average rainfall
		r	Rainfall excess rate
p	Pressure	r	Radius
p	Permeability	r^2	Correlation coefficient
p	Monthly daytime hours	r	Reaction rate
p_c	Pan coefficient	r_{avg}	Average street runoff rate
Q	Streamflow	S	Storage
$\underline{Q}$	Discharge	S'	Storage potential of soil
$\overline{Q}$	Average discharge	S	Weighted physical slope of the main sewer
Q_I	Interflow		
Q_i	ith flow in sequence	S	Potential maximum retention in ground
Q_j	jth annual flow		
Q_p	Peak discharge	S	Energy gradient
Q	Hydrograph ordinate value	S	Overland slope
Q_t	Discharge at time t	SAR	Sodium adsorption ratio
$Q(t)$	Surface runoff rate at time t	SDR	Sediment delivery ratio
Q_w	Overland flow discharge rate, also gutter depression flow (Appendix F)	S_c	Groundwater aquifer storage coefficient
		SY	Sediment yield
Q_{10}	10-yr peak flow rate	SS	Ratio of horizontal to vertical length of side slope
$Q_{2.33}$	Mean annual flood rate		
Q_{50}	50-yr flow rate	S_t	Lake storage at time t
Q_{100}	100-yr flow rate	S_s	Soil storage percentage
Q_i^0	Observed hydrograph flow rate	SL_d	Shoreline length at depth d
		SNM	Daily snowmelt
Q'	Discharge per unit width	S_n	Standard deviation of annual maximum rain depths
q	Specific discharge		
q	Discharge per unit area	S_p	Total storage potential

S_x	Roadway cross-slope	Vn	Product of velocity and
S_y	Specific yield		Manning's roughness
S_{yx}	Standard error of estimated		coefficient
	runoff	V_5	Volume of pond at 5-ft depth
s	Linear slope parameter	V_M	Minimum volume of pond
s^2	Variance	V_m	Volume of soil material
s	Standard deviation	V_R	Runoff volume from mean
T	Transpiration		storm
T	Topographic factor	V_P	Volume of pond
T_D	Dewpoint temperature	V_D	Volume of pond at depth D
T	Temperature	V_I	Volume of pond from
T_r	Recurrence interval		impervious area
T	Transmissivity	V_v	Volume of voids
T	Width of gutter flow	v	Darcy's velocity
T	Travel time overland	v_s	Settling velocity
T'	Temperature	v_s	Seepage velocity
	differential—snowmelt	v_0	Design settling velocity
TC	Total cost	W	Power
T_A	Return period, annual series	W	Withdrawal
T_a	Air temperature	W	Width of depressed area
t_b	Time base of the hydrograph	W	Width of watershed
t_c	Time of concentration	W	Width of overland flow
T_e	Temperature of evaporated	W_i	Weighted area
	water	W_0	Width of curb inlet opening
T_{max}	Daily maximum temperature	WT	Weighted flow
T_0	Water surface temperature	w	Rectangular channel width
T_p	Partial series return period	w_s	Average watershed slope
t_p	Time to peak	X	Random variable
t_r	Recession limit time	X	Weight of organics adsorbed
t	Time	x	Weighting factor
t	Frequency of occurrence	$\bar{x}$	First moment about the origin
t_e	Time of equilibrium of runoff		(mean)
	rate	Y	Stage elevation
t_l	Lag time	Y	Release flow
t_d	Pond detention time	y	Depth of flow
t_w	Hydraulic residence time	Z	Side slope of a channel
	(lakes)	Z_o	Optimal solution
U	Unit hydrograph ordinates	z	Elevation
U	Consumptive use of water	z	Standard deviation
U	Wind velocity	z_0	Roughness parameter
U	Stream order	α	Angle
u	Average wind velocity	α	Regression constant
u	Velocity in x direction	α	Kinematic parameter
V	Volume	α_p	Portion of advective energy
V	Volume of water	β	Regression coefficient
V	Aquifer volume	β	Fluid compressibility
$\overline{V}$	Average velocity	Γ	Gamma function
V	Velocity	γ	Specific weight

δ	Standard error parameter	σ	Standard deviation
θ_s	Soil water content at saturation	σ^2	Variance
		ν	Kinematic viscosity
θ_i	Initial soil water content	τ	Time parameter
Φ	Infiltration index	α	Loading coefficient
ρ_s	Snowpack density	β	Loading coefficient
μ	Mean value	Δ	Interevent time
μ	Dynamic viscosity	Ω	Pond empty rate
η	Effective soil porosity	λ	Multiplier
μ_r'	rth moment about the origin	λ	Exponential distribution constant
μ_s	Dynamic viscosity of saltwater	Ψ	Capillary suction pressure head
$\rho = \rho_w$	Density of water		

METRIC UNITS WITH ENGLISH EQUIVALENTS

Length

Metric Units

millimeter	(mm)	10 mm = cm
centimeter	(cm)	100 cm = m
meter	(m)	1000 m = km
kilometer	(km)	

English equivalents

meters	$\times$ 39.37	= inches	$\times$ 0.0254	= meters
meters	$\times$ 3.28	= feet	$\times$ 0.3049	= meters
kilometers	$\times$ 0.62	= miles	$\times$ 1.6129	= kilometers
millimeters	$\times$ 0.039	= inches	$\times$ 25.4	= millimeters
centimeters	$\times$ 0.394	= inches	$\times$ 2.54	= centimeters

Example: Convert 3 m to feet. 3 m $\times$ 3.28 ft/m = 9.84 ft

Area

Metric Units

square millimeter	(mm^2)	$10^2 \ mm^2 = cm^2$
square centimeter	(cm^2)	$10^4 \ cm^2 \ = m^2$
square meter	(m^2)	$10^6 \ m^2 \ \ = km^2$
hectare	(ha)	$10^2 \ ha \ \ = km^2$

English Equivalents

$mm^2 \times \quad 0.00155 = in^2 \ \times 645.16 \ = mm^2$

$cm^2 \ \times \quad 0.155 \ \ = in.^2 \times \quad 6.45 \ = cm^2$

$m^2 \ \ \times \ 10.764 \ \ = ft^2 \ \times \quad 0.093 = m^2$

$km^2 \ \times \quad 0.384 \ \ = mi^2 \times \quad 2.605 = km^2$

$km^2 \ \times 247.10 \quad = ac \ \times \quad 0.004 = km^2$

$ha \ \ \times \quad 2.471 \ \ \ = ac \ \times \quad 0.405 = ha$

$ha \ \ \times \quad 0.00386 = mi^2 \times 259 \quad \ \ = ha$

Mass

Metric Units

milligram	(mg)	$1000 \ mg = g$
gram	(g)	$1000 \ g \ \ = kg$
kilogram	(kg)	$1000 \ kg \ = 1 \ tonne \ (t)$

English Equivalents

$milligram \times \ 0.01543 = grains \ \times \ 64.809 = mg$

$gram \quad \ \times \ 0.0022 \ = pounds \times 453.6 \quad = gram$

$gram \quad \ \times 15.43 \quad = grains \ \times \quad 0.065 = gram$

$kilogram \ \times \ 2.205 \quad = pounds \times \quad 0.454 = kg$

$kilogram \ \times \ 0.0011 \ = ton \quad \ \times 907.20 \ = kg \times 0.06854 = slug$

$tonne \ (t) \times \ 1.1023 \ = ton \quad \ \times \quad 0.907 = tonne$

(using 2000 pounds ton (short ton))

Volume

Metric Units

cubic centimeter	(cm^3)	$10^6 \ cm^3 \ = m^3$
cubic meter	(m^3)	$10^3 L \quad = m^3$
liter	(L)	$10^3 \ cm^3 \ = L$

English Equivalents

$cm^3 \times \ 0.061 \quad \ \ = in.^3 \ \times 16.393 \quad \ \ = cm^3$

$m^3 \ \times 35.314 \quad = ft^3 \ \ \times \ 0.028 \quad \ \ = m^3$

$L \ \ \times \ 1.057 \quad \ \ = qt. \ \ \times \ 0.946 \quad \ \ = L$

English Equivalents (Continued)

L	$\times$ 0.264	= gal	$\times$ 3.788	= L
L	$\times$ 0.81(10^{-6})	= ac-ft	$\times$ 1.235(10^{6})	= L
m^3	$\times$ 0.41(10^{-3})	= SFD	$\times$ 2.45(10^{3})	= m^3

(second foot day)

Time

Metric Units = English Units

second	(sec)
day	(day)
year	(yr or a)

86,400 sec = 1 day

365 day = yr or a

(366 days every 4 years)

Force

Metric Unit with English Equivalent

newton   (N) $\times 0.22481$ = lb (weight) $\times 4.4482$ = Newtons

Commonly Used Conversion Factors

Linear Velocity

$$m/sec \ \times 3.280 = fps \times 0.305 = m/sec$$
$$km/sec \times 2.230 \times 10^3 = mph \times 0.448 \times 10^{-3} = km/sec$$
$$km/hr \ \times 0.621 = mph$$
$$km/hr \ \times 0.540 = knots$$

Flow or Discharge

$m^3/sec \times 15.850(10^3)$ = gpm $\times$ 0.063(10^{-3}) = m^3/sec

$m^3/sec \times 2.12(10^3)$ = cfm $\times$ 0.472(10^{-3}) = m^3/sec

$m^3/sec \times 35.314$ = cfs $\times$ 0.0283 = m^3/sec

L/sec $\times$ 15.850 = gpm $\times$ 0.063 = L/sec

$m^3/sec \times 22.82$ = MGD $\times$ 0.0438 = m^3/sec

L/day $\times$ 0.264 = GPD $\times$ 3.788 = L/day

Loading

kg/km $\times$ 3.576 = lb/mi $\times$ 0.280 = kg/km

kg/ha $\times$ 0.892 = lb/ac $\times$ 1.12 = kg/ha

kg/ha $\times$ 0.286 = ton/mi^2 $\times$ 3.50 = kg/ha

kg/ha $\times$ 0.446(10^{-3}) = ton/ac $\times$ 2.24(10^3) = kg/ha

kg/m^3 $\times$ 0.065 = lb/ft^3 $\times$ 15.38 = kg/m^3

m^3/m^2 $\times$ 3.28 = ft^3/ft^2 $\times$ 0.305 = m^3/m^2

Density

$kg/cm^3 \times 0.0624$ $= lb/ft^3 \times 16.026$ $= kg/cm^3$

$lb/gal \times 1.2 \times 10^5 = mg/L \times 8.33 \times 10^{-6} = lb/gal$

$kg/m^3 \times 0.065$ $= lb/ft^3 \times 15.38$ $= kg/m^3$

Commonly Used Conversions

Area

$43,560 \ ft^2 = 1 \ ac$

$4,840 \ \ yd^2 = 1 \ ac$

$144 \ \ \ \ in.^2 = 1 \ ft^2$

$640 \ \ \ \ ac \ = 1 \ mi^2$

Volume

$7.48 \ gal \ = 1 \ ft^3$

$1728 \ in.^3 = 1 \ ft^3$

$1 \ MGD \ = 694.4 \ gpm$

$8.34 \ lb \ \ = 1 \ gal \ (of \ water)$

$62.43 \ lb \ = 1 \ ft^3 \ (of \ water)$

Mass

$2000 \ lb = 1 \ ton$

$454 \ g \ \ = 1 \ lb$

$7000 \ gr = 1 \ lb$

$2240 \ lb = 1 \ long \ ton$

Other Conversions

Pressure

$2.307 \ ft \ H_2O = 1 \ lb/in.^2$

$2.036 \ in. \ Hg \ = 1 \ lb/in.^2$

$14.70 \ psia \ \ \ \ \ = 1 \ atm$

$29.92 \ in \ Hg \ \ \ = 1 \ atm$

$33.93 \ ft \ H_2O \ = 1 \ atm$

$76.0 \ cm \ Hg \ \ \ = 1 \ atm$

$0.205 \ kg/m^2 = 1 \ lb/ft^2$

Miscellaneous

$in. \cdot mi^2 \ \times \ 26.9 \ \ \ \ = SFD \ \ \ \ \ \times \ 2.45(10^3) \ = m^3$

$in. \cdot mi^2 \ \times \ 53.3 \ \ \ \ = ac\text{-}ft \ \ \ \ \times \ 1.235(10^3) = m^3$

$SFD/mi^2 \times \ 0.0372 = in. \ \ \ \ \ \ \times \ 25.4 \ \ \ \ \ \ \ \ = mm$

$CFS \ \ \ \ \times \ 0.992 \ = ac\text{-}in./hour \times 101.6 \ \ \ \ \ \ = m^3/hr$

$lb/gal \ \ \ \times 120(10^3) = mg/L$

Miscellaneous (Continued)

$g = 9.806 \text{ m/s}^2 = 32.174 \text{ ft/s}^2$

standard conditions 4°C, 706 mm Hg

ρ water $= 1.94 \text{ slugs/ft}^3 = 1000 \text{ kg/m}^3$ (4°C)

γ water $= 62.43 \text{ lb/ft}^3 = 9806 \text{ N/m}^3$ (4°C)

for permeability:

$1 \text{ ft/day} = 0.305 \text{ m/day} = 7.48 \text{ gal/day-ft}^2$

for transmissivity:

$1 \text{ ft}^2/\text{day} = 0.0929 \text{ m}^2/\text{day} = 7.48 \text{ gal/day-ft}$

Also

$\log_e 10 = 2.30259$

$e = 2.7828$

TABLE B.1   Physical Properties of Water

TEMPERATURE (°F)	SPECIFIC WEIGHT γ (lb/ft^3)	DENSITY ρ (slugs/ft^3)	VISCOSITY $\mu \times 10^5$ lb(s/ft^2)	KINEMATIC VISCOSITY $\nu \times 10^5$ (ft^2/s)
32	62.42	1.940	3.746	1.931
40	62.42	1.940	3.229	1.664
50	62.41	1.940	2.735	1.410
60	62.36	1.938	2.359	1.217
70	62.29	1.936	2.050	1.059
80	62.22	1.934	1.799	0.930
90	62.13	1.931	1.595	0.826
100	62.00	1.927	1.424	0.739
110	61.87	1.923	1.284	0.667

Source: Hydraulic Models ASCE Manual 25, New York, 1942.

TABLE B.2   SI Unit Prefixes

PREFIX	SYMBOL	MULTIPLES	PREFIX	SYMBOL	MULTIPLES
tera	T	10^{12}	deci	d	10^{-1}
giga	G	10^9	centi	c	10^{-2}
mega	M	10^6	milli	m	10^{-3}
kilo	k	10^3	micro	μ	10^{-6}
hecto	h	10^2	nano	n	10^{-9}
deka	da	10	pico	p	10^{-12}

TABLE B.3 Vapor Pressures of Water in Metric Units

TEMP. °C	SATURATION VAPOR PRESSURE	
	mm Hg	MILLIBARS
−15*	1.44	1.92
−10	2.15	2.87
−5	3.16	4.22
0	4.58	6.11
5	6.54	8.72
10	9.20	12.27
15	12.78	17.04
20	17.53	23.37
25	23.76	31.67
30	31.83	42.43
35	42.18	56.24
40	55.34	73.78
50	92.56	123.40
60	149.46	199.26
70	233.79	311.69
80	355.28	473.57
90	525.89	701.13
100	760.00	1013.25

*Value refers to conditions over ice; actual vapor pressure is slightly less.

Vapor Pressures of Water in English Units

TEMP. °F	SATURATION VAPOR PRESSURE	
	in Hg	MILLIBARS
0	0.045	1.54
10	0.070	2.38
20	0.100	3.40
32	0.180	6.11
40	0.248	8.39
50	0.362	12.27
60	0.522	17.66
70	0.739	25.03
80	1.032	34.96
90	1.422	48.15
100	1.933	65.47
120	3.448	116.75
140	5.884	199.26
160	9.656	326.98
180	15.295	517.95
200	23.468	794.72
212	29.921	1013.25

Note: 1 mm Hg = 1.334 mb; 1 mb = 0.0295 in. Hg; 1 lb. = 4.482 N.
Standard atmosphere (sea level):

101,325 N/m^2	29.92 in. Hg
1.013 bars	14.696 lb/in^2
760 mm Hg	10.33 m H$_2$O

TABLE B.4   Representative Ranges of Various Inorganic Constituents in Leachate from Sanitary Landfills

PARAMETER	REPRESENTATIVE RANGE (mg/L)
K^+	200–1000
Na^+	200–1200
Ca^{2+}	100–3000
Mg^+	100–1500
Cl^-	300–3000
SO_4^{2-}	10–1000
Alkalinity	500–10,000
Fe (total)	1–1000
Mn	0.01–100
Cu	<10
Ni	0.01–1
Zn	0.1–100
Pb	<5
Hg	<0.2
NO_3^-	0.1–10
NH_4^+	10–1000
P as PO_4	1–100
Organic nitrogen	10–1000
Total dissolved organic carbon	200–30,000
COD (chemical oxidation demand)	1000–90,000
Total dissolved solids	5000–40,000
pH	4–8

Adapted from Bedient, Rifai, and Newell (1994).

TABLE B.5　Formula Weights and Equivalent Weight of Ions Commonly Found in Water

ION	FORMULA WEIGHT	EQUIVALENT WEIGHT	ION	FORMULA WEIGHT	EQUIVALENT WEIGHT
Al^{3+}	27.0	9.0	Fe^{3+}	55.8	18.6
Ba^{++}	137.0	68.7	Pb^{++}	207.0	104.0
HCO_3^-	61.0	61.0	Li^+	6.94	6.94
Br^-	79.9	79.9	Mg^{++}	24.3	12.2
Ca^{++}	40.1	20	Mn^{++}	54.9	27.5
CO_3^{--}	60.0	30	Mn^{4+}	54.9	13.7
Cl^-	35.5	35.5	NO^{3-}	62.0	62.0
Cr^{6+}	52.0	8.67	PO_4^{3-}	95.0	31.7
Cu^{++}	63.6	31.8	K^+	39.1	39.1
F^-	19.0	19.0	Na^+	23.0	23.0
H^-	1.01	1.01	Sr^{++}	87.6	43.8
OH^-	17.0	17.0	SO_4^-	96.1	48.0
I^-	127.0	127.0	S^-	32.1	16.0
Fe^{++}	55.8	27.9	Zn^{++}	65.4	32.7

TABLE B.6 Variation of Relative Humidity in Percent with Temperature and Wet-Bulb Depression on the Fahrenheit Scale
Pressure = 30.00 in. = 1015.9 mb = 101.59 kPa

AIR TEMP., °F	WET BULB DEPRESSION, DEGREES														
	0	1	2	3	4	6	8	10	12	14	16	18	20	25	30
0	84	56	27												
5	86	63	40	16											
10	89	69	50	30	11										
15	91	74	58	42	26										
20	94	79	65	51	37	10									
25	96	84	71	59	47	24	1								
30	99	88	77	66	56	35	15								
35	100	91	81	72	63	45	27	10							
40	100	92	84	76	68	52	37	22	9						
45	100	93	85	78	71	57	44	31	19	6					
50	100	93	87	80	74	61	49	38	27	16	5				
55	100	94	88	82	76	65	54	43	33	24	14	5			
60	100	94	89	83	78	68	58	48	39	30	21	13	5		
65	100	95	90	85	80	70	61	52	44	35	28	20	13		
70	100	95	90	86	81	72	64	55	48	40	33	26	19	3	
75	100	96	91	87	82	74	66	58	51	44	37	31	24	10	
80	100	96	91	87	83	75	68	61	54	47	41	35	29	15	3
85	100	96	92	88	84	77	70	63	56	50	44	38	33	20	8
90	100	96	92	89	85	78	71	65	58	53	47	41	36	24	13
95	100	96	93	89	86	79	72	66	60	55	49	44	39	28	17
100	100	96	93	89	86	80	74	68	62	57	51	46	42	31	21

Source: U.S. Weather Bureau, Relative Humidity and Dew Point Table, TA 454-O-3E, September, 1965.

TABLE B.7   Variation of Relative Humidity in Percent with Temperature and Wet-Bulb Depression on the Celsius Scale

Pressure = 29.24 in. = 990 mb = 99.00 kPa

AIR TEMP., °C	WET BULB DEPRESSION, DEGREES															
	0	1	2	3	4	5	6	7	8	9	10	11	12	13	14	15
−10	91	60	31	2												
−8	93	65	39	13												
−6	94	70	46	23	0											
−4	96	74	53	32	11											
−2	98	78	58	39	21	3										
0	100	81	63	46	29	13										
2	100	84	68	52	37	22	7									
4	100	85	71	57	43	29	16									
6	100	86	73	60	48	35	24	11								
8	100	87	75	63	51	40	29	19	8							
10	100	88	77	66	55	44	34	24	15	6						
12	100	89	78	68	58	48	39	29	21	12	4					
14	100	90	79	70	60	51	42	34	26	18	10	3				
16	100	90	81	71	63	54	46	38	30	23	15	8				
18	100	91	82	73	65	57	49	41	34	27	20	14	7			
20	100	91	83	74	66	59	51	44	37	31	24	18	12	6		
22	100	92	83	76	68	61	54	47	40	34	28	22	17	11	6	
24	100	92	84	77	69	62	56	49	43	37	31	26	20	15	10	5
26	100	92	85	78	71	64	58	51	46	40	34	29	24	19	14	10
28	100	93	85	78	72	65	59	53	48	42	37	32	27	22	18	13
30	100	93	86	79	73	67	61	55	50	44	39	35	30	25	21	17

Source: "Radiosonde Observation Computation Tables," Dept. of Commerce–Dept. of Defense, Washington, June 1972.

TABLE B.8 Variation of Dewpoint with Temperature and Wet-Bulb Depression and of Saturation Vapor Pressure over Water with Temperature on the Fahrenheit Scale

Pressure = 30.00 in. = 1015.9 mb = 101.59 kPa

AIR TEMP., °F	SATURATION VAPOR PRESSURE MILLIBARS	SATURATION VAPOR PRESSURE IN Hg	WET-BULB DEPRESSION, DEGREES 0	1	2	3	4	6	8	10	12	14	16	18	20	25	30
0	1.52	0.045	−4	−12	−26												
5	1.91	0.056	2	−5	−14	−31											
10	2.40	0.071	7	2	−5	−15	−34										
15	2.99	0.088	13	8	3	−4	−14										
20	3.71	0.110	18	15	10	5	−2	−27									
25	4.58	0.135	24	21	17	13	8	−7									
30	5.63	0.166	30	27	24	20	16	6	−12								
35	6.89	0.203	35	33	30	27	24	16	5	−16							
40	8.39	0.248	40	38	35	33	30	24	16	4	−18						
45	10.17	0.300	45	43	41	39	36	31	24	16	5	−18					
50	12.27	0.362	50	48	46	44	42	37	32	25	17	5	−17				
55	14.75	0.436	55	53	51	50	48	43	39	33	27	18	7	−15			
60	17.66	0.522	60	58	57	55	53	49	45	40	35	29	20	9	−11		
65	21.07	0.622	65	63	62	60	59	55	51	47	42	37	31	23	12		
70	25.03	0.739	70	69	67	66	64	61	57	53	49	45	39	33	26	−14	
75	29.63	0.875	75	74	72	71	69	66	63	59	56	52	47	42	36	14	
80	34.96	1.032	80	79	77	76	74	72	68	65	62	58	54	50	45	29	−9
85	41.10	1.214	85	84	82	81	80	77	74	71	68	64	61	57	53	39	18
90	48.15	1.422	90	89	87	86	85	82	79	76	73	70	67	63	60	48	33
95	56.24	1.661	95	94	93	91	90	87	85	82	79	76	73	70	66	56	44
100	65.47	1.933	100	99	98	96	95	93	90	87	85	82	79	76	73	64	53

Source: U.S. Weather Bureau, Relative Humidity and Dew Point Table, TA 454-0-3E, September 1965.

TABLE B.9 Variation of Dewpoint with Temperature and Wet-Bulb Depression and of Saturation Vapor Pressure over Water with Temperature on the Celsius Scale

Pressure = 1013.2 mb = 29.92 in = 101.32 kPa

AIR TEMP., °C	SATURATION VAPOR PRESSURE MILLIBARS	SATURATION VAPOR PRESSURE IN Hg	WET-BULB DEPRESSION, DEGREES 0	1	2	3	4	5	6	7	8	9	10	11	12	13	14	15
−10	2.86	0.085	−11	−16	−24													
−8	3.35	0.099	−9	−13	−20	−33												
−6	3.91	0.115	−7	−11	−16	−24												
−4	4.55	0.134	−5	−8	−12	−19	−32											
−2	5.28	0.156	−2	−5	−9	−14	−22											
0	6.11	0.180	0	−3	−6	−11	−16	−27										
2	7.05	0.208	2	−1	−3	−7	−12	−19	−33									
4	8.13	0.240	4	2	−1	−4	−8	−13	−21	−47								
6	9.35	0.276	6	4	2	−1	−5	−9	−14	−23								
8	10.72	0.317	8	6	4	1	−2	−5	−9	−15	−26							
10	12.27	0.362	10	8	6	4	1	−2	−5	−10	−17	−29						
12	14.02	0.414	12	10	8	6	4	1	−2	−6	−11	−18	−34					
14	15.98	0.472	14	12	11	9	6	4	1	−2	−6	−11	−19					
16	18.17	0.532	16	14	13	11	9	7	4	1	−2	−6	−11					
18	20.63	0.609	18	16	15	13	11	9	7	4	2	−2	−6	−10				
20	23.37	0.690	20	19	17	15	14	12	10	7	5	2	−1	−4				
22	26.43	0.780	22	21	19	17	16	14	12	10	8	5	2	−1	−5			
24	29.83	0.881	24	23	21	20	18	16	15	13	11	8	6	3	−1	−5	−10	
26	33.61	0.992	26	25	23	22	20	19	18	15	13	11	9	6	4	0	−4	−9
28	37.80	1.116	28	27	25	24	22	21	19	18	16	14	12	10	7	4	1	−3
30	42.43	1.253	30	29	27	26	25	23	22	20	18	17	15	13	10	8	5	2
32	47.55	1.404	32	31	29	28	27	25	24	22	21	19	17	15	13	11	9	6
34	53.20	1.571	34	33	32	30	29	28	26	25	23	21	20	17	16	14	12	10
36	59.42	1.755	36	35	34	32	31	30	28	27	25	24	22	21	19	17	15	13
38	66.26	1.957	38	37	36	34	33	32	30	29	28	26	25	23	21	20	18	16
40	73.78	2.179	40	39	38	36	35	34	33	31	30	28	27	25	24	22	20	19

Source: U.S. National Weather Service, Marine Surface Observations, *Weather Bur. Handb.* 1, 1969.

NONDIMENSIONAL RAINFALL AND
INTENSITY-DURATION-FREQUENCY CURVES

TABLE C.1 NRCS Type II: Rainfall Distribution (24 hr)

NONDIMENSIONAL		NONDIMENSIONAL		NONDIMENSIONAL		NONDIMENSIONAL	
TIME	RAINFALL	TIME	RAINFALL	TIME	RAINFALL	TIME	RAINFALL
.000	.000	.260	.085	.521	.735	.781	.934
.010	.002	.271	.090	.531	.758	.792	.938
.021	.005	.281	.095	.542	.776	.802	.942
.031	.008	.292	.100	.552	.791	.813	.946
.042	.011	.302	.105	.563	.804	.823	.950
.052	.014	.313	.110	.573	.815	.833	.953
.063	.017	.323	.115	.583	.825	.844	.956
.073	.020	.333	.120	.594	.834	.854	.959
.083	.023	.344	.126	.604	.842	.865	.962
.094	.026	.354	.133	.615	.849	.875	.965
.104	.029	.365	.140	.625	.856	.885	.968
.115	.032	.375	.147	.635	.863	.896	.971
.125	.035	.385	.155	.646	.869	.906	.974
.135	.038	.396	.163	.656	.875	.917	.977
.146	.041	.406	.172	.667	.881	.927	.980
.156	.044	.417	.181	.677	.887	.938	.983
.167	.048	.427	.191	.688	.893	.948	.986
.177	.052	.438	.203	.698	.898	.958	.989
.188	.056	.448	.218	.708	.903	.969	.992
.198	.060	.458	.236	.719	.908	.979	.995
.208	.064	.469	.257	.729	.913	.990	.998
.219	.068	.479	.283	.740	.918	1.000	1.000
.229	.072	.490	.387	.750	.922		
.240	.076	.500	.663	.760	.926		
.250	.080	.510	.707	.771	.930		

TABLE C.2 NRCS Type III: Rainfall Distribution (24 hr)

NONDIMENSIONAL		NONDIMENSIONAL		NONDIMENSIONAL		NONDIMENSIONAL	
TIME	RAINFALL	TIME	RAINFALL	TIME	RAINFALL	TIME	RAINFALL
.000	.000	.260	.076	.521	.702	.781	.940
.010	.002	.271	.080	.531	.729	.792	.944
.021	.005	.281	.085	.542	.751	.802	.947
.031	.007	.292	.089	.552	.769	.813	.951
.042	.010	.302	.094	.563	.785	.823	.954
.052	.012	.313	.100	.573	.799	.833	.957
.063	.015	.323	.107	.583	.811	.844	.960
.073	.017	.333	.115	.594	.823	.854	.963
.083	.020	.344	.122	.604	.834	.865	.966
.094	.023	.354	.130	.615	.844	.875	.969
.104	.026	.365	.139	.625	.853	.885	.972
.115	.028	.375	.148	.635	.862	.896	.975
.125	.031	.385	.157	.646	.870	.906	.978
.135	.034	.396	.167	.656	.878	.917	.981
.146	.037	.406	.178	.667	.886	.927	.983
.156	.040	.417	.189	.677	.893	.938	.986
.167	.043	.427	.202	.688	.900	.948	.988
.177	.047	.438	.216	.698	.907	.958	.991
.188	.050	.448	.232	.708	.911	.969	.993
.198	.053	.458	.250	.719	.916	.979	.996
.208	.057	.469	.271	.729	.920	.990	.998
.219	.060	.479	.298	.740	.925	1.000	1.00
.229	.064	.490	.339	.750	.929		
.240	.068	.500	.500	.760	.933		
.250	.072	.510	.662	.771	.936		

TABLE C.3   Corps of Engineers Design Storms

Hr	(vol. ≈ 7.90 in.)		(vol. ≈ 9.00 in.)		(vol. ≈ 11.00 in.)	
	P_{inc}	ΣP	P_{inc}	ΣP	P_{inc}	ΣP
0.0	—	—	—	—	—	—
0.5	0.06	0.06	0.06	0.06	0.08	0.08
1.0	0.06	0.12	0.06	0.12	0.08	0.16
1.5	0.06	0.18	0.06	0.18	0.08	0.24
2.0	0.06	0.24	0.06	0.24	0.08	0.32
2.5	0.06	0.30	0.06	0.30	0.08	0.40
3.0	0.06	0.36	0.07	0.37	0.09	0.49
3.5	0.06	0.42	0.07	0.44	0.09	0.58
4.0	0.06	0.48	0.07	0.51	0.09	0.67
4.5	0.08	0.56	0.09	0.60	0.11	0.78
5.0	0.08	0.64	0.09	0.69	0.11	0.89
5.5	0.09	0.73	0.11	0.80	0.13	1.02
6.0	0.09	0.82	0.11	0.91	0.13	1.15
6.5	0.09	0.91	0.11	1.02	0.13	1.28
7.0	0.09	1.00	0.11	1.13	0.13	1.41
7.5	0.13	1.13	0.15	1.28	0.19	1.60
8.0	0.13	1.26	0.15	1.43	0.19	1.79
8.5	0.13	1.39	0.15	1.58	0.19	1.98

continued

TABLE C.3 *(Continued)*

Hr	(vol. ≈ 7.90 in.)		(vol. ≈ 9.00 in.)		(vol. ≈ 11.00 in.)	
	P_{inc}	ΣP	P_{inc}	ΣP	P_{inc}	ΣP
9.0	0.13	1.52	0.15	1.73	0.19	2.17
9.5	0.14	1.66	0.16	1.89	0.20	2.37
10.0	0.14	1.80	0.16	2.05	0.20	2.57
10.5	0.14	1.94	0.16	2.21	0.20	2.77
11.0	0.14	2.08	0.16	2.37	0.20	2.97
11.5	0.16	2.24	0.18	2.55	0.22	3.19
12.0	0.16	2.40	0.18	2.73	0.22	3.41
12.5	0.19	2.59	0.22	2.95	0.26	3.67
13.0	0.21	2.80	0.23	3.18	0.29	3.96
13.5	0.29	3.09	0.33	3.51	0.41	4.37
14.0	0.30	3.39	0.34	3.85	0.42	4.79
14.5	0.46	3.85	0.52	4.37	0.64	5.43
15.0	0.46	4.31	0.52	4.89	0.64	6.07
15.5	0.73	5.04	0.83	5.72	1.01	7.08
16.0	0.81	5.85	0.92	6.64	1.12	8.20
16.5	0.34	6.19	0.39	7.03	0.47	8.67
17.0	0.33	6.52	0.38	7.41	0.46	9.13
17.5	0.18	6.70	0.21	7.62	0.25	9.38
18.0	0.17	6.87	0.20	7.82	0.24	9.62
18.5	0.11	6.98	0.13	7.95	0.15	9.77
19.0	0.11	7.09	0.13	8.08	0.15	9.92
19.5	0.09	7.18	0.11	8.19	0.13	10.05
20.0	0.09	7.27	0.11	8.30	0.13	10.18
20.5	0.09	7.36	0.10	8.40	0.12	10.30
21.0	0.09	7.45	0.10	8.50	0.12	10.42
21.5	0.08	7.53	0.09	8.59	0.11	10.53
22.0	0.08	7.61	0.09	8.68	0.11	10.64
22.5	0.09	7.70	0.08	8.76	0.10	10.74
23.0	0.07	7.77	0.08	8.84	0.10	10.84
23.5	0.07	7.84	0.08	8.92	0.10	10.94
24.0	0.06	7.90	0.08	9.00	0.06	11.00

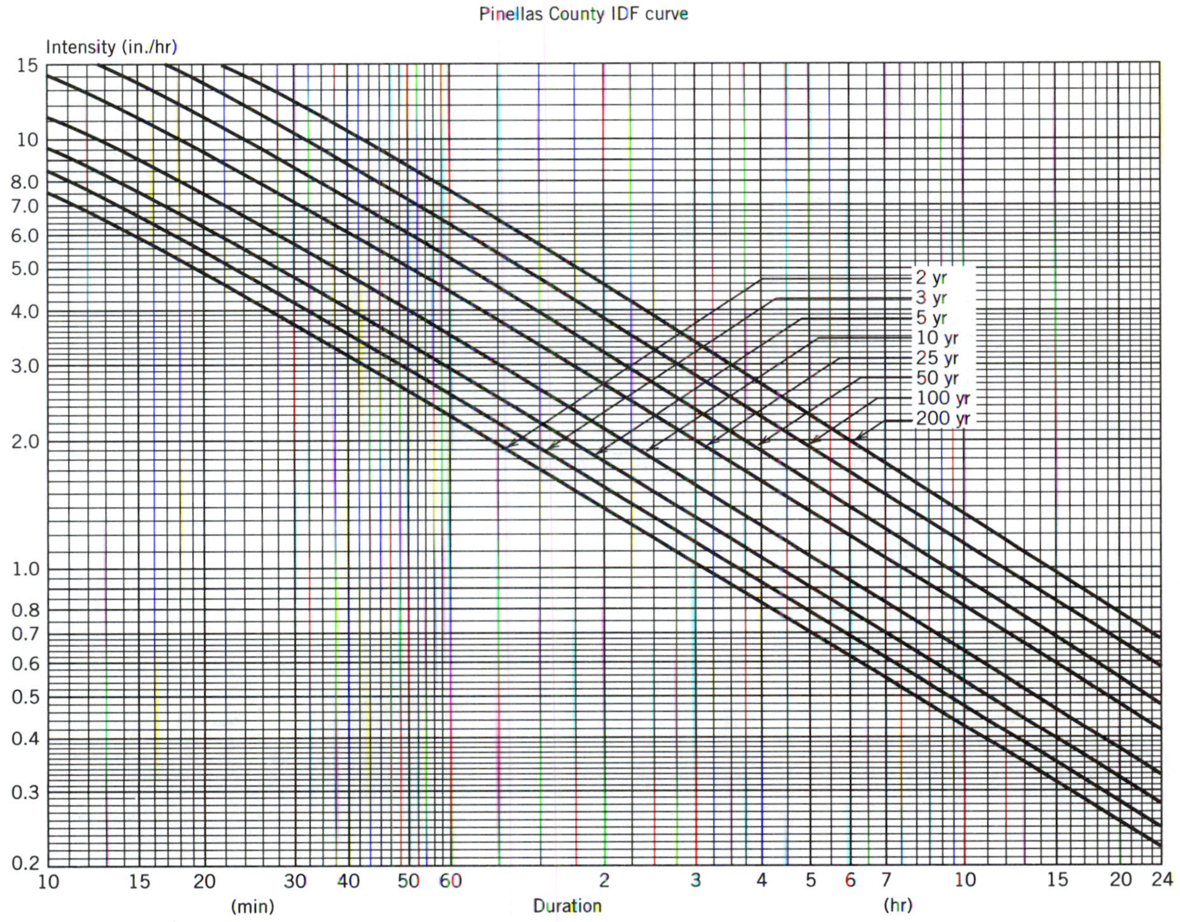

Pinellas County IDF curve

Intensity (in./hr)

FIGURE C.1 IDF curve for Tampa, Florida.

457

Orange County IDF curve

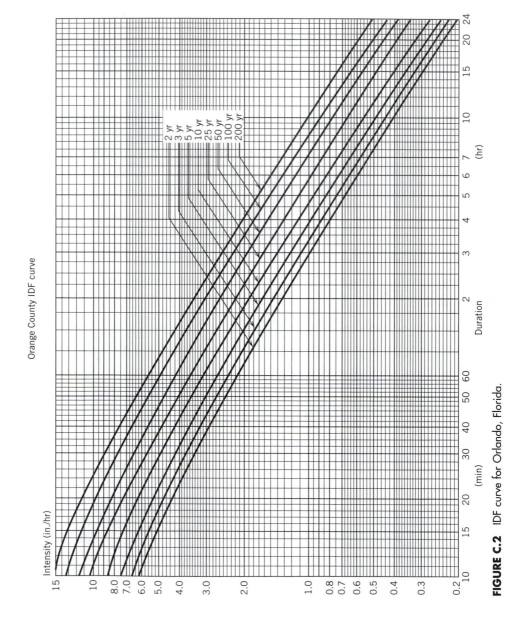

FIGURE C.2 IDF curve for Orlando, Florida.

Dade County IDF curve

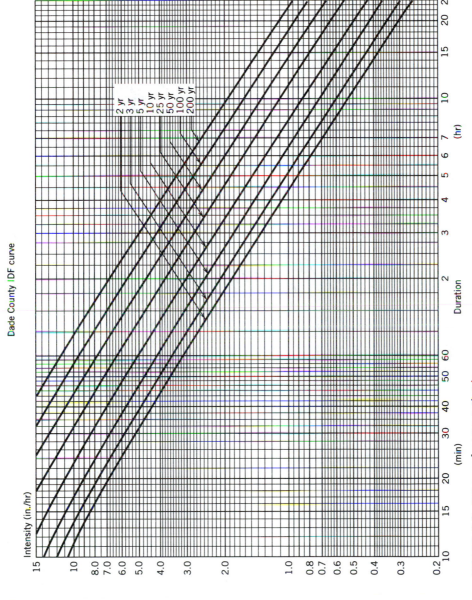

FIGURE C.3 IDF curve for Miami, Florida.

Duval County IDF curve

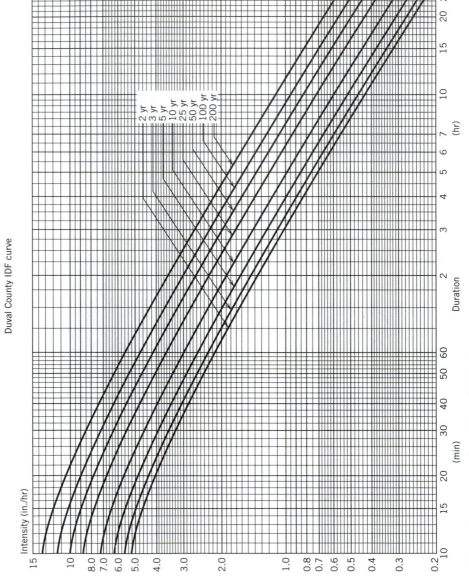

FIGURE C.4 IDF curve for Jacksonville, Florida.

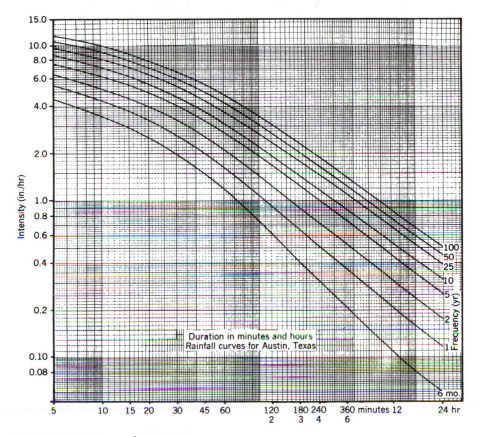

FIGURE C.5   IDF curve for Austin, Texas.

APPENDIX
D

STATISTICAL TABLES

TABLE D.1 Normal Distribution Function Table

$$F(z) = \frac{1}{\sqrt{2\pi}} \int_{-\infty}^{z} e^{-\frac{t^2}{2}} dt$$

z	.0	.0100	.0200	.0300	.0400	.0500	.0600	.0700	.0800	.0900
.0	.5000	.5040	.5080	.5120	.5160	.5199	.5239	.5279	.5319	.5359
.10	.5398	.5438	.5478	.5517	.5557	.5596	.5636	.5675	.5714	.5733
.20	.5793	.5832	.5871	.5910	.5948	.5987	.6026	.6064	.6103	.6141
.30	.6179	.6217	.6255	.6293	.6331	.6368	.6406	.6443	.6480	.6517
.40	.6554	.6591	.6628	.6664	.6700	.6736	.6772	.6808	.6844	.6879
.50	.6915	.6950	.6985	.7019	.7054	.7088	.7123	.7157	.7190	.7224
.60	.7257	.7291	.7324	.7356	.7389	.7422	.7454	.7486	.7517	.7549
.70	.7580	.7611	.7642	.7673	.7703	.7734	.7764	.7793	.7823	.7852
.80	.7881	.7910	.7939	.7967	.7995	.8023	.8051	.8078	.8106	.8133
.90	.8159	.8186	.8212	.8238	.8264	.8289	.8315	.8340	.8365	.8389
1.00	.8413	.8437	.8461	.8485	.8508	.8531	.8554	.8577	.8599	.8621
1.10	.8643	.8665	.8686	.8708	.8729	.8749	.8770	.8790	.8810	.8830
1.20	.8849	.8869	.8888	.8906	.8925	.8943	.8962	.8980	.8997	.9015
1.30	.9032	.9049	.9066	.9082	.9099	.9115	.9131	.9147	.9162	.9177
1.40	.9192	.9207	.9222	.9236	.9251	.9265	.9278	.9292	.9306	.9319

continued

TABLE D.1 (*Continued*)

z	.0	.0100	.0200	.0300	.0400	.0500	.0600	.0700	.0800	.0900
1.50	.9332	.9345	.9357	.9370	.9382	.9394	.9406	.9418	.9429	.9441
1.60	.9452	.9463	.9474	.9484	.9795	.9505	.9515	.9525	.9535	.9545
1.70	.9554	.9564	.9573	.9582	.9591	.9599	.9608	.9616	.9625	.9633
1.80	.9641	.9648	.9656	.9664	.9671	.9678	.9686	.9693	.9699	.9706
1.90	.9713	.9719	.9726	.9732	.9738	.9744	.9750	.9756	.9761	.9767
2.00	.9772	.9778	.9783	.9788	.9793	.9798	.9803	.9808	.9812	.9817
2.10	.9821	.9826	.9830	.9834	.9838	.9842	.9846	.9850	.9854	.9857
2.20	.9861	.9864	.9868	.9871	.9874	.9878	.9881	.9884	.9887	.9890
2.30	.9893	.9895	.9898	.9901	.9904	.9906	.9909	.9911	.9913	.9916
2.40	.9918	.9920	.9922	.9924	.9926	.9928	.9930	.9932	.9934	.9936
2.50	.9938	.9940	.9941	.9943	.9944	.9946	.9949	.9948	.9951	.9952
2.60	.9953	.9955	.9956	.9957	.9958	.9960	.9961	.9962	.9963	.9964
2.70	.9965	.9966	.9967	.9968	.9969	.9970	.9971	.9972	.9973	.9974
2.80	.9974	.9975	.9976	.9977	.9977	.9978	.9979	.9979	.9980	.9981
2.90	.9981	.9982	.9982	.9983	.9983	.9984	.9985	.9985	.9985	.9986
3.00	.9986	.9987	.9987	.9988	.9988	.9988	.9989	.9989	.9990	.9990
3.10	.9990	.9991	.9991	.9991	.9991	.9992	.9992	.9992	.9993	.9993
3.20	.9993	.9993	.9993	.9994	.9994	.9994	.9994	.9995	.9995	.9995
3.30	.9995	.9995	.9995	.9996	.9996	.9996	.9996	.9996	.9996	.9996
3.40	.9997	.9997	.9997	.9997	.9997	.9997	.9997	.9997	.9997	.9997

TABLE D.2 Binomial Distribution Function

$$B(x; N, p) = \sum_{k=0}^{x} \binom{N}{x} p^k (1-p)^{N-k}$$

N	x	.05	.10	.15	.20	.25	.30	.35	.40	.45	.50
2-											
	0	.9025	.8100	.7225	.6400	.5625	.4900	.4225	.3600	.3025	.2500
	1	.9975	.9900	.9775	.9600	.9375	.9100	.8775	.8400	.7975	.7500
3-											
	0	.8574	.7290	.6141	.5120	.4219	.3430	.2746	.2160	.1664	.1250
	1	.9928	.9720	.9393	.8960	.8437	.7840	.7183	.6480	.5748	.5000
	2	.9999	.9990	.9966	.9920	.9844	.9730	.9571	.9360	.9089	.8750
4-											
	0	.8145	.6561	.5220	.4096	.3164	.2401	.1785	.1296	.0915	.0625
	1	.9860	.9477	.8905	.8192	.7383	.6517	.5630	.4752	.3910	.3125
	2	.9995	.9963	.9880	.9728	.9492	.9163	.8735	.8208	.7585	.6875
	3	1.0000	.9999	.9995	.9984	.9961	.9919	.9850	.9744	.9590	.9375
5-											
	0	.7738	.5905	.4437	.3277	.2373	.1681	.1160	.0778	.0503	.0313
	1	.9774	.9185	.8352	.7373	.6328	.5282	.4284	.3370	.2562	.1875
	2	.9988	.9914	.9734	.9421	.8965	.8369	.7648	.6826	.5931	.5000
	3	1.0000	.9995	.9978	.9933	.9844	.9692	.9460	.9130	.8688	.8125
	4	1.0000	1.0000	.9999	.9997	.9990	.9976	.9947	.9898	.9815	.9688

continued

465

TABLE D.2 *(Continued)*

N	X	.05	.10	.15	.20	.25	.30	.35	.40	.45	.50
6-	0	.7351	.5314	.3771	.2621	.1780	.1176	.0754	.0467	.0277	.0156
	1	.9672	.8857	.7765	.6554	.5339	.4202	.3191	.2333	.1636	.1094
	2	.9978	.9842	.9527	.9011	.8306	.7443	.6471	.5443	.4415	.3438
	3	.9999	.9987	.9941	.9830	.9624	.9295	.8826	.8208	.7447	.6563
	4	1.0000	.9999	.9996	.9984	.9954	.9891	.9777	.9590	.9308	.8906
	5	1.0000	1.0000	1.0000	.9999	.9998	.9993	.9982	.9959	.9917	.9844
7-	0	.6983	.4783	.3206	.2097	.1335	.0824	.0490	.0280	.0152	.0078
	1	.9556	.8503	.7166	.5767	.4449	.3294	.2338	.1586	.1024	.0625
	2	.9962	.9743	.9262	.8520	.7564	.6471	.5323	.4199	.3164	.2266
	3	.9998	.9973	.9879	.9667	.9294	.8740	.8002	.7102	.6083	.5000
	4	1.0000	.9998	.9988	.9953	.9871	.9712	.9444	.9037	.8471	.7734
	5	1.0000	1.0000	.9999	.9996	.9987	.9962	.9910	.9812	.9643	.9375
	6	1.0000	1.0000	1.0000	1.0000	.9999	.9998	.9994	.9984	.9963	.9922
8-	0	.6634	.4305	.2725	.1678	.1001	.0576	.0319	.0168	.0084	.0039
	1	.9428	.8131	.6572	.5033	.3671	.2553	.1691	.1064	.0632	.0352
	2	.9942	.9619	.8948	.7969	.6785	.5518	.4278	.3154	.2201	.1445
	3	.9996	.9950	.9786	.9437	.8862	.8059	.7064	.5941	.4770	.3633
	4	1.0000	.9996	.9971	.9896	.9727	.9420	.8939	.8263	.7396	.6367
	5	1.0000	1.0000	.9998	.9988	.9958	.9887	.9747	.9502	.9115	.8555
	6	1.0000	1.0000	1.0000	.9999	.9996	.9987	.9964	.9915	.9819	.9648
	7	1.0000	1.0000	1.0000	1.0000	1.0000	.9999	.9998	.9993	.9983	.9961

n	k										
9-	0	.6302	.3874	.2316	.1342	.0751	.0404	.0207	.0101	.0046	.0020
	1	.9288	.7748	.5995	.4362	.3003	.1960	.1211	.0705	.0385	.0195
	2	.9916	.9470	.8591	.7382	.6007	.4628	.3373	.2318	.1495	.0898
	3	.9994	.9917	.9661	.9144	.8343	.7297	.6089	.4826	.3614	.2539
	4	1.0000	.9991	.9944	.9804	.9511	.9012	.8283	.7334	.6214	.5000
	5	1.0000	.9999	.9994	.9969	.9900	.9747	.9464	.9006	.8342	.7461
	6	1.0000	1.0000	1.0000	.9997	.9987	.9957	.9888	.9750	.9502	.9102
	7	1.0000	1.0000	1.0000	1.0000	.9999	.9996	.9986	.9962	.9909	.9805
	8	1.0000	1.0000	1.0000	1.0000	1.0000	1.0000	.9999	.9997	.9992	.9980
10-	0	.5987	.3487	.1969	.1074	.0563	.0282	.0135	.0060	.0025	.0010
	1	.9139	.7361	.5443	.3758	.2440	.1493	.0860	.0464	.0233	.0107
	2	.9885	.9298	.8202	.6778	.5256	.3828	.2616	.1673	.0996	.0547
	3	.9990	.9872	.9500	.8791	.7759	.6496	.5138	.3823	.2660	.1719
	4	.9999	.9984	.9901	.9672	.9219	.8497	.7515	.6331	.5044	.3770
	5	1.0000	.9999	.9986	.9936	.9803	.9527	.9051	.8338	.7384	.6230
	6	1.0000	1.0000	.9999	.9991	.9965	.9894	.9740	.9452	.8980	.8281
	7	1.0000	1.0000	1.0000	.9999	.9996	.9984	.9952	.9877	.9726	.9453
	8	1.0000	1.0000	1.0000	1.0000	1.0000	.9999	.9995	.9983	.9955	.9893
	9	1.0000	1.0000	1.0000	1.0000	1.0000	1.0000	1.0000	.9999	.9997	.9990

continued

467

TABLE D.2 *(Continued)*

N	x	.05	.10	.15	.20	.25	.30	.35	.40	.45	.50
11-	0	.5688	.3138	.1673	.0859	.0422	.0198	.0088	.0036	.0014	.0005
	1	.8981	.6974	.4922	.3221	.1971	.1130	.0606	.0302	.0139	.0059
	2	.9848	.9104	.7788	.6174	.4552	.3127	.2001	.1189	.0652	.0327
	3	.9984	.9815	.9306	.8389	.7133	.5696	.4256	.2963	.1911	.1133
	4	.9999	.9972	.9841	.9496	.8854	.7897	.6683	.5328	.3971	.2744
	5	1.0000	.9997	.9973	.9883	.9657	.9218	.8513	.7535	.6331	.5000
	6	1.0000	1.0000	.9997	.9980	.9924	.9784	.9499	.9006	.8262	.7256
	7	1.0000	1.0000	1.0000	.9998	.9988	.9957	.9878	.9707	.9390	.8867
	8	1.0000	1.0000	1.0000	1.0000	.9999	.9994	.9980	.9941	.9852	.9673
	9	1.0000	1.0000	1.0000	1.0000	1.0000	1.0000	.9998	.9993	.9978	.9941
	10	1.0000	1.0000	1.0000	1.0000	1.0000	1.0000	1.0000	1.0000	.9998	.9995
12-	0	.5404	.2824	.1422	.0687	.0317	.0138	.0057	.0022	.0008	.0002
	1	.8816	.6590	.4435	.2749	.1584	.0850	.0424	.0196	.0083	.0032
	2	.9804	.8891	.7358	.5583	.3907	.2528	.1513	.0834	.0421	.0193
	3	.9978	.9744	.9078	.7946	.6488	.4925	.3467	.2253	.1345	.0730
	4	.9998	.9957	.9761	.9274	.8424	.7237	.5833	.4382	.3044	.1938
	5	1.0000	.9995	.9954	.9806	.9456	.8822	.7873	.6652	.5269	.3872
	6	1.0000	.9999	.9993	.9961	.9857	.9614	.9154	.8418	.7393	.6128
	7	1.0000	1.0000	.9999	.9994	.9972	.9905	.9745	.9427	.8883	.8062
	8	1.0000	1.0000	1.0000	.9999	.9996	.9983	.9944	.9847	.9644	.9270
	9	1.0000	1.0000	1.0000	1.0000	1.0000	.9998	.9992	.9972	.9921	.9807
	10	1.0000	1.0000	1.0000	1.0000	1.0000	1.0000	.9999	.9997	.9989	.9968
	11	1.0000	1.0000	1.0000	1.0000	1.0000	1.0000	1.0000	1.0000	.9999	.9998

n	x										
13-	0	.5133	.2542	.1209	.0550	.0238	.0097	.0037	.0013	.0004	.0001
	1	.8646	.6213	.3983	.2336	.1267	.0637	.0296	.0126	.0049	.0017
	2	.9755	.8661	.6920	.5017	.3326	.2025	.1132	.0579	.0269	.0112
	3	.9969	.9658	.8820	.7473	.5843	.4206	.2783	.1686	.0929	.0461
	4	.9997	.9935	.9658	.9009	.7940	.6543	.5005	.3530	.2279	.1334
	5	1.0000	.9991	.9925	.9700	.9198	.8346	.7159	.5744	.4268	.2905
	6	1.0000	.9999	.9987	.9930	.9757	.9376	.8705	.7712	.6437	.5000
	7	1.0000	1.0000	.9998	.9988	.9944	.9818	.9538	.9023	.8212	.7095
	8	1.0000	1.0000	1.0000	.9998	.9990	.9960	.9874	.9679	.9302	.8666
	9	1.0000	1.0000	1.0000	1.0000	.9999	.9993	.9975	.9922	.9797	.9539
	10	1.0000	1.0000	1.0000	1.0000	1.0000	.9999	.9997	.9987	.9959	.9888
	11	1.0000	1.0000	1.0000	1.0000	1.0000	1.0000	1.0000	.9999	.9995	.9983
	12	1.0000	1.0000	1.0000	1.0000	1.0000	1.0000	1.0000	1.0000	1.0000	.9999
14-	0	.4877	.2288	.1028	.0440	.0178	.0068	.0024	.0008	.0002	.0001
	1	.8470	.5846	.3567	.1979	.1010	.0475	.0205	.0081	.0029	.0009
	2	.9699	.8416	.6479	.4481	.2811	.1608	.0839	.0398	.0170	.0065
	3	.9958	.9559	.8535	.6982	.5213	.3552	.2205	.1243	.0632	.0287
	4	.9996	.9908	.9533	.8702	.7415	.5842	.4227	.2793	.1672	.0898
	5	1.0000	.9985	.9885	.9561	.8883	.7805	.6405	.4859	.3373	.2120
	6	1.0000	.9998	.9978	.9884	.9617	.9067	.8164	.6925	.5461	.3953
	7	1.0000	1.0000	.9997	.9976	.9897	.9685	.9247	.8499	.7414	.6047

continued

TABLE D.2 *(Continued)*

N	x	.05	.10	.15	.20	.25	.30	.35	.40	.45	.50
	8	1.0000	1.0000	1.0000	.9996	.9978	.9917	.9757	.9417	.8811	.7880
	9	1.0000	1.0000	1.0000	1.0000	.9997	.9983	.9940	.9825	.9574	.9102
	10	1.0000	1.0000	1.0000	1.0000	1.0000	.9998	.9989	.9961	.9886	.9713
	11	1.0000	1.0000	1.0000	1.0000	1.0000	1.0000	.9999	.9994	.9978	.9935
	12	1.0000	1.0000	1.0000	1.0000	1.0000	1.0000	1.0000	.9999	.9997	.9991
	13	1.0000	1.0000	1.0000	1.0000	1.0000	1.0000	1.0000	1.0000	1.0000	.9999
15-	0	.4633	.2059	.0874	.0352	.0134	.0047	.0016	.0005	.0001	.0000
	1	.8290	.5490	.3186	.1671	.0802	.0353	.0142	.0052	.0017	.0005
	2	.9638	.8159	.6042	.3980	.2361	.1268	.0617	.0271	.0107	.0037
	3	.9945	.9444	.8227	.6482	.4613	.2969	.1727	.0905	.0424	.0176
	4	.9994	.9873	.9383	.8358	.6865	.5155	.3519	.2173	.1204	.0592
	5	.9999	.9978	.9832	.9389	.8516	.7216	.5643	.4032	.2608	.1509
	6	1.0000	.9997	.9964	.9819	.9434	.8689	.7548	.6098	.4522	.3036
	7	1.0000	1.0000	.9994	.9958	.9827	.9500	.8868	.7869	.6535	.5000
	8	1.0000	1.0000	.9999	.9992	.9958	.9848	.9578	.9050	.8182	.6964
	9	1.0000	1.0000	1.0000	.9999	.9992	.9963	.9876	.9662	.9231	.8491
	10	1.0000	1.0000	1.0000	1.0000	.9999	.9993	.9972	.9907	.9745	.9408
	11	1.0000	1.0000	1.0000	1.0000	1.0000	.9999	.9995	.9981	.9937	.9824
	12	1.0000	1.0000	1.0000	1.0000	1.0000	1.0000	.9999	.9997	.9989	.9963
	13	1.0000	1.0000	1.0000	1.0000	1.0000	1.0000	1.0000	1.0000	.9999	.9995
	14	1.0000	1.0000	1.0000	1.0000	1.0000	1.0000	1.0000	1.0000	1.0000	1.0000
16-	0	.4401	.1853	.0743	.0281	.0100	.0033	.0010	.0003	.0001	.0000

470

1	.8108	.5147	.2839	.1407	.0635	.0261	.0098	.0033	.0010	.0003
2	.9571	.7893	.5614	.3518	.1971	.0994	.0451	.0183	.0066	.0021
3	.9930	.9316	.7899	.5981	.4050	.2459	.1339	.0651	.0281	.0106
4	.9991	.9830	.9209	.7982	.6302	.4499	.2892	.1666	.0853	.0384
5	.9999	.9967	.9765	.9183	.8103	.6598	.4900	.3288	.1976	.1051
6	1.0000	.9995	.9944	.9733	.9204	.8247	.6881	.5272	.3660	.2272
7	1.0000	.9999	.9989	.9930	.9729	.9256	.8406	.7161	.5629	.4018
8	1.0000	1.0000	.9998	.9985	.9925	.9743	.9329	.8577	.7441	.5982
9	1.0000	1.0000	1.0000	.9998	.9984	.9929	.9771	.9417	.8759	.7728
10	1.0000	1.0000	1.0000	1.0000	.9997	.9984	.9938	.9809	.9514	.8949
11	1.0000	1.0000	1.0000	1.0000	1.0000	.9997	.9987	.9951	.9851	.9616
12	1.0000	1.0000	1.0000	1.0000	1.0000	1.0000	.9998	.9991	.9965	.9894
13	1.0000	1.0000	1.0000	1.0000	1.0000	1.0000	1.0000	.9999	.9994	.9979
14	1.0000	1.0000	1.0000	1.0000	1.0000	1.0000	1.0000	1.0000	.9999	.9997
15	1.0000	1.0000	1.0000	1.0000	1.0000	1.0000	1.0000	1.0000	1.0000	1.0000
0	.4181	.1668	.0631	.0225	.0075	.0023	.0007	.0002	.0000	.0000
1	.7922	.4818	.2525	.1182	.0501	.0193	.0067	.0021	.0006	.0001
2	.9497	.7618	.5198	.3096	.1637	.0774	.0327	.0123	.0041	.0012
3	.9912	.9174	.7556	.5489	.3530	.2019	.1028	.0464	.0184	.0064
4	.9988	.9779	.9013	.7582	.5739	.3887	.2348	.1260	.0596	.0245
5	.9999	.9953	.9681	.8943	.7653	.5968	.4197	.2639	.1471	.0717

17-

continued

TABLE D.2 (Continued)

N	x	.05	.10	.15	.20	.25	.30	.35	.40	.45	.50
	6	1.0000	.9992	.9917	.9623	.8929	.7752	.6188	.4478	.2902	.1662
	7	1.0000	.9999	.9983	.9891	.9598	.8954	.7872	.6405	.4743	.3145
	8	1.0000	1.0000	.9997	.9974	.9876	.9597	.9006	.8011	.6626	.5000
	9	1.0000	1.0000	1.0000	.9995	.9969	.9873	.9617	.9081	.8166	.6855
	10	1.0000	1.0000	1.0000	.9999	.9994	.9968	.9880	.9652	.9174	.8338
	11	1.0000	1.0000	1.0000	1.0000	.9999	.9993	.9970	.9894	.9699	.9283
	12	1.0000	1.0000	1.0000	1.0000	1.0000	.9999	.9994	.9975	.9914	.9755
	13	1.0000	1.0000	1.0000	1.0000	1.0000	1.0000	.9999	.9995	.9981	.9936
	14	1.0000	1.0000	1.0000	1.0000	1.0000	1.0000	1.0000	.9999	.9997	.9988
	15	1.0000	1.0000	1.0000	1.0000	1.0000	1.0000	1.0000	1.0000	1.0000	.9999
	16	1.0000	1.0000	1.0000	1.0000	1.0000	1.0000	1.0000	1.0000	1.0000	1.0000
18-	0	.3972	.1501	.0536	.0180	.0056	.0016	.0004	.0001	.0000	.0000
	1	.7735	.4503	.2241	.0991	.0395	.0142	.0046	.0013	.0003	.0001
	2	.9419	.7338	.4797	.2713	.1353	.0600	.0236	.0082	.0025	.0007
	3	.9891	.9018	.7202	.5010	.3057	.1646	.0783	.0328	.0120	.0038
	4	.9985	.9718	.8794	.7164	.5187	.3327	.1886	.0942	.0411	.0154
	5	.9998	.9936	.9581	.8671	.7174	.5344	.3550	.2088	.1077	.0481
	6	1.0000	.9988	.9882	.9487	.8610	.7217	.5491	.3743	.2258	.1189
	7	1.0000	.9998	.9973	.9837	.9431	.8593	.7283	.5634	.3915	.2403
	8	1.0000	1.0000	.9995	.9957	.9807	.9404	.8609	.7368	.5778	.4073
	9	1.0000	1.0000	.9999	.9991	.9946	.9790	.9403	.8653	.7473	.5927
	10	1.0000	1.0000	1.0000	.9998	.9988	.9939	.9788	.9424	.8720	.7597
	11	1.0000	1.0000	1.0000	1.0000	.9998	.9986	.9938	.9797	.9463	.8811
	12	1.0000	1.0000	1.0000	1.0000	1.0000	.9997	.9986	.9942	.9817	.9519

continued

13	1.0000	1.0000	1.0000	1.0000	1.0000	1.0000	.9999	.9993	.9962	.9846
14	1.0000	1.0000	1.0000	1.0000	1.0000	1.0000	1.0000	.9999	.9990	.9951
15	1.0000	1.0000	1.0000	1.0000	1.0000	1.0000	1.0000	1.0000	.9998	.9987
16	1.0000	1.0000	1.0000	1.0000	1.0000	1.0000	1.0000	1.0000	1.0000	.9997
17	1.0000	1.0000	1.0000	1.0000	1.0000	1.0000	1.0000	1.0000	1.0000	1.0000

19-

0	.3774	.1351	.0456	.0144	.0042	.0011	.0003	.0001	.0000	.0000
1	.7547	.4203	.1985	.0829	.0310	.0104	.0031	.0008	.0002	.0000
2	.9335	.7054	.4413	.2369	.1113	.0462	.0170	.0055	.0015	.0004
3	.9868	.8850	.6841	.4551	.2631	.1332	.0591	.0230	.0077	.0022
4	.9980	.9648	.8556	.6733	.4654	.2822	.1500	.0696	.0280	.0096
5	.9998	.9914	.9463	.8369	.6678	.4739	.2968	.1629	.0777	.0318
6	1.0000	.9983	.9837	.9324	.8251	.6655	.4812	.3081	.1727	.0835
7	1.0000	.9997	.9959	.9767	.9225	.8180	.6656	.4878	.3169	.1796
8	1.0000	1.0000	.9992	.9933	.9713	.9161	.8145	.6675	.4940	.3238
9	1.0000	1.0000	.9999	.9984	.9911	.9674	.9125	.8139	.6710	.5000
10	1.0000	1.0000	1.0000	.9997	.9977	.9895	.9653	.9115	.8159	.6762
11	1.0000	1.0000	1.0000	1.0000	.9995	.9972	.9886	.9648	.9129	.8204
12	1.0000	1.0000	1.0000	1.0000	.9999	.9994	.9969	.9884	.9658	.9165
13	1.0000	1.0000	1.0000	1.0000	1.0000	.9999	.9993	.9969	.9891	.9682
14	1.0000	1.0000	1.0000	1.0000	1.0000	1.0000	.9999	.9994	.9972	.9904

TABLE D.2 (Continued)

N	x	.05	.10	.15	.20	.25	.30	.35	.40	.45	.50
	15	1.0000	1.0000	1.0000	1.0000	1.0000	1.0000	1.0000	.9999	.9995	.9978
	16	1.0000	1.0000	1.0000	1.0000	1.0000	1.0000	1.0000	1.0000	.9999	.9996
	17	1.0000	1.0000	1.0000	1.0000	1.0000	1.0000	1.0000	1.0000	1.0000	1.0000
	18	1.0000	1.0000	1.0000	1.0000	1.0000	1.0000	1.0000	1.0000	1.0000	1.0000
20-	0	.3585	.1216	.0388	.0115	.0032	.0008	.0002	.0000	.0000	.0000
	1	.7358	.3917	.1756	.0692	.0243	.0076	.0021	.0005	.0001	.0000
	2	.9245	.6769	.4049	.2061	.0913	.0355	.0121	.0036	.0009	.0002
	3	.9841	.8670	.6477	.4114	.2252	.1071	.0444	.0160	.0049	.0013
	4	.9974	.9568	.8298	.6296	.4148	.2375	.1182	.0510	.0189	.0059
	5	.9997	.9887	.9327	.8042	.6172	.4164	.2454	.1256	.0553	.0207
	6	1.0000	.9976	.9781	.9133	.7858	.6080	.4166	.2500	.1299	.0577
	7	1.0000	.9996	.9941	.9679	.8982	.7723	.6010	.4159	.2520	.1316
	8	1.0000	.9999	.9987	.9900	.9591	.8867	.7624	.5956	.4143	.2517
	9	1.0000	1.0000	.9998	.9974	.9861	.9520	.8782	.7553	.5914	.4119
	10	1.0000	1.0000	1.0000	.9994	.9961	.9829	.9468	.8725	.7507	.5881
	11	1.0000	1.0000	1.0000	.9999	.9991	.9949	.9804	.9435	.8692	.7483
	12	1.0000	1.0000	1.0000	1.0000	.9998	.9987	.9940	.9790	.9420	.8684
	13	1.0000	1.0000	1.0000	1.0000	1.0000	.9997	.9985	.9935	.9786	.9423
	14	1.0000	1.0000	1.0000	1.0000	1.0000	1.0000	.9997	.9984	.9936	.9793
	15	1.0000	1.0000	1.0000	1.0000	1.0000	1.0000	1.0000	.9997	.9985	.9941
	16	1.0000	1.0000	1.0000	1.0000	1.0000	1.0000	1.0000	1.0000	.9997	.9987
	17	1.0000	1.0000	1.0000	1.0000	1.0000	1.0000	1.0000	1.0000	1.0000	.9998
	18	1.0000	1.0000	1.0000	1.0000	1.0000	1.0000	1.0000	1.0000	1.0000	1.0000
	19	1.0000	1.0000	1.0000	1.0000	1.0000	1.0000	1.0000	1.0000	1.0000	1.0000

TABLE D.3 *K Standard Deviates: Log Pearson Type III Positive Skew Values (+G)*

EXCEEDENCE PROBABILITY

G \ P	.999	.990	.975	.950	.900	.500	.100	.050	.025	.010	.001
.0	−3.090	−2.326	−1.960	−1.645	−1.282	.000	1.281	1.645	1.960	2.326	3.090
.1	−2.948	−2.253	−1.912	−1.616	−1.270	−0.166	1.292	1.673	2.007	2.400	3.233
.2	−2.808	−2.178	−1.864	−1.586	−1.258	−0.330	1.301	1.700	2.053	2.472	3.377
.3	−2.670	−2.104	−1.814	−1.555	−1.245	−0.500	1.309	1.726	2.098	2.544	3.521
.4	−2.533	−2.030	−1.764	−1.523	−1.231	−0.665	1.317	1.750	2.142	2.615	3.666
.5	−2.400	−1.954	−1.714	−1.491	−1.216	−0.830	1.323	1.774	2.185	2.686	3.810
.6	−2.688	−1.880	−1.662	−1.458	−1.200	−0.099	1.330	1.797	2.227	2.755	3.955
.7	−2.140	−1.806	−1.611	−1.423	−1.835	−0.116	1.333	1.818	2.268	2.823	4.100
.8	−2.017	−1.733	−1.560	−1.388	−1.656	−0.132	1.336	1.840	2.308	2.891	4.244
.9	−1.899	−1.660	−1.507	−1.353	−1.147	−0.148	1.339	1.858	2.346	2.957	4.388
1.0	−1.786	−1.588	−1.455	−1.317	−1.280	−0.164	1.340	1.877	2.383	3.022	4.531
1.5	−1.313	−1.256	−1.200	−1.130	−1.018	−0.240	1.333	1.951	2.552	3.330	5.233
2.0	−0.999	.989	−0.975	.949	−0.895	−0.307	1.303	1.995	2.688	3.605	5.907
2.5	−0.800	−0.799	−0.797	−0.790	−0.770	−0.360	1.250	2.012	2.793	3.845	6.548
3.0	−0.667	−0.666	−0.666	−0.665	−0.660	−0.396	1.180	2.003	2.867	4.051	7.152
4.0	−0.500	−0.500	−0.500	−0.499	−0.499	−0.413	1.000	1.920	2.933	4.367	8.253
5.0	−0.400	−0.400	−0.400	−0.400	−0.400	−0.379	.795	1.773	2.909	4.457	9.219

continued

TABLE D.3 *(Continued)*

G	.999	.990	.975	.950	.900	.500	.100	.050	.025	.010	.001
6.0	−0.333	−0.333	−0.333	−0.333	−0.333	−0.330	.589	1.585	2.817	4.687	10.068
7.0	−0.285	−0.285	−0.285	−0.285	−0.285	−0.285	.400	1.377	2.676	4.726	10.813
8.0	−0.250	−0.250	−0.250	−0.250	−0.250	−0.249	.239	1.163	2.500	4.705	11.468
.0	−3.090	−2.326	−1.960	−1.645	−1.281	.000	1.281	1.645	1.960	2.326	3.090
−0.1	−3.233	−2.400	−2.007	−1.673	−1.291	.166	1.270	1.616	1.912	2.252	2.948
−0.2	−3.377	−2.472	−2.053	−1.700	−1.301	.033	1.258	1.586	1.863	2.178	2.808
−0.3	−3.352	−2.544	−2.098	−1.726	−1.309	.050	1.245	1.555	1.814	2.104	2.670
−0.4	−3.666	−2.615	−2.142	−1.750	−1.317	.066	1.231	1.523	1.764	2.029	2.532
−0.5	−3.810	−2.685	−2.185	−1.774	−1.323	.083	1.216	1.491	1.713	1.954	2.400
−0.6	−3.956	−2.755	−2.220	−1.800	−1.330	.099	1.200	1.457	1.662	1.880	2.268
−0.7	−4.100	−2.823	−2.268	−1.818	−1.332	.116	1.183	1.423	1.611	1.806	2.350
−0.8	−4.244	−2.891	−2.307	−1.839	−1.336	.132	1.166	1.388	1.560	1.732	2.184
−0.9	−4.388	−2.957	−2.346	−1.858	−1.338	.148	1.147	1.353	1.507	1.660	2.030
−1.0	−4.531	−3.022	−2.383	−1.876	−1.340	.164	1.127	1.317	1.455	1.588	1.884
−1.5	−5.233	−3.330	−2.552	−1.950	−1.333	.240	1.018	1.130	1.200	1.256	1.312
−2.0	−5.907	−3.605	−2.689	−1.995	−1.303	.306	.894	.949	.975	.990	.999
−2.5	−6.548	−3.845	−2.793	−2.012	−1.250	.360	.770	.790	.797	.799	.800
−3.0	−7.152	−4.051	−2.867	−2.003	−1.180	.395	.660	.665	.666	.666	.666
−4.0	−8.252	−4.367	−2.933	−1.920	−1.000	.413	.499	.500	.500	.500	.500
−5.0	−9.219	−4.573	−2.909	−1.773	−0.795	.380	.400	.400	.400	.400	.400
−6.0	−10.068	−4.686	−2.817	−1.585	−0.590	.329	.333	.333	.333	.333	.333
−7.0	−10.813	−4.726	−2.676	−1.377	−0.400	.285	.285	.285	.285	.285	.285

TABLE D.4   K Standard Deviates: GUMBEL Distribution ($G = +1.14$)

SAMPLE SIZE n	RETURN PERIOD								
	5	10	15	20	25	50	75	100	1000
10	1.058	1.848	2.294	2.606	2.846	3.587	4.018	4.322	6.752
15	0.967	1.703	2.117	2.410	2.632	3.321	3.721	4.005	6.265
20	0.919	1.625	2.023	2.302	2.517	3.179	3.563	3.836	6.006
25	0.888	1.575	1.963	2.235	2.444	3.088	3.463	3.729	5.842
30	0.866	1.541	1.922	2.188	2.393	3.026	3.393	3.653	5.727
35	0.851	1.515	1.891	2.153	2.356	2.979	3.341	3.598	5.642
40	0.838	1.495	1.867	2.126	2.326	2.943	3.301	3.554	5.576
45	0.829	1.479	1.847	2.104	2.303	2.913	3.268	3.520	5.522
50	0.820	1.466	1.831	2.086	2.283	2.889	3.241	3.491	5.478
55	0.813	1.455	1.818	2.071	2.267	2.869	3.219	3.467	5.442
60	0.807	1.446	1.806	2.059	2.253	2.852	3.200	3.446	5.410
65	0.801	1.437	1.796	2.048	2.241	2.837	3.183	3.428	5.383
70	0.797	1.430	1.788	2.038	2.230	2.824	3.169	3.413	5.359
75	0.792	1.423	1.780	2.029	2.218	2.812	3.155	3.399	5.338
80	0.788	1.417	1.773	2.020	2.212	2.802	3.145	3.387	5.320
85	0.787	1.413	1.767	2.013	2.205	2.793	3.135	3.376	5.303
90	0.782	1.409	1.762	2.009	2.198	2.785	3.125	3.366	5.288
95	0.780	1.405	1.757	2.002	2.193	2.777	3.117	3.357	5.273
100	0.779	1.401	1.752	1.998	2.187	2.770	3.109	3.349	5.261
∞	0.719	1.305	1.635	1.866	2.044	2.592	2.911	3.137	4.936

TABLE D.5 Parameters δ for Standard Error of Normal Distribution

EXCEEDENCE PROBABILITY IN PERCENT						
50.0	20.0	10.0	4.0	2.0	1.0	.2
CORRESPONDING RETURN PERIOD IN YEARS						
2	5	10	25	50	100	500
1.0000	1.1637	1.3496	1.5916	1.7634	1.9253	2.2624

TABLE D.6 Parameter δ for Standard Error of Log-Normal Distribution

	EXCEEDENCE PROBABILITY IN PERCENT						
	50.0	20.0	10.0	4.0	2.0	1.0	0.2
	CORRESPONDING RETURN PERIOD IN YEARS						
COEFFICIENT OF VARIATION	2	5	10	25	50	100	500
.05	.9983	1.2162	1.4323	1.7105	1.9087	2.0968	2.4939
.10	.9932	1.2698	1.5222	1.8453	2.0766	2.2979	2.7714
.15	.9848	1.3241	1.6187	1.9956	2.2676	2.5298	3.0993
.20	.9733	1.3784	1.7211	2.1613	2.4819	2.7940	3.4820
.25	.9589	1.4323	1.8289	2.3423	2.7202	3.0917	3.9241
.30	.9420	1.4855	1.9417	2.5383	2.9829	3.4246	4.4305
.35	.9229	1.5378	2.0591	2.7496	3.2708	3.7942	5.0065
.40	.9021	1.5890	2.1811	2.9762	3.5845	4.2023	5.6574
.45	.8801	1.6389	2.3074	3.2184	3.9251	4.6508	6.3890
.50	.8575	1.6876	2.4382	3.4766	4.2935	5.1418	7.2076
.55	.8351	1.7351	2.5735	3.7514	4.6910	5.6774	8.1196
.60	.8138	1.7814	2.7134	4.0435	5.1190	6.2604	9.1322
.65	.7945	1.8266	2.8583	4.3535	5.5790	6.8934	10.2529
.70	.7784	1.8709	3.0085	4.6826	6.0729	7.5794	11.4897
.75	.7669	1.9143	3.1644	5.0316	6.6024	8.3217	12.8513
.80	.7615	1.9570	3.3264	5.4018	7.1698	9.1238	14.3468
.85	.7635	1.9991	3.4949	5.7945	7.7773	9.9894	15.9861
.90	.7746	2.0408	3.6705	6.2109	8.4272	10.9225	17.7796
.95	.7959	2.0821	3.8536	6.6524	9.1221	11.9272	19.7381
1.00	.8284	2.1232	4.0449	7.1206	9.8646	13.0081	21.8734

TABLE D.7   Parameter δ for Standard Error of Gumbel Extreme Value Distribution

SAMPLE SIZE n	EXCEEDENCE PROBABILITY IN PERCENT						
	50.0	20.0	10.0	4.0	2.0	1.0	0.2
	CORRESPONDING RETURN PERIOD IN YEARS						
	2	5	10	25	50	100	500
10	.9305	1.8540	2.6200	3.6275	4.3870	5.1460	6.9103
15	.9270	1.7695	2.4756	3.4083	4.1127	4.8173	6.4565
20	.9250	1.7249	2.3990	3.2919	3.9670	4.6427	6.2154
25	.9237	1.6968	2.3507	3.2183	3.8748	4.5322	6.0626
30	.9229	1.6772	2.3169	3.1667	3.8103	4.4547	5.9556
35	.9223	1.6627	2.2919	3.1286	3.7624	4.3973	5.8763
40	.9218	1.6514	2.2725	3.0990	3.7253	4.3528	5.8147
45	.9214	1.6424	2.2569	3.0752	3.6955	4.3171	5.7653
50	.9211	1.6350	2.2441	3.0555	3.6707	4.2874	5.7242
55	.9208	1.6288	2.2333	3.0390	3.6502	4.2626	5.6900
60	.9206	1.6235	2.2241	3.0249	3.6325	4.2414	5.6607
65	.9204	1.6190	2.2163	3.0130	3.6175	4.2234	5.6357
70	.9202	1.6149	2.2092	3.0022	3.6039	4.2071	5.6132
75	.9200	1.6114	2.2032	2.9929	3.5923	4.1932	5.5939
80	.9199	1.6083	2.1977	2.9846	3.5818	4.1806	5.5765
85	.9198	1.6055	2.1929	2.9771	3.5725	4.1694	5.5610
90	.9197	1.6030	2.1885	2.9704	3.5640	4.1592	5.5468
95	.9196	1.6007	2.1845	2.9643	3.5563	4.1500	5.5341
100	.9195	1.5986	2.1808	2.9586	3.5492	4.1414	5.5222

TABLE D.8 Parameter δ for Standard Error of Log-Pearson Type III Distribution

COEFFICIENT OF SKEW	EXCEEDENCE PROBABILITY IN PERCENT						
	50.0	20.0	10.0	4.0	2.0	1.0	0.2
	CORRESPONDING RETURN PERIOD IN YEARS						
	2	5	10	25	50	100	500
.0	1.0801	1.1698	1.3748	1.8013	2.1992	2.6369	3.7212
.1	1.0808	1.2006	1.4368	1.9092	2.3429	2.8174	3.9902
.2	1.0830	1.2310	1.4990	2.0229	2.4990	3.0181	4.3001
.3	1.0866	1.2610	1.5611	2.1414	2.6661	3.2373	4.6486
.4	1.0918	1.2906	1.6228	2.2639	2.8428	3.4732	5.0336
.5	1.0987	1.3200	1.6840	2.3898	3.0283	3.7247	5.4534
.6	1.1073	1.3493	1.7442	2.5182	3.2215	3.9905	5.9066
.7	1.1179	1.3786	1.8033	2.6486	3.4215	4.2695	6.3920
.8	1.1304	1.4083	1.8611	2.7802	3.6274	4.5607	6.9085
.9	1.1449	1.4386	1.9172	2.9123	3.8383	4.8631	7.4550
1.0	1.1614	1.4701	1.9717	3.0442	4.0532	5.1756	8.0303
1.1	1.1799	1.5032	2.0243	3.1751	4.2711	5.4969	8.6335
1.2	1.2003	1.5385	2.0751	3.3043	4.4909	5.8259	9.2631
1.3	1.2223	1.5767	2.1242	3.4311	4.7115	6.1613	9.9177
1.4	1.2457	1.6186	2.1718	3.5546	4.9319	6.5017	10.5959
1.5	1.2701	1.6649	2.2182	3.6741	5.1507	6.8456	11.2957
1.6	1.2951	1.7164	2.2640	3.7891	5.3669	7.1915	12.0155
1.7	1.3202	1.7741	2.3097	3.8989	5.5792	7.5378	12.7231
1.8	1.3450	1.8385	2.3562	4.0029	5.7865	7.8829	13.5064
1.9	1.3687	1.9104	2.4046	4.1008	5.9875	8.2252	14.2731
2.0	1.3907	1.9904	2.4560	4.1922	6.1812	8.5629	15.0508

TABLE D.9 Confidence Limit Deviate Values for Normal and Log-Normal Distributions

CONFIDENCE LEVEL	SYSTEMATIC RECORD LENGTH n	EXCEEDENCE PROBABILITY									
		.002	.010	.020	.040	.100	.200	.500	.800	.990
.05	10	4.862	3.981	3.549	3.075	2.355	1.702	.580	−.317	−1.563
	15	4.304	3.520	3.136	2.713	2.068	1.482	.455	−.406	−1.677
	20	4.033	3.295	2.934	2.534	1.926	1.370	.387	−.460	−1.749
	25	3.868	3.158	2.809	2.425	1.838	1.301	.342	−.497	−1.801
	30	3.755	3.064	2.724	2.350	1.777	1.252	.310	−.525	−1.840
	40	3.608	2.941	2.613	2.251	1.697	1.188	.266	−.656	−1.896
	50	3.515	2.862	2.542	2.188	1.646	1.146	.237	−.592	−1.936
	60	3.448	2.807	2.492	2.143	1.609	1.116	.216	−.612	−1.966
	70	3.399	2.765	2.454	2.110	1.581	1.093	.199	−.629	−1.990
	80	3.360	2.733	2.425	2.083	1.559	1.076	.186	−.642	−2.010
	90	3.328	2.706	2.400	2.062	1.542	1.061	.175	−.652	−2.026
	100	3.301	2.684	2.380	2.044	1.527	1.049	.166	−.662	−2.040

continued

TABLE D.9 *(Continued)*

CONFIDENCE LEVEL	SYSTEMATIC RECORD LENGTH n	EXCEEDENCE PROBABILITY									
		.002	.010	.020	.040	.100	.200	.500	.800	.990
.95	10	1.989	1.563	1.348	1.104	.712	.317	−.580	−1.702	−3.981
	15	2.121	1.677	1.454	1.203	.802	.406	−.455	−1.482	−3.520
	20	2.204	1.749	1.522	1.266	.858	.460	−.387	−1.370	−3.295
	25	2.264	1.801	1.569	1.309	.898	.497	−.342	−1.301	−3.158
	30	2.310	1.840	1.605	1.342	.928	.525	−.310	−1.252	−3.064
	40	2.375	1.896	1.657	1.391	.970	.565	−.266	−1.188	−2.941
	50	2.421	1.936	1.694	1.424	1.000	.592	−.237	−1.146	−2.862
	60	2.456	1.966	1.722	1.450	1.022	.612	−.216	−1.116	−2.807
	70	2.484	1.990	1.745	1.470	1.040	.629	−.199	−1.093	−2.765
	80	2.507	2.010	1.762	1.487	1.054	.642	−.186	−1.076	−2.733
	90	2.526	2.026	1.778	1.500	1.066	.652	−.175	−1.061	−2.706
	100	2.542	2.040	1.791	1.512	1.077	.662	−.166	−1.049	−2.684

DERIVATION OF EQUATIONS FOR SWALE DESIGN

E.1 TRIANGULAR SHAPED SWALE

Recall that the total infiltration volume (F) per unit time equals the rate of infiltration (f) times the contact area (A) between the water and the swale bottom. This infiltration volume is expressed as

$$F = fA(\Delta t) \tag{E.1}$$

Whereas the contact area can be defined as the product of the wetted perimeter (P) and the length of the swale (L).

$$A = L \times P \tag{E.2}$$

For design, the volume of infiltration equals the volume of runoff water that enters the system, or

$$F = Q(\Delta t) \tag{E.3}$$

where Q = flow rate of water in ft³/sec.

Substituting Equations (E.2) and (E.3) into Equation (E.1), we get

$$Q = L \times P \times f$$

and for solving L, we obtain

$$L = \frac{3600Q}{P \times f} \tag{E.4}$$

where 3600 is a conversion factor from cfs to ft³/hr.

The perimeter of the swale depends on the rate of discharge through the swale. Thus, a relationship between the two is necessary to be able to use Equation E.4.

For a triangular section, the depth of flow (D) can be defined as

$$D = \frac{P}{2\sqrt{1 + Z^2}} \tag{E.5}$$

483

where

P = wetted perimeter of the swale

Z = horizontal component of side slope Z, which means Z horizontal per one vertical

so that

$$P = 2D\sqrt{1 + Z^2} \tag{E.6}$$

and the cross-sectional area of the flow can be expressed as

$$A = ZD^2 \tag{E.7}$$

One of the most common equations to compute flow of water in open channels is the Manning's equation. When the U.S. customary system of units is used, Manning's equation can be expressed as

$$Q = \frac{1.486}{n} AR^{2/3}S^{1/2} \tag{E.8}$$

where

Q = average flow rate (cfs)

n = coefficient of roughness

R = hydraulic radius (ft)

S = longitudinal slope of swale

A = cross-sectional area of flow (ft^2)

For trapezoidal sections, the hydraulic radius can be defined as

$$R = \frac{D}{2} \frac{Z}{\sqrt{1 + Z^2}} \tag{E.9}$$

Therefore, substituting Equations E.9 and E.7 into Equation E.8, we obtain

$$Q = \frac{1.486}{n} ZD^2 \left(\frac{DZ}{2\sqrt{1 + Z^2}}\right)^{2/3} (S)^{1/2} \tag{E.10}$$

or

$$Q = \frac{1.486}{1.59n} \frac{Z^{5/3}S^{1/2}D^{8/3}}{(1 + Z^2)^{1/3}} \tag{E.11}$$

and

$$Q = \frac{Z^{5/3}S^{1/2}D^{8/3}}{1.073n(1 + Z^2)^{1/3}} \tag{E.12}$$

Solving for D, it is found that

$$D = \left[\frac{1.073Qn(1 + Z^2)^{1/3}}{Z^{5/3}S^{1/2}}\right]^{3/8} \tag{E.13}$$

Substituting the value of D expressed in Equation E.13 into Equation E.6, the wetted perimeter can be expressed as

$$P = 2(1 + Z^2)^{1/2}\left[\frac{1.073Qn(1 + Z^2)^{1/3}}{Z^{5/3}S^{1/2}}\right]^{3/8} \tag{E.14}$$

Finally, if one substitutes the value of P into Equation E.4 we get

$$L = \frac{3600Q}{2\left[\dfrac{1.073Qn(1 + Z^2)^{1/3}}{Z^{5/3}S^{1/2}}\right]^{3/8}(1 + Z^2)^{1/2} \times \dfrac{f}{12}} \tag{E.15}$$

Note that f is divided by 12 in order to be able to use the common units of in./hr. Then,

$$L = \frac{3600Q}{\dfrac{2.054Q^{3/8}n^{3/8}(1 + Z^2)^{1/8}(1 + Z^2)^{1/2}f}{12Z^{5/8}S^{3/16}}} \tag{E.16}$$

and

$$L = \frac{21,032QZ^{5/8}S^{3/16}}{Q^{3/8}(1 + Z^2)^{5/8}fn^{3/8}} \tag{E.17}$$

Finally, we obtain

$$L = \frac{21,032Q^{5/8}Z^{5/8}S^{3/16}}{n^{3/8}(1 + Z^2)^{5/8}f} \tag{E.18}$$

where
L = length of swale (ft)
Q = average flow rate to be percolated (cfs)
Z = horizontal distance per one foot of elevation change inside slope
S = longitudinal or flow slope
n = Manning's roughness coefficient
f = infiltration rate (in./hr)

Equation E.18 can be used to predict the necessary length of swale to percolate certain amounts of runoff waters when U.S. Customary system of units is used.

When the International System of units is used, Manning's equation is written in the form

$$Q = \frac{1}{n} A R^{2/3} S^{1/2} \tag{E.19}$$

and Equation E.14 would be expressed as

$$P = 2 \left[\frac{1.5874 Q n (1 + Z^2)^{1/3}}{Z^{5/3} S^{1/2}} \right]^{3/8} (1 + Z^2)^{1/2} \tag{E.20}$$

Substituting this value into Equation E.4, we get

$$L = \frac{3600 Q}{2 \left[\dfrac{1.5874 Q n (1 + Z^2)^{1/3}}{Z^{5/3} S^{1/2}} \right]^{3/8} (1 + Z^2)^{1/2} \dfrac{f}{100}} \tag{E.21}$$

Note that f is divided by 100 in order to use the infiltration rate in units of cm/hr. Then,

$$L = \frac{3600 Q}{\dfrac{2.378 Q^{3/8} (1 + Z^2)^{5/8} f n^{3/8}}{100 Z^{5/8} S^{3/16}}} \tag{E.22}$$

which can be expressed as

$$L = \frac{151{,}361 Q^{5/8} Z^{3/8} S^{3/16}}{n^{3/8} (1 + Z^2)^{5/8} f} \tag{E.23}$$

where
L = length of swale (m)
Q = average flow rate (m³/s)
S = longitudinal slope
n = Manning's roughness coefficient
Z = side slope
f = infiltration rate (cm/hr)

Equation E.23 can be used to calculate the necessary length of swale (L) to percolate the runoff (Q) when the International System of units is used.

E.2   TRAPEZOIDAL SHAPED SWALE

Applying Manning's equation for U.S. Customary System of units, we obtain

$$Q = \frac{1.486}{n} A R^{2/3} S^{1/2} \tag{E.24}$$

For a trapezoidal section, the cross-sectional area of flow is defined

$$A = BD + ZD^2 \tag{E.25}$$

where
B = bottom width of the swale (ft)
D = depth of flow (ft)
Z = side slope

and, the wetted perimeter (P) can be defined as

$$P = B + 2D\sqrt{1 + Z^2} \tag{E.26}$$

Substituting $R = A/P$ and Equations E.25 and E.26 into Equation E.24, it is found that

$$Q = \frac{1.486}{n}(BD + ZD^2)\left[\frac{(B + ZD)D}{B + 2D(1 + Z^2)^{1/2}}\right]^{2/3} S^{1/2} \tag{E.27}$$

which equals

$$Q = \frac{1.486}{n}\frac{D^{5/3}(B + ZD)^{5/3}}{(B + 2D\sqrt{1 + Z^2})^{2/3}} S^{1/2} \tag{E.28}$$

For design, the most efficient section should be used. For a trapezoidal section, the most efficient section is defined as

$$\frac{B}{D} = 2(\sqrt{1 + Z^2} - Z) \tag{E.29}$$

Solving for D we get

$$D = \frac{B}{2(\sqrt{1 + Z^2} - Z)} \tag{E.30}$$

Substituting

$$Q = \frac{1.486}{n}\frac{D^{5/3}\left[B + \dfrac{BZ}{2(\sqrt{1 + Z^2} - Z)}\right]^{5/3}}{\left[B + \dfrac{2B\sqrt{1 + Z^2}}{2(\sqrt{1 + Z^2} - Z)}\right]^{2/3}} S^{1/2} \tag{E.31}$$

Solving for D we get

$$D = \left\{ \frac{nQ\left[B + \dfrac{B\sqrt{1 + Z^2}}{(\sqrt{1 + Z^2} - Z)} \right]^{2/3}}{1.486\left[B + \dfrac{BZ}{2(\sqrt{1 + Z^2} - Z)} \right]^{5/3}} \right\}^{3/5} \tag{E.32}$$

and

$$D = \frac{Q^{3/5}\left(B + \dfrac{B\sqrt{1 + Z^2}}{(\sqrt{1 + Z^2} - Z)} \right)^{2/5} n^{3/5}}{1.268\left[B + \dfrac{BZ}{2(\sqrt{1 + Z^2} - Z)} \right]} \tag{E.33}$$

Substituting in Equation E.26 we get

$$P = B + \frac{2Q^{3/5}n^{3/5}\left[\dfrac{B\sqrt{1 + Z^2}}{B + (\sqrt{1 + Z^2} - Z)} \right]^{2/5}}{1.268\left[B + \dfrac{BZ}{2(\sqrt{1 + Z^2} - Z)} \right]} (1 + Z^2)^{1/2} \tag{E.34}$$

$$D = \left\{ \frac{1.068nQ(1 + Z^2)^{1/3}}{S^{1/2}Z^{2/3}[2(1 + Z^2)^{1/2} - Z]} \right\}^{3/8} \tag{E.35}$$

Substituting the value of D in Equation E.35 into Equation E.26 we get

$$P = B + 2\left\{ \frac{1.068nQ(1 + Z^2)^{1/3}}{S^{1/2}Z^{2/3}[2(1 + Z^2)^{1/2} - Z]} \right\}^{3/8} (1 + Z^2)^{1/2} \tag{E.36}$$

Recall from the triangular section derivation that

$$L = \frac{Q(3600)}{P \times f} \tag{E.37}$$

which, for units of in./hr for the infiltration rate, can be expressed as

$$L = \frac{3600Q}{P \times \dfrac{f}{12}} \tag{E.38}$$

or

$$L = \frac{43,200Q}{P \times f} \tag{E.39}$$

Substituting the value of wetted perimeter defined in Equation E.36 into Equation E.39 we obtain

$$L = \frac{43,200Q}{\left(B + 2\left\{\dfrac{1.068nQ(1 + Z^2)^{1/3}}{S^{1/2}Z^{2/3}2[(1 + Z^2)^{1/2} - Z]}\right\}^{3/8}(1 + Z^2)^{1/2}\right)f} \tag{E.40}$$

where

L = length of swale (ft)
B = bottom width of swale (ft)
Q = average flow rate (cfs)
n = Manning's roughness coefficient
Z = side slope, horizontal component
S = longitudinal slope
f = infiltration rate of swale (in./hr)

If the International System of units is used, the length of swale necessary to percolate the runoff (Q) can be expressed as

$$L = \frac{360,000Q}{\left(B + 2\left\{\dfrac{nQ(1 + Z^2)^{1/3}}{S^{1/2}Z^{2/3}2[(1 + Z^2)^{1/2} - Z]}\right\}^{3/8}(1 + Z^2)^{1/2}\right)f} \tag{E.41}$$

where

L = length of swale (m)
Q = average flow rate (cfs)
f = infiltration rate (cm/hr)
n = Manning's roughness coefficient
Z = side slope, horizontal component
S = longitudinal slope
B = bottom width of swale (m)

GUTTERS, INLETS, AND ROADWAYS

F.1 INTRODUCTION

This appendix aids in the calculation of peak discharges and time of concentration for all small watersheds composed primarily of impervious surfaces (i.e. roadways). The variables of major concern are the size of water shed, rainfall intensities, impervious materials, and the geometry of the roadway. The geometry variables include details on drainage slope, cross-slope, gutter cross sections, and inlet designs. These variables determine the depth of flow, velocity, and flow discharge, which then determines time of concentration.

Rainfall intensity to be used for estimating maximum discharge is specified by an estimate of the time of concentration. If the time of concentration were about 10 min, the IDF curve will specify the intensity corresponding to a duration of 10 min and a specific return frequency (say 1 in 10 yr). Since the watershed area is relatively easy to define and the roadway watershed area is impervious, the peak discharge can be estimated using the rational formula. However, it may be necessary to check the assumption of the time of concentration because it affects the choice of rainfall intensity. Since there are many details of a hydraulic nature, not hydrologic, that affect roadway and gutter design, the details are outlined in this appendix.

F.2 GUTTER DESIGN

Rainfall excess from a roadway watershed either is by overland flow or intercepted by some form of a gutter. Typical gutter shapes are of three types: Curb, V shaped, and curb with depressed gutter. All three cross-sections are shown in Figure F.1. Typical commercial area design is done using the curb type, whereas residential areas use both the curb and no-curb V shapes. The no-curb type is primarily used where swales (ditches that both infiltrate and transport) are used in a drainage plan, and usually average flow velocity is less than 2 fps.

The variables used to determine flow rate (Q) for a gutter section are:

Longitudinal slope (ft/ft)	S
Cross-slope (ft/ft)	S_x
Width of flow (ft)	T
Depth of gutter flow (ft)	d
Width of depressed area (ft)	W

For a curb design with a depressed gutter area, two flow rates are calculated and designated as Q_s for standard area flow rate and Q_w for the depressed area. The total flow is the sum of Q_s and Q_w.

Since gutter flow is a form of open channel flow, a modified form of Manning's equation can be used. For such conditions, the hydraulic radius is adjusted for the very shallow flow. The modified formula for the curb type gutter is (Izzard, 1946):

$$Q = (0.56/n)S_x^{5/3}S^{1/2}T^{8/3} \tag{F.1}$$

where

Q = gutter flow rate (cfs)
S_x = roadway cross-slope (ft/ft)
S = longitudinal slope (ft/ft)
T = width of flow (ft)
n = Manning's coefficient, see Table F.1

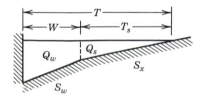

1. Curb with straight cross-slope

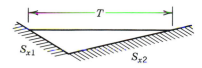

2. No curb, V-shaped gutter

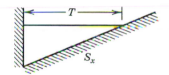

3. Curb with depressed gutter

FIGURE F.1 Typical gutter sections. *Source:* The Federal Highway Administration, 1984.

TABLE F.1   Manning's *n* Values for Street and Pavement Gutters

TYPE OF GUTTER OR PAVEMENT	MANNING'S *n*
Concrete gutter, troweled finish	.012
Asphalt pavement	
Smooth texture	.013
Rough texture	.016
Concrete gutter with asphalt pavement	
Smooth	.013
Rough	.015
Concrete pavement	
Float finish	.014
Broom finish	.016
For gutters with small slope, where sediment may accumulate, increase above values of *n* by	.002

Note: Estimates are by the Federal Highway Administration.
Source: Department of Transportation, Federal Highway Administration (1984).

Manning's coefficients were tabulated by the U.S. Department of Transportation (1984). Various charts have been developed to aid in solving Manning's equation, one is shown in Figure F.2. Given longitudinal slope and cross-slope with an estimate of the water profile depth and width, and for a type of material, velocity (V) can be estimated. The quantity (Vn) refers to the product of velocity and Manning's coefficient. The solution to the example given in Figure F.2 is $V = 2$ ft/sec and $Vn = 0.032$ ft/sec.

For a depressed gutter or a section with composite cross slopes, use Figures F.1 and F.2 to calculate the flow Q_s, then use Figure F.3 to estimate the relative fraction of flow in the depressed section or

$$E_0 = \frac{Q_w}{Q} \tag{F.2}$$

where

E_0 = relative flow fraction
Q_w = flow rate in depressed area (cfs)
Q = total flow rate (cfs)

The total discharge is calculated from

$$Q = Q_s + Q_w \tag{F.3}$$

but $Q_w = E_0 Q$; thus

$$Q = Q_s + E_0 Q$$

and

$$Q = \frac{Q_s}{(1 - E_0)} \tag{F.4}$$

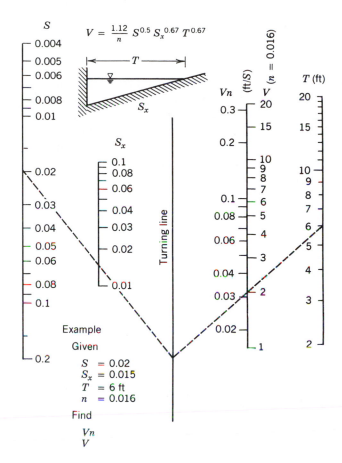

$$V = \frac{1.12}{n} S^{0.5} S_x^{0.67} T^{0.67}$$

Example

Given

$S = 0.02$
$S_x = 0.015$
$T = 6$ ft
$n = 0.016$

Find

Vn
V

FIGURE F.2 Velocity in triangular gutter sections. *Source:* The Federal Highway Administration, 1984.

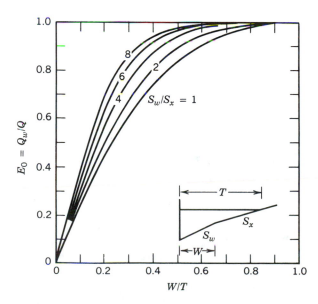

FIGURE F.3 The ratio of frontal flow to total gutter flow. *Source:* The Federal Highway Administration, 1984.

The depth of gutter flow and width of gutter flow are related by

$$d = TS_x \qquad\qquad (F.5)$$

From the top of a continuous longitudinal slope to the end or bottom, the watershed area grows, and if the roadway width to the center line or cross-sectional slope peak remains constant, the area increases linearly. As the area increases, the depth and width of flow increases. Velocity of flow in the gutter is not constant but will vary with the depth and width of flow. If the watershed area varies linearly with gutter distance, an average velocity may be at first estimated as the velocity of flow at the midpoint in the watershed. This velocity is used to estimate time of concentration, and thus intensity of rainfall. Gutter design may be iterative. However, the gutter area must be capable of transporting the flow rate for any condition, otherwise overtopping or flooding will occur. Inlets to capture the water are designed to eliminate flooding.

F.3   INLET DESIGN

There are essentially two types of inlets for the collection of stormwater: a curb opening and a gutter opening. These openings can also be used in combination. The typical curb and gutter inlets with grate are shown in Figure F.4. The curb inlet is formed by an opening in the curb face and generally has a depressed gutter section in front. The gutter inlet is formed by an opening in the gutter which is covered by a metal grate. These are usually square or rectangular in shape. However, in some applications, a slotted pipe that allows drainage to enter continuously along its

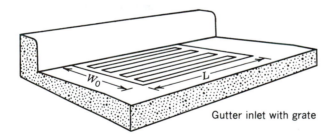

Gutter inlet with grate

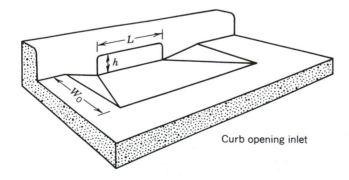

Curb opening inlet

FIGURE F.4   Perspective views of gutter and curb-opening inlets. *Source:* The Federal Highway Administration, 1984.

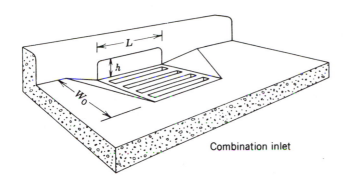

Combination inlet

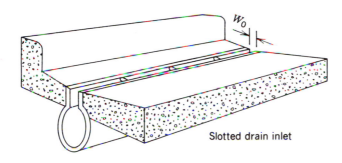

Slotted drain inlet

FIGURE F.5 Perspective views of combination and slotted drain inlets. *Source:* The Federal Highway Administration, 1984.

longitudinal axis is used (see Figure F.5). The combination inlet is also shown in Figure F.5.

Curb-opening inlets allow a capture of a specific maximum quantity and flow rate. They operate as weirs up to a depth equal to the opening height (h), then the inlet operates as an orifice at about 1.4 times the opening height. There is a transition section of flow from the weir to orifice flow rate. The weir flow equation is

$$Q_i = 2.3(L + 1.8W_o)d^{1.5} \tag{F.6}$$

where

Q_i = capture flow rate of a curb opening (cfs)
 L = length of curb opening (ft)
W_o = width of curb opening (ft)
 d = depth at curb (ft) and $d \leq h + d \leq h + a$ where h = opening height (ft) and a = depressing depth (ft).

Without a depressed gutter, the weir flow equation for the curb opening is

$$Q_i = 2.3Ld^{1.5} \qquad d \leq h \tag{F.7}$$

The orifice equation for the curb opening is

$$Q_i = 0.67A\left[2g\left(d - \frac{h}{2}\right)\right]^{1/2} \tag{F.8}$$

where

$A = hL$ = area of opening (ft^2)
g = acceleration due to gravity (32.2 ft/sec^2)

Capture efficiency is

$$E = \left(\frac{Q_i}{Q}\right)(100) \tag{F.9}$$

where

E = efficiency (%)
Q = total gutter flow (cfs)

Carryover is defined as

$$Q_c = Q - Q_i \tag{F.10}$$

where Q_c = bypass or carry over flow (cfs). For a slotted pipe inlet with slot widths greater than 1.75 in., the length of inlet required for complete interception is

$$L_T = 0.6Q^{0.42}S^{0.3}\left(\frac{1}{nS_x}\right)^{0.6} \tag{F.11}$$

where L_T = length of slotted pipe inlet (ft). The interception capacity of a gutter inlet can in very general terms be defined as

$$Q_i = 3.0P_G d^{1.5} \tag{F.12}$$

where P_G = perimeter of a grate (ft).

There are many different grate configurations that affect the "catch" efficiency. Empirical design procedures for interception capacity and efficiencies for seven grate types, slotted inlets, curb inlets, and combination ones are given in greater detail by the U.S. Federal Highway Administration, FHWA (1984). Also presented are additional details on roadway geometry, embankment, inlets, bridge deck inlets, un-steady-state flow, and IDF curve development. Roadway designers are encouraged to review the FHWA publication and similar ones.

F.4   REFERENCE

Federal Highway Administration. 1984. HEC No. 12, U.S. Department of Transportation, McLean, VA.

Izzard, C. F., "Hydraulics of Runoff from Developed Surfaces," *Proceedings, Highway Research Board,* Vol. 26, 1946, pp. 129–150.

DERIVATION OF THE FUNCTIONAL FORMS OF THE CONTINUOUS TIME CONVOLUTION FORMULAS

For this model, rainfall excess is convoluted with a routing function that can be changed with time to produce an output hydrograph. The model is derived from a mass balance, and it provides for nonlinearity of the routing function (impulse response) in the watershed. Another advantage of the routing response function is that it includes only one parameter to be estimated. This simplicity has enabled the convolution integral to be solved, thus allowing for time continuous solution of the model. Also, the model has been improved by representing the rainfall excess with straight line increasing or decreasing functions over any time interval to reflect more accurately the constantly changing amount of rainfall excess during a storm.

G.1 ROUTING FUNCTION

There have been several response functions developed by different hydrologists for the convolution integral, but most of these functions are based on empirical equations involving many parameters. The theory begins with the assumption that a linear relationship exists between the amount of storage and the outflow rate as given by

$$Q = kS \tag{G.1}$$

or in differential form

$$\frac{dQ}{dt} = k\left(\frac{dS}{dt}\right) \tag{G.2}$$

where
Q = outflow rate (L^3/t)
S = storage volume (L^3)
k = storage coefficient (t^{-1})
t = time (t)

where k is a type of friction factor reflecting the watershed's resistance to flow.

By using this relationship and a mass balance (continuity equation) for a watershed, the convolutional integral with the routing function can be derived in the following manner. The continuity equation for a watershed is

$$\text{input} - \text{output} \pm \text{generation} = \text{accumulation}$$

Assuming there is no generation in the watershed then

$$r(t) - Q(t) = \frac{dS}{dt} \tag{G.3}$$

where $r(t)$ = input (rainfall excess rate).

Substitute the storage–outflow relationship (Equation G.2) into the continuity equation (Equation G.3), and there results

$$r(t) - Q(t) = \left(\frac{1}{k}\right)\left(\frac{dQ}{dt}\right) \tag{G.4}$$

Multiply both sides of Equation G.4 by the exponential function, e^{kt}, and we get

$$e^{kt}\frac{dQ}{dt} + ke^{kt}Q(t) = ke^{kt}r(t) \tag{G.5}$$

By definition:

$$\frac{d}{dt}(Qe^{kt}) = e^{kt}\frac{dQ}{dt} + ke^{kt}Q(t) \tag{G.6}$$

By substituting Equation G.5 into Equation G.6 we get:

$$\frac{d}{dt}(Qe^{kt}) = ke^{kt}r(t) \tag{G.7}$$

Separate the variables:

$$d(Qe^{kt}) = ke^{kt}r(t)\,dt \tag{G.8}$$

Integrate both sides:

$$\int d(Qe^{kt}) = \int ke^{kt}r(t)\,dt \tag{G.9}$$

Substitute τ (a dummy variable of time) for t in both sides of the equation where

$$t = \tau \qquad dt = d\tau$$

so when $t = 0$, $\tau = 0$, and when $t = t$, $\tau = t$:

$$d(Qe^{k\tau}) = ke^{k\tau}r(\tau)\,d\tau \tag{G.10}$$

Make the integrals definite by using the initial conditions at $t = 0$, $Q = Q_0$ and at $t = t$, $Q = Q(t)$:

$$\int_{Q_0,0}^{Q(t),t} d(Qe^{k\tau}) = \int_0^t ke^{k\tau}r(\tau)\,d\tau \tag{G.11}$$

By solving the left-hand side integral:

$$Q(t)e^{kt} - Q_0 = \int_0^t ke^{k\tau}r(\tau)\,d\tau \tag{G.12}$$

Q_0 is the runoff at $t = 0$ (also called base flow). If this flow is separated from the runoff caused by rainfall, then Q_0 is taken out of the equation and

$$Q(t)e^{kt} = \int_0^t ke^{k\tau}r(\tau)\,d\tau \tag{G.13}$$

where $Q(t)$ is the outflow rate excluding base flow. Dividing both sides of the equation by e^{kt} yields

$$Q(t) = e^{-kt}\int_0^t ke^{k\tau}r(\tau)\,d\tau \tag{G.14}$$

which equals

$$Q(t) = \int_0^t ke^{-(t-\tau)k}r(\tau)\,d\tau \tag{G.15}$$

Equation G.15 is the convolution equation using the assumption of an exponential routing response. The rainfall excess function is $r(\tau)$ and $ke^{-(t-\tau)k}$ is the exponential routing function.

SELECTED HYDROLOGIC SOIL CLASSIFICATIONS

The soil scientists of the U.S. Department of Agriculture, Natural Resources Conservation Service (NRCS), have classified over 4000 soil types. Most areas of the United States have been classified and reported in "county by county soil surveys" readily available from the NRCS. The tables in this appendix give a representative example of this classification. County Survey soil maps should be consulted for additional detail.

TABLE H.1 Soil Names and Hydrologic Classifications

Aaberg	C	Aberdeen	D	Abscota	B
Aastad	B	Abes	D	Absher	D
Abac	D	Abilene	C	Absted	D
Abajo	C	Abington	B	Acacio	C
Abbott	D	Abiqua	C	Academy	C
Abbottstown	C	Abo	B/C	Acadia	D
Abcal	D	Abok	D	Acana	D
Abegg	B	Abra	C	Acasco	D
Abela	B	Abraham	B	Aceitunas	B
Abell	B	Absarkee	C	Acel	D

TABLE H.1   *(Continued)*

Acker	B	Aeneas	B	Akela	C
Ackmen	B	Aetna	B	Aladdin	B
Acme	C	Afton	D	Alae	A
Aco	B	Agar	B	Alaeloa	B
Acolita	B	Agassiz	D	Alaga	A
Acoma	C	Agate	D	Alakai	D
Acove	C	Agawam	B	Alama	B
Acree	C	Agency	C	Alamance	B
Acrelane	C	Ager	D	Alamo	D
Acton	B	Agner	B	Alamosa	C
Acuff	B	Agnew	B/C	Alapaha	D
Acworth	B	Agnus	B	Alapai	A
Acy	C	Agua	B	Alban	B
Ada	B	Agua Dulce	C	Albano	D
Adair	D	Agua Fria	C	Albany	C
Adams	A	Aguadilla	B	Albaton	D
Adamson	B	Agualt	B	Albee	C
Adamstown		Agueda	B	Albemarle	B
Adamsville	C	Aguilita	B	Albertville	C
Adaton	D	Aguirre	D	Albia	C
Adaven	D	Agustin	B	Albion	B
Addielou	C	Ahatone	D	Albrights	C
Addison	D	Ahl	C	Alcalde	C
Addy	C	Ahlstrom	C	Alcester	B
Ade	A	Ahmeek	B	Alcoa	B
Adel	A	Aholt	D	Alcona	B
Adelaide	D	Ahtanum	C	Alcova	B
Adelanto	B	Ahwahnee	C	Alda	C
Adelino	B	Aibonito	C	Aldax	D
Adelphia	C	Aiken	B/C	Alden	D
Adena	C	Aikman	D	Alder	B
Adger	D	Ailey	B	Alderdale	C
Adilis	A	Ainakea	B	Alderwood	C
Adirondack		Airmont	C	Aldino	C
Adiv	B	Airotsa	B	Aldwell	C
Adjuntas	C	Airport	D	Aleknagik	B
Adkins	B	Aits	B	Alemeda	C
Adler	C	Ajo	C	Alex	B
Adolph	D	Akaka	A	Alexandria	C
Adrian	A/D	Akaska	B	Alexis	B

continued

TABLE H.1   (Continued)

Alford	B	Alpon	B	Ammon	B
Algansee	B	Alpowa	B	Amole	C
Algerita	B	Alps	C	Amor	B
Algiers	C/D	Alsea	B	Amos	C
Algoma	B/D	Alspaugh	C	Amsden	B
Alhambra	B	Alstad	B	Amsterdam	B
Alice	A	Alstown	B	Amtoft	D
Alicel	B	Altamont	D	Amy	D
Alicia	B	Altavista	C	Anacapa	B
Alida	B	Altdorf	D	Anahuac	D
Alikchi	B	Altmar	B	Anamite	D
Aline	A	Alto	C	Anapra	B
Alko	D	Altoga	C	Anasazi	B
Allagash	B	Alton	B	Anatone	D
Allard	B	Altus	B	Anaverde	D
Allegheny	B	Altvan	B	Anawalt	D
Allemands	D	Alum	B	Ancho	B
Allen	B	Alusa	D	Anchor Bay	D
Allendale	C	Alvin	B	Anchor Point	D
Allens Park	B	Alvira	C	Anchorage	A
Allensville	C	Alviso	D	Anclote	D
Allentine	D	Alvor	C	Anco	C
Allenwood	B	Amador	D	Anderly	C
Allessio	B	Amagon	D	Anders	C
Alley	C	Amalu	D	Anderson	B
Alliance	B	Amana	B	Andes	C
Alligator	D	Amargosa	D	Andorinia	C
Allis	D	Amarillo	B	Andover	D
Allison	C	Amasa	B	Andreen	B
Allouez	C	Amberson		Andreeson	C
Alloway		Amboy	C	Andres	B
Almac	B	Ambraw	C	Andrews	C
Almena	C	Amedee	A	Aned	D
Almont	D	Amelia	B	Aneth	A
Almy	B	Amenia	B	Angelica	D
Aloha	C	Americus	A	Angelina	B/D
Alonso	B	Ames	C	Angelo	C
Alovar	C	Amesha	B	Angie	C
Alpena	B	Amherst	C	Angle	A
Alpha	C	Amity	C	Anglen	B

continued

TABLE H.1   *(Continued)*

Angola	C	Appling	B	Arlando	B
Angostura	B	Apron	B	Arle	B
Anhalt	D	Apt	C	Arling	D
Aniak	D	Aptakisic	B	Arlington	C
Anita	D	Araby		Arloval	C
Ankeny	A	Arada	C	Armagh	D
Anlauf	C	Aransas	D	Armijo	D
Annabella	B	Arapien	C	Armington	D
Annandale	C	Arave	D	Armjohee	D
Anniston	B	Araveton	B	Armour	B
Anoka	A	Arbela	C	Armster	C
Anones	C	Arbone	B	Armstrong	D
Ansari	D	Arbor	B	Arnegard	B
Ansel	B	Arbuckle	B	Arnhart	C
Anselmo	A	Arcata	B	Arnheim	C
Anson	B	Arch	B	Arno	D
Ant Flat	C	Archabal	B	Arnold	B
Antelope Springs	C	Archer	C	Arnot	C/D
Antero	C	Archin	C	Arny	A
Antho	B	Arco	B	Arosa	C
Anthony	B	Arcola	C	Arp	C
Antigo	B	Ard	C	Arrington	B
Antilon	B	Arden	B	Arritola	D
Antioch	D	Ardenvoir	B	Arrolime	C
Antler	C	Ardilla	C	Arron	D
Antoine	C	Ardostook		Arrow	B
Antrobus	B	Aredale	B	Arrowsmith	B
Anty	B	Arena	C	Arroyo Seco	B
Anvik	B	Arenales		Arta	C
Anway	B	Arendsa	A	Artois	C
Anza	B	Arendtsville	B	Arvada	D
Anziano	C	Arenzville	B	Arvana	C
Apache	D	Argonaut	D	Arveson	D
Apakuie	A	Arguello	B	Arvilla	B
Apishapa	C	Argyle	B	Arzell	C
Apison	B	Arho	B	Asa	B
Apopka	A	Ariel	C	Asbury	B
Appian	C	Arizo	A	Aschoff	B
Applegate	C	Arkabutla	C	Ash Springs	C
Appleton	C	Arkport	B	Ashby	C

continued

TABLE H.1   *(Continued)*

Ashcroft	B	Atsion	C	Azarman	C
Ashdale	B	Atterberry	B	Azeltine	B
Ashe	B	Attewan	A	Azfield	B
Ashkum	C	Attica	B	Aztalan	B
Ashlar	B	Attleboro		Aztec	B
Ashley	A	Atwater	B	Azule	C
Ashton	B	Atwell	C/D	Azwell	B
Ashue	B	Atwood	B	Babb	A
Ashuelot	C	Au Gres	B	Babbington	B
Ashwood	C	Aubbeenaubbe	B	Babcock	C
Askew	C	Auberry	C/D	Babylon	A
Aso	C	Auburn	D	Baca	C
Asotin	C	Auburndale	B	Bach	D
Aspen	B	Audian	C	Bachus	C
Aspermont	B	Augsburg	B	Backbone	A
Assalon	B	Augusta	C	Baculan	A
Assinniboine	B	Auld	D	Badenaugh	B
Assumption	B	Aura	B	Badger	C
Astatula	A	Aurora	C	Badgerton	B
Astor	A/D	Austin	C	Bado	D
Astoria	B	Austwell	D	Badus	C
Atascadero	C	Auxvasse	D	Bagard	C
Atascosa	D	Auzqui	B	Bagdad	B
Atco	B	Ava	C	Baggott	D
Atencio	B	Avalanche	B	Bagley	B
Atepic	D	Avalon	B	Bahem	B
Athelwold	B	Avery	B	Baile	D
Athena	B	Avon	C	Bainville	C
Athens	B	Avonburg	D	Baird Hollow	C
Atherly	B	Avondale	E	Bajura	D
Atherton	B/D	Awbrey	D	Bakeoven	D
Athmar	C	Axtell	D	Baker	C
Athol	B	Ayar	D	Baker Pass	B
Atkinson	B	Aycock	B	Balaam	A
Atlas	D	Ayr	B	Balch	D
Atlee	C	Ayres	D	Balcom	B
Atmore	B/D	Ayrshire	C	Bald	C
Atoka	C	Aysees	B	Balder	C
Aton	B	Ayun	B	Baldock	B/C
Atrypa	C	Azaar	C	Baldwin	D

continued

TABLE H.1   *(Continued)*

Baldy	B	Barishman	C	Bassfield	B
Bale	C	Barker	C	Bassler	D
Ballard	B	Barkerville	C	Bastian	D
Baller	D	Barkley	B	Bastrop	B
Ballinger	C	Barlane	D	Bata	A
Balm	B/C	Barling	C	Batavia	B
Balman	B/C	Barlow	B	Bates	B
Balon	B	Barnard	D	Bath	C
Baltic	D	Barnes	B	Batterson	D
Baltimore	B	Barneston	B	Battle Creek	C
Balto	D	Barney	A	Batza	D
Bamber	B	Barnhardt	B	Baudette	B
Bamforth	B	Barnstead		Bauer	C
Bancas	B	Barnum	B	Baugh	B/C
Bancroft	B	Baron	C	Baxter	B
Bandera	B	Barrada	D	Baxtervillle	B
Bango	C	Barrett	D	Bayamon	B
Bangston	A	Barrington	B	Bayard	A
Bangur	B	Barron	B	Baybord	D
Bankard	A	Barronett	C	Bayerton	C
Banks	A	Barrows	D	Baylor	D
Banner	C	Barry	D	Bayshore	B/C
Bannerville	C/D	Barstow	B	Bayside	C
Bannock	B	Barth	C	Bayucos	D
Banquete	D	Bartine	C	Baywood	A
Barabou	B	Bartle	D	Bazette	C
Baraga	C	Bartley	C	Bazile	B
Barbary	D	Barton	B	Bead	C
Barboor	B	Bartonflat	B	Beadle	C
Barbourville	B	Bascom	B	Beales	A
Barclay	C	Basehor	D	Bear Basin	B
Barco	B	Bashaw	D	Bear Creek	C
Barcus	B	Basher	B	Bear Lake	C
Bard	D	Basile	D	Bear Prairie	C
Barden	C	Basin	C	Beardall	C
Bardley	C	Basinger	C	Bearden	D
Barela	C	Basket	C	Beardstown	A
Barfield	D	Bass	A	Bearmough	B
Barfuss	B	Bassel	B	Bearpaw	B
Barge	C	Bassett	B	Bearskin	D

continued

TABLE H.1   *(Continued)*

Beasley	C	Belfast	B	Benz	D
Beasun	C	Belfield	B	Beotia	B
Beaton	C	Belfore	B	Beowawe	D
Beatty	C	Belgrade	B	Bercail	C
Beaucoup	B	Belinda	D	Berda	B
Beauford	D	Belknap	C	Berea	C
Beaumont	D	Bellamy	C	Bereniceton	B
Beauregard	C	Bellavista	D	Berent	A
Beausite	B	Belle	B	Bergland	D
Beauvais	B	Bellefontaine		Bergstrom	B
Beaverton	A	Bellicum	B	Berino	B
Beck	C	Bellingham	C	Berkeley	
Becker	B	Bellpine	C	Berks	C
Becket	C	Belmont	B	Berkshire	B
Beckley	B	Belmore	B	Berlin	C
Beckton	D	Belt	D	Bermaldo	B
Beckwith	C	Belted	D	Bermesa	C
Beckwourth	B	Belton	C	Bermudian	B
Becreek	B	Beltrami	B	Bernal	D
Bedford	C	Beltsville	C	Bernard	D
Bedington	B	Beluga	D	Bernardino	C
Bedner	C	Belvoir	C	Bernardston	C
Beebe	A	Ben Hur	B	Bernhill	B
Beecher	C	Ben Lumond	B	Bernice	A
Beechy		Benclare	C	Berning	C
Beehive	B	Benevola	C	Berrendos	D
Beek	C	Benewah	C	Berryland	D
Beenom	D	Benfirld	C	Bertelson	B
Beezar	B	Benge	B	Berthoud	B
Begay	B	Benin	D	Bertie	C
Begoshian	C	Benito	D	Bertolotti	B
Behanin	B	Benjamin	D	Bertrand	B
Behemotosh	B	Benman	A	Berville	D
Behring	D	Benndale	B	Beryl	B
Beirman	D	Bennett	C	Bessemer	B
Bejucos	B	Bennington	D	Bethany	C
Belcher	D	Benoit	D	Bethel	D
Belden	D	Benson	C/D	Betteravia	C
Belding	B	Benteen	B	Betts	B
Belen	C	Bentonville	C	Beulah	B

continued

TABLE H.1   *(Continued)*

Bevent	B	Birchwood	C	Blackwater	D
Beverly	B	Birdow	B	Blackwell	B/D
Bew	D	Birds	C	Bladen	D
Bewleyville	B	Birdsall	D	Blago	D
Bewlin	D	Birdsboro	B	Blaine	B
Bexar	C	Birdsley	D	Blair	C
Bezzant	B	Birkbeck	B	Blairton	C
Bibb	B/D	Bisbee	A	Blake	C
Bibon	A	Biscay	C	Blakeland	A
Bickelton	B	Bishop	B/C	Blakeney	C
Bickleton	C	Bisping	B	Blakeport	B
Bickmore	C	Bissell	B	Blalock	D
Bicondoa	C	Bisti	C	Blamer	C
Biddeford	D	Bit	D	Blanca	B
Biddleman	C	Bitter Spring	C	Blanchard	A
Bidman	C	Bitteron	A	Blanchester	B/D
Bidwell	B	Bitterroot	C	Bland	C
Bieber	D	Bitton	B	Blandford	C
Bienville	A	Bixby	B	Blanding	B
Big Blue	D	Bjork	C	Blaney	B
Big Horn	C	Blachly	C	Blanket	C
Big Timber	D	Black Butte	C	Blanton	A
Bigel	A	Black Canyon	D	Blanyon	C
Bigelow	C	Black Mountain	D	Blasdell	A
Bigetty	C	Black Ridge	D	Blasingame	C
Biggs	A	Blackburn	B	Blazon	D
Biggsville	B	Blackcap	A	Blencoe	C
Bignell	B	Blackett	B	Blend	D
Bigwin	D	Blackfoot	B/C	Blendon	B
Bijou	A	Blackhall	D	Blethen	B
Billett	A	Blackhawk	D	Blevins	B
Billings	C	Blackleaf	B	Blevinton	B/D
Bindle	B	Blackleed	A	Blichton	D
Binford	B	Blacklock	D	Bliss	D
Bingham	B	Blackman	C	Blockton	C
Binnsville	D	Blackoak	C	Blodgett	A
Bins	B	Blackpipe	C	Blomford	B
Binton	C	Blackrock	B	Bloom	C
Bippus	B	Blackston	B	Bloomfield	A
Birch	A	Blacktail	B	Blooming	B

continued

TABLE H.1   *(Continued)*

Bloor	D	Bombay	B	Bosco	B
Blossom	C	Bon	B	Bosket	B
Blount	C	Bonaccord	D	Bosler	B
Blountville	C	Bonaparte	A	Bosque	B
Blucher	C	Bond	D	Boss	D
Blue Earth	D	Bondranch	D	Boston	C
Blue Lake	A	Bondurant	B	Bostwick	B
Blue Star	B	Bone	D	Boswell	D
Bluebell	C	Bong	B	Bosworth	D
Bluejoint	B	Bonham	C	Botella	B
Bluepoint	B	Bonifay	A	Bothwell	C
Bluewing	B	Bonilla	B	Bottineau	C
Bluffdale	C	Bonita	D	Bottle	A
Bluffton	D	Bonn	D	Boulder	B
Bluford	D	Bonner	B	Boulder Lake	D
Bly	B	Bonnet	B	Boulder Point	B
Blythe	D	Bonneville	B	Boulflat	D
Boardtree	C	Bonnick	A	Bourne	C
Bobs	D	Bonnie	D	Bow	C
Bobtail	B	Bonsall	D	Bowbac	C
Bock	B	Bonta	C	Bowbells	B
Bodell	D	Bonti	C	Bowdoin	D
Bodenburg	B	Booker	D	Bowdre	C
Bodine	B	Boomer	B	Bowers	C
Boel	A	Boone	A	Bowie	B
Boelus	A	Boonesboro	B	Bowman	B/D
Boesel	B	Boonton	C	Bowmansville	C
Boettcher	C	Booth	C	Boxelder	C
Bogan	C	Boracho	C	Boxwell	C
Bogart	B	Borah	A/D	Boy	A
Bogue	D	Borda	D	Boyce	B/D
Bohannon	C	Bordeaux	B	Boyd	D
Bohemian	B	Borden	B	Boyer	B
Boistfort	C	Border	B	Boynton	
Bolar	C	Bornstedt	C	Boysag	D
Bold	B	Borrego	C	Boysen	D
Boles	C	Borup	B	Bozarth	C
Bolivar	B	Borvant	D	Boze	B
Bolivia	B	Borza	C	Bozeman	A
Bolton	B	Bosanko	D	Braceville	C

continued

TABLE H.1   *(Continued)*

Bracken	D	Bremen	B	Brinkerton	D
Brackett	C	Bremer	B	Briscot	B
Brad	D	Bremo	C	Brite	C
Braddock	C	Brems	A	Britton	C
Bradenton	B/D	Brenda	C	Brizam	A
Brader	D	Brennan	B	Broad	C
Bradford	B	Brenner	C/D	Broad Canyon	B
Bradshaw	B	Brent	C	Broadalbin	C
Bradway	D	Brenton	B	Broadax	B
Brady	B	Brentwood	B	Broadbrook	C
Bradyville	C	Bresser	B	Broadhead	C
Braham	B	Brevard	B	Broadhurst	D
Brainerd	B	Brevort	B	Brock	D
Brallier	D	Brewer	C	Brockliss	C
Bram	B	Brewster	D	Brockman	C
Bramard	B	Brewton	C	Brocko	B
Bramble	C	Brickel	C	Brockport	D
Bramwell	C	Brickton	C	Brockton	D
Brand	D	Bridge	C	Brockway	B
Brandenburg	A	Bridgehampton	B	Brody	C
Brandon	B	Bridgeport	B	Broe	B
Brandywine	C	Bridger	A	Brogan	B
Branford	B	Bridgeson	B/C	Brogdon	B
Brantford	B	Bridget	B	Brolliar	D
Branyon	D	Bridgeville	B	Bromo	B
Brashear	C	Bridgport	B	Bronaugh	B
Brassfield	B	Briedwell	B	Broncho	B
Bratton	B	Brief	B	Bronson	B
Bravane	D	Briensburg		Bronte	C
Braxton	C	Briggs	A	Brooke	C
Braymill	A/D	Briggsdale	C	Brookfield	B
Brays	D	Briggsville	C	Brookings	B
Brayton	C	Brighton	A/D	Brooklyn	D
Brazito	A	Brightwood	C	Brookside	C
Brazos	A	Brill	B	Brookston	B/C
Brea	B	Brim	C	Brooksville	D
Breckenridge	D	Brimfield	C/D	Broomfield	D
Brecknock	B	Brimley	B	Broseley	B
Breece	B	Brinegar	B	Bross	B
Bregar	D	Brinkert	C	Broughton	D

continued

TABLE H.1   *(Continued)*

Broward	C	Bukreek	B	Buse	B
Brownell	B	Bull Run	B	Bush	B
Brownfield	A	Bull Trail	B	Bushnell	C
Brownlee	B	Bullion	D	Bushvalley	D
Broyles	B	Bullrey	B	Buster	C
Bruce	D	Bully	B	Butano	C
Bruffy	C	Bumgard	B	Butler	D
Bruin	C	Buncombe	A	Butlertown	C
Bruneel	B/C	Bundo	B	Butte	C
Bruno	A	Bundyman	C	Butterfield	C
Brunt	C	Bunejug	C	Button	C
Brush		Bunker	D	Buxin	D
Brussett	B	Bunselmeier	C	Buxton	C
Bryan	A	Buntingville	B/C	Byars	D
Brycan	B	Bunyan	B	Bynum	C
Bryce	D	Burbank	A	Byron	A
Bucan	D	Burch	B	Caballo	B
Buchanan	C	Burchard	B	Cabarton	D
Buchenau	C	Burchell	B/C	Cabba	C
Bucher	C	Burdett	C	Cabbart	D
Buckhouse	A	Buren	C	Cabezon	D
Buckingham		Burgess	C	Cabin	C
Buckland	C	Burgi	B	Cabinet	C
Bucklebar	B	Burgin	D	Cable	D
Buckley	B/C	Burke	C	Cabo Rojo	C
Bucklon	D	Burkhardt	B	Cabot	D
Buckner	A	Burleigh	D	Cacapon	B
Buckney	A	Burleson	D	Cache	D
Bucks	B	Burlington	A	Cacique	C
Buckskin	C	Burma		Caddo	D
Bucoda	C	Burmester	D	Cadeville	D
Budd	B	Burnac	C	Cadmus	B
Bude	C	Burnette	B	Cadoma	D
Buell	B	Burnham	D	Cador	C
Buena Vista	B	Burnside	B	Cagey	C
Buff Peak	C	Burnsville	B	Caguabo	D
Buffington	B	Burnt Lake	B	Cagwin	B
Buffmeyer	B	Burris	D	Cahaba	B
Buick	C	Burt	D	Cahill	B
Buist	B	Burton	B	Cahone	C

continued

TABLE H.1   (Continued)

Cahto	C	Camargo	B	Canoncito	B
Caid	B	Camarillo	B/D	Canova	B/D
Cairo	D	Camas	A	Cantala	B
Cajalco	C	Camascreek	B/D	Canton	B
Cajon	A	Cambridge	C	Cantril	B
Calabar	D	Camden	B	Cantua	B
Calabasas	B	Cameron	D	Canutio	B
Calais	C	Cameyville	C	Canyon	D
Calamine	D	Camgeny	C	Capac	B
Calapooya	C	Camillus	B	Capay	D
Calawah	B	Camp	B	Cape	D
Calco	C	Campbell	B/C	Cape Fear	D
Calder	D	Camphora	B	Capers	D
Caldwell	B	Campia	B	Capillo	C
Caleast	C	Campo	C	Caples	C
Caleb	B	Campone	B/C	Capps	B
Calera	C	Campspass	C	Capshaw	C
Calhi	A	Campus	B	Capulin	B
Calhoun	D	Camroden	C	Caputa	C
Calico	D	Cana	C	Caraco	C
Califon	D	Canaan	C/D	Caralampi	B
Calimus	B	Canadian	B	Carbo	C
Calita	B	Canadice	D	Carbol	D
Caliza	B	Canandaigua	D	Carbondale	D
Calkins	C	Canaseraga	C	Carbury	B
Callabo	C	Canaveral	C	Carcity	D
Callahan	C	Canburn	D	Cardiff	B
Calleguas	D	Cande	B	Cardington	C
Callings	C	Candelero	C	Cardon	D
Calloway	C	Cane	C	Carey	B
Calmar	B	Caneadea	D	Carey Lake	B
Calneva	C	Caneek	B	Careytown	D
Calouse	B	Canel	B	Cargill	C
Calpine	B	Canelo	D	Caribe	B
Calvert	D	Caney	C	Caribel	B
Calverton	C	Canez	B	Caribou	B
Calvin	C	Canfield	C	Carlin	D
Calvista	D	Canisted	C	Carlinton	B
Cam	B	Canninger	D	Carlisle	A/D
Camaguey	D	Cannon	B	Carlotta	B

continued

TABLE H.1   *(Continued)*

Carlow	D	Tusel	C	Ulricher	B
Carlsbad	C	Tuskeego	C	Ulupalakua	B
Carlsborg	A	Tusler	B	Uly	B
Carlson	C	Tusquitee	B	Ulysses	B
Carlton	B	Tustin	B	Uma	A
Carmi	B	Tustumena	B	Umapine	B/C
Carnasaw	C	Tuthill	B	Umiat	D
Carnegie	C	Tutni	B	Umikoa	B
Carnero	C	Tutwiler	B	Umil	D
Carney	D	Tuxedo		Umnak	B
Caroline	C	Tuxekan	B	Umpa	B
Carr	B	Twin Creek	B	Umpqua	B
Carrisalitos	D	Twining	C	Una	D
Carrizo	A	Twisp	B	Unadilla	B
Carsitas	A	Two Dot	C	Unaweep	B
Carsley	C	Tybo	D	Uncom	B
Carso	D	Tyee	D	Uncompangre	D
Carson	D	Tygart	D	Uneeda	B
Carstairs	B	Tyler	D	Ungers	B
Carstump	C	Tyndall	B/C	Union	C
Cart	B	Tyner	A	Uniontown	B
Cartagena	D	Tyrone	C	Unionville	C
Cartecay	C	Tyson	C	Unisun	C
Caruso	C	Uana	D	Updike	D
Caruthersville	B	Ubar	C	Upsal	C
Carver	A	Ubly	B	Upsata	A
Carwile	D	Ucola	D	Upshur	C
Turnbow	C	Ucolo	C	Upton	C
Turner	B	Ucopia	B	Uracca	B
Turnerville	B	Uddlpho	C	Urbana	C
Turney	B	Udel	D	Urbo	D
Turrah	D	Uffens	D	Urich	D
Turret	B	Ugak	D	Urne	B
Turria	C	Uhland	B	Ursine	D
Turson	B/C	Uhlig	B	Urtah	C
Tuscan	D	Uinta	B	Urwil	D
Tuscarawas	C	Ukiah	C	Usal	B
Tuscarora	C	Ulen	B	Ushar	B
Tuscola	B	Ullda	B	Usine	B
Tuscumbia	D	Ulm	B	Uska	D

continued

TABLE H.1   *(Continued)*

Utaline	B	Vandergrift	C	Venice	D
Ute	C	Vanderhoff	D	Venlo	D
Utica	A	Vanderlip	A	Venus	B
Utley	B	Vanet	D	Verboort	D
Utuado	B	Vang	B	Verde	C
Uvada	D	Vanhorn	B	Verdel	D
Uvalde	C	Vannoy	B	Verdella	D
Uwala	B	Vanoss	B	Verdico	D
Vacherie	C	Vantage	C	Verdigris	B
Vader	B	Varco	C	Verdun	D
Vado	B	Varelum	C	Vergennes	D
Vaiden	D	Varick	D	Verhalen	D
Vailton	B	Varina	C	Vermejo	D
Valby	C	Varna	C	Vernal	B
Valco	C	Varro	B	Vernalis	B
Valdez	B/C	Varysburg	B	Vernia	A
Vale	B	Vashti	C	Vernon	D
Valencia	B	Vasquez	B	Verona	C
Valent	A	Vassalboro	D	Vesser	C
Valentine	A	Vassar	B	Veston	D
Valera	C	Vastine	C	Vetal	A
Valkaria	B/D	Vaucluse	C	Veteran	B
Vallan	D	Vaughnsville	C	Veyo	D
Vallecitos	C/D	Vayas	D	Via	B
Valleono	B	Veal	B	Vian	B
Vallers	C	Veazie	B	Viboras	D
Valmont	C	Vebar	B	Viborg	B
Valmy	B	Vecont	D	Vickery	C
Valois	B	Vega	C	Vicksburg	B
Vamer	D	Vega Alta	C	Victor	A
Van Buren		Vega Baja	C	Victoria	D
Van Dusen	B	Vekol	D	Victory	B
Van Nostern	B	Velda	B	Vicu	D
Van Wagoner	D	Velma	B	Vida	B
Vanajo	D	Velva	B	Vidrine	C
Vananda	D	Vena	C	Vieja	D
Vance	C	Venango	C	Vienna	B
Vanda	D	Venator	D	Vieques	B
Vandalia	C	Veneta	C	View	C
Vanderdasson	D	Venezia	D	Vigar	C

continued

TABLE H.1   (Continued)

Vigo	D	Volkmar	B	Wahtigup	B
Vigus	C	Volney	B	Wahtum	D
Viking	D	Volperie	C	Waiaha	D
Vil	D	Voltaire	D	Waiakoa	C
Vilas	A	Volumer	D	Waialeale	D
Villa Grove	B	Volusia	C	Waialua	B
Villars	B	Vona	B	Waiawa	D
Villy	D	Vore	B	Waihuna	D
Vina	B	Vrodman	B	Waikaloa	B
Vincennes	C	Vulcan	C	Waikane	B
Vincent	C	Vylach	D	Waikapu	B
Vineyard	C	Wabanica	D	Waikomo	D
Vingo	B	Wabash	D	Wailuku	B
Vining	C	Wabasha	D	Waimea	B
Vinita	C	Wabassa	B/D	Wainee	B
Vinland	C	Wabek	B	Wainola	A
Vinsad	C	Waca	C	Waipahu	C
Vint	B	Wacota	B	Waiska	B
Vinton	B	Wacousta	C	Waits	B
Vira	C	Wadams	B	Wake	D
Viraton	C	Waddell	B	Wakeen	B
Virden	C	Waddups	B	Wakefield	B
Virgil	B	Wadell	B	Wakeland	B/D
Virgin Peak	D	Wadena	B	Wakonda	C
Virgin River	D	Wadesboro	B	Wakulla	A
Virtue	C	Wadleigh	D	Walcott	B
Visalia	B	Wadmalaw	D	Waldeck	C
Vista	C	Wadsworth	C	Waldo	D
Vives	B	Wages	B	Waldron	D
Vivi	B	Wagner	D	Waldroup	D
Vlasaty	C	Wagram	A	Wales	B
Voca	C	Waha	C	Walford	C
Vodermaier	B	Wahee	D	Walke	C
Voladora	B	Wahiawa	B	Wall	B
Volco	D	Wahikuli	B	Walla Walla	B
Volente	C	Wahkeena	B	Wallace	B
Volga	D	Wahkiacus	B	Waller	B/D
Volin	B	Wahluke	B	Wallington	C
Volinia	B	Wahmonie	D	Wallis	B
Volke	C	Wahpeton	C	Wallkill	C/D

continued

TABLE H.1   *(Continued)*

Wallman	C	Warrenton	B/D	Waukee	B
Wallowa	C	Warrior		Waukegan	B
Wallpack	C	Warsaw	B	Waukena	D
Wallrock	B/C	Warsing	B	Waukon	B
Wallsburg	D	Warwick	A	Waumbek	B
Wallson	B	Wasatch	A	Waurika	D
Walpole	C	Wasepi	B	Wauseon	B/D
Walsh	B	Washburn		Waverly	B/D
Walshville	D	Washde	C	Wawaka	C
Walters	A	Washington	B	Waycup	B
Walton	C	Washougal	B	Wayden	D
Walum	B	Washtenaw	C/D	Wayland	C/D
Walvan	B	Wasidja	B	Wayne	B
Wamba	B/C	Wasilla	D	Waynesboro	B
Wamic	B	Wassaic	B	Wayside	
Wampsville	B	Watab	C	Wea	B
Wanatah	B	Watauga	B	Weaver	C
Wanblee	D	Watchaug	B	Webb	C
Wando	A	Watchung	D	Weber	B
Wanetta	A	Waterboro		Webster	C
Wanilla	C	Waterbury	D	Wedekind	D
Wann	A	Waterino	C	Wedertz	C
Wapal	B	Waters	C	Wedge	A
Wapato	C/D	Watkins	B	Wedowee	D
Wapello	B	Watkins Ridge	B	Weed	B
Wapinitia	B	Wato	B	Weeding	A/C
Wapping	B	Watopa	B	Weedmark	B
Wapsie	B	Watrous	B	Weeksville	B/D
Warba	B	Watseka	C	Weepon	D
Ward	D	Watson	C	Wehadkee	D
Wardboro	A	Watsonia	D	Weikert	C/D
Wardell	D	Watsonville	D	Weimer	D
Warden	B	Watt	D	Weinbach	C
Wardwell	C	Watton	C	Weir	D
Ware	B	Waubay	B	Weirman	B
Wareham	C	Waubeek	B	Weiser	C
Warm Springs	C	Waubonsie	B	Weishaupt	D
Warman	D	Wauchula	B/D	Weiss	A
Warners	A/D	Waucoma	B	Weitchpec	B
Warren		Wauconda	B	Welaka	A

continued

TABLE H.1 (Continued)

Welby	B	Westville	B	Whitewood	C
Welch	C	Wethersfield	C	Whitley	B
Weld	C	Wethey	B/C	Whitlock	B
Welda	C	Wetterhorn	C	Whitman	D
Weldon	D	Wetzel	D	Whitney	B
Welduna	B	Weymouth	B	Whitore	A
Weller	C	Whakana	B	Whitsol	B
Wellington	D	Whalan	B	Whitson	D
Wellman	B	Wharton	C	Whitwell	C
Wellner	B	Whatcom	C	Wholan	C
Wellsboro	C	Whately	D	Wibaux	C
Wellston	B	Wheatley	D	Wichita	C
Wellsville	B	Wheatridge	C	Wichup	D
Welring	D	Wheatville	B	Wickersham	B
Wemple	B	Wheeler	B	Wickett	C
Wenas	B/C	Wheeling	B	Wickiup	C
Wenatcher	C	Whelchel	B	Wickliffe	D
Wendel	B/C	Whellon	D	Wickman	B
Wenham		Whetstone	B	Wicksburg	B
Wenona	C	Whidbey	C	Widta	B
Wentworth	B	Whippany	C	Widtsoe	C
Werlow	C	Whipstock	C	Wiehl	C
Werner	B	Whirlo	B	Wien	D
Weso	C	Whit	B	Wiggleton	B
Wessel	B	Whitaker	C	Wigton	A
Westbrook	D	Whitcomb	C	Wilbraham	C
Westbury	C	White Bird	C	Wilbur	C
Westcreek	B	White House	C	Wilco	C
Westerville	C	White Store	D	Wilcox	D
Westfall	C	White Swan	C	Wilcoxson	C
Westfield		Whitecap	D	Wildcat	D
Westford		Whitefish	B	Wilder	B
Westland	B/D	Whiteford	B	Wilderness	C
Westminster	C/D	Whitehorse	B	Wildrose	D
Westmore	B	Whitelake	B	Wildwood	D
Westmoreland	B	Whitelaw	B	Wiley	C
Weston	D	Whiteman	D	Wilkes	C
Westphalia	B	Whiterock	D	Wilkeson	C
Westplain	C	Whitesburg	C	Wilkins	D
Westport	A	Whitewater	B	Will	D

continued

TABLE H.1   (Continued)

Willacy	B	Winger	C	Wolfesen	C
Willakenzie	C	Wingville	B/D	Wolfeson	C
Willamar	D	Winifred	C	Wolford	B
Willamette	B	Wink	B	Wolftever	C
Willapa	C	Winkel	D	Wolverine	A
Willard	B	Winkleman	C	Wood River	D
Willette	A/D	Winkler	A	Woodbine	B
Willhand	B	Winlo	D	Woodbridge	C
Williams	B	Winlock	C	Woodburn	C
Williamsburg	B	Winn	C	Woodbury	D
Williamson	C	Winnebago	B	Woodcock	B
Willis	C	Winnemucca	B	Woodenville	C
Willits	B	Winneshiek	B	Woodglen	D
Willoughby	B	Winnett	D	Woodhall	B
Willow Creek	B	Winona	D	Woodhurst	A
Willowdale	B	Winooski	B	Woodinville	C/D
Willows	D	Winston	A	Woodly	B
Willwood	A	Winters	C	Woodlyn	C/D
Wilmer	C	Wintersburg	C	Woodmansie	B
Wilpar	D	Winterset	C	Woodmere	B
Wilson	D	Winthrop	A	Woodrock	C
Wiltshire	C	Wintoner	C	Woodrow	C
Winans	B/C	Winu	C	Woods Cross	D
Winberry	D	Winz	C	Woodsfield	C
Winchester	A	Wishard	A	Woodside	A
Winchuck	C	Wisheylu	C	Woodson	D
Wind River	B	Wishkam	C	Woodstock	C/D
Winder	B/D	Wiskam	C	Woodstown	C
Windham	B	Wisner	D	Woodward	B
Windmill	B	Witbeck	D	Woolman	B
Windom	B	Witch	D	Woolper	C
Windsor	A	Witham	D	Woolsey	C
Windthorst	C	Withee	C	Wooskow	B/C
Windy	C	Witt	B	Woosley	C
Wineg	C	Witzel	D	Wooster	C
Winema	C	Woden	B	Woostern	B
Winetti	B	Wolcottsburg		Wooten	A
Winfield	C	Woldale	C/D	Worcester	B
Wing	D	Wolf	B	Worf	D
Wingate	B	Wolf Point	D	Work	C

continued

TABLE H.1   *(Continued)*

Worland	B	Yamhill	C	Yuba	D
Worley	C	Yampa	C	Yuko	C
Wormser	C	Yamsay	D	Yukon	D
Worock	B	Yana	B	Yunes	D
Worsham	D	Yancy	C	Yunque	C
Worth	C	Yardley	C	Zaar	D
Worthen	B	Yates	D	Zaca	D
Worthing	D	Yauco	C	Zacharias	B
Worthington	C	Yawdim	D	Zachary	D
Wortman	C	Yawkey	C	Zafra	B
Wrentham	C	Yaxon	B	Zahill	B
Wright	C	Yeary	C	Zahl	B
Wrightman	C	Yeates Hollow	C	Zaleski	C
Wrightsville	D	Yegen	B	Zalla	A
Wunjey	B	Yelm	B	Zamora	B
Wurtsboro	C	Yenrab	A	Zane	C
Wyalusing	D	Yeoman	B	Zaneis	B
Wyard	B	Yesum	B	Zanesville	C
Wyarno	B	Yetull	A	Zanone	C
Wyatt	C	Yoder	B	Zapata	C
Wyeast	C	Yokohl	D	Zavala	B
Wyeville	C	Yollabolly	D	Zavco	C
Wygant	C	Yolo	B	Zeb	B
Wykoff	B	Yologo	D	Zeesix	C
Wyman	B	Yomba	C	Zell	B
Wymoose	D	Yomont	B	Zen	C
Wymore	C	Yoncalla	C	Zenda	C
Wynn	B	Yonges	D	Zenia	B
Wyo	B	Yonna	B/D	Zeniff	B
Wyocena	B	Yordy	B	Zeona	A
Xavier	B	York	C	Ziegler	C
Yacolt	B	Yorkville	D	Zigweid	B
Yahara	B	Yost	C	Zillah	B/C
Yahola	B	Youga	B	Zim	D
Yaki	D	Youman	C	Zimmerman	A
Yakima	B	Youngston	B	Zing	C
Yakus	D	Yourame	A	Zinzer	B
Yallani	B	Yovimpa	D	Zion	C
Yalmer	B	Ysidora	D	Zipp	C/D
Yamac	B	Yturbide	A	Zita	B

continued

TABLE H.1   *(Continued)*

Zoar	C	Zufelt	B/D	Zunhall	B/C
Zoate	D	Zukan	D	Zuni	D
Zohner	B/D	Zumbro	B	Zurich	B
Zook	C	Zumwalt	C	Zwingle	D
Zorravista	A	Zundell	B/C		

Source: U.S. Department of Agriculture, Soil Conservation Service 1986. *Urban Hydrology For Small Watersheds.* TR55, Washington, D.C. (June).
A blank hydrological soil group indicates the soil group has not been determined. Two soil groups such as B/C indicates the drained/undrained situation.

STORMWATER PROGRAMS

To assist in calculations, a number of computer programs are included with this text. The programs were written by Dr. Ronald D. Eaglin and Dr. Marty Wanielista. These programs were specifically written to accompany this text and to illustrate the principles of hydrology.

I.1 SYSTEM REQUIREMENTS

The stormwater programs are available on a disk included in the back of the book. The programs must be installed on your system prior to using them. The programs require Microsoft Windows™ operating system 3.1 or greater (the programs will work with Microsoft Windows™ 95 or NT). The Microsoft Windows™ operating system was specifically chosen for a number of reasons. The number of host platforms which support Microsoft Windows™ 3.0 and the common user interface used by Microsoft Windows™ makes programs written for this operating system easy to use. You will need a 386 machine or greater with at least 2 MB of RAM and 2.5 MB of hard disk space. You will also need a 3.5-in. floppy drive capable of using high-density 1.44 MB disks. If you wish to install the package on a machine with only a 5.25-in. high-density drive, the contents of the disk can be copied to a 5.25-in. high-density floppy and installed from that floppy. The programs cannot be installed from a low-density floppy.

I.2 INSTALLATION

To install the programs, put the disk in the A or B drive. Using Windows™ Program Manager select **File, Run** and type **A:SETUP** or **B:SETUP** in the command line edit field in the Run window and hit the **Enter** key (or click the **OK** button). The programs will be installed to your hard disk, and a program group will be created on your system. Once the programs have been installed, the hydrology programs group will be added to Program Manager, as shown in Figure I.1.

I.3 SMADA-STORMWATER MANAGEMENT AND DESIGN AID

SMADA is a Windows program designed to generate watershed hydrographs and route hydrographs through ponds using inventory routing. SMADA is designed to be

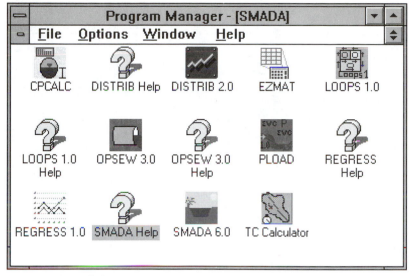

FIGURE I.1   Hydrology programs group.

easy to use and also powerful. Features can be accessed from pull-down menus or using the button tool bar. As the mouse moves over buttons, the status bar gives the user a message indicating the function of the buttons. The SMADA main screen is shown in Figure I.2.

There are four main windows in SMADA corresponding to watershed, rainfall, hydrograph, and pond. These dialog boxes are used to enter information into the program, to perform calculation, and to present results. Output can be obtained in text format or graphical output can also be obtained.

I.3.1   Watershed Input

The Watershed dialog box can be accessed from the Watershed pull-down menu or by clicking the **Watershed** button in the tool bar. The Watershed button is the first button after the word edit in the tool bar. The Watershed Input window is shown in Figure I.3. In the Watershed window you can input watershed information and select

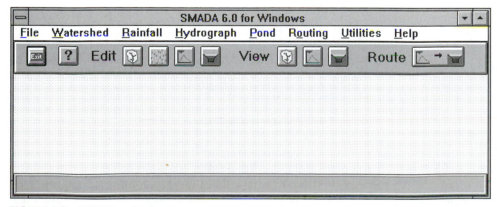

FIGURE I.2   SMADA main screen.

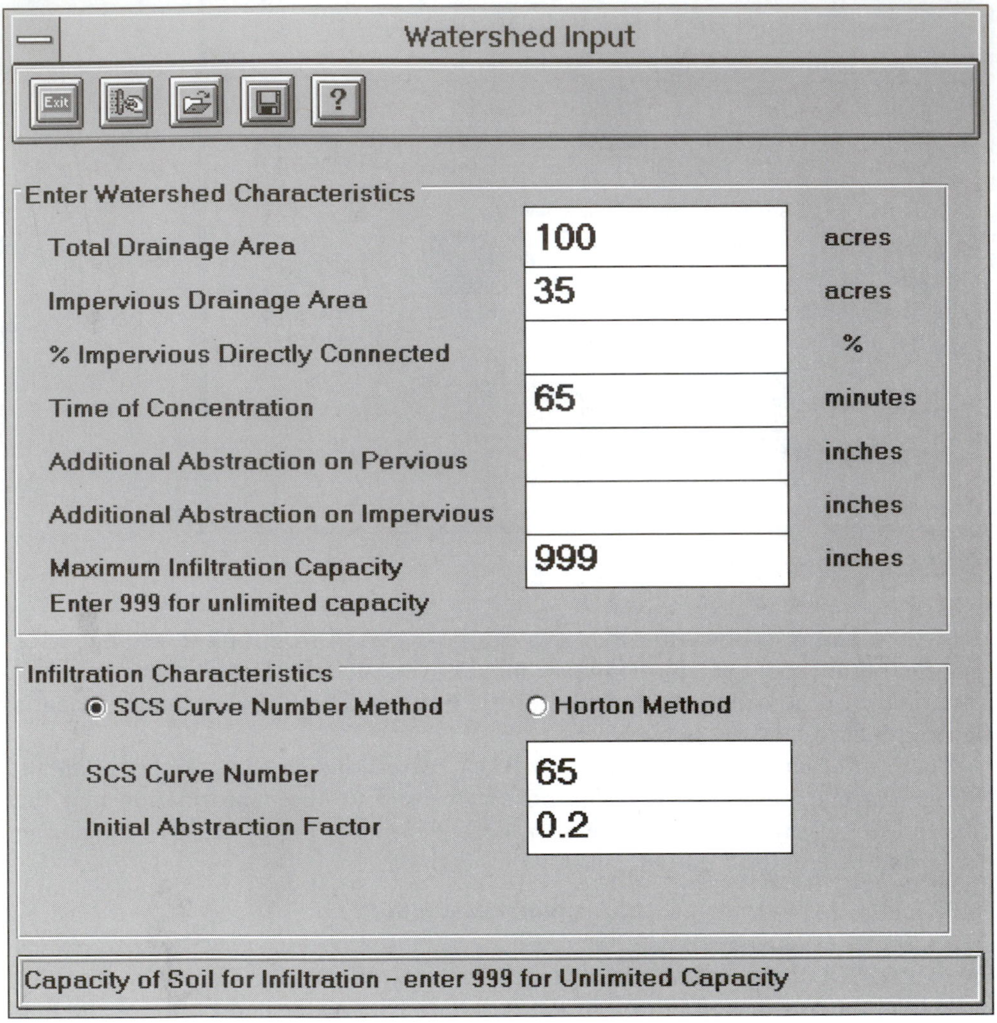

FIGURE I.3 Watershed Input window for NRCS method.

an infiltration method. When done you should click the **Exit** button, you may also save a watershed, retrieve a watershed or clear all watershed information from within this window. You can also access the online help at any time by clicking the **Question Mark** button. Double-clicking the dash in the upper-left corner will exit you from the window *without saving* any of the entered information. Exit the window this way if you wish to exit without saving any information.

I.3.2 *Rainfall Properties*

The Rainfall Properties window contains a number of options that are used for a dimensionless curve or to enter rainfall information by hand. The Rainfall Properties window is shown in Figure I.4. The time step you enter at this point will be the time step the program uses to calculate the hydrograph in later steps. A total rainfall must be entered if you wish to use a dimensionless curve. In the case of a user-entered

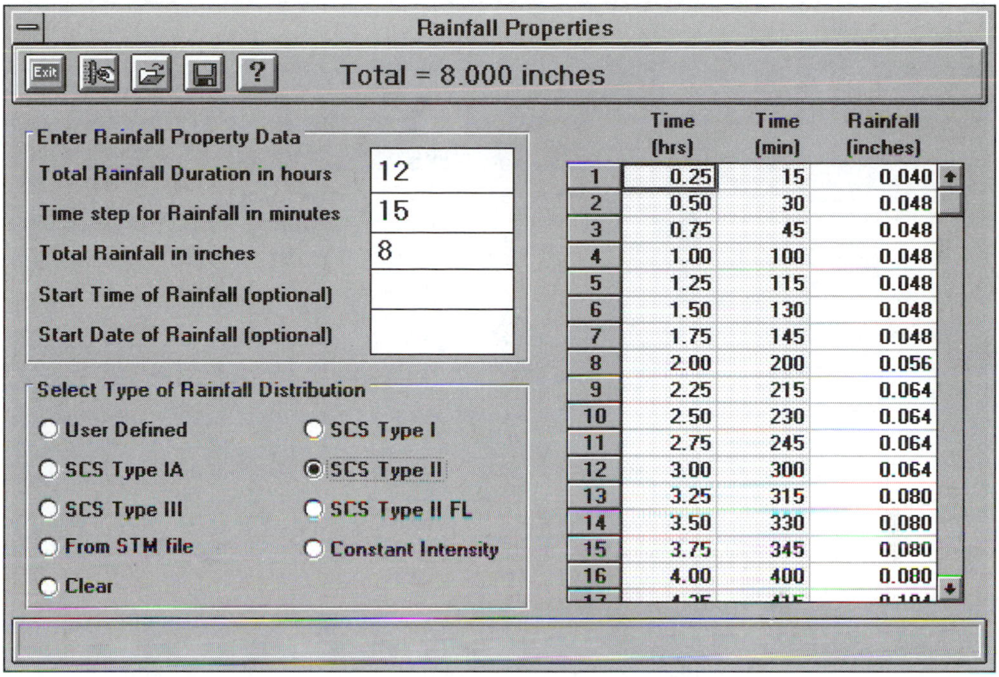

FIGURE I.4   Rainfall Properties window.

curve, the total will be calculated and displayed as you enter the information. You must enter a total duration and a time step for a user-defined rainfall. Click the **Exit** button to exit the dialog box and keep your changes.

I.3.3   Hydrograph Generation

The Hydrograph window allows for the calculation of hydrographs using a number of methods. The Hydrograph Generation window is shown in Figure I.5. The hydrograph is calculated by selecting the hydrograph method and clicking the **Generate Hydrograph** button. To use the unit hydrograph method, the unit hydrograph must be entered. This can be done using the **Create** option in the Unit Hydrograph menu under the Hydrograph menu. The NRCS method requires the input of a peak attenuation factor. Common peak attenuation factors of 484 and 256 can be used with the NRCS procedure by clicking the correct buttons. The calculated hydrograph will appear in the spreadsheet control to the right of the window. A plot of the calculated hydrograph will also appear in the lower-left-hand corner of the window. Saving the hydrograph by clicking the save button will also save the associated rainfall and watershed. Opening a hydrograph clicking the **Open** button will also open the associated watershed and rainfall. Click the **Exit** button to exit this window. *Hint:* Any of the contents of the hydrograph (or any) spreadsheet control can be copied to the clipboard by highlighting it with the mouse and pressing **Control-Insert.** Highlighting the fields with the mouse can be done by holding down the left mouse button and dragging the mouse over the fields. This information can then be pasted directly into any other Windows application.

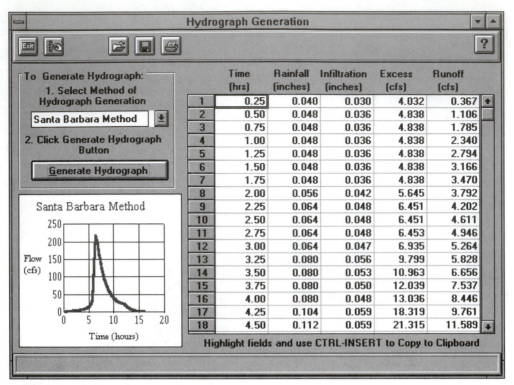

FIGURE I.5   Hydrograph Generation window.

I.3.4   Pond Design

To route this hydrograph through a pond, a stage storage discharge relationship must be entered or calculated. This can be done using the Pond Design window shown in Figure I.6. The pond window contains a spreadsheet for direct entry of pond stage storage discharge information. Also contained are a number of buttons to assist in calculations. The Pond Shape button (7th from left) allows the user to perform automatic calculations of area from stage using a square or circular geometry and a given constant side slope. The Calculate Storage from Area-Stage button (8th from left) will automatically calculate the storage for all entered area and stage information. Two (2) weirs can be placed automatically on the pond using the two weir buttons (5th and 6th from left). Weir choices include 60 degree V-notch, 90 degree V-notch, sharp-crested rectangular, broad-crested rectangular, and orifice weir. Discharge will automatically be calculated from the weir information and placed into the spreadsheet control. To route the current hydrograph to the pond, click the **Route** button (9th from left). The current hydrograph will be routed through the pond using 1-minute time steps. Output will be reported on the time step for which rainfall was entered.

I.3.5   Text and Plot Viewer

Output from SMADA can be displayed through the Text Output window. From this window, text can be sent to the printer, a file, or the clipboard. Plots of routing, hydrographs, and rainfall information can also be displayed, sent to the clipboard, or

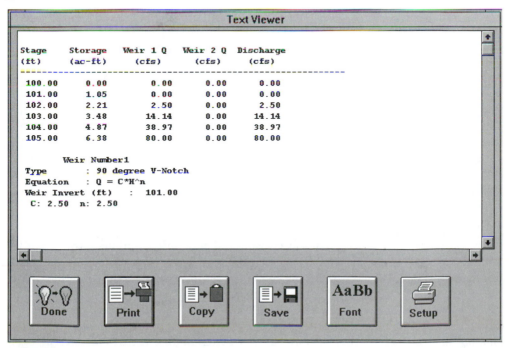

FIGURE I.6 Pond Design window.

Stage (ft)	Storage (ac-ft)	Weir 1 Q (cfs)	Weir 2 Q (cfs)	Discharge (cfs)
100.00	0.00	0.00	0.00	0.00
101.00	1.05	0.00	0.00	0.00
102.00	2.21	2.50	0.00	2.50
103.00	3.48	14.14	0.00	14.14
104.00	4.87	38.97	0.00	38.97
105.00	6.38	80.00	0.00	80.00

Weir Number 1
Type   : 90 degree V-Notch
Equation   : $Q = C*H^n$
Weir Invert (ft)   : 101.00
 C: 2.50 n: 2.50

FIGURE I.7 Text Viewer with Pond Display.

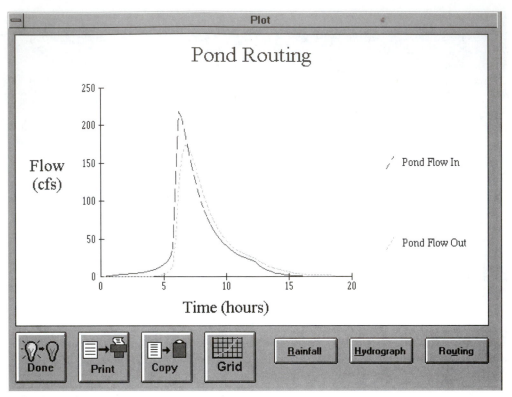

FIGURE I.8   Plot window with display of Pond Routing.

sent to the printer. This is performed from the Plot window. The Text Viewer and the Plot window are shown in Figures I.7 and I.8.

I.4   DISTRIB AND REGRESS STATISTICS PROGRAMS

Two statistics programs are included to help you solve complex hydrology problems statistical distributions analysis and regression analysis. These programs are DISTRIB and REGRESS. DISTRIB fits univariate data to a theoretical distribution and predicts values based on the fit and a user-given return period. REGRESS performs a number of regression fits for bivariate (x, y) data. Most of the work can be performed from each program's main window.

I.4.1   REGRESS

To use REGRESS, enter the data in the spreadsheet control and click the curve fit form you desire. The main REGRESS window is shown in Figure I.9. The data will automatically plot, and the curve fit coefficients will output at the bottom of the window. To output information from REGRESS, select **Regression** from the Statistics menu. A number of output choices will appear. Select the output you desire. If the form chosen has an 'n' parameter, it can be solved using a nonlinear iterative search. This option will search a range of values of n searching for the best fit. All fits will be reported. This option is available under **Statistics, Regression, Iterative Search.**

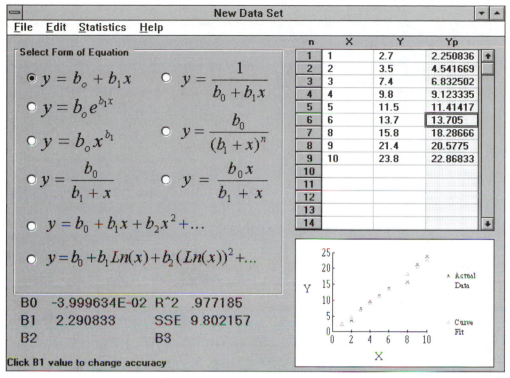

FIGURE I.9   REGRESS Main window.

I.4.2 DISTRIB

DISTRIB works in much the same manner as REGRESS. DISTRIB handles only univariate data. The DISTRIB main window is shown in Figure I.10. Enter the univariate data in the spreadsheet control and click the desired distribution fit. The program will output the results to the screen. For the conversion of partial series data to an annual series, the program allows for the input of an annualization factor. This factor is the number of partial series points entered divided by the number of years these points represent. It allows you to perform annual return period predictions using partial series data. The Annualization Factor option is in the Edit menu.

I.5—PLOAD POLLUTANT LOADING PROGRAM

PLOAD is a program which calculates pollutant loadings from a watershed. For the program to calculate these pollutant loadings, the program must first open a file containing this information for various land uses; a landuse file. When you start the program, you will be asked if you wish to use a landuse file. The only time you should answer no is if you plan to create your own landuse file. These landuse files have the default extension LUF. A sample landuse file called LANDUSE.LUF is included with the program. As you use PLOAD you may create more landuse files. These files can be created using a text editor or using the PLOAD program. PLOAD allows you to open and save two types of files; landuse and analysis. A landuse file, as mentioned

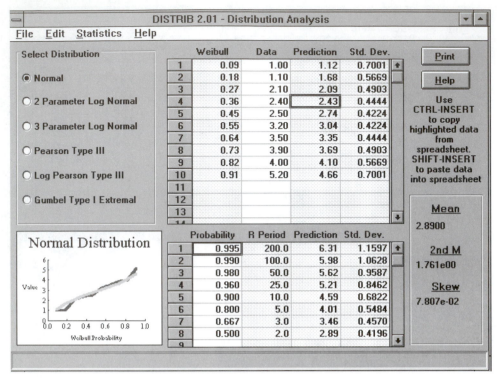

FIGURE I.10   DISTRIB main screen.

above, contains default loading information for various pollutants and land uses. An analysis file contains predictions for pollutant loading for a watershed.

I.5.1   Performing a Pollutant Analysis

To perform an analysis, click on **Pollutant Load** in the Calculate menu. A number of dialog box screens will appear. Pollutant loadings are calculated on the basis of mass per time. To calculate the load, a total rainfall and the time frame for that rainfall must be entered. This is done in the Enter Rainfall for Pollutant Analysis window. When finished click the **Done** button. The Enter Watershed Information dialog box appears. This box uses a number of Spin buttons. A Spin button is a button that has an up and a down arrow on it. To view the possible selections click the **up** or the **down portion** of the **Spin** button. Five (5) watersheds can be added together in a single analysis. The possible land uses are read from the landuse file and will vary with each file. The default land use file [LANDUSE.LUF] has six land uses in it; half-acre urban residential, urban commercial, pasture, woodland, citrus, and highway. To add a land use to your analysis select the watershed number using the Spin button, select the landuse using the Spin button, and enter an area. The program uses the rational method ($Q = CiA$) for determining runoff, and the default rational coefficient can be modified for the analysis. The flowrate (L^3/T) calculated is then multiplied by the default pollutant concentration (M/L^3) to yield pollutant load (M/T). The Text Viewer displays the results of the analysis. Figure I.11 shows the PLOAD main screen. Figures I.12 and I.13 show the input screens for performing a pollutant analysis. Figure I.14 shows the output of this analysis.

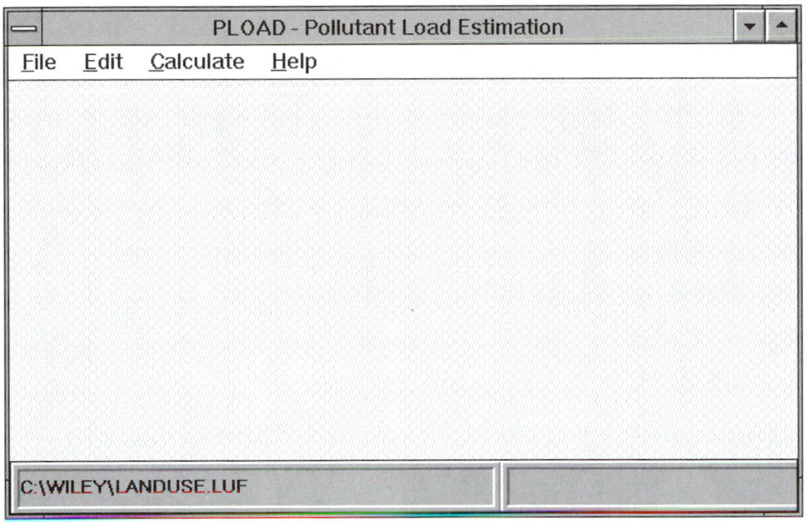

FIGURE I.11   PLOAD main screen.

I.5.2   Modifying Land Use Information

Landuse information can be modified by selecting the **Landuse/Pollutant Information** option in the Edit menu. From the Landuse/Pollutant dialog box the default pollutant loadings for each pollutant can be modified. Landuses can be added clicking the **Add Landuse** button, and pollutants can be added clicking the **Add Pollutant** button. For each landuse added a default concentration must be added for each pollutant. For each pollutant added a default concentration must be added for each landuse. Figure I.15 shows the Landuse and Pollutant editing window. If you wish to keep this information for later, select the **Save, Landuse** option in the File menu. If you save the modified landuse with the name LANDUSE, the original landuse file will be overwritten, so be sure to select a different name for your landuse file. The format for the landuse

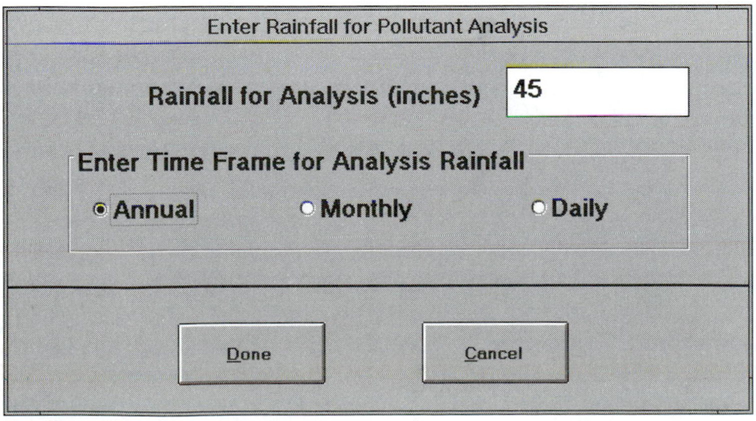

FIGURE I.12   Enter Rainfall for Pollutant Analysis.

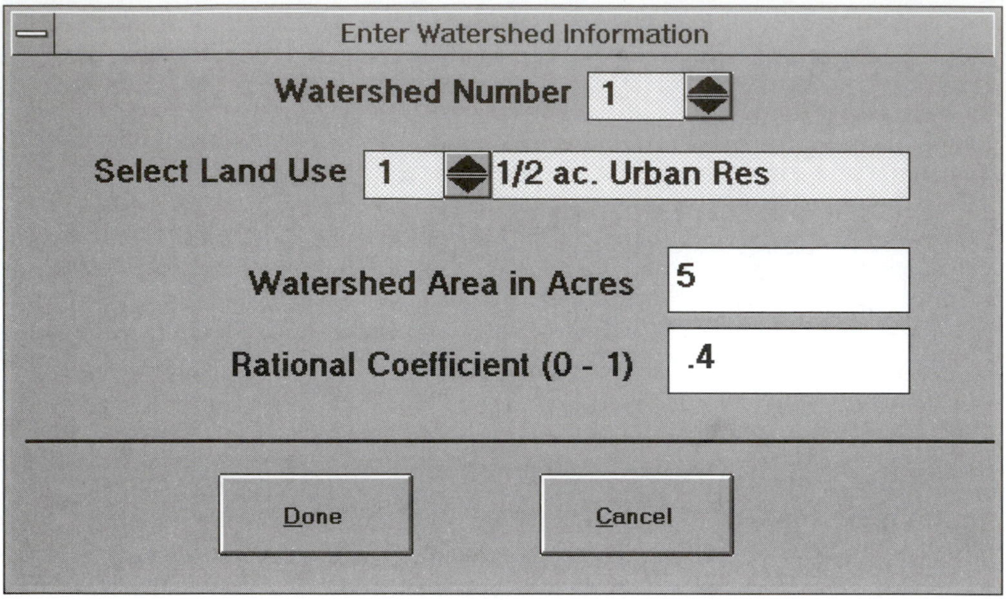

FIGURE I.13 The Enter Watershed Information dialog box.

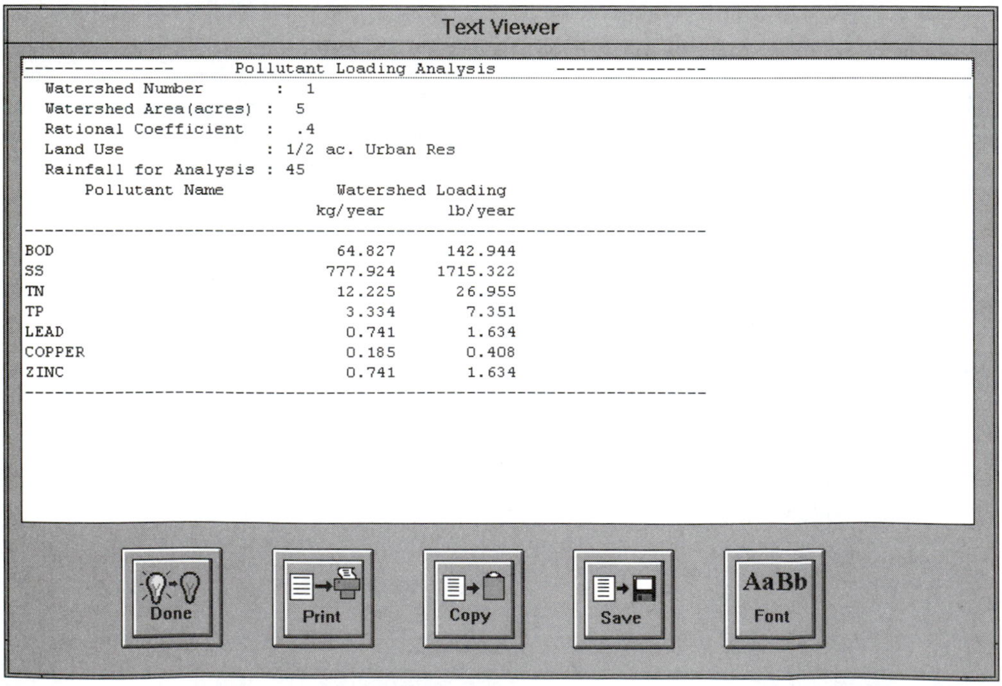

FIGURE I.14 Text Viewer with Results of Pollutant Analysis.

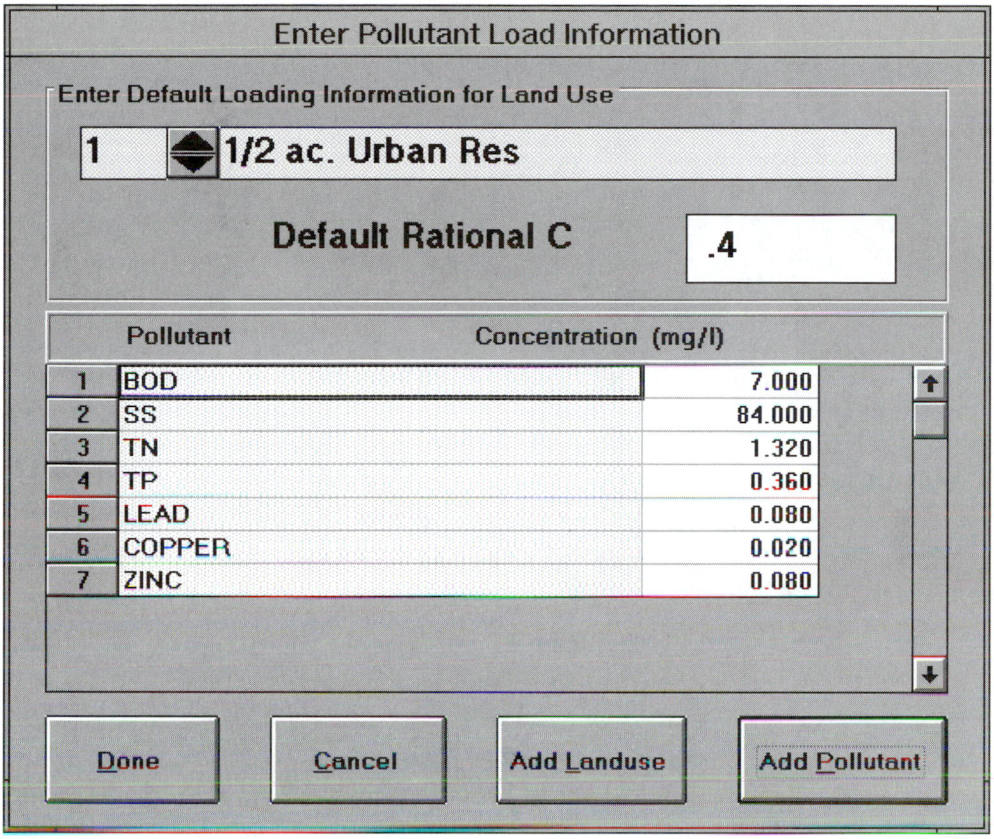

FIGURE I.15   Enter Pollutant Load Information dialog box.

file is one entry per line with the following entries;

   Integer—Number of Landuses

repeat the following for each landuse

   String—Name of Landuse
   Integer—Number of pollutants

repeat the following for number of pollutants

   String—name of pollutant
   Float—default pollutant concentration

To observe an example of this format, open the file LANDUSE.LUF with any text editor.

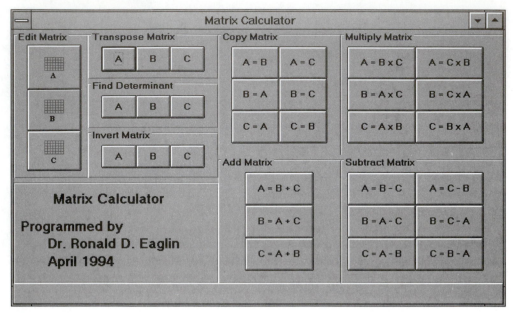

FIGURE I.16   EZMAT main screen.

I.6   EZMAT MATRIX CALCULATOR

The programs included with this book contain EZMAT-Matrix Calculator. This program calculates and displays addition, subtraction, multiplication, transposing, determinant, and inversion of matrices. The program allows for the manipulation of three matrices simultaneously. These matrices are labeled A, B, and C. Each matrix can be edited using Edit Matrix button, as shown in Figure I.16. Clicking the **Edit Matrix** button will bring up the Matrix Editor for the Matrix A, B, or C, as shown in Figure I.17. This editor looks very much like a spreadsheet with a number of buttons on the top. These buttons from left to right are: Exit and Keep Changes, Clear Matrix Values, Open a Matrix File (*.MTX), Save a Matrix File, Invert Displayed Matrix, and Transpose Displayed Matrix. Transpose and invert can also be performed from

Matrix Editor

Enter Matrix Values

	A	B	C	D	E	F	G
1	1.00	2.00	21.00	32.00			
2	3.00	4.00	3.00	8.00			
3	7.00	3.00	6.00	11.00			
4	11.00	34.00	73.00	9.00			
5							
6							
7							
8							

FIGURE I.17   Matrix Editor screen.

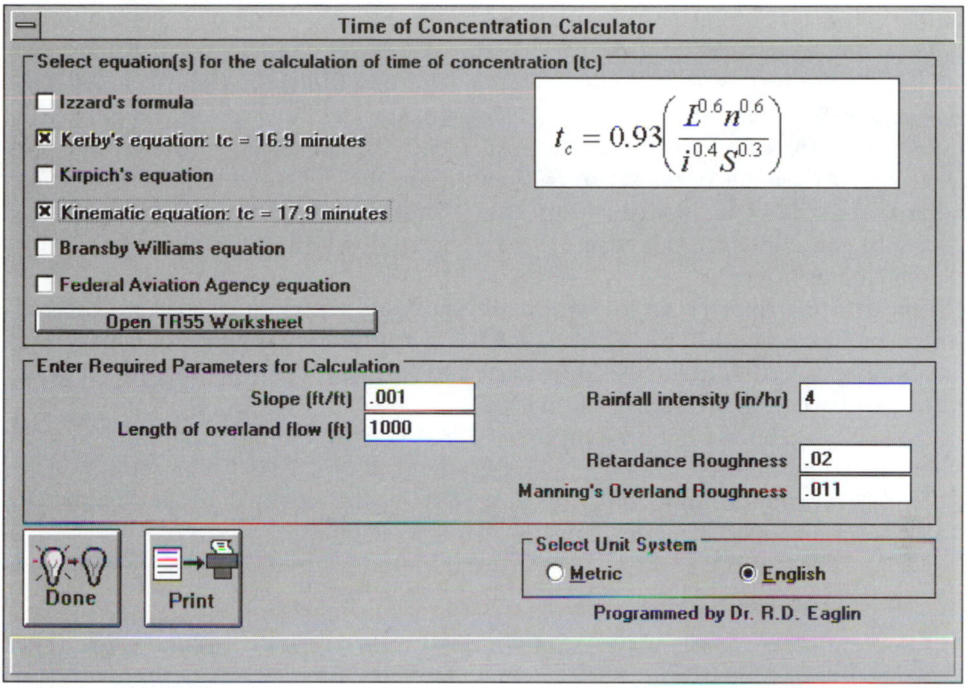

FIGURE I.18 TC Calculator main screen.

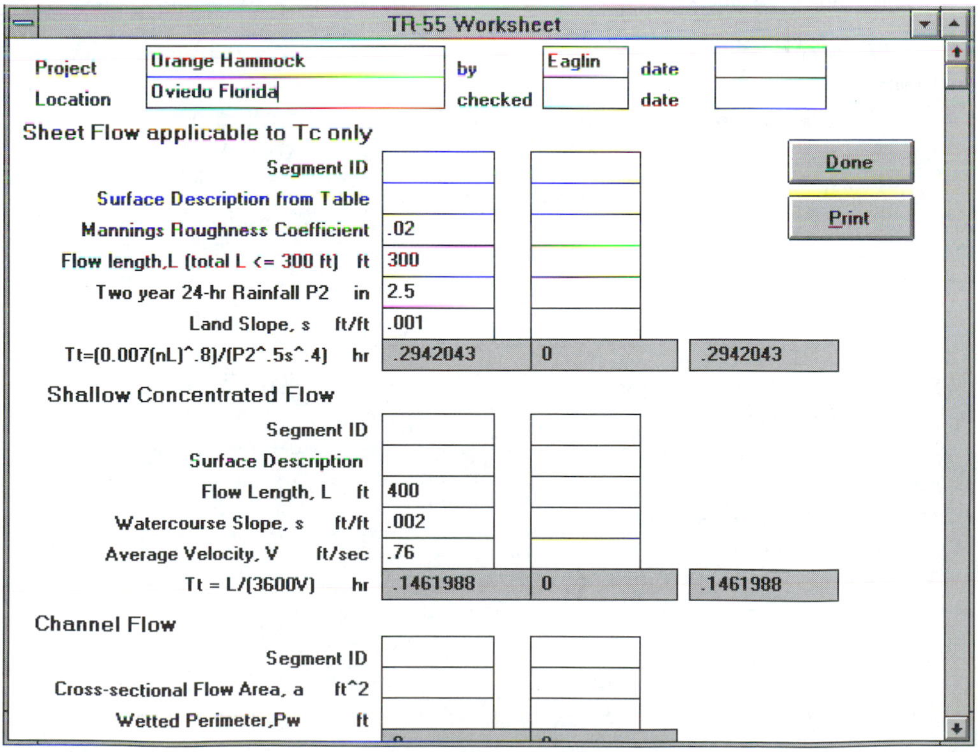

FIGURE I.19 The TR-55 Worksheet option of the TC Calculator.

the main screen. The Matrix Calculator works much in the same way as a handheld calculator. If you wish to multiply two matrices ($A \times B$) you can enter the first one as A, the second one as B, and then click the **Multiply Matrix** button ($C = A \times B$). The results will be put into matrix C. This matrix can then be viewed using the Edit Matrix button for C. EZMAT has no print facility for matrices; however, matrices can be copied into any Windows spreadsheet using the Windows clipboard by highlighting the matrix in the editor and pressing **Control-Insert.** This will then copy the matrix into the clipboard, allowing it to be inserted into a spreadsheet or a word processor for printing.

Note: If a matrix is inverted twice, it should yield the original matrix. This will not necessarily be the case because of roundoff error and the number of calculations involved. The roundoff error will be reflected in the final result which will be close, but not exactly equal to the original matrix.

Note: As an exercise enter any square (3×3 or 4×4) matrix in A, Set $B = A$ clicking the **Copy** button in the main screen. Invert matrix B and then multiply $C = A \times B$ and view the resulting matrix C. C should be very nearly equal to the identity matrix (1's on diagonal, 0's elsewhere).

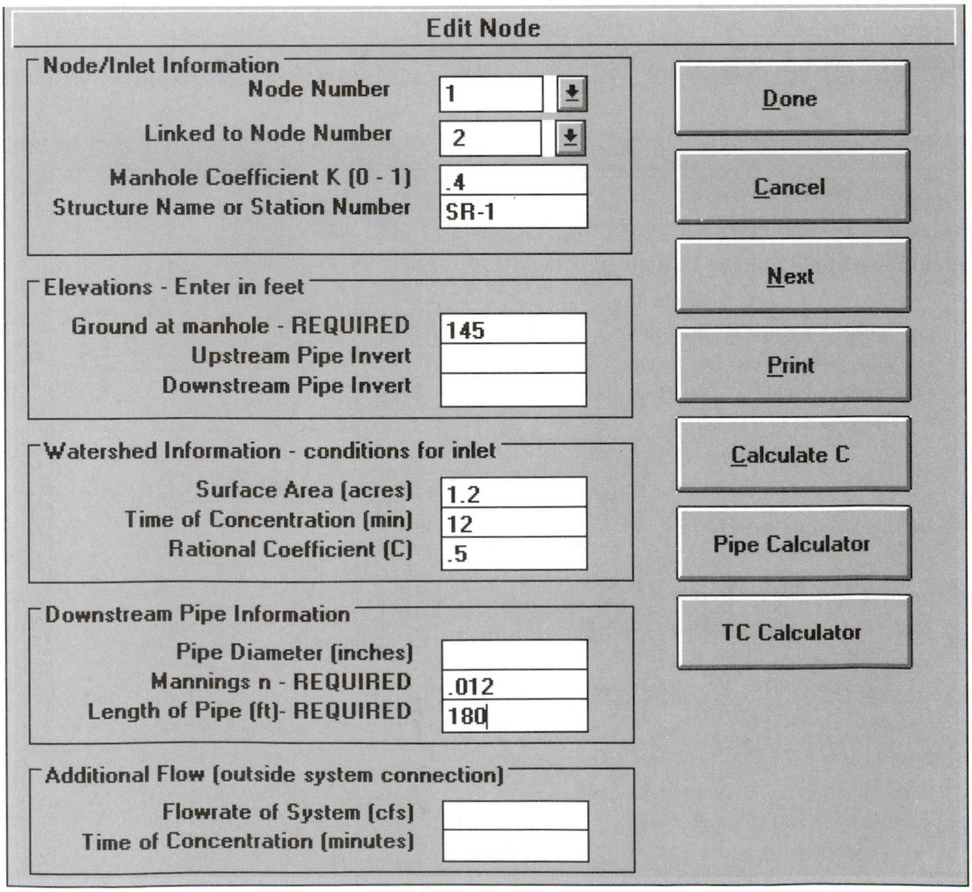

FIGURE I.20 Node Entry screen of the OPSEW Storm Sewer design program.

I.7   TC CALCULATOR

The Time of Concentration Calculator allows calculations for time of concentration for each of the equations in the text and compares the results of each. Checking any of the equations in the main window will cause the required data to be enabled in the lower half of the window shown in Figure I.18. If all the necessary data are entered, the time of concentration will be automatically calculated and the results displayed. Useful information is displayed to assist in selecting coefficients and viewing the equations. This information is displayed based on the location of the cursor. The results of each selected equation can be printed.

The TR-55 calculator can be used by clicking the **Open TR-55 Calculator** button on the main window. This option will open a second window which is similar to the worksheet outlined in Technical Release 55, *Urban Hydrology for Small Watersheds*, Soil Conservation Service, June 1986. This worksheet is shown in Figure I.19.

I.8   OPSEW-STORM SEWER DESIGN

To assist in the design of storm sewer, a computer program, OPSEW, specifically designed for this purpose is included with the text. The use of OPSEW helps specify the size of pipes and the depth of trench cut to meet flow and cover specifications. The user can pick various designs and compare their cost and effectiveness. The OPSEW program uses the concept of a link and a node. A node is a representation

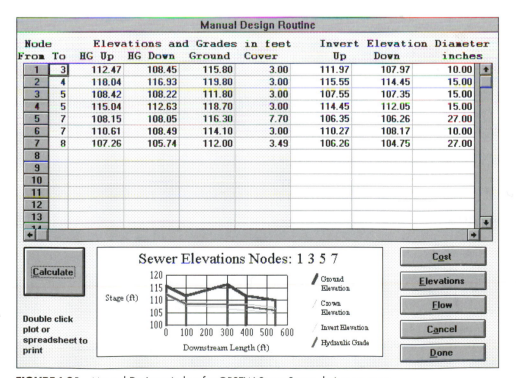

FIGURE I.21   Manual Design window for OPSEW Storm Sewer design program.

of a manhole, junction, or a storm sewer inlet. Every node except the outlet has a downstream 'link.' A link is a length of pipe which conveys water from node. An example of the node entry screen is shown in Figure I.20. All the required data for each node in the storm sewer system must be entered, and each node must be numbered such that downstream nodes have a higher number than the nodes which are upstream of them. When all the required data for the OPSEW program is entered, the Manual Design window (Figure I.21) can be used. The automatic design routines may also be used to give a good starting point for design. These routines are in the Design menu and will calculate pipe diameters and/or pipe inverts for your storm sewer system.

Extensive on-line documentation is available from the OPSEW program by clicking the **F1** key or from the Help menu. Because the OPSEW program is so extensive, the reader is referred to this documentation when using the OPSEW program. A number of example problems are included in this documentation and can be printed using the Windows™ Help engine.

SUPPLEMENTAL HYDROLOGY ASSIGNMENTS

This appendix contains suggested assignments to assist in understanding hydrologic processes. These assignments may be used as laboratories or as extended homework. They are structured so that the assignment numbers correspond with the chapter covered. The authors have used them as laboratory assignments giving the students additional group and individual help. For some chapters more than one assignment is presented. The instructor may wish to provide specific data or collect data for use with these assignments. The USGS also maintains a historical database of water resources data on the worldwide web at address *http://h2o.usgs.gov/,* which may be used as supplemental data with many of these assignments.

HYDROLOGY ASSIGNMENT 1
PROGRAMMING / SIMPSON'S RULE

Objective

A. To learn fundamental programming concepts and an application in Hydrology.

B. To determine the area under a curve using programming methods.

Equipment

A. Computer with spreadsheet software.

B. Programming Language (BASIC, FORTRAN, C, etc. . . .).

Procedure

A. Plot flow versus time for given set of data (any hydrograph may be used).

B. Determine total volume of flow using a spreadsheet and Simpson's rule.

C. Determine total volume of flow by writing a short computer program. The program should accept flow and time as inputs and should give total volume as output.

Calculations

A. Area under curve for a single time step

$$V_i = \left(\frac{q_i + q_{i+1}}{2} \right) \Delta t$$

B. Total volume under curve

$$V_T = \sum_{i=1}^{i=n-1} V_i$$

Where q_i is the rate of flow for each step, V_i is the volume for each step. Note, be careful with units or any appropriate conversion factors.

HYDROLOGY ASSIGNMENT 2
STREAM FLOW/PROBABILITY DISTRIBUTIONS

Objective

To demonstrate the use of probability distributions in developing flood predictions.

Equipment

Library, DISTRIB computer program, REGRESS computer program, spreadsheet software, the Internet connection.

Procedure

A. Using the *Water Resources Data* available from United States Geological Survey select three gaging sites for a river reach on any selected river. Select a river and sites where at least 10 yr of data exist. Record 10–15/yr maximum flow and gage height values at each site along with information concerning location of gaging stations including contributing area at each gage. (Note: historical data is available from the USGS on the internet at address *http:// wwwdwatcm.wr.usgs.gov/historical.htm/*).

B. Using the data collected and the DISTRIB for Windows program, predict the 10-, 25-, and 100-yr return period maximum flows for the three stations.

C. Using REGRESS, develop a relationship (or use any reported relationship) between flow and gage height for the three stations. Be sure to investigate exponential and power relationships between flow and gage height. Using this relationship and the predictions from part B, also predict the maximum gage stage at each of the gaging stations.

Report

Your report should include the following;

1. Location map including the general area of the gaging stations.
2. A tabulation of the year, maximum stage, and maximum flow used in developing the distribution.
3. A tabulation (for each station) of the return period and the predicted stage and flow for that return period.
4. The results of the curve fit between maximum stage and flow, including a plot of actual and predicted data.

HYDROLOGY ASSIGNMENT 3A
PRECIPITATION OVER A DRAINAGE BASIN

Objective

To illustrate differences between several methods of calculating average precipitation over a drainage basin (catchment or watershed) due to a precipitation event.

Equipment

Area measurement device: Planimeter, Computer with Digitizer (AutoCAD).

Procedure

Utilize data for that drainage basin shown in Figure J-1.

Calculations

Calculate the mean areal depth by three methods:

1. *Mean station method*   Calculate the mean areal depth of precipitation over the drainage basin (i.e., mean areal depth is computed as the simple arithmetic mean of the precipitation at the selected stations).
2. *Theissen method*   By this method it is assumed that the precipitation at any point is equal to the precipitation at the nearest station, including stations outside the drainage basin. This method attempts to allow for the nonuniform distribution of gages by providing a weighing factor for each gage.

    The stations are plotted on a map, and connecting lines drawn between adjacent stations. Perpendicular bisectors of these connecting lines form polygons around each station. The sides of each polygon are the boundaries of the effective area assigned to each station. The area of each polygon is determined by planimetry and may be expressed as a percentage of the total area.

    The weighted average rainfall for the total area is computed by multiplying the precipitation at each station by its assigned percentage of area and totaling the contribution of each polygon.

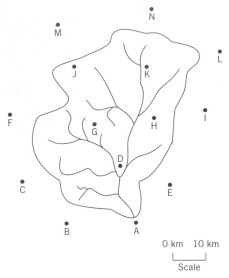

FIGURE J.1   Watershed for Hydrology assignment 3A.

Station	A	B	C	D	E	F	G
Precip. (cm)	5.5	3.2	4.0	6.0	4.9	5.1	7.0
Station	H	I	J	K	L	M	N
Precip. (cm)	5.8	4.0	3.5	3.0	2.0	2.5	1.5

3. *Isohyetal method*   Construct a isohyetal map, (i.e., lines or contours of equal precipitation for the drainage basin). The average precipitation for an area is computed by weighing the average precipitation between successive isohyets by the area between the isohyets totaling these products, and dividing by the total area.

Report

Discuss the accuracy of your results. Which method has the potential for being the most accurate? Why?

HYDROLOGY ASSIGNMENT 3B
DEVELOPMENT OF AN IDF CURVE

Objective

A. To use probability distributions for predictive analysis.

B. To demonstrate the procedures used in the generation of an IDF curve.

Equipment

DISTRIB Computer program, REGRESS Computer Program

Procedure

A. Data from a National Oceanic and Atmospheric Administration (NOAA) rainfall station has been analyzed and separated into storms of differing durations. These durations are 15 and 30 minutes, and 1, 2, 4, 6, 10, and 12 hr. These data are provided in Table J.1. A minimum interevent dry period of 4 hr was used to generate the data in Table J.1.

Using DISTRIB for Windows, enter the data from each duration. (Note: Rainfall data for various regions of the country is available from web address *http://www.esdim.noaa.gov/*). Using the available distributions, determine which distribution best fits the data. The available distributions are: Normal, 2 Parameter Log Normal, 3 Parameter Log Normal, Type I Extremal (Gumbel), Pearson Type III, and Log Pearson Type III. You should pick 1 distribution to fit *all* the data sets. This distribution will be used for the predictive analysis.

B. Using the distribution chosen in part A develop a prediction for each event duration for 2-, 5-, 10-, 25-, and 100-yr return periods. You should have 9*6 or 54 predictions. It would be advisable to organize these predictions in a spreadsheet by duration and return period. (Hint: the cumulative probability for a 10-yr return period is $1-\frac{1}{10}$ or 0.9). Plot these on a log–log plot with intensity on the y axis and duration on the x axis.

TABLE J.1   For Each Duration (in hours), the Yearly Maximum Rainfall Intensity in Inches Per Hour

	DURATIONS								
YEAR	0.25	0.5	1	2	4	6	10	12	24
1973	0.4	0.5	0.7	1.0	1.2	1.3	1.3	1.4	1.8
1974	1.1	1.7	2.1	3.7	5.7	5.9	6.4	6.4	7.8
1975	0.9	1.6	1.9	2.9	3.4	3.7	4.0	4.2	5.9
1976	1.1	1.8	3.0	3.8	5.4	6.1	7.7	7.9	12.0
1977	1.0	1.8	2.9	3.2	3.2	3.2	3.2	3.2	3.2
1978	0.7	1.1	1.5	1.9	1.9	2.3	2.6	2.6	2.6
1979	1.0	1.6	1.9	2.9	3.4	3.4	3.4	3.4	4.5
1980	0.6	1.1	1.6	2.2	2.4	2.6	3.2	3.8	3.9
1981	0.7	1.1	3.7	3.7	3.7	3.7	3.7	3.7	3.7
1982	1.1	1.9	2.8	3.5	3.7	4.4	5.3	5.4	5.4
1983	0.8	1.1	1.6	2.2	3.5	3.7	3.9	3.9	3.9
1984	1.0	1.1	15.2	15.2	15.2	15.2	15.2	15.2	15.2
1985	1.0	1.3	1.7	1.9	2.1	2.9	3.7	4.6	6.0
1986	0.8	1.0	1.5	2.1	2.4	3.0	3.5	3.6	5.0
1987	1.7	2.9	4.1	4.3	4.3	4.3	4.3	4.3	4.3
1988	1.0	1.5	1.8	3.2	4.0	4.8	6.8	7.0	8.4
1989	1.0	1.9	3.5	3.5	3.5	3.5	3.5	3.5	3.5

C. For each return period, fit the data to the curve using the REGRESS program (i = intensity in inches per hour, D = duration in minutes)

$$i = \frac{a}{b + D}$$

and report the constants (a and b) for each return period. Plot the equation on a log–log scale.

Report

The report should include the following: Sample Distribution Plot, Distribution Chosen, Array of Predictions, Plot of Predictions, Curve Fit Coefficients, Plot of Curve fits.

HYDROLOGY ASSIGNMENT 4A
CONSUMPTIVE USE

Discussion

The terms *evapotranspiration* and *consumptive use* are frequently used interchangeably. However, they are synonymous only if evapotranspiration is used to indicate the amount of water consumed (used up and lost by change from a liquid state to a vapor state) in evaporation and transpiration in raising plants.

Evaporation is the process by which water accumulated on the land surfaces (including that held in surface depressions, water bodies, vegetation, etc.) is converted into the vapor state and returned to the atmosphere.

Transpiration is the process by which plants transfer water from the root zone to the leaf surface, where it eventually evaporates into the atmosphere.

Evapotranspiration is the process by which water in the land surface, soil, and vegetation is converted into the vapor state and returned to the atmosphere. Evapotranspiration is usually taken to be the sum of evaporation and transpiration.

Case I

During a precipitation event we are not normally concerned with evaporation and transpiration, but initial abstractions (which later become lost to evaporation) and infiltration (which may later become a part of evaporation and transpiration) in determining rainfall excess.

Case II

During a precipitation event with temporary storage (e.g., retention or detention ponds) evaporation may become an important item in cases where it is desired to maintain the pond in the dry state insofar as possible.

Case III

Inversely, during a normal stream flow condition, we are not concerned with initial abstractions but with infiltration, evaporation, and transpiration (if there is significant vegetation along the stream bank).

Case IV

In a situation involving "permanent storage," good reservoir management would require knowledge about rainfall, evaporation, infiltration, and possibly transpiration. Further, use of the "storage" for productive purposes (consumptive use) such as irrigation, water supply, and similar uses would require careful analysis of the needs of humans, plants, and animals for the water—i.e., we must have good *consumptive use* information.

See Chapter 4 of the text and related material.

Objective

To obtain reliable data on consumptive use for a variety of vegetation types.

Equipment

Library.

Procedure

A. Review the various indices for reference material on evaporation, transpiration, evapotranspiration, and related topics.

B. Review pertinent journals, i.e., ASCE Proceedings (esp. Hydraulics, and Irrigation and Drainage); Water Resources Research; Climatological Documents; ASCE and AGU Transactions, etc.

C. Review appropriate books containing material related to required topics.

D. Provide *consumptive use* data and references on the vegetative type of your choice. Include:

1. Values for k = crop coefficient (see Blaney–Criddle).
2. Values for the transpiration ratio (ratio of the weight of water transpired by a plant during the growing season to the weight of dry matter produced by the plant, excluding roots).
3. Record how date was determined by the original investigator (field experiment, water budget calculations, laboratory procedures, etc.).
4. Record the location (latitude and longitude) and elevation if known or given.
5. Record other pertinent research data if given (e.g., insolation, wind movement, temperature, soil type, etc.).

Calculations

Using the Blaney–Criddle equation (See Chapter 4 of text) calculate an estimated consumptive use requirement for the vegetation type of your choice during a typical growing season. Assume an 80-acre plot. Please identify all assumptions about length of growing season, solar time, location, climatic zone, etc. pertinent to your calculation.

HYDROLOGY ASSIGNMENT 4B
EVAPORATION PROBLEM

Objective

To obtain a valid estimate for the lake evaporation rate for a local water budget study.

TABLE J.2   Data for Three Central Florida Weather Stations for Month of June

STATION	PRECIP. (in.)	EVAP. (in.)	WIND (mi.)	AIR (max.)	TEMPERATURE (min.)	SOIL (max.)	TEMPERATURE (min.)
Lisbon	9.59	7.34	868	88.8	69.4		
Lake Alfred	2.59	8.51	1289	90.5	69.2	94.5	75.9
Vero Beach	4.93	7.04	998	88.5	71.9		

Procedure

A. Using the methods outlined in Chapter 4 of the text estimate the pan evaporation for a specific location which is equidistant from the locations of the supplied data. Compare the results with local field data.

B. Convert pan evaporation into lake evaporation and use this data to estimate the volume of water lost by two lakes; one 21.22 acres and the other 10 acres.

C. Use Equation 4.6 and the given meteorological data to confirm your estimate of lake evaporation.

Local conditions coincident with your field study and pan evaporation measurements were:

$$\text{Mean pan water temperature} = 63°F$$

$$\text{Mean air temperature} = 57°F$$

$$\text{Mean wind speed} = 9 \text{ mph}$$

TABLE J.3   Short Term Evaporation Pan Measurements at Field Site

DATE	TIME	ELAPSED TIME Hr:Min	PAN RDG in	P in	E in	ACCUM. E in
6/6	10:00	0	3.370	0	0	0
	15:35	5:35	3.257	0	.113	.113
	21:00	11:00	4.253	1.48	.484	.597
6/8	9:15	23:15	4.262	0	(−).009	.588
	9:15	23:15	2.776	new datum		
	12:00	26:00	2.763	0	.013	.601
	14:35	28:35	2.730	0	.032	.633
	15:50	29:50	3.161	.45	.019	.652
6/9	10:00	48:00	3.140	0	.021	.673
	15:00	53:00	3.035	0	.105	.778
6/10	14:30	76:30	3.027	.08	.088	.866
6/11	15:00	101:00	3.282	.50	.255	1.121
6/12	8:00	118:00	3.157	T	.135	1.256

HYDROLOGY ASSIGNMENT 5A
INFILTRATION USING NRCS CURVE NUMBER

Objective

A. To understand the use of a spreadsheet in Hydrology
B. To determine the relationship between NRCS curve number, rainfall excess, infiltration, and rainfall.

Equipment

Computer with spreadsheet software.

Procedure

A. Using the equations;

$$S' = \frac{1000}{CN} - 10$$

$$R = \frac{(P - 0.2S')^2}{P + 0.8S'} \qquad \text{for P} > 0.2S'$$

$$R = 0 \qquad \text{for P} \leq 0.2S'$$

where
CN = NRCS Curve Number
  P = Precipitation (in)
  R = Rainfall Excess (in)

Plot rainfall excess versus precipitation for curve numbers of 30, 50, 70, 80, 90, and 100. Precipitation (on the x axis) should vary from 0.0 to 10.0 in.

B. For a 6-hr storm with constant rainfall intensity, calculate cumulative infiltration versus time using half hour increments. Use a spreadsheet to perform the calculations of rainfall excess and infiltration. The total storm precipitation is 4.5 in. Perform the calculation for the curve numbers of 30, 50, 70, 80, 90, and 100. Calculate both the incremental infiltration and the total infiltration. Plot the cumulative infiltration for all six curve numbers on the same plot (cumulative infiltration as y versus time as x).

C. For the problem in part B calculate an infiltration rate (ΔInfiltration/ΔTime) for the storm duration. Plot this infiltration rate versus time for the six curve numbers. What does the area under this curve represent? How are the curves in part B and part C related. How does the plot of cumulative infiltration relate to a plot of cumulative runoff? How does the plot of infiltration rate relate to a plot of runoff rate? (Hint: remember the rainfall intensity or rainfall rate is constant).

HYDROLOGY ASSIGNMENT 5B
DETERMINATION OF TIME OF CONCENTRATION

Objective

A. To demonstrate the use of TR-55 methodology in the determination of time of concentration.

B. Comparison of different methods for the determination of time of concentration.

Equipment

TCCALC Computer Program, Scale, USGS Topography map

Procedure

A. Using the supplied excerpt of a USGS Topographical map Figure J.2 delineate the boundaries of the watershed.

B. Using the topographic data and land cover data to estimate the time of concentration for this watershed.

HYDROLOGY ASSIGNMENT 5C
INFILTRATION WITH A DOUBLE RING INFILTROMETER

Objective

To estimate infiltration rates with the use of a double-ring infiltrometer.

Equipment

Double-ring infiltrometer with measuring column.

Procedure

A. Set up a double-ring infiltration apparatus in an undisturbed area. Both the inner and outer rings should be sunk about 10 cm into the soil, or until they cannot be sunk further.

B. Fill the inner and outer rings to a nominal depth and record this depth (head) of water.

C. Keep this constant head in both the inner and outer rings by adding water continuously or at regular time intervals. Continue this procedure until equal volumes of water are added for each time interval. Record the volume of water added to the inner ring for each time interval.

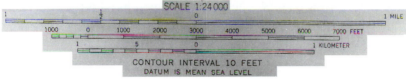

CONTOUR INTERVAL 10 FEET
DATUM IS MEAN SEA LEVEL

FIGURE J.2   USGS Topographical map excerpt for Hydrology Assignment 5B.

Calculations

Develop a Horton equation for the infiltration data collected in this experiment. Plot the results of the developed equation versus the actual data and compare the results.

Data

If no data are available for this experiment, the following data may be substituted. The double-ring infiltrometer had a constant head of 6 in. The diameter of the inner ring was 12 in. Time is in minutes, volume is in liters.

TABLE J.4 Sample Data for Hydrology Assignment 5C

TIME	VOLUME	TIME	VOLUME
1	0.64	40	1.65
2	0.54	50	1.29
4	0.67	60	1.03
8	0.48	70	0.80
10	0.40	80	0.62
20	2.52	90	0.50
30	2.00		

HYDROLOGY ASSIGNMENT 6A
INVESTIGATION OF HYDROGRAPH PROCEDURES

Objective

A. To demonstrate the methodology used to generate a watershed hydrograph.
B. To investigate the effect of time step and watershed parameters in hydrograph generation.

Equipment

SMADA Computer Program

Procedure

A watershed has a total area of 100 acres, 75 acres are pervious and 25 acres are impervious. Half of this impervious area is directly connected. The time of concentration of this watershed is 60 min. The pervious region has a NRCS curve number of 70, the impervious region can be assumed to have a curve number of 98.

A. Using an NRCS Type II rainfall of duration 6 hr, and total volume of 4 in., generate hydrographs for the watershed with time steps varying from 5 min to 1 hr. Enter information about total and impervious areas, use the curve number for the pervious region only in the calculations. Record the peak of the hydrograph for each time step using both the NRCS 256 and the NRCS 484 hydrograph generation procedures.
B. Calculate a composite curve number for the watershed. With this calculated composite curve number redo the analyses performed in part A assuming the watershed is homogeneous (the entire watershed is pervious with the calculated composite curve number).

Discussion

Discuss the accuracy of both methods of hydrograph generation (using separate routing and using a composite curve number). Which methodology is most accurate? What effect does time step have on the hydrograph accuracy? Make an organized table of your results.

HYDROLOGY ASSIGNMENT 7A
SOIL EROSION PROBLEM

Objective

To evaluate the probable soil erosion from two similar plots under differing cropping use (e.g., a woodlot and a citrus grove), to determine an estimate of the amount of sediment delivered annually to a nearby stream.

Procedure

A. Utilize the RUSLE equation (Equation 7.26);

$$A = RKLSCP$$

where
A = soil loss in tons/acre/year
R = soil erosivity factor
K = soil type factor
L = length factor
S = slope factor
C = crop management factor
P = practice factor

to estimate the soil loss in tons per acre per year. Reliable information must be obtained about each parameter.

  1. *R Factor:* Isoerodent maps are available from which a site specific value for the R factor may be determined. Alternately R can be calculated as the product of storm kinetic energy and maximum 30-min intensity and summed for all storms for a year. These are then averaged over a number of years. The equation

$$R = \frac{EI_{max}}{100} \qquad \text{where } E = 916 + (331)\log_{10}I_{30}$$

where
  R = soil erosivity factor
  E = average storm kinetic energy
  I_{30} = 30 min storm intensity
  I_{max} = maximum 30-min storm intensity

  2. *K Factor:* The K factor is readily determined from a nomograph if the soil characteristics are known. This nomograph is shown in Figure J.3. To find soil conditions for your local area you can consult your local county branch of the NRCS. Soil information is also accessible on the worldwide website at address: http://www.nrcs.usda.gov/

  3. *LS Factor:* The slope length factors are commonly combined into one topographic factor and may be calculated from the equation;

$$LS = \left(\frac{L}{72.6}\right)^m (65.41 \sin^2\theta + 4.56 \sin\theta + 0.065)$$

	Watershed 1	Watershed 2
Type	woodlot	citrus grove
Cover	pine and oak trees	citrus trees
Canopy	75% cover	25% cover
Ground	50% cover	10% cover
Fall height	6.5 ft	13 ft
C Factor	0.036	0.31
P Factor	1.0	0.5

FIGURE J.3   Soil erodibility nomograph (see Wishmeier and Smith, 1978) of Agricultural Handbook No. 537.

where

LS = slope length factor of soil loss equation

m = 0.5 for slopes greater than 5%

= 0.4 for slopes between 3.5% and 4.5%

= 0.3 for slopes between 1% and 3%

= 0.2 for slopes less than 1%

θ = slope as an angle.

4. *C Factor:* The crop management factor is sensitive to canopy cover, ground cover, and raindrop fall height.

5. *P Factor:* Practice factor is determined by comparing the soil loss under a given practice or combination of practices to that under normal tillage (up slope) of fallow plots.

B. Calculate soil loss for two watersheds using the given information. Both watersheds have a total area of 80 acres, a maximum 30-min rainfall intensity of 5.8 in./hr, storm kinetic energy (E/100) of 77.58 ft-tons/acre. The watershed soil type for both watersheds is Chandler fine sand consisting of 50% silt and very fine sand, 10% sand, and 2% organic material. It is of fine granular structure and drains rapidly. Both watersheds have a slope of 2% and length of 300 ft.

HYDROLOGY ASSIGNMENT 8A
INVENTORY ROUTING

Objective

To use the inventory equation to develop a pond design for flow routing.

Equipment

SMADA computer program, Spreadsheet, AutoCAD or other CAD program

Procedure

A. Develop hydrographs for the following Orlando area watershed: Use a 25-yr, 12-hr storm (using an IDF curve from Appendix C) with the NRCS Type II distribution and 15-min-time increments. The groundwater table is average 6 ft below the surface (limited soil storage capacity).

Precondition (woods, fair condition)

Total area	100 acres	Impervious area	5 acres
Soil type for pervious region—loam		Time of concentration	72 min

Postcondition (partially developed)

Impervious area	25 acres, 20 acres directly connected to outlet.
Time of concentration	57 min

B. A pond must be designed so that the first 1.5 in. of runoff from the impervious region must be stored. The pond inventory must also be designed so that the peak flowrate of the postcondition does not exceed the peak flowrate of the precondition. This can be done by properly placing a weir at the correct invert. Use a sharp crested rectangular weir for your pond. The pond side slope should not exceed 6 : 1 (horizontal : vertical). The pond **permanent pool** must be 6 ft below the ground due to GWT constraints. Determine the stage–storage–discharge relationship for your pond based on your pond design.

C. Using the stage–storage–discharge calculated for your pond use SMADA to route the postcondition watershed to your pond. Routing is done in SMADA by first **defining** all your nodes (3 total: 1. Watershed → 2. Pond → 3. Outlet) and then linking the nodes (1 → 2 → 3). **IF** your postcondition peak discharge exceeds the precondition peak discharge, then you must return to your pond design and either make the pond larger or the weir smaller. A good starting point for the area of the pond bottom is 5% of watershed area.

Report

A. A plan view of your pond is required. This can be done using AutoCAD or it can be drawn by hand. A side view is also required indicating the following: permanent pool elevation, storage volume, Weir invert elevation, Weir size and type, peak stage during design storm, volume in pond during peak design storm stage, and pond side slope.

B. A plot of both pre- and postcondition hydrographs is required. Peak flowrate and difference in runoff volumes between the storms should be noted.

HYDROLOGY ASSIGNMENT 8B
HYDRAULIC ROUTING / STORM SEWER DESIGN

Objective

To learn how to use hydraulic routing techniques in a practical application.

Equipment

Spreadsheet software, OPSEW Computer Program.

	S-60	S-60A
Inlet area (acres)	0.408	0.660
Rational coefficient	0.90	0.85
Length of pipe (ft)	110	220
Ground elevation (ft)	128.0	110.0
Invert elevation (ft)	unset	unset
Downstream invert (ft)	unset	unset
Time of concentration (min)	10.0	10.0
Pipe Diameter (in)	unset	unset
Manning's n	0.02	0.02

Information

FIGURE J.4   Schematic of sewer system for Hydrology Assignment 8B.

Procedure

A. Using OPSEW, design a conveyance system for the following sewer system (Figure J.4). Design for the 10-yr storm and full flow in pipes. The hydraulic grade must be kept 12 in. below the grate elevation at S-60, which is at the ground elevation. At least 3 ft of cover must be maintained above the pipes. The ground elevation at the outlet is 103.0.

Output

A. Your results should contain the following at each node: pipe diameter, length of pipe, drainage area, watershed runoff, cumulative time of concentration, time of flow in pipe, total design flowrate in pipe, invert elevations, hydraulic grade, and flow velocity in pipe.

HYDROLOGY LABORATORY 8C
CHANNEL ROUTING

Objective

To demonstrate the methods used in Muskingum stream routing.

Equipment

SMADA computer program, spreadsheet software.

Procedure

A. It is proposed to use a rectangular channel to transport the runoff from a 25-yr, 6-hr storm to a regional stormwater facility. The watershed is 650 ac, t_c = 145 minutes, 12% Impervious of which 35% is directly connected. The watershed is primarily type B soils and pasture and contains a lake capable of storing 32 ac-ft of water before discharge. The discharge is at an elevation of 123 ft and the stormwater facility has a surface elevation of 97 ft. The facility is 1.7 miles from the watershed. Generate a hydrograph (NRCS 484) for the watershed.

B.   Using the average flowrate from the hydrograph, design a channel to transport the water from the watershed to the stormwater facility. The channel will have a Manning's n of 0.03. For a rectangular channel hydraulic radius (R), cross-sectional area (A), and velocity (V) can be calculated as shown by calculating depth (d) for the average flow in a rectangular channel of width W. This depth should not exceed 8 ft.

$$R = \frac{A}{P_w} = \frac{Wd}{W + 2d}, Q = \frac{1.486}{n} AR^{2/3}S^{1/2}, V = \frac{Q}{A}$$

C.   Using a Muskingum routing scheme, calculate a storage time constant (K) for the reach. Using this K and a c of both 0.2 and 0.4, perform a routing of the watershed hydrograph (I) to produce an outlet hydrograph (O) going INTO the stormwater facility. The equations are

$$c_0 = \frac{-Kc + 0.5\,\Delta t}{K - Kc + 0.5\,\Delta t}, c_1 = \frac{Kc + 0.5\,\Delta t}{K - Kc + 0.5\,\Delta t}, c_2 = \frac{K - Kc - 0.5\,\Delta t}{K - Kc + 0.5\,\Delta t}$$

$$c_0 + c_1 + c_2 = 1.0$$

$$O_{i+1} = c_0 I_{i+1} + c_1 I_i + c_2 O_i$$

HYDROLOGY ASSIGNMENT 10A
DESIGN OF STORMWATER CONTROL SYSTEM

Objective

To apply the principles of Hydrology to perform a design of a stormwater control system.

Equipment

SMADA computer program, spreadsheet software, AutoCAD or other drafting software.

Discussion

The shown sports arena at a local university (Figure J.5) requires a parking area. The precondition is as shown. All runoff from roadways and impervious areas flows to the pond in the front of the arena which is currently at capacity. All runoff from impervious areas flows to one of two wetland areas. Develop a precondition hydrograph for flow from the impervious areas into the wetland system. The soil is type A, and the area is primarily palmetto and pine forest. The groundwater table is at 104 ft throughout the property. For scaling purposes the distance along the straight lengths of the track next to the arena are 150 yd.

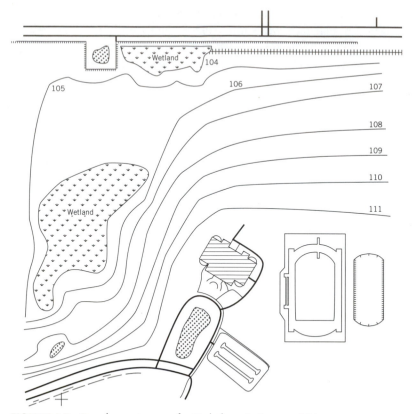

FIGURE J.5   Sample sports-arena for Hydrology Assignment 10A.

The parking area should provide at least 1600 parking spaces (use 400 sf per space). All marked wetland areas *must* have at least a 100-ft buffer between them and any impervious surfaces. The parking lot should be accessible.

Calculations

A. Design a storm sewer system for the parking area. The flow in at any manhole must be less than 3 cfs during the design sewer system storm (10-yr, Zone 7). The sewer system will outlet into a wet detention pond.

B. Design a wet detention system for the parking lot. The system should have a storage volume of at least 1.5 in. over impervious area and be capable of attenuating the peak runoff from a 25-yr, 24-hr NRCS Type II storm to the precondition. Pond layout and cross sections, pre- and post-hydrographs with plots should be included in your report.

HYDROLOGY ASSIGNMENT 10B
CASE STUDY

Objective

To use hydrologic principles in a practical case study.

Procedure

Using the water budget methods discussed in the text analyze the 650-acre Little Joe Creek watershed. Assume that for the study of a 100-acre surface water reservoir and surrounding lands the following data are known for a 12-mon period: precipitation is 40 in., evaporation pan data is 58 in., controlled outflow from the reservoir is 5 cfs to satisfy water rights downstream.

The temporary system gaging station for a small perennial surface stream showed an average discharge of 3 cfs into the reservoir during the year. The stream channel of slope 0.02 ft/ft is underlain by a sand and gravel deposit of cross-sectional area 10,000 ft^2 and of permeability 500 gpd/ft^2 which transmits another 112 acre-ft/yr as underflow.

Several farmers irrigate by pumping water from the reservoir. There are 300 acres of pasture (consumptive use of 2.16 ft/yr) and 200 acres of alfalfa (consumptive use of 3.51 ft/yr). Neither the pasture or the alfalfa drain into the reservoir. Adjacent to the watershed and feeding from the reservoir are 50 acres of bulrushes which use 9.6 ft/yr.

A.   Consider the entire 650 acres as a single hydrologic system. For a 12-month period determine (in consistent volume units of acre-feet):

- The reservoir evaporation
- The controlled outflow
- The precipitation over the watershed
- The gaged inflow
- The groundwater flow (underflow)
- The total consumptive use
- The total evapotranspiration

B.   How much of the reservoir volume could be 'saved' if the bulrushes could be drained and the 50 acres converted to pasture.

C.   There are other ungaged tributaries of area 320 acres which drain into the reservoir. The possibility also exists that springs exist in the reservoir which also add water

TABLE J.5   Annual Total Precipitation for a Sample Weather Station

YEAR	PRECIPITATION	YEAR	PRECIPITATION
1971	49.27	1981	43.91
1972	52.18	1982	55.39
1973	59.10	1983	40.91
1974	54.02	1984	52.65
1975	39.61	1985	56.70
1976	55.95	1986	51.35
1977	50.13	1987	47.40
1978	44.28	1988	57.37
1979	57.92	1989	51.05
1980	41.45	1990	40.00

to the reservoir. Estimate the ungaged inflow if the reservoir surface rose 60 in. during the 12-mon period.

D. A weather station at a nearby airport recorded the following total precipitation values for the years recorded.

Also an Index of wetness versus average annual discharge per square mile is essentially linear (plot the following data on semilog paper).

cfs/mi²	Index
0.40	0.7
5.00	1.3

You are the engineer in charge at a local water district office. A small municipality in the vicinity seeks a permit to use 3 MGD from the reservoir. What would be your decision and why?

INDEX

Limited Use License Agreement

This is the John Wiley and Sons, Inc. (Wiley) limited use License Agreement, which governs your use of any Wiley proprietary software products (Licensed Program) and User Manual (s) delivered with it.

Your use of the Licensed Program indicates your acceptance of the terms and conditions of this Agreement. If you do not accept or agree with them, you must return the Licensed Program unused within 30 days of receipt or, if purchased, within 30 days, as evidenced by a copy of your receipt, in which case, the purchase price will be fully refunded.

License: Wiley hereby grants you, and you accept, a non-exclusive and non-transferable license, to use the Licensed Program on the following terms and conditions only:

a. You have been granted an Individual Software License and you may use the Licensed Program on a single personal computer for your own personal use only.
b. A backup copy or copies may be made but all such backup copies are subject the terms and conditions of this agreement.
c. You may not make or distribute unauthorized copies of the Licensed Program, create by decompilation, or otherwise, the source code of the Licensed Program, or use, copy, modify, or transfer the Licensed Program in whole or in part, except as expressly permitted by this Agreement.
d. A backup copy or copies may be made only as provided by the User Manual(s), except as expressly permitted by this Agreement.

If you transfer possession of any copy or modification of the Licensed Program to any third party, your license is automatically terminated. Such termination shall be in addition to and not in lieu of any equitable, civil, or other remedies available to Wiley.

Term: This License Agreement is effective until terminated. You may terminate it at any time by destroying the Licensed Program with all copies made (with or without authorization).

This Agreement will also terminate upon the conditions discussed elsewhere in this Agreement, or if you fail to comply with any term or condition of this Agreement. Upon such termination, you agree to destroy the Licensed Program and any copies made (with or without authorization) of either.

Wiley's Rights: You acknowledge that all rights (including without limitation, copyrights, patents and trade secrets) in the Licensed Program (including without limitation, the structure, sequence, organization, flow, logic, source code, object code and all means and forms of operation of the Licensed Program) are the sole and exclusive property of Wiley. By accepting this Agreement, you do not become the owner of the Licensed Program, but you do have the right to use it in accordance with the provision of this Agreement. You agree to protect the Licensed Program from unauthorized use, reproduction, or distribution. You further acknowledge that the Licensed Program contains valuable trade secrets and confidential information belonging to Wiley. You may not disclose any component of the Licensed Program, whether or not in machine readable form, except as expressly provided in this Agreement.

THIS LIMITED WARRANTY IS IN LIEU OF ALL OTHER WARRANTIES, EXPRESSED OR IMPLIED, INCLUDING WITHOUT LIMITATION, ANY WARRANTIES OR MERCHANTIBILITY OR FITNESS FOR A PARTICULAR P'

EXCEPT AS SPECIFIED ABOVE, THE LICENSED PROGRAM IS FURNISHED BY WILEY ON AN "AS IS" BASIS AND WITHOUT WARRANTY AS TO THE PERFORMANCE OR RESULTS YOU MAY OBTAIN USING THE LICENSED PROGRAM. THE ENTIRE RISK IS TO THE RESULTS OR PERFORMANCE, AND THE COST OF ALL NECESSARY SERVICING, REPAIR, OR CORRECTION OF THE LICENSED PROGRAM IS ASSUMED BY YOU.

IN NO EVENT WILL WILEY BE LIABLE TO YOU FOR ANY DAMAGES, INCLUDING LOST PROFITS, LOST SAVINGS, OR OTHER INCIDENTAL OR CONSEQUENTIAL DAMAGES ARISING OUT OF THE USE OR INABILITY TO USE THE LICENSED PROGRAM EVEN IF WILEY OR AN AUTHORIZED WILEY DEALER HAS BEEN ADVISED OF THE POSSIBILITY OF SUCH DAMAGES.

THIS LIMITED WARRANTY GIVES YOU SPECIFIC LEGAL RIGHTS. YOU MAY HAVE OTHERS BY OPERATION OF LAW WHICH VARIES FROM STATE TO STATE. IF ANY OF THE PROVISIONS OF THIS AGREEMENT ARE INVALID UNDER ANY APPLICABLE STATUE OR RULE OF LAW, THEY ARE TO THAT EXTENT DEEMED OMITTED.

This Agreement represents the entire agreement between us and supersedes any proposals or prior agreements, oral or written, and any other communication between us relating to the subject matter of this Agreement.

This Agreement will be governed and construed as if wholly entered into and performed within the State of New York.

You acknowledge that you have read this Agreement, and agree to bound by its terms and conditions.